Valery G. Romanovski
Douglas S. Shafer

The Center and Cyclicity Problems:
A Computational Algebra Approach

Birkhäuser
Boston • Basel • Berlin

Valery G. Romanovski
Center for Applied Mathematics
and Theoretical Physics
University of Maribor
Krekova 2
2000 Maribor, Slovenia
valery.romanovsky@uni-mb.si

Douglas S. Shafer
Department of Mathematics
University of North Carolina
Charlotte, NC 28025
USA
dsshafer@uncc.edu

ISBN 978-0-8176-4726-1 eISBN 978-0-8176-4727-8
DOI 10.1007/978-0-8176-4727-8

Library of Congress Control Number: PCN applied for

Mathematics Subject Classification (2000): 34C07, 37G15, 37G05, 34C23, 34C14, 34-01, 37-01, 13-01, 14-01

Cover designed by Alex Gerasev.

Printed on acid-free paper.

Springer is part of Springer Science+Business Media (www.springer.com)

To our families and teachers.

Preface

The primary object of study in this book is small-amplitude periodic solutions of two-dimensional autonomous systems of ordinary differential equations,

$$\dot{x} = P(x,y), \qquad \dot{y} = Q(x,y),$$

for which the right-hand sides are polynomials. Such systems are called polynomial systems. If the origin is an isolated singularity of a polynomial (or real analytic) system, and if there does not exist an orbit that tends to the singularity, in either forward or reverse time, with a definite limiting tangent direction, then the singularity must be either a *center*, in which case there is a neighborhood of the origin in which every orbit except the origin is periodic, or a *focus*, in which case there is a neighborhood of the origin in which every orbit spirals towards or away from the origin. The problem of distinguishing between a center and a focus for a given polynomial system or a family of such systems is known as the Poincaré center problem or the center-focus problem. Although it dates from the end of the 19th century, it is completely solved only for linear and quadratic systems ($\max\{\deg(P),\deg(Q)\}$ equal to 1 or 2, respectively) and a few particular cases in families of higher degree.

Relatively simple analysis shows that when the matrix of the linearization of the system at the the singular point has eigenvalues with nonzero real parts, the singular point is a focus. If, however, the real parts of the eigenvalues are zero then the type of the singular point depends on the nonlinear terms of polynomials in a nontrivial way. A general method due to Poincaré and Lyapunov reduces the problem to that of solving an infinite system of polynomial equations whose variables are parameters of the system of differential equations. That is, the center-focus problem is reduced to the problem of finding the variety of the ideal generated by a collection of polynomials, called the focus quantities of the system.

A second problem, called the cyclicity problem, is to estimate the number of limit cycles, that is, isolated periodic solutions, that can bifurcate from a center or focus when the coefficients of the system of differential equations are perturbed by an arbitrarily small amount, but in such a way as to remain in a particular family of systems, for example in the family of all quadratic polynomial systems if the original system was quadratic. This problem is a part of the still unresolved 16th Hilbert problem and is often called the local 16th Hilbert problem. In fact, in order to find an upper bound for the cyclicity of a center or focus in a polynomial system it is sufficient to obtain a basis for the above-mentioned ideal of focus quantities. Thus the study of these two famous problems in the qualitative theory of differential equations can be carried out through the study of polynomial ideals, that is, through the study of an object of commutative algebra.

Recent decades have seen a surge of interest in the center and cyclicity problems. Certainly an important reason for this is that the resolution of these problems involves extremely laborious computations, which nowadays can be carried out using powerful computational facilities. Applications of concepts that could not be utilized even 30 years ago are now feasible, often even on a personal computer, because of advances in the mathematical theory, in the computer software of computational algebra, and in computer technology. This book is intended to give the reader a thorough grounding in the theory, and explains and illustrates methods of computational algebra, as a means of approaching the center-focus and cyclicity problems.

The methods we present can be most effectively exploited if the original real system of differential equations is properly complexified; hence, the idea of complexifying a real system, and more generally working in a complex setting, is one of the central ideas of the text. Although the idea of extracting information about a real system of ordinary differential equations from its complexification goes back to Lyapunov, it is still relatively scantily used. Our belief that it deserves exposition at the level of a textbook has been a primary motivation for this work. In addition to that, it has appeared to us that by and large specialists in the qualitative theory of differential equations are not well versed in these new methods of computational algebra, and conversely that there appears to be a general lack of knowledge on the part of specialists in computational algebra about the possibility of an algebraic treatment of these problems of differential equations. We have written this work with the intention of trying to help to draw together these two mathematical communities.

Thus, the readers we have had in mind in writing this work have been graduate students and researchers in nonlinear differential equations and computational algebra, and in fields outside mathematics in which the investigation of nonlinear oscillation is relevant. The book is designed to be suitable for use as a primary textbook in an advanced graduate course or as a supplementary source for beginning graduate courses. Among other things, this has meant motivating and illustrating the material with many examples, and including a great many exercises, arranged in the order in which the topics they cover appear in the text. It has also meant that we have given complete proofs of a number of theorems that are not readily available in the current literature and that we have given much more detailed versions of proofs that were written for specialists. All in all, researchers working in the theory of limit cycles of polynomial systems should find it a valuable reference resource, and because it is self-contained and written to be accessible to nonspecialists, researchers in other fields should find it an understandable and helpful introduction to the tools they need to study the onset of stable periodic motion, such as ideals in polynomial rings and Gröbner bases.

The first two chapters introduce the primary technical tools for this approach to the center and cyclicity problems, as well as questions of linearizability and isochronicity that are naturally investigated in the same manner. The first chapter lays the groundwork of computational algebra. We give the main properties of ideals in polynomial rings and their affine varieties, explain the concept of Gröbner bases, a key component of various algorithms of computational algebra, and provide explicit algorithms for elimination and implicitization problems and for basic operations on ideals in polynomial rings and on their varieties. The second chapter begins with the main theorems of Lyapunov's second method, theorems that are aimed at the investigation of the stability of singularities (in this context often termed equilibrium points) by means of Lyapunov functions. We then cover the basics of the theory of normal forms of ordinary differential equations, including an algorithm for the normalization procedure and a criterion for convergence of normalization transformations and normal forms.

Chapter 3 is devoted to the center problem. We describe how the concept of a center can be generalized to complex systems, in order to take advantage of working over the algebraically closed field $\mathbb{C}$ in place of $\mathbb{R}$. This leads to the study of the variety, in the space of parameters of the system, that corresponds to systems with a center, which is called the center variety. We present an efficient computational algorithm for computing the focus quantities, which are the polynomials that define the center variety. Then we describe two main mechanisms for proving the existence of a center in a polynomial system, Darboux integrability and time-reversibility, thereby completing the description of all the tools needed for this method of approach to the center-focus problem. This program and its efficiency are demonstrated by applying it to resolve the center problem for the full family of quadratic systems and for one particular family of cubic systems. In a final section, as a complement to the rest of the chapter, particularly aspects of symmetry, the important special case of Liénard systems is presented.

If all solutions in a neighborhood of a singular point are periodic, then a question that arises naturally is whether all solutions have the same period. This is the so-called isochronicity problem that has attracted study from the time of Huygens and the Bernoullis. In Chapter 4 we present a natural generalization of the concept of isochronicity to complex systems of differential equations, the idea of linearizability. We then introduce and develop methods for investigating linearizability in the complex setting.

As indicated above, one possible mechanism for the existence of a center is time-reversibility of the system. Chapter 5 presents an algorithm for computing all time-reversible systems within a given polynomial family. This takes on additional importance because in all known cases the set of time-reversible systems forms exactly one component of the center variety. The algorithm is derived using the study of invariants of the rotation group of the system and is a nice application of that theory and the algebraic theory developed in Chapter 1.

The last chapter is devoted to the cyclicity problem. We describe Bautin's method, which reduces the study of cyclicity to finding a basis of the ideal of focus quantities, and then show how to obtain the solution for the cyclicity problem in the case that the ideal of focus quantities is radical. In the case that the ideal generated by the first few focus quantities is not radical, the problem becomes much more difficult; at present there is no algorithmic approach for its treatment. Nevertheless we present a particular family of cubic systems for which it is possible, using Gröbner basis calculations, to obtain a bound on cyclicity. Finally, as a further illustration of the applicability of the ideas developed in the text, we investigate the problem of the maximum number of cycles that can maintain the original period of an isochronous center in $\mathbb{R}^2$ when it is perturbed slightly within the collection of centers, the so-called problem of bifurcation of critical periods.

Specialists perusing the table of contents and the bibliography will surely miss some of their favorite topics and references. For example, we have not mentioned methods that approach the center and cyclicity problems based on the theory of resultants and triangular decomposition, and have not treated the cyclicity problem specifically in the important special case of Liénard systems, such as we did for the center problem. We are well aware that there is much more that could be included, but one has to draw the line somewhere, and we can only say that we have made choices of what to include and what to omit based on what seemed best to us, always with an eye to what we hoped would be most valuable to the readers of this book.

The first author acknowledges the financial support of this work by the Slovenian Research Agency. We thank all those with whom we consulted on various aspects of this work, especially Vladimir Basov, Carmen Chicone, Freddy Dumortier, Maoan Han, Evan Houston, and Dongming Wang.

Maribor, Charlotte
May 2008

Valery G. Romanovski
Douglas S. Shafer

Contents

List of Tables

Notation and Conventions

$\mathbb{N}$	the set of natural numbers $\{1, 2, 3, \ldots\}$
$\mathbb{N}_0$	$\mathbb{N} \cup \{0\}$
$\mathbb{Z}$	the ring of integers
$\mathbb{Q}$	the field of rational numbers
$\mathbb{R}$	the field of real numbers
$\mathbb{C}$	the field of complex numbers
$A \subset B$	A is a subset of B, $A = B$ allowed
$A \subsetneqq B$	A is a proper subset of B
$A \setminus B$	elements that are in A and are not in B

See the Index of Notation beginning on p. 323 for a full list of notation.

Chapter 1
Polynomial Ideals and Their Varieties

As indicated in the Preface, solutions of the fundamental questions addressed in this book, the center and cyclicity problems, are expressed in terms of the sets of common zeros of collections of polynomials in the coefficients of the underlying family of systems of differential equations. These sets of common zeros are termed *varieties*. They are determined not so much by the specific polynomials themselves as by larger collections of polynomials, the so-called ideals that the original collections of polynomials generate. In the first section of this chapter we discuss these basic concepts: polynomials, varieties, and ideals. An ideal can have more than one set of generating polynomials, and a fundamental problem is that of deciding when two ideals, hence the varieties they determine, are the same, even though presented by different sets of generators. To address this and related isssues, in Sections 1.2 and 1.3 we introduce the concept of a Gröbner basis and certain fundamental techniques and algorithms of computational algebra for the study of polynomial ideals and their varieties. The last section is devoted to the decomposition of varieties into their simplest components and shows how this decomposition is connected to the structure of the generating ideals. For a fuller exposition of the concepts presented here, the reader may consult [1, 18, 23, 60].

1.1 Fundamental Concepts

A *polynomial* in variables $x_1, x_2, \dots, x_n$ with coefficients in a field k is a formal expression of the form

$$f = \sum_{\alpha \in S} a_\alpha \mathbf{x}^\alpha, \tag{1.1}$$

where S is a finite subset of $\mathbb{N}_0^n$, $a_\alpha \in k$, and for $\alpha = (\alpha_1, \alpha_2, \dots, \alpha_n)$, $\mathbf{x}^\alpha$ denotes the *monomial* $x_1^{\alpha_1} x_2^{\alpha_2} \cdots x_n^{\alpha_n}$. In most cases of interest k will be $\mathbb{Q}$, $\mathbb{R}$, or $\mathbb{C}$. The product $a_\alpha \mathbf{x}^\alpha$ is called a *term* of the polynomial f. The set of all polynomials in the variables $x_1, \dots, x_n$ with coefficients in k is denoted by $k[x_1, \dots, x_n]$. With the natural and well-known addition and multiplication, $k[x_1, \dots, x_n]$ is a commutative ring. The

V.G. Romanovski, D.S. Shafer, *The Center and Cyclicity Problems*,
DOI 10.1007/978-0-8176-4727-8_1,

full degree of a monomial $\mathbf{x}^\alpha$ is the number $|\alpha| = \alpha_1 + \cdots + \alpha_n$. The *full degree* of a term $a_\alpha \mathbf{x}^\alpha$ is the full degree of the monomial $\mathbf{x}^\alpha$. The *full degree* of a polynomial f as in (1.1), denoted by $\deg(f)$, is the maximum of $|\alpha|$ among all monomials (with nonzero coefficients a_α, of course) of f.

If a field k and a natural number n are given, then we term the set

$$k^n = \{(a_1, \ldots, a_n) : a_1, \ldots, a_n \in k\}$$

n-dimensional affine space. If f is the polynomial in (1.1) and $(a_1, \ldots, a_n) \in k^n$, then $f(a_1, \ldots, a_n)$ will denote the element $\sum_\alpha a_\alpha a_1^{\alpha_1} \cdots a_n^{\alpha_n}$ of k. Thus, to any polynomial $f \in k[x_1, \ldots, x_n]$ is associated the function $f : k^n \to k$ defined by

$$f : (a_1, \ldots, a_n) \mapsto f(a_1, \ldots, a_n).$$

This ability to consider polynomials as functions defines a kind of duality between the algebra and geometry of affine spaces. In the case of an arbitrary field k this interconnection between polynomials and functions on affine spaces can hold some surprises. For example, the statements "f is the zero polynomial" (all coefficients a_α are equal to zero) and "f is the zero function" ($f|_{k^n} \equiv 0$) are not necessarily equivalent (see Exercise 1.1). However, we will work mainly with the infinite fields $\mathbb{Q}$, $\mathbb{R}$, and $\mathbb{C}$, for which the following two statements show that our naive intuition is correct.

Proposition 1.1.1. *Let k be an infinite field and $f \in k[x_1, \ldots, x_n]$. Then f is the zero element of $k[x_1, \ldots, x_n]$ (that is, all coefficients a_α of f are equal to zero) if and only if $f : k^n \to k$ is the zero function.*

Proof. Certainly if every coefficient of the polynomial f is the zero polynomial then the corresponding function is the zero function. We must establish the converse:

$$\text{If } f(a_1, \ldots, a_n) = 0 \text{ for all } (a_1, \ldots, a_n) \in k^n, \text{ then } f \text{ is the zero polynomial.} \quad (1.2)$$

We will do this by induction on the number of variables in the polynomial ring.

Basis step. For $n = 1$, the antecedent in (1.2) means that either (i) f is the zero polynomial or (ii) $\deg(f)$ is defined and at least 1 and f has infinitely many roots. It is well known, however (Exercise 1.2), that every polynomial $f \in k[x]$ for which $\deg(f) = s > 0$ has at most s roots. Hence only alternative (i) is possible, so (1.2) holds for $n = 1$.

Inductive step. Suppose (1.2) holds in the ring $k[x_1, \ldots, x_p]$ for $p = 1, 2, \ldots, n-1$. Let $f \in k[x_1, \ldots, x_n]$ be such that the antecedent in (1.2) holds for f. We can write f in the form

$$f = \sum_{j=0}^{m} g_j(x_1, \ldots, x_{n-1}) x_n^j$$

for some finite m, and will show that g_j is the zero polynomial for each j, $1 \le j \le m$. This will imply that f is the zero polynomial. Thus fix any $a = (a_1, \ldots, a_{n-1}) \in k^{n-1}$ and define $f_a \in k[x_n]$ by

$$f_a = \sum_{j=0}^{m} g_j(a_1,\ldots,a_{n-1})x_n^j.$$

By hypothesis, $f_a(a_n) = 0$ for all $a_n \in k$. Hence, by the induction hypothesis, f_a is the zero polynomial; that is, its coefficients $g_k(a_1,\ldots,a_{n-1})$ are equal to zero for all $j, 0 \leq j \leq m$. But $(a_1,\ldots,a_{n-1})$ was an arbitrary point in k^{n-1}, hence the evaluation function corresponding to g_j is the zero function for $j = 1,\ldots,m$, which, by the induction hypothesis, implies that g_j is the zero polynomial for $j = 1,\ldots,m$, as required. Thus the proposition holds. □

The proposition yields the following result.

Corollary 1.1.2. *If k is an infinite field and f and g are elements of $k[x_1,\ldots,x_n]$, then $f = g$ in $k[x_1,\ldots,x_n]$ if and only if the functions $f : k^n \to k$ and $g : k^n \to k$ are equal.*

Proof. Suppose f and g in $k[x_1,\ldots,x_n]$ define the same function on k^n. Then $f - g$ is the zero function. Hence, by Proposition 1.1.1, $f - g$ is the zero polynomial in $k[x_1,\ldots,x_n]$, so that $f = g$ in $k[x_1,\ldots,x_n]$. The converse is clear. □

Throughout this chapter, unless otherwise indicated k will denote an arbitrary field. The main geometric object of study in this chapter is what is called an *affine variety* in k^n, defined as follows.

Definition 1.1.3. Let k be a field and let $f_1,\ldots,f_s$ be (finitely many) elements of $k[x_1,\ldots,x_n]$. The *affine variety* defined by the polynomials $f_1,\ldots,f_s$ is the set

$$\mathbf{V}(f_1,\ldots,f_s) = \{(a_1,\ldots,a_n) \in k^n \ : \ f_j(a_1,\ldots,a_n) = 0 \text{ for } 1 \leq j \leq s\}.$$

An *affine variety* is a subset V of k^n for which there exist finitely many polynomials such that $V = \mathbf{V}(f_1,\ldots,f_s)$. A *subvariety* of V is a subset of V that is itself an affine variety.

In other words, the affine variety $\mathbf{V}(f_1,\ldots,f_s) \subset k^n$ is the set of solutions of the system

$$f_1 = 0,\ f_2 = 0,\ \ldots,\ f_s = 0 \tag{1.3}$$

of finitely many polynomial equations in k^n. Of course, this set depends on k and could very well be empty: $\mathbf{V}(x^2 + y^2 + 1) = \varnothing$ for $k = \mathbb{R}$ but not for $k = \mathbb{C}$, while $\mathbf{V}(x^2 + y^2 + 1, x, y) = \varnothing$ no matter what k is, since k is a field.

The following proposition gives an important property of affine varieties. The proof is left as Exercise 1.3, in which the reader is asked to prove in addition that the arbitrary (that is, possibly infinite) intersection of affine varieties is still an affine variety.

Proposition 1.1.4. *If $V \subset k^n$ and $W \subset k^n$ are affine varieties, then $V \cup W$ and $V \cap W$ are also affine varieties.*

It is easy to see that, given an affine variety V, the collection of polynomials $\{f_1,\ldots,f_s\}$ such that $V=\mathbf{V}(f_1,\ldots,f_s)$ is not unique, and thus cannot be uniquely recovered from the point set V. For example, for any a and b in k, $a\neq 0$, it is apparent that $\mathbf{V}(f_1,\ldots,f_s)=\mathbf{V}(af_1+bf_2,f_2,\ldots,f_s)$. See also Example 1.1.13 and Proposition 1.1.11. In order to connect a given variety with a particular collection of polynomials, we need the concept of an ideal, the main algebraic object of study in this chapter.

Definition 1.1.5. An *ideal* of $k[x_1,\ldots,x_n]$ is a subset I of $k[x_1,\ldots,x_n]$ satisfying
(a) $0\in I$,
(b) if $f,g\in I$ then $f+g\in I$, and
(c) if $f\in I$ and $h\in k[x_1,\ldots,x_n]$, then $hf\in I$.

Let $f_1,\ldots,f_s$ be elements of $k[x_1,\ldots,x_n]$. We denote by $\langle f_1,\ldots,f_s\rangle$ the set of all linear combinations of $f_1,\ldots,f_s$ with coefficients from $k[x_1,\ldots,x_n]$:

$$\langle f_1,\ldots,f_s\rangle=\left\{\sum_{j=1}^{s}h_jf_j\;:\;h_1,\ldots,h_s\in k[x_1,\ldots,x_n]\right\}. \tag{1.4}$$

It is easily seen that the set $\langle f_1,\ldots,f_s\rangle$ is an ideal in $k[x_1,\ldots,x_n]$. We call $\langle f_1,\ldots,f_s\rangle$ the *ideal generated by the polynomials* $f_1,\ldots,f_s$, and the polynomials themselves *generators* of I. A generalization of this idea that will be important later is the following: if F is any nonempty subset of $k[x_1,\ldots,x_n]$ (possibly infinite), then we let $\langle f : f\in F\rangle$ denote the set of all *finite* linear combinations of elements of F with coefficients from $k[x_1,\ldots,x_n]$. (Occasionally we will abbreviate the notation to just $\langle F\rangle$.) Then $\langle f : f\in F\rangle$ is also an ideal, the *ideal generated by the elements of* F, which are likewise called its generators (Exercise 1.4; see Exercise 1.38). An arbitrary ideal $I\subset k[x_1,\ldots,x_n]$ is called *finitely generated* if there exist polynomials $f_1,\ldots,f_s\in k[x_1,\ldots,x_n]$ such that $I=\langle f_1,\ldots,f_s\rangle$; the set $f_1,\ldots,f_s$ is called a *basis* of I. The concept of an ideal arises in the context of arbitrary commutative rings. In that setting an ideal need not be finitely generated, but in a polynomial ring over a field it must be:

Theorem 1.1.6 (Hilbert Basis Theorem). *If k is a field, then every ideal in the polynomial ring $k[x_1,\ldots,x_n]$ is finitely generated.*

For a proof of the Hilbert Basis Theorem the reader is referred to [1, 60, 132, 195].

Corollary 1.1.7. *Every ascending chain of ideals $I_1\subset I_2\subset I_3\subset\cdots$ in a polynomial ring over a field k stabilizes. That is, there exists $m\geq 1$ such that for every $j>m$, $I_j=I_m$.*

Proof. Let $I_1\subset I_2\subset I_3\subset\cdots$ be an ascending chain of ideals in $k[x_1,\ldots,x_n]$ and set $I=\cup_{j=1}^{\infty}I_j$, clearly an ideal in $k[x_1,\ldots,x_n]$. By the Hilbert Basis Theorem there exist $f_1,\ldots,f_s$ in $k[x_1,\ldots,x_n]$ such that $I=\langle f_1,\ldots,f_s\rangle$. Choose any $N\in\mathbb{N}$ such that $F=\{f_1,\ldots,f_s\}\subset I_N$, and suppose that $g\in I_p$ for some $p\geq N$. Since $g\in I$ and F is a basis for I, there exist $h_1,\ldots,h_s\in k[x_1,\ldots,x_n]$ such that $g=h_1f_1+\cdots+h_sf_s$.

But then because $F \subset I_N$ and I_N is an ideal, $g \in I_N$. Thus $I_p \subset I_N$, and the ascending chain has stabilized by I_N. □

Rings in which every strictly ascending chain of ideals stabilizes are called *Noetherian* rings. The Hilbert Basis Theorem and its corollary hold under the milder condition that k be only a commutative Noetherian ring. Some condition is necessary, though, which is why in the statements above we explicitly included the condition that k be a field, which is enough for our puposes.

Occasionally we will find that it is important not to distinguish between two polynomials whose difference lies in a particular ideal I. Thus, we define a relation on $k[x_1,\ldots,x_n]$ by saying that f and g are related if $f-g \in I$. This relation is an equivalence relation (Exercise 1.5) and is the basis for the following definition.

Definition 1.1.8. Let I be an ideal in $k[x_1,\ldots,x_n]$. Two polynomials f and g in $k[x_1,\ldots,x_n]$ are congruent modulo I, denoted $f \equiv g \bmod I$, if $f-g \in I$. The set of equivalence classes is denoted $k[x_1,\ldots,x_n]/I$.

As a simple example, if in $\mathbb{R}[x]$ we take $I = \langle x \rangle$, then $f \equiv g \bmod I$ precisely when $f(x) - g(x) = xh(x)$ for some polynomial h. Hence f and g are equivalent if and only if they have the same constant term.

If for $f \in k[x_1,\ldots,x_n]$ the equivalence class of f is denoted $[f]$, then for any f_1 and f_2 in $[f]$ and for any g_1 and g_2 in $[g]$, $(f_1+g_1)-(f_2+g_2) \in I$ and $f_1g_1 - f_2g_2 \in I$. We conclude that an addition and multiplication are defined on $k[x_1,\ldots,x_n]/I$ by $[f]+[g] = [f+g]$ and $[f][g] = [fg]$, which give it the structure of a ring (Exercise 1.6).

Suppose $f_1,\ldots,f_s \in k[x_1,\ldots,x_n]$ and consider system (1.3), whose solution set is the affine variety $V = \mathbf{V}(f_1,\ldots,f_s)$. The reader may readily verify that for any $\mathbf{a} \in k^n$, $\mathbf{a} \in V$ if and only if $f(\mathbf{a}) = 0$ for *every* $f \in I = \langle f_1,\ldots,f_s \rangle$. V is the set of common zeros of the full (typically infinite) set I of polynomials. Moreover, given the ideal I, as the following proposition states, the particular choice of generators is unimportant; the same variety will be determined. Thus, it is the *ideal* that determines the variety, and not the particular collection of polynomials $f_1,\ldots,f_s$.

Proposition 1.1.9. *Let $f_1,\ldots,f_s$ and $g_1,\ldots,g_m$ be bases of an ideal $I \in k[x_1,\ldots,x_n]$, that is, $I = \langle f_1,\ldots,f_s \rangle = \langle g_1,\ldots,g_m \rangle$. Then $\mathbf{V}(f_1,\ldots,f_s) = \mathbf{V}(g_1,\ldots,g_m)$.*

The straightforward proof is left to the reader.

We have seen how a finite collection of polynomials defines a variety. Conversely, given a variety V, there is naturally associated to it an ideal. As already noted, the collection of polynomials in a system (1.3) for which V is the solution set is not unique, and neither is the ideal they generate, although any such ideal has the property that V is precisely the subset of k^n on which every element of the ideal vanishes. The ideal naturally associated to V is the one given in the following definition.

Definition 1.1.10. Let $V \subset k^n$ be an affine variety. The *ideal of the variety V* is the set

$$\mathbf{I}(V) = \{f \in k[x_1,\ldots,x_n] \ : \ f(a_1,\ldots,a_n) = 0 \text{ for all } (a_1,\ldots,a_n) \in V\}.$$

In Exercise 1.7 the reader is asked to show that $\mathbf{I}(V)$ *is* an ideal in $k[x_1,\ldots,x_n]$, even if V is not a variety, but simply an arbitrary subset of k^n. (See also the discussion following Theorem 1.3.18.)

The ideal naturally associated to a variety V bears the following relation to the family of ideals that come from the polynomials in any system of equations that define V.

Proposition 1.1.11. *Let $f_1,\ldots,f_s$ be elements of $k[x_1,\ldots,x_n]$. Then the set inclusion $\langle f_1,\ldots,f_s\rangle \subset \mathbf{I}(\mathbf{V}(f_1,\ldots,f_s))$ always holds, but could be strict.*

Proof. Let $f \in \langle f_1,\ldots,f_s\rangle$. Then there exist $h_1,\ldots,h_s \in k[x_1,\ldots,x_n]$ such that $f = h_1f_1+\cdots+h_sf_s$. Since $f_1,\ldots,f_s$ all vanish on $\mathbf{V}(f_1,\ldots,f_s)$, so does f, so $f \in \mathbf{I}(\mathbf{V}(f_1,\ldots,f_s))$. The demonstration that the inclusion can be strict is given by Example 1.1.13. □

When V is not just a subset of k^n but a variety, the ideal $\mathbf{I}(V)$ naturally determined by V uniquely determines V:

Proposition 1.1.12. *Let V and W be affine varieties in k^n. Then*
1. $V \subset W$ if and only if $\mathbf{I}(W) \subset \mathbf{I}(V)$.
2. $V = W$ if and only if $\mathbf{I}(W) = \mathbf{I}(V)$.

Proof. (1) Suppose $V \subset W$. Then any polynomial that vanishes on W also vanishes on V, so $\mathbf{I}(W) \subset \mathbf{I}(V)$. Suppose conversely that $\mathbf{I}(W) \subset \mathbf{I}(V)$. Choose any collection $\{h_1,\ldots,h_s\} \subset k[x_1,\ldots,x_n]$ such that $W = \mathbf{V}(h_1,\ldots,h_s)$, which must exist, since W is a variety. Then for $1 \le j \le s$, $h_j \in \mathbf{I}(W) \subset \mathbf{I}(V)$, so that if $\mathbf{a} \in V$, then $h_j(\mathbf{a}) = 0$. That is, if $\mathbf{a} \in V$, then $\mathbf{a} \in \mathbf{V}(h_1,\ldots,h_s) = W$, so $V \subset W$.

Statement (2) is an immediate consequence of statement (1). □

Example 1.1.13. Let $V = \{(0,0)\} \subset \mathbb{R}^2$. Then $\mathbf{I}(V)$ is the set of all polynomials in two variables without constant term. We will express V as $\mathbf{V}(f_1,f_2)$ in two different ways. Choosing $f_1 = x$ and $f_2 = y$, $V = \mathbf{V}(f_1,f_2)$ and $I = \langle x,y\rangle$ is the same ideal as $\mathbf{I}(V)$. Choosing $f_1 = x^2$ and $f_2 = y$, $V = \mathbf{V}(f_1,f_2)$, but $J = \langle x^2,y\rangle$ is the set of elements of $\mathbb{R}[x,y]$, every term of which contains x^2 or y; hence $J \subsetneq \mathbf{I}(V)$. Note that both I and J have the property that V is precisely the set of common zeros of all their elements.

Denote by $\mathbb{V}$ the set of all affine varieties of k^n and by $\mathbb{I}$ the set of all polynomial ideals in $k[x_1,\ldots,x_n]$. Then Definition 1.1.10 defines a map

$$\mathbf{I} : \mathbb{V} \to \mathbb{I}. \tag{1.5}$$

Because every ideal I of $k[x_1,\ldots,x_n]$ has a finite basis (Theorem 1.1.6), so that $I = \langle f_1,\ldots,f_s\rangle$, and because the variety defined using any basis of I is the same as that defined using any other (Proposition 1.1.9), there is also a natural map from $\mathbb{I}$ to $\mathbb{V}$ defined by

$$\mathbf{V} : \mathbb{I} \to \mathbb{V} : \langle f_1,\ldots,f_s\rangle \mapsto \mathbf{V}(f_1,\ldots,f_s). \tag{1.6}$$

That is, for an ideal I in $k[x_1,\dots,x_n]$, $\mathbf{V}(I) = \mathbf{V}(f_1,\dots,f_s)$ for any finite collection of polynomials satisfying $I = \langle f_1,\dots,f_s\rangle$. Thus the symbol $\mathbf{V}$ will be doing double duty, since we will continue to write $\mathbf{V}(f_1,\dots,f_s)$ in place of the more cumbersome $\mathbf{V}(\langle f_1,\dots,f_s\rangle)$. The following theorem establishes some properties of the maps $\mathbf{I}$ and $\mathbf{V}$. (See also Theorem 1.3.15.)

Theorem 1.1.14. *For any field k, the maps* $\mathbf{I}$ *and* $\mathbf{V}$ *are inclusion-reversing.* $\mathbf{I}$ *is one-to-one (injective) and* $\mathbf{V}$ *is onto (surjective). Furthermore, for any variety* $V \subset k^n$, $\mathbf{V}(\mathbf{I}(V)) = V$.

Proof. In Exercise 1.8 the reader is asked to show that the maps $\mathbf{I}$ and $\mathbf{V}$ are inclusion-reversing. Now let an affine variety $V = \mathbf{V}(f_1,\dots,f_s)$ of k^n be given. Since $\mathbf{I}(V)$ is the collection of *all* polynomials that vanish on V, if $\mathbf{a} \in V$, then every element of $\mathbf{I}(V)$ vanishes at $\mathbf{a}$, so $\mathbf{a}$ is in the set of common zeros of $\mathbf{I}(V)$, which is $\mathbf{V}(\mathbf{I}(V))$. Thus, $V \subset \mathbf{V}(\mathbf{I}(V))$. For the reverse inclusion, by the definition of $\mathbf{I}(V)$, $f_j \in \mathbf{I}(V)$, $1 \le j \le s$; hence, $\langle f_1,\dots,f_s\rangle \subset \mathbf{I}(V)$. Since $\mathbf{V}$ is inclusion-reversing, $\mathbf{V}(\mathbf{I}(V)) \subset \mathbf{V}(\langle f_1,\dots,f_s\rangle) = \mathbf{V}(f_1,\dots,f_s) = V$.

Finally, $\mathbf{I}$ is one-to-one because it has a left inverse, and $\mathbf{V}$ is onto because it has a right inverse. □

1.2 The Ideal Membership Problem and Gröbner Bases

One of the main problems of computational algebra is the *Ideal Membership Problem*, formulated as follows.

> **Ideal Membership Problem.** Let $I \subset k[x_1,\dots,x_n]$ be an ideal and let f be an element of $k[x_1,\dots,x_n]$. Determine whether or not f is an element of I.

We first consider the polynomial ring with one variable x. One important feature of this ring is the existence of the Division Algorithm: given two polynomials f and g in $k[x]$, $g \neq 0$, there exist unique elements q and r of $k[x]$, the *quotient* and *remainder*, respectively, of f upon division by g, such that $f = qg + r$, and either $r = 0$ or $\deg(r) < \deg(g)$. To *divide* f by g is to express f as $f = qg + r$. We say that g *divides* f if $r = 0$, and write it as $g \mid f$. As outlined in Exercises 1.9–1.12, the greatest common divisor of two polynomials in $k[x]$ is defined, is easily computed using the Euclidean Algorithm, and can be used in conjunction with the Hilbert Basis Theorem to show that every ideal in $k[x]$ is generated by a single element. (An ideal generated by a single element is called a *principal* ideal, and a ring in which every ideal is principal is a *principal ideal domain*). The Ideal Membership Problem is then readily solved: given an ideal I and a polynomial f, we first find a generator g for I, then divide f by g; $f \in I$ if and only if $g \mid f$.

In polynomial rings of several variables, we want to follow an analogous procedure for solving the Ideal Membership Problem: performing a division and examining a remainder. Matters are more complicated, however. In particular, in general ideals are not generated by just one polynomial, so we have to formulate a procedure for dividing a polynomial f by a set F of polynomials, and although there is a way to generalize the Division Algorithm to do this for elements of $k[x_1,\ldots,x_n]$, a complication arises in that the remainder under the division is not necessarily unique.

To describe the division algorithm in $k[x_1,\ldots,x_n]$, we must digress for several paragraphs to introduce the concepts of a term ordering and of reduction of a polynomial modulo a set of polynomials, along with attendant terminology. We first of all specify an ordering on the terms of the polynomials. In the case of one variable there is the natural ordering according to degree. In the multivariable case there are different orders that can be used. We will define the general concept of a *term order* and a few of the most frequently used term orders. Observe that because of the one-to-one correspondence between monomials $\mathbf{x}^\alpha = x_1^{\alpha_1}x_2^{\alpha_2}\cdots x_n^{\alpha_n}$ and n-tuples $\alpha = (\alpha_1,\ldots,\alpha_n) \in \mathbb{N}_0^n$, it is sufficient to order elements of $\mathbb{N}_0^n$ (for, as in the one-variable case, the actual coefficients of the terms play no role in the ordering). Underlying this correspondence, of course, is the assumption of the ordering $x_1 > x_2 > \cdots > x_n$ of the variables themselves.

Recall that a *partial order* $\succ$ on a set S is a binary relation that is reflexive ($a \succ a$ for all $a \in S$), antisymmetric ($a \succ b$ and $b \succ a$ only if $a = b$), and transitive ($a \succ b$ and $b \succ c$ implies $a \succ c$). A *total order* $>$ on S is a partial order under which any two elements can be compared: for all a and b in S, either $a = b$, $a > b$, or $b > a$.

Definition 1.2.1. A *term order* on $k[x_1,\ldots,x_n]$ is a total order $>$ on $\mathbb{N}_0^n$ having the following two properties:
(a) for all α, β, and γ in $\mathbb{N}_0^n$, if $\alpha > \beta$, then $\alpha+\gamma > \beta+\gamma$; and
(b) $\mathbb{N}_0^n$ is *well-ordered* by $>$: if S is any nonempty subset of $\mathbb{N}_0^n$, then there exists a smallest element μ of S (for all $\alpha \in S\setminus\{\mu\}$, $\alpha > \mu$).

The monomials $\{\mathbf{x}^\alpha : \alpha \in \mathbb{N}_0\}$ are then ordered by the ordering of their exponents, so that $\mathbf{x}^\alpha > \mathbf{x}^\beta$ if and only if $\alpha > \beta$. Note that while we speak of the term order $>$ as being on $k[x_1,\ldots,x_n]$, we are not actually ordering all elements of $k[x_1,\ldots,x_n]$, but only the monomials, hence the individual terms of the polynomials that comprise $k[x_1,\ldots,x_n]$; this explains the terminology *term order*. The terminology *monomial order* is also widely used.

A sequence α_j in $\mathbb{N}_0^n$ is *strictly descending* if, for all j, $\alpha_j > \alpha_{j+1}$ and $\alpha_j \neq \alpha_{j+1}$. Such a sequence *terminates* if it is finite.

Proposition 1.2.2. *A total order $>$ on $\mathbb{N}_0^n$ well-orders $\mathbb{N}_0^n$ if and only if each strictly descending sequence of elements of $\mathbb{N}_0^n$ terminates.*

Proof. If there exists a strictly descending sequence $\alpha_1 > \alpha_2 > \alpha_3 > \cdots$ that does not terminate, then $\{\alpha_1,\alpha_2,\ldots\}$ is a nonempty subset of $\mathbb{N}_0^n$ with no minimal element, and $>$ does not well-order $\mathbb{N}_0^n$.

Conversely, if $>$ does not well-order $\mathbb{N}_0^n$, then there exists a nonempty subset A of $\mathbb{N}_0^n$ that has no minimal element. Let α_1 be an arbitrary element of A. It is not minimal; hence there exists $\alpha_2 \in A$, $\alpha_2 \neq \alpha_1$, such that $\alpha_1 > \alpha_2$. Continuing the process, we get a strictly descending sequence that does not terminate. □

We now define the three most commonly used term orders; in Exercise 1.16 we ask the reader to verify that they indeed meet the conditions in Definition 1.2.1. Addition and rescaling in $\mathbb{Z}^n$ are performed componentwise: for α, $\beta \in \mathbb{Z}^n$ and $p \in \mathbb{Z}$, the jth entry of $\alpha + p\beta$ is the jth entry of α plus p times the jth entry of β. The word "graded" is sometimes used where we use the word "degree."

Definition 1.2.3. Let $\alpha = (\alpha_1, \dots, \alpha_n)$ and $\beta = (\beta_1, \dots, \beta_n)$ be elements of $\mathbb{N}_0^n$.
(a) Lexicographic Order. Define $\alpha >_{\text{lex}} \beta$ if and only if, reading left to right, the first nonzero entry in the n-tuple $\alpha - \beta \in \mathbb{Z}^n$ is positive.
(b) Degree Lexicographic Order. Define $\alpha >_{\text{deglex}} \beta$ if and only if

$$|\alpha| = \sum_{j=1}^{n} \alpha_j > |\beta| = \sum_{j=1}^{n} \beta_j \qquad \text{or} \qquad |\alpha| = |\beta| \text{ and } \alpha >_{\text{lex}} \beta .$$

(c) Degree Reverse Lexicographic Order. Define $\alpha >_{\text{degrev}} \beta$ if and only if either $|\alpha| > |\beta|$ or $|\alpha| = |\beta|$ and, reading right to left, the first nonzero entry in the n-tuple $\alpha - \beta \in \mathbb{Z}^n$ is negative.

For example, if $\alpha = (1,4,4,2)$ and $\beta = (1,2,6,2)$, then α is greater than β with respect to all three orders. Note in particular that this example shows that degrev is not simply the reverse of deglex.

When a term order $>$ on $k[x_1, \dots, x_n]$ is given, we write $a_\alpha \mathbf{x}^\alpha > a_\beta \mathbf{x}^\beta$ if and only if $\alpha > \beta$. We reiterate that the definitions above are based on the presumed ordering $x_1 > \cdots > x_n$ of the variables. This ordering must be explicitly identified when non-subscripted variables are in use. For instance, if in $k[x,y]$ we choose $y > x$, then $y^5 >_{\text{lex}} x^9$ (since $(5,0) >_{\text{lex}} (0,9)$) and $xy^4 >_{\text{deglex}} x^2y^3$ (since $4+1 = 3+2$ and $(4,1) >_{\text{lex}} (3,2)$), and we will typically write these latter two terms as y^4x and y^3x^2 to reflect the underlying ordering of the variables themselves.

Fixing a term order $>$ on $k[x_1, \dots, x_n]$, any nonzero $f \in k[x_1, \dots, x_n]$ may be written in the *standard form*, with respect to $>$,

$$f = a_1\mathbf{x}^{\alpha_1} + a_2\mathbf{x}^{\alpha_2} + \cdots + a_s\mathbf{x}^{\alpha_s}, \tag{1.7}$$

where $a_j \neq 0$ for $j = 1, \dots, s$, $\alpha_i \neq \alpha_j$ for $i \neq j$ and $1 \leq i, j \leq s$, and where, with respect to the specified term order, $\alpha_1 > \alpha_2 > \cdots > \alpha_s$.

Definition 1.2.4. Let a term order on $k[x_1, \dots, x_n]$ be specified and let f be a nonzero element of $k[x_1, \dots, x_n]$, written in the standard form (1.7).
(a) The *leading term* $\text{LT}(f)$ of f is the term $\text{LT}(f) = a_1\mathbf{x}^{\alpha_1}$.
(b) The *leading monomial* $\text{LM}(f)$ of f is the monomial $\text{LM}(f) = \mathbf{x}^{\alpha_1}$.
(c) The *leading coefficient* $\text{LC}(f)$ of f is the coefficient $\text{LC}(f) = a_1$.

The concept of division of single-variable polynomials has an obvious generalization to the case of division of one monomial by another: we say that a monomial $\mathbf{x}^\alpha = x_1^{\alpha_1} \cdots x_n^{\alpha_n}$ *divides* a monomial $\mathbf{x}^\beta = x_1^{\beta_1} \cdots x_n^{\beta_n}$, written $\mathbf{x}^\alpha \mid \mathbf{x}^\beta$, if and only if $\beta_j \geq \alpha_j$ for all j, $1 \leq j \leq n$. In such a case the notation $\mathbf{x}^\beta / \mathbf{x}^\alpha$ denotes the monomial

$x_1^{\beta_1-\alpha_1}\cdots x_n^{\beta_n-\alpha_n}$. In $k[x_1,\ldots,x_n]$, to divide a polynomial f by nonzero polynomials $\{f_1,\ldots,f_s\}$ means to represent f in the form

$$f = u_1 f_1 + \cdots + u_s f_s + r,$$

where $u_1,\ldots,u_s, r \in k[x_1,\ldots,x_n]$, and either $r = 0$ or $\deg(r) \le \deg(f)$ (the inequality is not strict). The most important part of this expression is the remainder r, not the weights u_j, for the context in which we intend to apply the division concept is that the f_i are generators of an ideal I, and we want the division to produce a zero remainder r if and only if f is in I.

We must first specify a term order on $k[x_1,\ldots,x_n]$. The main idea then of the algorithm for the division is the same as in the one-variable case: we reduce the leading term of f (as determined by the specified term order) by multiplying some f_j by an appropriate term and subtracting. We will describe the procedure in detail, but to understand the motivation for the following definition, recall that in the familiar one-variable case of polynomial long division, in the first pass through the algorithm dividing, for example, $f = 6x^3 + \cdots$ (lower-order terms omitted) by $g = 7x^2 + \cdots$, we compare the leading terms, multiply g by $\frac{6}{7}x = \frac{6x^3}{7x^2} = \frac{\mathrm{LT}(f)}{\mathrm{LT}(g)}$, and then subtract the product from f to obtain a polynomial $h = f - \frac{6}{7}xg$ that satisfies $\deg(h) < \deg(f)$.

Definition 1.2.5. (a) For f, g, $h \in k[x_1,\ldots,x_n]$ with $g \neq 0$, we say that f *reduces to h modulo g in one step*, written as

$$f \xrightarrow{g} h,$$

if and only if $\mathrm{LM}(g)$ divides a nonzero term X that appears in f and

$$h = f - \frac{X}{\mathrm{LT}(g)} g. \tag{1.8}$$

(b) For $f, f_1,\ldots,f_s, h \in k[x_1,\ldots,x_n]$ with $f_j \neq 0$, $1 \le j \le s$, letting $F = \{f_1,\ldots,f_s\}$, we say that f *reduces to h modulo F*, written as

$$f \xrightarrow{F} h,$$

if and only if there exist a sequence of indices $j_1, j_2, \ldots, j_m \in \{1,\ldots,s\}$ and a sequence of polynomials $h_1,\ldots,h_{m-1} \in k[x_1,\ldots,x_n]$ such that

$$f \xrightarrow{f_{j_1}} h_1 \xrightarrow{f_{j_2}} h_2 \xrightarrow{f_{j_3}} \cdots \xrightarrow{f_{j_{m-1}}} h_{m-1} \xrightarrow{f_{j_m}} h.$$

Remark 1.2.6. Applying part (a) of the definition repeatedly shows that if $f \xrightarrow{F} h$, then there exist $u_j \in k[x_1,\ldots,x_n]$ such that $f = u_1 f_1 + \cdots + u_s f_s + h$. Hence, by Definition 1.1.8 f, reduces to h modulo $F = \{f_1,\ldots,f_s\}$ only if $f \equiv h \bmod \langle f_1,\ldots,f_s\rangle$. The converse is false, as shown by Example 1.2.12.

Example 1.2.7. We illustrate each part of Definition 1.2.5.
(a) In $\mathbb{Q}[x,y]$ with $x > y$ and the term order deglex, let $f = x^2y + 2xy - 3x + 5$ and

$g = xy + 6y^2 - 4x$. If the role of X is played by the leading term x^2y in f, then

$$h = f - \frac{x^2y}{xy}(xy + 6y^2 - 4x) = -6xy^2 + 4x^2 + 2xy - 3x + 5,$$

so $f \xrightarrow{g} h$ and $\text{LM}(h) < \text{LM}(f)$. If the role of X is played by the term $2xy$ in f, then

$$\tilde{h} = f - \frac{2xy}{xy}(xy + 6y^2 - 4x) = x^2y - 12y^2 + 5x + 5,$$

so $f \xrightarrow{g} \tilde{h}$ and $\text{LT}(\tilde{h}) = \text{LT}(f)$. In either case we remove the term X from f and replace it with a term that is smaller with respect to deglex.
(b) In $\mathbb{Q}[x,y]$ with $y > x$ and the term order deglex, let $f = y^2x + y^2 + 3y$, $f_1 = yx + 2$, and $f_2 = y + x$. Then

$$y^2x + y^2 + 3y \xrightarrow{f_1} y^2 + y \xrightarrow{f_2} -yx + y \xrightarrow{f_2} x^2 + y,$$

so $f \xrightarrow{\{f_1,f_2\}} x^2 + y$.

Definition 1.2.8. Suppose $f, f_1, \ldots, f_s \in k[x_1, \ldots, x_n]$, $f_j \neq 0$ for $1 \leq j \leq s$, and let $F = \{f_1, \ldots, f_s\}$.

(a) A polynomial $r \in k[x_1, \ldots, x_n]$ is *reduced with respect to F* if either $r = 0$ or no monomial that appears in the polynomial r is divisible by any element of the set $\{\text{LM}(f_1), \ldots, \text{LM}(f_s)\}$.

(b) A polynomial $r \in k[x_1, \ldots, x_n]$ is a *remainder for f with respect to F* if $f \xrightarrow{F} r$ and r is reduced with respect to F.

The Multivariable Division Algorithm is the direct analogue of the procedure used to divide one single-variable polynomial by another. To divide f by the ordered set $F = \{f_1, \ldots, f_s\}$, we proceed iteratively, at each step performing a familiar polynomial long division using one element of F. Typically, the set F of divisors is presented to us in no particular order, so as a preliminary we must order its elements in some fashion; the order selected can affect the final result. At the first step in the actual division process, the "active divisor" is the first element of F, call it f_j, whose leading term divides the leading term of f; at this step we replace f by the polynomial h of (1.8) when $X = \text{LT}(f)$ and $g = f_j$, thereby reducing f somewhat using f_j. At each succeeding step, the active divisor is the first element of F whose leading term divides the leading term of the current polynomial h; at this step we similarly reduce h somewhat using the active divsior. If at any stage no division is possible, then the leading term of h is added to the remainder, and we try the same process again, continuing until no division is possible at all. By Exercise 1.17, building up the remainder successively is permissible. An explicit description of the procedure is given in Table 1.1 on page 12. In the next theorem we will prove that the algorithm works correctly to perform the reduction $f \xrightarrow{F} r$ and generate the components of the expression $f = u_1 f_1 + \cdots + u_s f_s + r$, where r is a remainder for f with respect to F (thus showing that a remainder always exists), but first we present an example.

Multivariable Division Algorithm

Input:

$f \in k[x_1, \ldots, x_n]$
ordered set $F = \{f_1, \ldots, f_s\} \subset k[x_1, \ldots, x_n] \setminus \{0\}$

Output:

$u_1, \ldots, u_s, r \in k[x_1, \ldots, x_n]$ such that
1. $f = u_1 f_1 + \cdots + u_s f_s + r$,
2. r is reduced with respect to $\{f_1, \ldots, f_s\}$, and
3. $\max(\mathrm{LM}(u_1)\mathrm{LM}(f_1), \ldots, \mathrm{LM}(u_s)\mathrm{LM}(f_s), \mathrm{LM}(r)) = \mathrm{LM}(f)$

Procedure:

$u_1 := 0;\ \ldots,\ u_s := 0;\ r := 0;\ h := f$
WHILE $h \neq 0$ DO
  IF
  There exists j such that $\mathrm{LM}(f_j)$ divides $\mathrm{LM}(h)$
  THEN
  For the least j such that $\mathrm{LM}(f_j)$ divides $\mathrm{LM}(h)$
  $u_j := u_j + \dfrac{\mathrm{LT}(h)}{\mathrm{LT}(f_j)}$
  $h := h - \dfrac{\mathrm{LT}(h)}{\mathrm{LT}(f_j)} f_j$
  ELSE
  $r := r + \mathrm{LT}(h)$
  $h := h - \mathrm{LT}(h)$

Table 1.1 The Multivariable Division Algorithm

Example 1.2.9. In $\mathbb{Q}[x,y]$ with $x > y$ and the term order lex, we apply the algorithm to divide $f = x^2y + xy^3 + xy^2$ by the polynomials $f_1 = xy + 1$ and $f_2 = y^2 + 1$, ordered f_1 then f_2.

The two panels in Table 1.2 on page 14 show the computation in tabular form and underscore the analogy with the one-variable case. The top panel shows three divsions by f_1, at which point no division (by either divisor) is possible. The leading term $-x$ is sent to the remainder, and the process is restarted. Division by f_1 is impossible, but the bottom panel shows one further division by f_2, and then all remaining terms are sent to the remainder. Therefore,

$$f = u_1 f_1 + u_2 f_2 + r = (x + y^2 + y) f_1 + (-1) f_2 + (-x - y + 1).$$

That is, the quotient is $\{u_1, u_2\} = \{x + y^2 + y, -1\}$ and the remainder is $-x - y + 1$. (In general, the role of the divisor on each step will alternate between the f_j, so that in a hand computation the full table will contain more than s panels, and successive dividends when f_j is the active divisor must be added to obtain u_j.)

Now let us go through exactly the same computation by means of an explicit application of the Multivariable Division Algorithm. That is, we will follow the instructions presented in Table 1.1 in a step-by-step fashion.

First pass:
$\mathrm{LM}(f_1) \mid \mathrm{LM}(h)$ but $\mathrm{LM}(f_2) \nmid \mathrm{LM}(h)$
f_1 is least
$u_1 = 0 + \frac{x^2y}{xy} = x$
$h = (x^2y + xy^3 + xy^2) - x(xy+1) = xy^3 + xy^2 - x$

Second pass:
$\mathrm{LM}(f_1) \mid \mathrm{LM}(h)$ and $\mathrm{LM}(f_2) \mid \mathrm{LM}(h)$
f_1 is least
$u_1 = x + \frac{xy^3}{xy} = x + y^2$
$h = (xy^3 + xy^2 - x) - y^2(xy+1) = xy^2 - x - y^2$

Third pass:
$\mathrm{LM}(f_1) \mid \mathrm{LM}(h)$ and $\mathrm{LM}(f_2) \mid \mathrm{LM}(h)$
f_1 is least
$u_1 = x + y^2 + \frac{xy^2}{xy} = x + y^2 + y$
$h = (xy^2 - x - y^2) - y(xy+1) = -x - y^2 - y$

Fourth pass:
$\mathrm{LM}(f_1) \nmid \mathrm{LM}(h)$ and $\mathrm{LM}(f_2) \nmid \mathrm{LM}(h)$
$r = 0 + (-x) = -x$
$h = (-x - y^2 - y) - (-x) = -y^2 - y$

Fifth pass:
$\mathrm{LM}(f_1) \nmid \mathrm{LM}(h)$ but $\mathrm{LM}(f_2) \mid \mathrm{LM}(h)$
f_2 is least
$u_2 = 0 + \frac{-y^2}{y} = -1$
$h = (-y^2 - y) - (-1)(y^2+1) = -y + 1$

Sixth pass:
$\mathrm{LM}(f_1) \nmid \mathrm{LM}(h)$ and $\mathrm{LM}(f_2) \nmid \mathrm{LM}(h)$
$r = -x + (-y) = -x - y$
$h = (-y+1) - (-y) = 1$

Seventh pass:
$\mathrm{LM}(f_1) \nmid \mathrm{LM}(h)$ and $\mathrm{LM}(f_2) \nmid \mathrm{LM}(h)$
$r = -x - y + 1$
$h = 1 - 1 = 0$

A summary statement in the language of Definition 1.2.5 for these computations is the string of reductions and equalities

$$
\begin{aligned}
f \xrightarrow{f_1} h_1 \xrightarrow{f_1} h_2 \xrightarrow{f_1} h_3 = h_4 + (-x) \xrightarrow{f_2} h_5 + (-x) &= h_6 + (-x-y) \\
&= h_7 + (-x-y-1) = -x-y-1
\end{aligned}
$$

or, more succinctly, $f \xrightarrow{\{f_1, f_2\}} -x-y-1$.

$$
\begin{array}{r|l|l|l}
 & x + y^2 + y & \longleftarrow [u_1] & r: \\
xy + 1 & x^2y + xy^3 + xy^2 & \longleftarrow [f] & \\
 & x^2y \qquad\qquad + x & & \\
\hline
 & \quad xy^3 + xy^2 - x & \longleftarrow [h_1] & \\
 & \quad xy^3 \qquad\quad + y^2 & & \\
\hline
 & \qquad xy^2 - x - y^2 & \longleftarrow [h_2] & \\
 & \qquad xy^2 \qquad\quad + y & & \\
\hline
 & \qquad\quad - x - y^2 - y & \longleftarrow [h_3] & \\
 & & & \\
 & -1 & \longleftarrow [u_2] & \\
y^2 + 1 & -y^2 - y & \longleftarrow [h_4] & -x \\
 & -y^2 \qquad - 1 & & \\
\hline
 & \quad - y + 1 & \longleftarrow [h_5] & \\
 & \qquad\quad 1 & \longleftarrow [h_6] & -x-y \\
 & \qquad\quad 0 & \longleftarrow [h_7] & -x-y+1
\end{array}
$$

Table 1.2 The Computations of Example 1.2.9

Theorem 1.2.10. *Let an ordered set $F = \{f_1, \dots, f_s\} \subset k[x_1, \dots, x_n] \setminus \{0\}$ of nonzero polynomials and a polynomial $f \in k[x_1, \dots, x_n]$ be given. The Multivariable Division Algorithm produces polynomials $u_1, \dots, u_s, r \in k[x_1, \dots, x_n]$ such that*

$$f = u_1 f_1 + \cdots + u_s f_s + r, \tag{1.9}$$

where r is a remainder for f with respect to F and

$$\mathrm{LM}(f) = \max(\mathrm{LM}(u_1)\mathrm{LM}(f_1), \dots, \mathrm{LM}(u_s)\mathrm{LM}(f_s), \mathrm{LM}(r)), \tag{1.10}$$

where $\mathrm{LM}(u_j)\mathrm{LM}(f_j)$ is not present in (1.10) if $u_j = 0$.

Proof. The algorithm certainly produces the correct result $u_j = 0$, $1 \le j \le s$, and $r = f$ in the special cases that $f = 0$ or that no leading term in any of the divisors divides any term of f. Otherwise, after the first pass through the WHILE loop for which the IF statement is true, exactly one polynomial u_j is nonzero, and we have

$$\begin{aligned}\max[\mathrm{LM}(u_1)\mathrm{LM}(f_1),\ldots,\mathrm{LM}(u_s)\mathrm{LM}(f_s),\mathrm{LM}(h)]\\ = \max[\mathrm{LM}(u_1)LM(f_1),\ldots,\mathrm{LM}(u_s)\mathrm{LM}(f_s)],\end{aligned} \tag{1.11}$$

which clearly holds at every succeeding stage of the algorithm. Consequently, (1.10) holds at that and every succeeding stage of the algorithm, hence holds when the algorithm terminates.

At every stage of the algorithm, $f = u_1 f_1 + \cdots + u_s f_s + r + h$ holds. Because the algorithm halts precisely when $h = 0$, this implies that on the last step (1.9) holds. Moreover, since at each stage we add to r only terms that are not divisible by $\mathrm{LT}(f_j)$ for any j, $1 \leq j \leq s$, r is reduced with respect to F, and thus is a remainder of f with respect to F.

To show that the algorithm must terminate, let $h_1, h_2, \ldots$ be the sequence of polynomials produced by the successive values of h upon successive passes through the WHILE loop. The algorithm fails to terminate only if for every $j \in \mathbb{N}$ there is a jth pass through the loop, hence an $h_j \neq 0$. Then $\mathrm{LM}(h_j)$ exists for each $j \in \mathbb{N}$, and the sequence $\mathrm{LM}(h_1), \mathrm{LM}(h_2), \ldots$ satisfies $\mathrm{LM}(h_{j+1}) < \mathrm{LM}(h_j)$ and $\mathrm{LM}(h_{j+1}) \neq \mathrm{LM}(h_j)$, which contradicts Proposition 1.2.2. □

If we change the order of the polynomials in Example 1.2.9, dividing first by f_2 and then by f_1, then the quotient and remainder change to $\{xy + x, x - 1\}$ and $-2x + 1$, respectively (Exercise 1.21). Thus we see that, unlike the situation in the one-variable case, the quotient and remainder are not unique. They depend on the ordering of the polynomials in the set of divisors as well as on the term order chosen for the polynomial ring (see Exercise 1.22). But what is even worse from the point of view of solving the Ideal Membership Problem is that, as the following examples show, it is even possible that, keeping the term order fixed, there exist an element of the ideal generated by the divisors whose remainder can be zero under one ordering of the divisors and different from zero under another, or even different from zero no matter how the divisors are ordered.

Example 1.2.11. In the ring $\mathbb{R}[x,y]$ fix the lexicographic term order with $x > y$ and consider the polynomial $f = x^2y + xy + 2x + 2$. When we use the Multivariable Division Algorithm to reduce the polynomial f modulo the ordered set $\{f_1 = x^2 - 1, f_2 = xy + 2\}$, we obtain

$$f = yf_1 + f_2 + (2x + y).$$

Since the corresponding remainder, $2x + y$, is different from zero, we might conclude that f is not in the ideal $\langle f_1, f_2 \rangle$. If, however, we change the order of the divisors so that f_2 is first, we obtain

$$f = 0 \cdot f_1 + (x+1)f_2 + 0 = (x+1)f_2, \tag{1.12}$$

so that $f \in \langle f_1, f_2 \rangle$ after all.

Example 1.2.12. In the ring $\mathbb{R}[x,y]$ fix the lexicographic term order with $x > y$. Then $2y = 1 \cdot (x+y) + (-1) \cdot (x-y) \in \langle x+y, x-y \rangle$, but because $\mathrm{LT}(x+y) = \mathrm{LT}(x-y) = x$

does not divide $2y$ the remainder of $2y$ with respect to $\{x+y, x-y\}$ is unique and is $2y$. Thus, for either ordering of the divisors, the Multivariable Division Algorithm produces this nonzero remainder.

We see then that we have lost the tool that we had in polynomial rings of one variable for resolving the Ideal Membership Problem. Fortunately, not all is lost. While the Multivariable Division Algorithm cannot be improved in general, it has been discovered that if we use a certain special generating set for our ideals, then it is still true that $f \in \langle f_1, \ldots, f_s \rangle$ if and only if the remainder in the Division Algorithm is equal to zero, and we are able to decide the Ideal Membership Problem. Such a special generating set for an ideal is called a *Gröbner basis* or a *standard basis*. It is one of the primary tools of computational algebra and is the basis of numerous algorithms of computational algebra and algebraic geometry. To motivate the definition of a Gröbner basis we discuss Example 1.2.11 again. It showed that in the ring $k[x,y]$, under lex with $x > y$, for $f = x^2y + xy + 2x + 2$, $f_1 = x^2 - 1$, and $f_2 = xy + 2$,

$$f \xrightarrow{\{f_1, f_2\}} 2x + y.$$

But by (1.12), $f \in \langle f_1, f_2 \rangle$, so the remainder $r = 2x + y$ must also be in $\langle f_1, f_2 \rangle$. The trouble is that the leading term of r is not divisible by either $\mathrm{LM}(f_1)$ or $\mathrm{LM}(f_2)$, and this is what halts the division process in the Multivariable Division Algorithm. So the problem is that the ideal $\langle f_1, f_2 \rangle$ contains elements that are not divisible by a leading term of either element of the particular basis $\{f_1, f_2\}$ of the ideal.

If, for any ideal I, we had a basis B with the special property that the leading term of every polynomial in I was divisible by the leading term of some element of B, then the Multivariable Division Algorithm would provide an answer to the Ideal Membership Problem: a polynomial f is in the ideal I if and only if the remainder of f upon division by elements of B, in any order, is zero. This the idea behind the concept of a Gröbner basis of an ideal, and we use this special property as the defining characteristic of a Gröbner basis.

Definition 1.2.13. A *Gröbner basis* (also called a *standard basis*) of an ideal I in $k[x_1, \ldots, x_n]$ is a finite nonempty subset $G = \{g_1, \ldots, g_m\}$ of $I \setminus \{0\}$ with the following property: for every nonzero $f \in I$, there exists $g_j \in G$ such that $\mathrm{LT}(g_j) \mid \mathrm{LT}(f)$.

It is implicit in the definition that we do not consider the concept of a Gröbner basis G for the zero ideal, nor will we need it. See Section 5.2 of [18] for this more general situation, in which G must be allowed to be empty. Note that the requirement that the set G actually be a basis of the ideal I does not appear in the definition of a Gröbner basis but is a consequence of it (Theorem 1.2.16). Note also that whether or not a set G forms a Gröbner basis of an ideal I depends not only on the term order in use, but also on the underlying ordering of the variables. See Exercise 1.23.

With a Gröbner basis we again have the important property of uniqueness of the remainder, which we had in $k[x]$ and which we lost in the multivariable case for division by an arbitrary set of polynomials:

Proposition 1.2.14. *Let G be a Gröbner basis for a nonzero ideal I in $k[x_1,\ldots,x_n]$ and $f \in k[x_1,\ldots,x_n]$. Then the remainder of f with respect to G is unique.*

Proof. Suppose $f \xrightarrow{G} r_1$ and $f \xrightarrow{G} r_2$ and both r_1 and r_2 are reduced with respect to G. Since $f - r_1$ and $f - r_2$ are both in I, $r_1 - r_2 \in I$. By Definition 1.2.8, certainly $r_1 - r_2$ is reduced with respect to G. But then by Definition 1.2.13 it is immediate that $r_1 - r_2 = 0$, since it is in I. □

Definition 1.2.15. Let I be an ideal and f a polynomial in $k[x_1,\ldots,x_n]$. To *reduce f modulo the ideal I* means to find the unique remainder of f upon division by some Gröbner basis G of I. Given a nonzero polynomial g, to *reduce f modulo g* means to reduce f modulo the ideal $\langle g\rangle$.

Proposition 1.2.14 ensures that once a Gröbner basis is selected, the process is well-defined, although the remainder obtained depends on the Gröbner basis specified. We will see when this concept is applied in Section 3.7 that this ambiguity is not important in practice.

Let S be a subset of $k[x_1,\ldots,x_n]$ (possibly an ideal). We denote by $\mathrm{LT}(S)$ the set of leading terms of the polynomials that comprise S and by $\langle \mathrm{LT}(S)\rangle$ the ideal generated by $\mathrm{LT}(S)$ (the set of all finite linear combinations of elements of $\mathrm{LT}(S)$ with coefficients in $k[x_1,\ldots,x_n]$). The following theorem gives the main properties of Gröbner bases. We remind the reader that the expression $f \xrightarrow{F} h$ means that there is some sequence of reductions using the unordered set F of divisors that leads from f to h, which is not necessarily a remainder of f with respect to F. This is in contrast to the Multivariable Division Algorithm, in which F must be ordered, and the particular order selected determines a unique sequence of reductions from f to a remainder r.

Theorem 1.2.16. *Let $I \subset k[x_1,\ldots,x_n]$ be a nonzero ideal, let $G = \{g_1,\ldots,g_s\}$ be a finite set of nonzero elements of I, and let f be an arbitrary element of $k[x_1,\ldots,x_n]$. Then the following statements are equivalent:*

(i) G is a Gröbner basis for I;

(ii) $f \in I \Leftrightarrow f \xrightarrow{G} 0$;

(iii) $f \in I \Leftrightarrow f = \sum_{j=1}^{s} u_j g_j$ and $\mathrm{LM}(f) = \max_{1\le j\le s}(\mathrm{LM}(u_j)\mathrm{LM}(g_j))$;

(iv) $\langle \mathrm{LT}(G)\rangle = \langle \mathrm{LT}(I)\rangle$.

Proof. (i) $\Rightarrow$ (ii). Let any $f \in k[x_1,\ldots,x_n]$ be given. By Theorem 1.2.10 there exists $r \in k[x_1,\ldots,x_n]$ such that $f \xrightarrow{G} r$ and r is reduced with respect to G. If $f \in I$, then $r \in I$; hence, by the definition of Gröbner basis and the fact that r is reduced with respect to G, $r = 0$. Conversely, if $f \xrightarrow{G} 0$, then obviously $f \in I$.

(ii) $\Rightarrow$ (iii). Suppose $f \in I$. Then by (ii) there is a sequence of reductions

$$f \xrightarrow{g_{j_1}} h_1 \xrightarrow{g_{j_2}} h_2 \xrightarrow{g_{j_3}} \cdots \xrightarrow{g_{j_{m-1}}} h_{m-1} \xrightarrow{g_{j_m}} 0$$

which yields $f = u_1 g_1 + \cdots + u_2 g_2$ for some $u_j \in k[x_1,\ldots,x_n]$. Exactly as described in the first paragraph of the proof of Theorem 1.2.10, an equality analogous to (1.11) holds at each step of the reduction and thus the equality in (iii) holds. The reverse implication in (iii) is immediate.

(iii) $\Rightarrow$ (iv). $G \subset I$, hence $\langle \mathrm{LT}(G) \rangle \subset \langle \mathrm{LT}(I) \rangle$ is always true, so we must verify that $\langle \mathrm{LT}(I) \rangle \subset \langle \mathrm{LT}(G) \rangle$ when (iii) holds. The inclusion $\langle \mathrm{LT}(I) \rangle \subset \langle \mathrm{LT}(G) \rangle$ is implied by the implication: if $f \in I$, then $\mathrm{LT}(f) \in \langle \mathrm{LT}(G) \rangle$. Hence suppose $f \in I$. Then it follows immediately from the condition on $\mathrm{LM}(f)$ in (iii) that $\mathrm{LT}(f) = \sum_j \mathrm{LT}(u_j)\mathrm{LT}(g_j)$, where the summation is over all indices j such that $\mathrm{LM}(f) = \mathrm{LM}(u_j)\mathrm{LM}(g_j)$. Therefore $\langle \mathrm{LT}(I) \rangle \subset \langle \mathrm{LT}(G) \rangle$ holds.

(iv) $\Rightarrow$ (i). For $f \in I$, $\mathrm{LT}(f) \in \mathrm{LT}(I) \subset \langle \mathrm{LT}(I) \rangle = \langle \mathrm{LT}(G) \rangle$, so there exist $h_1, \ldots, h_m \in k[x_1, \ldots, x_n]$ such that

$$\mathrm{LT}(f) = \sum_{j=1}^{m} h_j \mathrm{LT}(g_j). \tag{1.13}$$

Write $\mathrm{LT}(f) = c x_1^{\beta_1} \cdots x_n^{\beta_n}$ and $\mathrm{LT}(g_1) = a_1 x_1^{\alpha_1} \cdots x_n^{\alpha_n}$. If there exists an index u such that $\alpha_u > \beta_u$, then every term in $h_1 \mathrm{LT}(g_1)$ has exponent on x_u exceeding β_u, hence must cancel out in the sum (1.13). Similarly for g_2 through g_m, implying that there must exist an index j so that $\mathrm{LT}(g_j) = a_j x_1^{\gamma_1} \cdots x_n^{\gamma_n}$ has $\gamma_u < \beta_u$ for $1 \le u \le n$, which is precisely the statement that $\mathrm{LT}(g_j) \mid \mathrm{LT}(f)$, so G is a Gröbner basis for I. □

The selection of a Gröbner basis of an ideal I provides a solution to the Ideal Membership Problem:

Ideal Membership Problem: Solution. If G is a Gröbner basis for an ideal I, then $f \in I$ if and only if $f \xrightarrow{G} 0$.

However, it is not at all clear from Definition 1.2.13 that every ideal must actually possess a Gröbner basis. We will now demonstrate the existence of a Gröbner basis for a general ideal by giving a constructive algorithm for producing one. The algorithm is due to Buchberger [27] and either it or some modification of it is a primary component of a majority of the algorithms of computational algebra and algebraic geometry. A key ingredient in the algorithm is the so-called S-polynomial of two polynomials. To get some feel for why it should be relevant, consider the ideal I generated by the set $F = \{f_1, f_2\} \subset \mathbb{R}[x,y]$, $f_1 = x^2y + y$ and $f_2 = xy^2 + x$. Then $f = yf_1 - xf_2 = y^2 - x^2 \in I$, but under lex or deglex with $x > y$, neither $\mathrm{LT}(f_1)$ nor $\mathrm{LT}(f_2)$ divides $\mathrm{LT}(f)$, showing that F is not a Gröbner basis of I. The polynomial f is the S-polynomial of f_1 and f_2 and clearly was constructed using their leading terms in precisely a fashion that would produce a cancellation that led to the failure of their leading terms to divide its leading term.

Definition 1.2.17. Let f and g be nonzero elements of $k[x_1, \ldots, x_n]$, $\mathrm{LM}(f) = \mathbf{x}^\alpha$ and $\mathrm{LM}(g) = \mathbf{x}^\beta$. The *least common multiple* of $\mathbf{x}^\alpha$ and $\mathbf{x}^\beta$, denoted $\mathrm{LCM}(\mathbf{x}^\alpha, \mathbf{x}^\beta)$, is the monomial $\mathbf{x}^\gamma = x_1^{\gamma_1} \cdots x_n^{\gamma_n}$ such that $\gamma_j = \max(\alpha_j, \beta_j)$, $1 \le j \le n$, and (with the same notation) the *S-polynomial* of f and g is the polynomial

$$S(f,g) = \frac{\mathbf{x}^\gamma}{\mathrm{LT}(f)} f - \frac{\mathbf{x}^\gamma}{\mathrm{LT}(g)} g.$$

The following lemma, whose proof is left as Exercise 1.24 (or see [1, §1.7]), describes how any polynomial f created as a sum of polynomials whose leading terms all cancel in the formation of f must be a combination of S-polynomials computed pairwise from the summands in f.

Lemma 1.2.18. *Fix a term order $<$ on $k[x_1,\dots,x_n]$. Suppose M is a monomial and $f_1,\dots,f_s$ are elements of $k[x_1,\dots,x_n]$ such that $\mathrm{LM}(f_j) = M$ for all j, $1 \le j \le s$. Set $f = \sum_{j=1}^{s} c_j f_j$, where $c_j \in k$. If $\mathrm{LM}(f) < M$, then there exist $d_1,\dots,d_{s-1} \in k$ such that*

$$f = d_1 S(f_1,f_2) + d_2 S(f_2,f_3) + \cdots + d_{s-1} S(f_{s-1},f_s). \tag{1.14}$$

In writing (1.14) we have used the fact that $S(f,g) = -S(g,f)$ to streamline the expression. We also use this fact in the following theorem, expressing the condition on $S(g_i,g_j)$ only where we know it to be at issue: for $i \neq j$. The theorem is of fundamental importance because it provides a computational method for determining whether or not a given set of polynomials is a Gröbner basis for the ideal that they generate.

Theorem 1.2.19 (Buchberger's Criterion). *Let I be a nonzero ideal in $k[x_1,\dots,x_n]$ and let $<$ be a fixed term order on $k[x_1,\dots,x_n]$. A generating set $G = \{g_1,\dots,g_s\}$ is a Gröbner basis for I with respect to $<$ if and only if $S(g_i,g_j) \xrightarrow{G} 0$ for all $i \neq j$.*

Proof. If G is a Gröbner basis, then by (ii) in Theorem 1.2.16, $f \xrightarrow{G} 0$ for all $f \in I$, including $S(g_i,g_j)$. Conversely, suppose that all S-polynomials of the g_j reduce to 0 modulo G and let

$$f = \sum_{j=1}^{s} h_j g_j \tag{1.15}$$

be an arbitrary element of I. We need to show that there exists an index j_0 for which $\mathrm{LT}(g_{j_0})$ divides $\mathrm{LT}(f)$. The choice of the set of polynomials h_j in the representation (1.15) of f is not unique, and to each such set there corresponds the monomial

$$M = \max_{1 \le j \le s} (\mathrm{LM}(h_j)\mathrm{LM}(g_j)) \ge \mathrm{LM}(f). \tag{1.16}$$

The set $\mathbf{M}$ of all such monomials is nonempty, hence it has a least element M_0 (since the term order is a well-order). We will show that for any $M \in \mathbf{M}$, if $M > \mathrm{LM}(f)$, then there is an element of $\mathbf{M}$ that is smaller than M, from which it follows that $M_0 = \mathrm{LM}(f)$, completing the proof.

Thus let M be an element of $\mathbf{M}$ for which $M \neq \mathrm{LM}(f)$. Let $\{h_1,\dots,h_s\}$ be a collection satisfying (1.15) and giving rise to M, and let

$$J = \{j \in \{1,\dots,s\} : \mathrm{LM}(h_j)\mathrm{LM}(g_j) = M\}.$$

For $j \in J$, we write $h_j = c_j M_j + \text{lower terms}$. Let $g = \sum_{j \in J} c_j M_j g_j$. Then

$$f = g + \tilde{g}, \tag{1.17}$$

where $\mathrm{LM}(\tilde{g}) < \mathrm{LM}(g)$. Thus $\mathrm{LM}(M_j g_j) = M$ for all $j \in J$, but $\mathrm{LM}(g) < M$. By Lemma 1.2.18, g is a combination of S-polynomials,

$$g = \sum_{i,j \in J} d_{ij} S(M_i g_i, M_j g_j), \tag{1.18}$$

with $d_{ij} \in k$.

We now compute these S-polynomials. Since the least common multiple of all pairs of leading terms of the $M_i g_i$ is M, using the definition of S-polynomials we obtain

$$S(M_i g_i, M_j g_j) = \frac{M}{M_{ij}} S(g_i, g_j),$$

where $M_{ij} = \mathrm{LCM}(\mathrm{LM}(g_i), \mathrm{LM}(g_j))$, which divides M because M_i divides M, M_j divides M, and M is the least common multiple of M_i and M_j. By hypothesis, $S(g_i, g_j) \xrightarrow{G} 0$ whenever $i \neq j$, hence $S(M_i g_i, M_j g_j) \xrightarrow{G} 0$ (Exercise 1.18), so by the definition of reduction modulo G,

$$S(M_i g_i, M_j g_j) = \sum_t h_{ijt} g_t, \tag{1.19}$$

and

$$\max_{1 \le t \le s} (\mathrm{LM}(h_{ijt}) \mathrm{LM}(g_t) = \mathrm{LM}(S(M_i g_i, M_j g_j)) < \max(\mathrm{LM}(M_i g_i), \mathrm{LM}(M_j g_j)) = M.$$

Substituting (1.19) into (1.18) and then (1.18) into (1.17), we obtain $f = \sum_{j=1}^{s} h'_j g_j$, where $\max_{1 \le j \le s}(\mathrm{LT}(h'_j)\mathrm{LT}(g_j)) < M$ (recall that $d_{ij} \in k$), as required. □

Example 1.2.20. (Continuation of Example 1.2.11.) Once again consider the ideal I in $k[x,y]$ generated by the polynomials $f_1 = x^2 - 1$ and $f_2 = xy + 2$. As before we use lex with $x > y$. If f and g are in I, then certainly $S(f,g)$ is. Beginning with the basis $\{f_1, f_2\}$ of I, we will recursively add S-polynomials to our basis until, based on Buchberger's Criterion, we achieve a Gröbner basis. Thus we initially set $G := \{f_1, f_2\}$ and, using $\mathrm{LCM}(x^2, xy) = x^2 y$, compute the S-polynomial

$$S(f_1, f_2) = \frac{x^2 y}{x^2}(x^2 - 1) - \frac{x^2 y}{xy}(xy + 2) = -2x - y.$$

Since $-2x - y$ is obviously reduced with respect to $\{f_1, f_2\}$, it must be added to G, which becomes $G := \{f_1, f_2, -2x - y\}$. We then compute the S-polynomial for every pair of polynomials in the new G and reduce it, if possible. We know already that $S(f_1, f_2) = -2x - y$; hence, we compute

$$S(f_1, -2x - y) = \frac{x^2}{x^2}(x^2 - 1) - \frac{x^2}{-2x}(-2x - y) = -\frac{1}{2}(xy + 2) \xrightarrow{G} 0$$
$$S(f_2, -2x - y) = \frac{xy}{xy}(xy + 2) - \frac{xy}{-2x}(-2x - y) = -\frac{1}{2}y^2 + 2.$$

Since $-\frac{1}{2}y^2+2$ is reduced with respect to G, it is added to G, which then becomes $G := \{f_1, f_2, -2x-y, -\frac{1}{2}y^2+2\}$. Further computations show that the S-polynomial of every pair of polynomials now in G is in $\langle G\rangle$. Therefore a Gröbner basis for $I = \langle f_1, f_2\rangle$ is $\{f_1, f_2, -2x-y, -\frac{1}{2}y^2+2\}$.

This example illustrates the algorithm in Table 1.3, based on Buchberger's Criterion, for computing a Gröbner basis of a polynomial ideal. (It is implicit in the algorithm that we order the set G at each stage in order to apply the Multivariable Division Algorithm.)

Buchberger's Algorithm

Input:

A set of polynomials $\{f_1,\dots,f_s\} \in k[x_1,\dots,x_n] \setminus \{0\}$

Output:

A Gröbner basis G for the ideal $\langle f_1,\dots,f_s\rangle$

Procedure:

$G := \{f_1,\dots,f_s\}$.
Step 1. For each pair $g_i, g_j \in G$, $i \neq j$, compute the S-polynomial $S(g_i,g_j)$ and apply the Multivariable Division Algorithm to compute a remainder r_{ij}:
$S(g_i,g_j) \xrightarrow{G} r_{ij}$
IF
All r_{ij} are equal to zero, output G
ELSE
Add all nonzero r_{ij} to G and return to Step 1.

Table 1.3 Buchberger's Algorithm

Although it is easy to understand why the algorithm produces a Gröbner basis for the ideal $\langle f_1,\dots,f_s\rangle$, it is not immediately obvious that the algorithm must terminate after a finite number of steps. In fact, its termination is a consequence of the Hilbert Basis Theorem (Theorem 1.1.6), as we now show.

Theorem 1.2.21. *Buchberger's Algorithm produces a Gröbner basis for the nonzero ideal $I = \langle f_1,\dots,f_s\rangle$.*

Proof. Since the original set G is a generating set for I, and at each step of the algorithm we add to G polynomials from I, certainly at each step G remains a generating set for I. Thus, if at some stage all remainders r_{ij} are equal to zero, then by Buchberger's criterion (Theorem 1.2.19) G is a Gröbner basis for I. Therefore we need only prove that the algorithm terminates.

To this end, let $G_1, G_2, G_3, \dots$ be the sequence of sets produced by the successive values of G upon successive passes through the algorithm. If the algorithm does not terminate, then we have a strictly increasing infinite sequence $G_j \subsetneq G_{j+1}$, $j \in \mathbb{N}$. Each set G_{j+1} is obtained from the set G_j by adjoining to G_j at least one polynomial h of I, where h is a nonzero remainder with respect to G_j of an S-polynomial $S(f_1, f_2)$ for $f_1, f_2 \in G_j$. Since h is reduced with respect to G_j, its leading term is not divisible by the leading term of any element of G_j, so that $\mathrm{LT}(h) \notin \langle \mathrm{LT}(G_j) \rangle$. Thus we obtain the strictly ascending chain of ideals

$$\langle \mathrm{LT}(G_1) \rangle \subsetneq \langle \mathrm{LT}(G_2) \rangle \subsetneq \langle \mathrm{LT}(G_3) \rangle \subsetneq \cdots ,$$

in contradiction to Corollary 1.1.7. □

Even if a term order is fixed, an imprecision in the computation of a Gröbner basis arises because the Multivariable Division Algorithm can produce different remainders for different orderings of polynomials in the set of divisors. Thus the output of Buchberger's Algorithm is not unique. Also, as a rule the algorithm is inefficient in the sense that the basis that it produces contains more polynomials than are necessary. We can eliminate the superfluous polynomials from the basis using the following fact, which follows immediately from the definition of a Gröbner basis.

Proposition 1.2.22. *Let G be a Gröbner basis for $I \subset k[x_1, \ldots, x_n]$. If $g \in G$ and* $\mathrm{LT}(g) \in \langle \mathrm{LT}(G \setminus \{g\}) \rangle$, *then $G \setminus \{g\}$ is also a Gröbner basis for I.*

Proof. Exercise 1.27. □

In particular, if $g \in G$ is such that $\mathrm{LT}(g') \mid \mathrm{LT}(g)$ for some other element g' of G, then g may be discarded.

Definition 1.2.23. A Gröbner basis $G = \{g_1, \ldots, g_m\}$ is called *minimal* if, for all $i, j \in \{1, \ldots, m\}$, $\mathrm{LC}(g_i) = 1$ and for $j \neq i$, $\mathrm{LM}(g_i)$ does not divide $\mathrm{LM}(g_j)$.

Theorem 1.2.24. *Every nonzero polynomial ideal has a minimal Gröbner basis.*

Proof. By the Hilbert Basis Theorem (Theorem 1.1.6) there exists a finite basis $\{f_1, \ldots, f_s\}$ of I. From the Gröbner basis G obtained from $\{f_1, \ldots, f_s\}$ by Buchberger's Algorithm (Theorem 1.2.21) discard all those polynomials g such that $\mathrm{LT}(g) \in \langle \mathrm{LT}(G \setminus \{g\}) \rangle$, then rescale the remaining polynomials to make their leading coefficients equal to 1. By Proposition 1.2.22 the resulting set is a minimal Gröbner basis. □

Proposition 1.2.25. *Any two minimal Gröbner bases G and G' of an ideal I of the ring $k[x_1, \ldots, x_n]$ have the same set of leading terms:* $\mathrm{LT}(G) = \mathrm{LT}(G')$. *Thus they have the same number of elements.*

Proof. Write $G = \{g_1, \ldots, g_s\}$ and $G' = \{g'_1, \ldots, g'_t\}$. Because $g_1 \in I$ and G' is a Gröbner basis, there exists an index i such that

$$\mathrm{LT}(g'_i) \mid \mathrm{LT}(g_1) . \tag{1.20}$$

Because $g'_i \in I$ and G is a Gröbner basis, there exists an index j such that

$$\mathrm{LT}(g_j) \mid \mathrm{LT}(g_i'), \tag{1.21}$$

whence $\mathrm{LT}(g_j) \mid \mathrm{LT}(g_1)$. Since G is minimal we must have $j = 1$. But then (1.20) and (1.21) taken together force $\mathrm{LT}(g_i') = \mathrm{LT}(g_1)$. Re-index G' so that $g_i' = g_1'$, and repeat the argument with g_1 replaced by g_2 (noting that $LT(g_1) \nmid LT(g_2)$, by minimality of G), obtaining existence of an index i such that $\mathrm{LT}(g_i') = \mathrm{LT}(g_2)$. Applying the same argument for g_3 through g_s, we obtain the fact that $t \geq s$ and, suitably indexed, $\mathrm{LT}(g_i') = \mathrm{LT}(g_i)$ for $1 \leq i \leq s$. Reversing the roles of G and G', we have $t \leq s$, and in consequence the fact that $\mathrm{LT}(G') = \mathrm{LT}(G)$.

By minimality, there is a one-to-one correspondence between leading terms of G and its elements, and similarly for G', so G and G' have the same number of elements. □

Although the minimal basis obtained by the procedure described in the proof of Theorem 1.2.24 can be much smaller than the original Gröbner basis provided by Buchberger's Algorithm, nevertheless it is not necessarily unique (see Exercise 1.28). Fortunately, we can attain uniqueness with one additional stipulation concerning the Gröbner basis.

Definition 1.2.26. A Gröbner basis $G = \{g_1, \ldots, g_m\}$ is called *reduced* if, for all i, $1 \leq i \leq m$, $\mathrm{LC}(g_i) = 1$ and no term of g_i is divisible by any $\mathrm{LT}(g_j)$ for $j \neq i$.

That is, the Gröbner basis is reduced if each of its elements g is monic and is reduced with respect to $G \setminus \{g\}$ (Definition 1.2.8). A quick comparison with Definition 1.2.23 shows that every reduced Gröbner basis is minimal.

Theorem 1.2.27. *Fix a term order. Every nonzero ideal $I \subset k[x_1, \ldots, x_n]$ has a unique reduced Gröbner basis with respect to this order.*

Proof. Let G be any minimal Gröbner basis for I, guaranteed by Theorem 1.2.24 to exist. For any $g \in G$, replace g by its remainder r upon division of g by elements of $G \setminus \{g\}$ (in some order), to form the set $H = (G \setminus \{g\}) \cup \{r\}$. Then $\mathrm{LT}(r) = \mathrm{LT}(g)$, since, by minimality of G, for no $g' \in G$ does $\mathrm{LT}(g') \mid \mathrm{LT}(g)$. Thus $\mathrm{LT}(H) = \mathrm{LT}(G)$, so $\langle \mathrm{LT}(H) \rangle = \langle \mathrm{LT}(G) \rangle$. Then by Theorem 1.2.16 H is also a Gröbner basis for I. It is clear that it is also minimal. Since r is a remainder of g with respect to $G \setminus \{g\}$, by definition it is reduced with respect to $G \setminus \{g\} = H \setminus \{r\}$. Applying this procedure to each of the finitely many elements of G in turn yields a reduced Gröbner basis for the ideal I.

Turning to the question of uniqueness, let two reduced Gröbner bases G and G' for I be given. By Proposition 1.2.25, $\mathrm{LT}(G) = \mathrm{LT}(G')$, so that for any $g \in G$ there exists $g' \in G'$ such that

$$\mathrm{LT}(g) = \mathrm{LT}(g'). \tag{1.22}$$

By the one-to-one correspondence between the elements of a minimal Gröbner basis and its leading terms, to establish uniqueness it is sufficient to show that $g = g'$. Thus consider the difference $g - g'$. Since $g - g' \in I$, by Theorem 1.2.16

$$g - g' \xrightarrow{G} 0. \tag{1.23}$$

On the other hand, $g - g'$ is already reduced with respect to G: for by (1.22) the leading terms of g and g' have cancelled, no other term of g is divisible by any element of $\mathrm{LT}(G)$ (minimality of G), and no other term of g' is divisible by any element of $\mathrm{LT}(G') = \mathrm{LT}(G)$ (minimality of G' and (1.22) again); that is,

$$g - g' \xrightarrow{G} g - g'. \tag{1.24}$$

Hence, from (1.23) and (1.24) by Proposition 1.2.14, we conclude that $g - g' = 0$, as required. □

Recall that by Theorem 1.2.16, a Gröbner basis G of an ideal I provides a solution to the Ideal Membership Problem: $f \in I$ if and only if $f \xrightarrow{G} 0$. An analogous problem is to determine whether or not two ideals $I, J \subset k[x_1, \ldots, x_n]$, each expressed in terms of a specific finite collection of generators, are the same ideal. Theorem 1.2.27 provides a solution that is easily implemented:

> **Equality of Ideals.** Nonzero ideals I and J in $k[x_1, \ldots, x_n]$ are the same ideal if and only if I and J have the same reduced Gröbner basis with respect to a fixed term order.

Buchberger's Algorithm is far from the most efficient way to compute a Gröbner basis. For practical use there are many variations and improvements, which the interested reader can find in [1, 18, 60], among other references. The point is, however, that such algorithms have now made it feasible to actually compute special generating sets for large ideals. Almost any readily available computer algebra system, such as Mathematica, Maple, or REDUCE, will have built-in routines for computing Gröbner bases with respect to lex, deglex, and degrev. In the following section we will repeatedly assert that such and such a collection is a Gröbner basis for a particular ideal. The reader is encouraged to use a computer algebra system to duplicate the calculations on his own.

1.3 Basic Properties and Algorithms

In this section we present some basic facts concerning polynomial ideals that will be important for us later. Based on these properties of polynomial ideals we develop algorithms for working constructively with them and interpret these algorithms from a geometric point of view, meaning that we will see which operations on affine varieties correspond to which operations on ideals, and conversely. Buchberger's Algorithm is an fundamental component of all the algorithms below. In our presentation we use for the most part the notation in [60], to which we refer the reader for more details and further results.

To begin the discussion, recall that in solving a system of linear equations, an effective method is to reduce the system to an equivalent one in which initial strings of variables are missing from some of the equations, then work "backwards" from constraints on a few of the variables to the full solution. The first few results in this section are a formalization of these ideas and an investigation into how they generalize to the full nonlinear situation (1.3).

Definition 1.3.1. Let I be an ideal in $k[x_1,\ldots,x_n]$ (with the implicit ordering of the variables $x_1 > \cdots > x_n$) and fix $\ell \in \{0,1,\ldots,n-1\}$. The *$\ell$th elimination ideal* of I is the ideal $I_\ell = I \cap k[x_{\ell+1},\ldots,x_n]$. Any point $(a_{\ell+1},\ldots,a_n) \in \mathbf{V}(I_\ell)$ is called a *partial solution* of the system $\{f=0 : f \in I\}$.

The following theorem is most helpful in investigating solutions of systems of polynomial equations. We will use it several times in later sections.

Theorem 1.3.2 (Elimination Theorem). *Fix the lexicographic term order on the ring $k[x_1,\ldots,x_n]$ with $x_1 > x_2 > \cdots > x_n$ and let G be a Gröbner basis for an ideal I of $k[x_1,\ldots,x_n]$ with respect to this order. Then for every ℓ, $0 \le \ell \le n-1$, the set*

$$G_\ell := G \cap k[x_{\ell+1},\ldots,x_n]$$

is a Gröbner basis for the ℓth elimination ideal I_ℓ.

Proof. Fix ℓ and suppose that $G = \{g_1,\ldots,g_s\}$. We may assume that the g_j are ordered so that exactly the first u elements of G lie in $k[x_{\ell+1},\ldots,x_n]$, that is, that $G_\ell = \{g_1,\ldots,g_u\}$. Since $G_\ell \subset I_\ell$, Theorem 1.2.16 is applicable and implies that G_ℓ is a Gröbner basis of I_ℓ provided $\langle \mathrm{LT}(I_\ell)\rangle = \langle \mathrm{LT}(G_\ell)\rangle$. It is certainly true that $\langle \mathrm{LT}(G_\ell)\rangle \subset \langle \mathrm{LT}(I_\ell)\rangle$. The reverse inclusion means that if $f \in I_\ell$, then $\mathrm{LT}(f)$ is a linear combination of leading terms of elements of G_ℓ. Since we are working with monomials, this is true if and only if $\mathrm{LT}(f)$ is divisible by $\mathrm{LT}(g_j)$ for some index $j \in \{1,\ldots,u\}$. (See the reasoning in the last part of the proof of Theorem 1.2.16.) Given $f \in I$, since G is a Gröbner basis for I, there exists a $g_j \in G$ such that $\mathrm{LT}(f)$ is divisible by $\mathrm{LT}(g_j)$. But f is also in I_ℓ, so f depends on only $x_{\ell+1},\ldots,x_n$, hence $LT(g_j)$ depends on only $x_{\ell+1},\ldots,x_n$. Because we are using the term order lex with $x_1 > x_2 > \cdots > x_n$, this implies that $g_j = g_j(x_{\ell+1},\ldots,x_n)$, that is, that $g_j \in G_\ell$. □

The Elimination Theorem provides an easy way to eliminate a group of variables from a polynomial system. Moreover, it provides a way to find all solutions of a polynomial system in the case that the solution set is finite, or in other words, to find the variety of a polynomial ideal in the case that the variety is *zero-dimensional*. We will build our discussion around the following example.

Example 1.3.3. Let us find the variety in $\mathbb{C}^3$ of the ideal $I = \langle f_1, f_2, f_3, f_4\rangle$, where

$$\begin{aligned}
f_1 &= y^2 + x + z - 1\\
f_2 &= x^2 + 2yx + 2zx - 4x - 3y + 2yz - 3z + 3\\
f_3 &= z^2 + x + y - 1\\
f_4 &= 2x^3 + 6zx^2 - 5x^2 - 4x - 7y + 4yz - 7z + 7,
\end{aligned}$$

that is, the solution set of the system

$$f_1 = 0, \quad f_2 = 0, \quad f_3 = 0, \quad f_4 = 0. \tag{1.25}$$

Under lex and the ordering $x > y > z$, a Gröbner basis for I is $G = \{g_1, g_2, g_3\}$, where $g_1 = x + y + z^2 - 1$, $g_2 = y^2 - y - z^2 + z$, and $g_3 = z^4 - 2z^3 + z^2$. Thus system (1.25) is equivalent to the system

$$\begin{aligned} x + y + z^2 - 1 &= 0 \\ y^2 - y - z^2 + z &= 0 \\ z^4 - 2z^3 + z^2 &= 0. \end{aligned} \tag{1.26}$$

System (1.26) is readily solved. The last equation factors as $z^2(z-1)^2 = 0$, yielding $z = 0$ or $z = 1$. Inserting these values successively into the second equation determines the corresponding values of y, and inserting pairs of (y,z)-values into the first equation determines the corresponding values of x, yielding the solution

$$V = \mathbf{V}(f_1, f_2, f_3, f_4) = \{(-1,1,1), (0,0,1), (0,1,0), (1,0,0)\} \subset \mathbb{C}^3, \tag{1.27}$$

so V is just a union of four points in $\mathbb{C}^3$ (and of course it is also a variety in $\mathbb{R}^3$ and in $\mathbb{Q}^3$).

Careful examination of the example shows that system (1.26) has a form similar to the row-echelon form for systems of linear equations (and indeed row reduction is a process that leads to a Gröbner basis). As in the linear case the procedure for solving system (1.25) consists of two steps: (1) the forward or elimination step, and (2) the backward or extension step.

In the first step we eliminated from system (1.25) the variables x and y, in the sense that we transformed our system into an equivalent system that contains a polynomial g_3 that depends only on the single variable z, and which is therefore easy to solve (although in general we might be unable to find the exact roots of such a polynomial if its degree is greater than four). Once we found the roots of g_3, which is the generator of the second elimination ideal I_2, we came to the second part of the procedure, the extension step. Here, using the roots of g_3, we first found roots of the polynomial g_2 (which, along with g_3, generates the first elimination ideal I_1), and then we found all solutions of system (1.26), which is equivalent to our original system.

The example shows the significance of the choice of term order for the elimination step, because in fact the original polynomials $F = \{f_1, \ldots, f_4\}$ also form a Gröbner basis for I, but with respect to the term order deglex with $y > x > z$. Unlike the Gröbner basis $G = \{g_1, g_2, g_3\}$, the Gröbner basis F does not contain a univariant polynomial.

Now consider this example from a geometrical point of view. The map

$$\pi_m : \mathbb{C}^n \to \mathbb{C}^{n-m} : (a_1, \ldots, a_n) \mapsto (a_{m+1}, \ldots, a_n)$$

is called the *projection* of $\mathbb{C}^n$ onto $\mathbb{C}^{n-m}$. For the variety (1.27), the projection $\pi_1(V) = \{(1,1), (0,1), (1,0), (0,0)\}$ is the variety of the first elimination ideal and

$\pi_2(V) = \{1,0\}$ is the variety of the second elimination ideal. So in solving the system, we first found the variety of the second elimination ideal I_2, extended it to the variety of the first elimination ideal I_1, and finally extended the latter variety to the variety of the original ideal $I = \langle f_1, f_2, f_3, f_4 \rangle$.

Unfortunately, the backward step, the extension of partial solutions, does not always work. In the example above, substitution of each root of g_j into the preceding polynomials yields a system that has solutions. The next example shows that this is not always the case. That is, it is not always possible to extend a partial solution to a solution of the original system.

Example 1.3.4. Let V be the solution set in $\mathbb{C}^3$ of the system

$$xy = 1, \quad xz = 1. \tag{1.28}$$

The reduced Gröbner basis of $I = \langle xy-1, xz-1 \rangle$ with respect to lex with $x > y > z$ is $\{xz-1, y-z\}$. Thus the first elimination ideal is $I_1 = \langle y-z \rangle$. The variety of I_1 is the line $y = z$ in the (y,z)-plane. That is, the partial solutions corresponding to I_1 are $\{(a,a) : a \in \mathbb{C}\}$. Any partial solution (a,a) for which $a \neq 0$ can be extended to the solution $(1/a, a, a)$ of (1.28). The partial solution $(0,0)$ cannot, which corresponds to the fact that the point $(0,0) \in \mathbf{V}(I_1)$ has no preimage in V under the projection π_1. The projection $\pi_1(V)$ of V onto $\mathbb{C}^2$ (regarded as the (y,z)-plane) is not the line $y = z$, but is this line with the point $(y,z) = (0,0)$ deleted, hence is not a variety (Exercise 1.30).

Note that the situation is the same if we consider the ideal I as an ideal in $\mathbb{R}[x,y,z]$. In this case we can easily sketch V, which is a hyperbola in the plane $y = z$ in $\mathbb{R}^3$, and we see from the picture that the projection of the hyperbola onto the plane $x = 0$ is the line $y = z$ with the origin deleted.

The next theorem gives a sufficient condition for the possibility of extending a partial solution of a polynomial system to the complete solution.

Theorem 1.3.5 (Extension Theorem). *Let $I = \langle f_1, \ldots, f_s \rangle$ be a nonzero ideal in the ring $\mathbb{C}[x_1, \ldots, x_n]$ and let I_1 be the first elimination ideal for I. Write the generators of I in the form $f_j = g_j(x_2, \ldots, x_n) x_1^{N_j} + \tilde{g}_i$, where $N_j \in \mathbb{N}_0$, $g_j \in \mathbb{C}[x_2, \ldots, x_n]$ are nonzero polynomials, and $\tilde{g}_j$ are the sums of terms of f_j of degree less than N_j in x_1. Consider a partial solution $(a_2, \ldots, a_n) \in \mathbf{V}(I_1)$. If $(a_2, \ldots, a_n) \notin \mathbf{V}(g_1, \ldots, g_s)$, then there exists a_1 such that $(a_1, a_2, \ldots, a_n) \in \mathbf{V}(I)$.*

The reader can find a proof, which uses the theory of resultants, in [60, Chap. 3, §6]. Note that the polynomials in Example 1.3.4 do not satisfy the condition of the theorem.

An important particular case of the Extension Theorem is that in which the leading term of at least one of the f_j, considered as polynomials in x_1, is a constant.

Corollary 1.3.6. *Let $I = \langle f_1, \ldots, f_s \rangle$ be an ideal in $\mathbb{C}[x_1, \ldots, x_n]$ and let I_1 be the first elimination ideal for I. Suppose there exists an f_j that can be represented in the form $f_j = c x_1^N + \tilde{g}$, where $N \in \mathbb{N}$, $c \in \mathbb{C} \setminus \{0\}$, and $\tilde{g}$ is the sum of the terms of f_j*

of degree less than N in x_1. Then for any partial solution $(a_2,\dots,a_n) \in \mathbf{V}(I_1)$, there exists $a_1 \in \mathbb{C}$ such that $(a_1,a_2,\dots,a_n) \in \mathbf{V}(I)$.

Assuming now that the ideal I in the theorem or in the corollary is an elimination ideal I_k, we obtain a method that will sometimes guarantee that a partial solution $(a_{k+1},\dots,a_n)$ can be extended to a complete solution of the system. For we first use the theorem (or corollary) to see if we can be sure that the original partial solution can be extended to a partial solution $(a_k,a_{k+1},\dots,a_n)$, then we apply the theorem again to see if we can be sure that this partial solution extends, and so on.

A restatement of Corollary 1.3.6 in geometric terms is the following. In Exercise 1.31 the reader is asked to supply a proof.

Corollary 1.3.7. *Let I and f_j be as in Corollary 1.3.6. Then the projection of $\mathbf{V}(I)$ onto the last $n-1$ components is equal to the variety of I_1; that is, $\pi_1(V) = \mathbf{V}(I_1)$.*

We have seen in Example 1.3.4 that the projection of a variety in k^n onto k^{n-m} is not necessarily a variety. The following theorem describes more precisely the geometry of the set $\pi_\ell(V)$ and its relation to the variety of I_ℓ. For a proof see, for example, [60, Chap. 3, §2].

Theorem 1.3.8 (Closure Theorem). *Let $V = \mathbf{V}(f_1,\dots,f_s)$ be an affine variety in $\mathbb{C}^n$ and let I_ℓ be the ℓth elimination ideal for the ideal $I = \langle f_1,\dots,f_s\rangle$. Then*
1. $\mathbf{V}(I_\ell)$ is the smallest affine variety containing $\pi_\ell(V) \subset \mathbb{C}^{n-\ell}$, and
2. if $V \neq \varnothing$, then there is an affine variety $W \subsetneq \mathbf{V}(I_\ell)$ such that $\mathbf{V}(I_\ell) \setminus W \subset \pi_\ell(V)$.

According to the Fundamental Theorem of Algebra every element of $\mathbb{C}[x]$ with $\deg(f) \geq 1$ has at least one root. The next theorem is a kind of analogue for the case of polynomials of several variables. It shows that any system of polynomials in several variables that is not equivalent to a constant has at least one solution over $\mathbb{C}$. We will need the following lemma concerning homogeneous polynomials. (A polynomial f in several variables is *homogeneous* if the full degree of every term of f is the same.) The idea is that in general it is possible that f not be the zero element of $k[x_1,\dots,x_n]$ yet g defined by $g(x_2,\dots,x_n) = f(1,x_2,\dots,x_n)$ be the zero in $k[x_2,\dots,x_n]$. An example is $f(x,y) = x^2y - xy$. The lemma shows that this phenomenon does not occur if f is homogeneous.

Lemma 1.3.9. *Let $f(x_1,\dots,x_n) \in k[x_1,\dots,x_n]$ be a homogeneous polynomial. Then $f(x_1,\dots,x_n)$ is the zero polynomial in $k[x_1,\dots,x_n]$ if and only if $f(1,x_2,\dots,x_n)$ is the zero polynomial in $k[x_2,\dots,x_n]$.*

Proof. If $f(x_1,\dots,x_n)$ is the zero polynomial, then certainly $f(1,x_2,\dots,x_n)$ is the zero polynomial as well. For the converse, let $f \in k[x_1,\dots,x_n]$ be a nonzero homogeneous polynomial with $\deg(f) = N$. Without loss of generality we assume that all similar terms of $f(x_1,\dots,x_n)$ are collected, so that the nonzero terms of f are different and satisfy the condition

$$a_\alpha x^\alpha \neq a_\beta x^\beta \text{ implies } \alpha \neq \beta\,. \tag{1.29}$$

Then for any two terms $c_1 x_2^{\beta_2} \cdots x_n^{\beta^n}$ and $c_2 x_2^{\gamma_2} \cdots x_n^{\gamma^n}$ in $f(1, x_2, \ldots, x_n)$ for which $c_1 c_2 \neq 0$, it must be true that $(\beta_2, \ldots, \beta_n) \neq (\gamma_2, \ldots, \gamma_n)$, else

$$c_1 x_1^{(N-(\beta_2+\cdots+\beta_n))} x_2^{\beta_2} \cdots x_n^{\beta^n} \qquad \text{and} \qquad c_2 x_1^{(N-(\gamma_2+\cdots+\gamma_n))} x_2^{\gamma_2} \cdots x_n^{\gamma^n}$$

are terms in f that violate condition (1.29). That is, no nonzero terms in $f(1, \ldots, x_n)$ are similar, hence none is lost (there is no cancellation) when terms in $f(1, \ldots, x_n)$ are collected. Since f is not the zero polynomial, there is a nonzero term $c x_1^{\alpha_1} \cdots x_n^{\alpha_n}$ in f, which yields a corresponding nonzero term $c x_2^{\alpha_2} \cdots x_n^{\alpha_n}$ in $f(1, x_2, \ldots, x_n)$, which is therefore not the zero polynomial. □

Theorem 1.3.10 (Weak Hilbert Nullstellensatz). *If I is an ideal in $\mathbb{C}[x_1, \ldots, x_n]$ such that $\mathbf{V}(I) = \varnothing$, then $I = \mathbb{C}[x_1, \ldots, x_n]$.*

Proof. To prove the theorem it is sufficient to show that $1 \in I$. We proceed by induction on the number of variables in our polynomial ring.

Basis step. Suppose that $n = 1$, so the polynomial ring is $\mathbb{C}[x]$, and let an ideal I for which $\mathbf{V}(I) = \varnothing$ be given. By Exercise 1.12 there exists $f \in \mathbb{C}[x]$ such that $I = \langle f \rangle$. The hypothesis $\mathbf{V}(I) = \varnothing$ means that f has no roots in $\mathbb{C}$, hence by the Fundamental Theorem of Algebra we conclude that $f = a \in \mathbb{C} \setminus \{0\}$. It follows that $I = \langle f \rangle = \mathbb{C}[x]$.

Inductive step. Suppose that the statement of the theorem holds in the case of $n - 1 \geq 1$ variables. Let I be an ideal in $\mathbb{C}[x_1, x_2, \ldots, x_n]$ such that $\mathbf{V}(I) = \varnothing$. By the Hilbert Basis Theorem (Theorem 1.1.6) there exist finitely many polynomials f_j such that $I = \langle f_1, \ldots f_s \rangle$. Set $N = \deg(f_1)$. If $N = 0$, then f_1 is constant, so that $I = \mathbb{C}[x_1, \ldots, x_n]$, as required. If $N \geq 1$, then for any choice of $(a_2, \ldots, a_n) \in \mathbb{C}^{n-1}$ the linear transformation

$$\begin{aligned} x_1 &= y_1 \\ x_2 &= y_2 + a_2 y_1 \\ &\vdots \\ x_n &= y_n + a_n y_1 \end{aligned} \tag{1.30}$$

induces a mapping

$$T : \mathbb{C}[z_1, \ldots, z_n] \to \mathbb{C}[z_1, \ldots, z_n] : f \mapsto \tilde{f} \tag{1.31}$$

defined by $\tilde{f}(y_1, \ldots, y_n) = f(y_1, y_2 + a_2 y_1, \ldots, y_n + a_n y_1)$. Let $g \in \mathbb{C}[z_1, \ldots, z_{n-1}]$ be defined by

$$\tilde{f}_1(y_1, \ldots, y_n) = f_1(y_1, y_2 + a_2 y_1, \ldots, y_n + a_n y_1) = g(a_2, \ldots, a_n) y_1^N + h, \tag{1.32}$$

where h is the sum of all terms of f_1 whose degree in y_1 is less than N. If we rewrite f_1 in the form $f_1 = u_N + u_{N-1} + \cdots + u_0$, where, for each j, u_j is a homogeneous polynomial of degree j in $x_1, \ldots, x_n$, then $g(a_2, \ldots, a_n) = u_N(1, a_2, \ldots, a_n)$. By Lemma 1.3.9, $g(a_2, \ldots, a_n)$ is a nonzero polynomial, hence there exist $\tilde{a}_2, \ldots, \tilde{a}_n$ such that $g(\tilde{a}_2, \ldots, \tilde{a}_n) \neq 0$. Fix a choice of such $\tilde{a}_2, \ldots, \tilde{a}_n$ to define T in (1.31).

Clearly $\tilde{I} := \{\tilde{f} : f \in I\}$ is an ideal in $\mathbb{C}[y_1,\ldots,y_n]$ and $\mathbf{V}(\tilde{I}) = \varnothing$, since the linear transformation given by (1.30) is invertible.

To prove the theorem it is sufficient to show that $1 \in \tilde{I}$ (this inclusion implies that $1 \in I$ because the transformation (1.31) does not change constant polynomials). From (1.32) and Corollary 1.3.7 we conclude that $\mathbf{V}(\tilde{I}_1) = \pi_1(\mathbf{V}(\tilde{I}))$. Therefore

$$\mathbf{V}(\tilde{I}_1) = \pi_1(\mathbf{V}(\tilde{I})) = \pi_1(\varnothing) = \varnothing.$$

Thus by the induction hypothesis $\tilde{I}_1 = \mathbb{C}[y_2,\ldots,y_n]$. Hence $1 \in \tilde{I}_1 \subset \tilde{I}$, the statement of the theorem thus holds for the case of n variables, and the proof is complete. □

The Weak Hilbert Nullstellensatz provides a way for checking whether or not a given polynomial system

$$f_1 = f_2 = \cdots = f_s = 0 \tag{1.33}$$

has a solution over $\mathbb{C}$, for in order to find out if there is a solution to (1.33) it is sufficient to compute a reduced Gröbner basis G for $\langle f_1,\ldots,f_s\rangle$ using any term order:

Existence of Solutions of a System of Polynomial Equations. System (1.33) has a solution over $\mathbb{C}$ if and only if the reduced Gröbner basis G for $\langle f_1,\ldots,f_s\rangle$ with respect to any term order on $\mathbb{C}[x_1,\ldots,x_n]$ is different from $\{1\}$.

Example 1.3.11. Consider the system

$$f_1 = 0, \quad f_2 = 0, \quad f_3 = 0, \quad f_4 = 0, \quad f_5 = 0, \tag{1.34}$$

where f_1, f_2, f_3, and f_4 are from Example 1.3.3 and $f_5 = 2x^3 + xy - z^2 + 3$. A computation shows that any reduced Gröbner basis of $\langle f_1,\ldots,f_5\rangle$ is $\{1\}$. Therefore system (1.34) has no solutions.

If we are interested in solutions of (1.33) over a field k that is not algebraically closed, then a Gröbner basis computation gives a definite answer only if it is equal to $\{1\}$. For example, if there is no solution in $\mathbb{C}^n$, then certainly there is no solution in $\mathbb{R}^n$. In general, if $\{1\}$ is a Gröbner basis for $\langle f_1,\ldots,f_s\rangle$ over a field k, then system (1.33) has no solutions over k.

We now introduce a concept, the *radical* of an ideal I, that will be of fundamental importance in the procedure for identifying the variety $\mathbf{V}(I)$ of I.

Definition 1.3.12. Let $I \subset k[x_1,\ldots,x_n]$ be an ideal. The radical of I, denoted $\sqrt{I}$, is the set

$$\sqrt{I} = \{f \in k[x_1,\ldots,x_n] : \text{ there exists } p \in \mathbb{N} \text{ such that } f^p \in I\}.$$

An ideal $J \subset k[x_1,\ldots,x_n]$ is called a *radical ideal* if $J = \sqrt{J}$.

In Exercise 1.32 the reader is asked to verify that $\sqrt{I}$ *is* an ideal, and in Exercise 1.33 to show that $\sqrt{I}$ determines the same affine variety as I:

$$\mathbf{V}(\sqrt{I}) = \mathbf{V}(I). \tag{1.35}$$

Example 1.3.13. Consider the set of ideals $I^{(p)} = \langle (x-y)^p \rangle$, $p \in \mathbb{N}$. All these ideals define the same variety V, which is the line $y = x$ in the plane k^2. It is easy to see that $I^{(1)} \supsetneq I^{(2)} \supsetneq I^{(3)} \supsetneq \cdots$ and that $\sqrt{I^{(p)}} = I^{(1)}$ for every index p, so the only radical ideal among the $I^{(p)}$ is $I^{(1)} = \langle x - y \rangle$.

This example indicates that it is the radical of an ideal that is fundamental in picking out the variety that it and other ideals may determine. The next theorem and Proposition 1.3.16 say this more precisely.

Theorem 1.3.14 (Strong Hilbert Nullstellensatz). *Let $f, f_1, \ldots, f_s$ be elements of $\mathbb{C}[x_1, \ldots, x_n]$. Then $f \in \mathbf{I}(\mathbf{V}(f_1, \ldots, f_s))$ if and only if there exists $p \in \mathbb{N}$ such that $f^p \in \langle f_1, \ldots, f_s \rangle$. In other words, for any ideal I in $\mathbb{C}[x_1, \ldots, x_n]$,*

$$\sqrt{I} = \mathbf{I}(\mathbf{V}(I)). \tag{1.36}$$

Proof. If $f^p \in \langle f_1, \ldots, f_s \rangle$, then f^p vanishes on the variety $\mathbf{V}(f_1, \ldots, f_s)$. Hence, because f and f^p have the same zero sets, f itself vanishes on $\mathbf{V}(f_1, \ldots, f_s)$. That is, $f \in \mathbf{I}(\mathbf{V}(f_1, \ldots, f_s))$.

For the reverse inclusion, suppose f vanishes on $\mathbf{V}(f_1, \ldots, f_s)$. We must show that there exist $p \in \mathbb{N}$ and $h_1, \ldots, h_s \in \mathbb{C}[x_1, \ldots, x_n]$ such that

$$f^p = \sum_{j=1}^{s} h_j f_j. \tag{1.37}$$

To do so we expand our polynomial ring by adding a new variable w, and consider the ideal

$$J = \langle f_1, \ldots, f_s, 1 - wf \rangle \subset \mathbb{C}[x_1, \ldots, x_n, w].$$

We claim that $\mathbf{V}(J) = \varnothing$. For consider any point $\tilde{A} = (a_1, \ldots, a_n, a_{n+1}) \in \mathbb{C}^{n+1}$, and let A be the projection of $\tilde{A}$ onto the subspace comprised of the first n coordinates, so that $A = (a_1, \ldots, a_n)$. Then either (i) $A \in \mathbf{V}(f_1, \ldots, f_s)$ or (ii) $A \notin \mathbf{V}(f_1, \ldots, f_s)$.

In case (i) $f(A) = 0$ so that $1 - wf$ is equal to 1 at $\tilde{A}$. But then it follows that $\tilde{A} \notin \mathbf{V}(f_1, \ldots, f_s, 1 - wf) = \mathbf{V}(J)$. In case (ii) there exists an index j, $1 \leq j \leq s$, such that $f_j(A) \neq 0$. We can consider f_j as an element of $\mathbb{C}[x_1, \ldots, x_n, w]$ that does not depend on w. Then $f_j(\tilde{A}) \neq 0$, which implies that $\tilde{A} \notin \mathbf{V}(J)$. Since $\tilde{A}$ was an arbitrary point of $\mathbb{C}^{n+1}$, the claim that $\mathbf{V}(J) = \varnothing$ is established.

By the Weak Hilbert Nullstellensatz, Theorem 1.3.10, $1 \in J$, so that

$$1 = \sum_{j=1}^{s} q_j(x_1, \ldots, x_n, w) f_j + q(x_1, \ldots, x_n, w)(1 - wf) \tag{1.38}$$

for some elements $q_1, \ldots, q_s, q$ of the ring $\mathbb{C}[x_1, \ldots, x_n, w]$. If we replace w by $1/f$ and multiply by a sufficiently high power of f to clear denominators, we then obtain (1.37), as required. □

Remark. Theorems 1.3.10 and 1.3.14 remain true with $\mathbb{C}$ replaced by any algebraically closed field, but not for $\mathbb{C}$ replaced by $\mathbb{R}$ (Exercise 1.34).

For any field k the map $\mathbf{I}$ actually maps into the set of radical ideals in $k[x_1,\ldots,x_n]$. It is a direct consequence of the Strong Nullstellensatz that over $\mathbb{C}$, if we restrict the domain of $\mathbf{V}$ to just the radical ideals, then it becomes one-to-one and is an inverse for $\mathbf{I}$.

Theorem 1.3.15. *Let k be a field and let $\mathbb{H} \subset \mathbb{I}$ denote the set of all radical ideals in $k[x_1,\ldots,x_n]$. Then $\mathbf{I} : \mathbb{V} \to \mathbb{H}$; that is, $\mathbf{I}(V)$ is a radical ideal in $k[x_1,\ldots,x_n]$. When $k = \mathbb{C}$, $\mathbf{V}|\mathbb{H}$ is an inverse of $\mathbf{I}$; that is, if $I \in \mathbb{H}$, then $\mathbf{I}(\mathbf{V}(I)) = I$.*

Proof. To verify that $\mathbf{I}(V)$ is a radical ideal, suppose $f \in k[x_1,\ldots,x_n]$ is such that $f^p \in \mathbf{I}(V)$. Then for any $\mathbf{a} \in V$, $f^p(\mathbf{a}) = 0$. That is, $(f(\mathbf{a}))^p = 0$, yielding $f(\mathbf{a}) = 0$, since k is a field. Because $\mathbf{a}$ is an arbitrary point of V, f vanishes at all points of V, so $f \in \mathbf{I}(V)$. By Theorem 1.1.14, the equality $\mathbf{V}(\mathbf{I}(V)) = V$ always holds. If $k = \mathbb{C}$, Theorem 1.3.14 applies, and for a radical ideal I equation (1.36) reads $I = \sqrt{I} = \mathbf{I}(\mathbf{V}(I))$. □

Part of the importance for us of the concept of the radical of an ideal is that when the field k in question is $\mathbb{C}$, it completely characterizes when two ideals determine the same affine variety:

Proposition 1.3.16. *Let I and J be ideals in $\mathbb{C}[x_1,\ldots,x_n]$. Then $\mathbf{V}(I) = \mathbf{V}(J)$ if and only if $\sqrt{I} = \sqrt{J}$.*

Proof. If $\mathbf{V}(I) = \mathbf{V}(J)$, then by (1.35), $\mathbf{V}(\sqrt{I}) = \mathbf{V}(\sqrt{J})$, hence by Theorem 1.3.15 we obtain $\sqrt{I} = \sqrt{J}$.

Conversely, if $\sqrt{I} = \sqrt{J}$, then $\mathbf{V}(\sqrt{I}) = \mathbf{V}(\sqrt{J})$ and by (1.35), $\mathbf{V}(I) = \mathbf{V}(J)$. □

It is a difficult computational problem to compute the radical of a given ideal (unless the ideal is particularly simple, as was the case in Example 1.3.13). However, examining the proof of the Theorem 1.3.14 we obtain the following method for checking whether or not a given polynomial belongs to the radical of a given ideal.

Theorem 1.3.17. *Let k be an arbitrary field and let $I = \langle f_1,\ldots,f_s\rangle$ be an ideal in $k[x_1,\ldots,x_n]$. Then $f \in \sqrt{I}$ if and only if $1 \in J := \langle f_1,\ldots,f_s, 1-wf\rangle \subset k[x_1,\ldots,x_n,w]$.*

Proof. If $1 \in J$, then from (1.38) and the discussion surrounding it we obtain that $f^p \in I$ for some $p \in \mathbb{N}$. Conversely, if $f \in \sqrt{I}$, then by the definition of $\sqrt{I}$, $f^p \in I \subset J$ for some $p \in \mathbb{N}$. Because $1 - wf \in J$ as well,

$$1 = w^p f^p + (1 - w^p f^p) = w^p f^p + (1 + wf + \cdots + w^{p-1} f^{p-1})(1 - wf) \in J,$$

since w^p, 1, $wf, \ldots, w^{p-1} f^{p-1} \in k[x_1,\ldots,x_n,w]$ and J is an ideal. □

This theorem provides the simple algorithm presented in Table 1.4 on page 33 for deciding whether or not a polynomial f lies in $\sqrt{\langle f_1,\ldots,f_s\rangle}$.

Radical Membership Test
Input:
$f, f_1, \ldots, f_s \in k[x_1, \ldots, x_n]$
Output:
"Yes" if $f \in \sqrt{\langle f_1, \ldots, f_s \rangle}$ or "No" if $f \notin \sqrt{\langle f_1, \ldots, f_s \rangle}$
Procedure:
1. Compute a reduced Gröbner basis G for $\langle f_1, \ldots, f_s, 1 - wf \rangle \subset k[x_1, \ldots, x_n, w]$ 2. IF $G = \{1\}$ THEN "Yes" ELSE "No"

Table 1.4 The Radical Membership Test

Proposition 1.3.16 and the Radical Membership Test together give us a means for checking whether or not the varieties of two given ideals $I = \langle f_1, \ldots, f_s \rangle$ and $J = \langle g_1, \ldots, g_t \rangle$ in $\mathbb{C}[x_1, \ldots, x_n]$ are equal. Indeed, by Exercise 1.35, $\sqrt{I} = \sqrt{J}$ if and only if $f_j \in \sqrt{J}$ for all j, $1 \le j \le s$, and $g_j \in \sqrt{I}$ for all j, $1 \le j \le t$. Thus by Proposition 1.3.16, $\mathbf{V}(I) = \mathbf{V}(J)$ if and only if the Radical Membership Test shows that $f_j \in \sqrt{J}$ for all j and $g_j \in \sqrt{I}$ for all j.

In succeeding chapters we will study the so-called center varieties of polynomial systems of ordinary differential equations. The importance of the Radical Membership Test for our studies of the center problem is that it gives a simple and efficient tool for checking whether or not the center varieties obtained by different methods or different authors are the same. Note, however, that the varieties in question must be complex, since the identification procedure is based on Proposition 1.3.16, which does not hold when $\mathbb{C}$ is replaced by $\mathbb{R}$.

Now we turn to the question of the relationship between operations on ideals and the corresponding effects on the affine varieties that they define, and between properties of ideals and properties of the corresponding varieties. Hence let I and J be two ideals in $k[x_1, \ldots, x_n]$. The *intersection* of I and J is their set-theoretic intersection

$$I \cap J = \{f \in k[x_1, \ldots, x_n] : f \in I \text{ and } f \in J\},$$

and the *sum* of I and J is

$$I + J := \{f + g : f \in I \text{ and } g \in J\}.$$

In Exercise 1.36 the reader is asked to show that $I \cap J$ and $I + J$ are again ideals in $k[x_1, \ldots, x_n]$ and to derive a basis for the latter. (See also Exercise 1.38. For a basis of $I \cap J$, see Proposition 1.3.25 and Table 1.5.) The intersection of ideals interacts well with the process of forming the radical of an ideal:

$$\sqrt{I \cap J} = \sqrt{I} \cap \sqrt{J}, \tag{1.39}$$

and with the mapping $\mathbf{I}$ defined by (1.5):

$$\mathbf{I}(A) \cap \mathbf{I}(B) = \mathbf{I}(A \cup B). \tag{1.40}$$

The reader is asked to prove these equalities in Exercise 1.39. Both intersection and sum mesh nicely with the mapping $\mathbf{V}$ defined by (1.6):

Theorem 1.3.18. *If I and J are ideals in $k[x_1, \ldots, x_n]$, then*
1. $\mathbf{V}(I + J) = \mathbf{V}(I) \cap \mathbf{V}(J)$ and
2. $\mathbf{V}(I \cap J) = \mathbf{V}(I) \cup \mathbf{V}(J)$.

Proof. Exercise 1.40 (or see, for example, [60]). □

In Definition 1.1.10 we defined the ideal $\mathbf{I}(V)$ of an affine variety. As the reader showed in Exercise 1.7, the fact that V was an affine variety was not needed in the definition, which could be applied equally well to any subset S of k^n to produce an ideal $\mathbf{I}(S)$. (However, the equality $\mathbf{V}(\mathbf{I}(V)) = V$ need no longer hold; see Exercise 1.41.) Thus for any set S in k^n we define $\mathbf{I}(S)$ as the set of all elements of $k[x_1, \ldots, x_n]$ that vanish on S. Then $\mathbf{I}(S)$ is an ideal in $k[x_1, \ldots, x_n]$, and the first few lines in the proof of Theorem 1.3.15 show that it is actually a radical ideal. There is also an affine variety naturally attached to any subset of k^n.

Definition 1.3.19. The *Zariski closure* of a set $S \subset k^n$ is the smallest variety containing S. It will be denoted by $\bar{S}$.

By part (b) of Exercise 1.3 the Zariski closure is well-defined: one merely takes the intersection of every variety that contains S. Another characterization of the Zariski closure will be given in Proposition 1.3.21. By Exercise 1.3, Proposition 1.1.4, and the obvious facts that $S \subset \bar{S}$, $\overline{(\bar{S})} = \bar{S}$, and $\bar{\varnothing} = \varnothing$, the mapping from the power set of k^n into itself defined by $S \to \bar{S}$ is a Kuratowski closure operation, hence can be used to define a topology on k^n (Theorem 3.7 of [200]), the *Zariski topology*. When k is $\mathbb{R}$ or $\mathbb{C}$ with the usual topology, the Zariski closure of a set S is a topologically closed set that contains S. If $C\ell(S)$ is the topological closure of S, then $S \subset C\ell(S) \subset \bar{S}$; the second inclusion, like the first, could be strict. See Exercise 1.42. The observation contained in the following proposition, which will prove useful later, underscores the analogy between the Zariski closure and topological closure. We have stated it as we have in order to emphasize this analogy.

Proposition 1.3.20. *Suppose P denotes a property that can be expressed in terms of polynomial conditions $f_1 = \cdots = f_s = 0$ and that it is known to hold on a subset $V \setminus S$ of a variety V. If $\overline{V \setminus S} = V$, then P holds on all of V.*

Proof. Let $W = \mathbf{V}(f_1,\ldots,f_s)$. Then the set of all points of k^n on which property P holds is the affine variety W, our assumption is just that $V \setminus S \subset W$, and we must show that $V \subset W$. If $V \cap S = \varnothing$, then $V \setminus S = V$, so the desired conclusion follows automatically from the assumption $V \setminus S \subset W$. If $V \cap S \neq \varnothing$, then $V \setminus S \neq V$, so the set $V \setminus S$ is not itself a variety, since it is not equal to the smallest variety $\overline{V \setminus S} = V$ that contains it. But W *is* a variety and it *does* contain $V \setminus S$, hence W contains the smallest variety that contains $V \setminus S$, namely V, as required. □

Here is the characterization of the Zariski closure of a set mentioned earlier.

Proposition 1.3.21. *Let S be a subset of k^n. The Zariski closure of S is equal to $\mathbf{V}(\mathbf{I}(S))$:*

$$\bar{S} = \mathbf{V}(\mathbf{I}(S)). \tag{1.41}$$

Proof. To establish (1.41) we must show that if W is any affine variety that contains S, then $\mathbf{V}(\mathbf{I}(S)) \subset W$. Since $S \subset W$, $\mathbf{I}(W) \subset \mathbf{I}(S)$, therefore by Theorem 1.1.14 $\mathbf{V}(\mathbf{I}(S)) \subset \mathbf{V}(\mathbf{I}(W)) = W$. □

Proposition 1.3.22. *For any set $S \subset k^n$,*

$$\mathbf{I}(\bar{S}) = \mathbf{I}(S). \tag{1.42}$$

Proof. Since $S \subset \bar{S}$, Theorem 1.1.14 yields $\mathbf{I}(\bar{S}) \subset \mathbf{I}(S)$, so we need only show that

$$\mathbf{I}(S) \subset \mathbf{I}(\bar{S}). \tag{1.43}$$

If (1.43) fails, then there exist a polynomial f in $\mathbf{I}(S)$ and a point $\mathbf{a} \in \bar{S} \setminus S$ such that $f(\mathbf{a}) \neq 0$. Let V be the affine variety defined by $V = \bar{S} \cap \mathbf{V}(f)$. Then $\mathbf{a} \in \bar{S}$ but $\mathbf{a} \notin V$ (since $f(\mathbf{a}) \neq 0$), so $V \subsetneq \bar{S}$. Because $S \subset \bar{S}$ and $S \subset \mathbf{V}(f)$ (since $f \in \mathbf{I}(S)$), $S \subset V$. Thus $S \subset V \subsetneq \bar{S}$, in contradiction to the fact that $\bar{S}$ is the smallest variety that contains the set S. □

Just above we defined the sum and intersection of polynomial ideals. From Theorem 1.3.18 we see that the first of these ideal operations corresponds to the intersection of their varieties, while the other operation corresponds to the union of their varieties. In order to study the set theoretic difference of varieties we need to introduce the notion of the quotient of two ideals.

Definition 1.3.23. Let I and J be ideals in $k[x_1,\ldots,x_n]$. Their *ideal quotient* $I : J$ is the ideal

$$I : J = \{f \in k[x_1,\ldots,x_n] : fg \in I \text{ for all } g \in J\}.$$

In Exercise 1.43 we ask the reader to prove that $I : J$ is indeed an ideal in $k[x_1,\ldots,x_n]$.

The next theorem shows that the formation of the quotient of ideals is an algebraic operation that closely corresponds to the geometric operation of forming the difference of two varieties. It is useful in the context of applications of Proposition 1.3.20, even when the set S is not itself a variety, for then we search for a variety $V_1 \subset S$ for which $\overline{V \setminus V_1} = V$ still holds.

Theorem 1.3.24. *If I and J are ideals in $k[x_1,\dots,x_n]$, then*

$$\overline{\mathbf{V}(I)\setminus\mathbf{V}(J)}\subset\mathbf{V}(I:J), \tag{1.44}$$

where the overline indicates Zariski closure. If $k=\mathbb{C}$ and I is a radical ideal, then

$$\overline{\mathbf{V}(I)\setminus\mathbf{V}(J)}=\mathbf{V}(I:J).$$

Proof. We first show that

$$I:J\subset\mathbf{I}(\overline{\mathbf{V}(I)\setminus\mathbf{V}(J)}). \tag{1.45}$$

Take any $f\in I:J$ and $\mathbf{a}\in\mathbf{V}(I)\setminus\mathbf{V}(J)$. Because $\mathbf{a}\notin\mathbf{V}(J)$ there exists $\tilde{g}\in J$ such that $\tilde{g}(\mathbf{a})\neq 0$. Since $f\in I:J$ and $\tilde{g}\in J$, $f\tilde{g}\in I$. Because $\mathbf{a}\in\mathbf{V}(I)$, $f(\mathbf{a})\tilde{g}(\mathbf{a})=0$, hence $f(\mathbf{a})=0$. Thus $f(\mathbf{a})=0$ for any $\mathbf{a}\in\mathbf{V}(I)\setminus\mathbf{V}(J)$, which yields the inclusion $f\in\mathbf{I}(\mathbf{V}(I)\setminus\mathbf{V}(J))$. Since, by Proposition 1.3.22, $\mathbf{I}(\mathbf{V}(I)\setminus\mathbf{V}(J))=\mathbf{I}(\overline{\mathbf{V}(I)\setminus\mathbf{V}(J)})$, (1.45) holds, and applying Theorem 1.1.14 we see that (1.44) is true as well.

To prove the reverse inclusion when $k=\mathbb{C}$ and I is a radical ideal, first suppose $h\in\mathbf{I}(\mathbf{V}(I)\setminus\mathbf{V}(J))$. Then for any $g\in J$, hg vanishes on $\mathbf{V}(I)$, hence by Theorem 1.3.14, $hg\in\sqrt{I}$, and because $I=\sqrt{I}$, $hg\in I$. Thus we have shown that

$$h\in\mathbf{I}(\mathbf{V}(I)\setminus\mathbf{V}(J)) \text{ implies } hg\in I \text{ for all } g\in J. \tag{1.46}$$

Now fix any $\mathbf{a}\in\mathbf{V}(I:J)$. By the definition of $\mathbf{V}(I:J)$, this means that if h is any polynomial with the property that $hg\in I$ for all $g\in J$, then $h(\mathbf{a})=0$:

$$hg\in I \text{ for all } g\in J \text{ implies } h(\mathbf{a})=0. \tag{1.47}$$

Thus if $h\in\mathbf{I}(\mathbf{V}(I)\setminus\mathbf{V}(J))$, then by (1.46) the antecedent in (1.47) is true, hence $h(\mathbf{a})=0$, meaning that $\mathbf{a}\in\mathbf{V}(\mathbf{I}(\mathbf{V}(I)\setminus\mathbf{V}(J)))$. Taking into account Proposition 1.3.21,

$$\mathbf{V}(I:J)\subset\mathbf{V}(\mathbf{I}(\mathbf{V}(I)\setminus\mathbf{V}(J)))=\overline{\mathbf{V}(I)\setminus\mathbf{V}(J)}. \quad \square$$

Suppose I and J are ideals in $k[x_1,\dots,x_n]$ for which we have explicit finite bases, say $I=\langle f_1,\dots,f_u\rangle$ and $J=\langle g_1,\dots,g_v\rangle$. We wish to use these bases to obtain bases for $I+J$, $I\cap J$, and $I:J$. For $I+J$ the answer is simple and is presented in Exercise 1.36. To solve the problem for $I\cap J$ we need the expression for $I\cap J$ in terms of an ideal in a polynomial ring with one additional indeterminate as given by the following proposition. The equality in the proposition is proved by treating its constitutents only as sets. See Exercise 1.37 for the truth of the more general equality of ideals that it suggests.

Proposition 1.3.25. *If $I=\langle f_1,\dots,f_u\rangle$ and $J=\langle g_1,\dots,g_v\rangle$ are ideals in $k[x_1,\dots,x_n]$, then*

$$I\cap J=\langle tf_1,\dots,tf_u,(1-t)g_1,\dots,(1-t)g_v\rangle\cap k[x_1,\dots,x_n].$$

Proof. If $f\in I\cap J$, then there exist $h_j,h'_j\in k[x_1,\dots,x_n]$ such that

$$f=tf+(1-t)f=t(h_1f_1+\cdots+h_uf_u)+(1-t)(h'_1g_1+\cdots+h'_vg_v).$$

Conversely, if there exist $h_j, h'_j \in k[t, x_1, \ldots, x_n]$ such that

$$f(\mathbf{x}) = h_1(\mathbf{x},t)tf_1(\mathbf{x}) + \cdots + h_u(\mathbf{x},t)tf_u(\mathbf{x}) \\ + h'_1(\mathbf{x},t)(1-t)g_1(\mathbf{x}) + \cdots + h'_v(\mathbf{x},t)(1-t)g_v(\mathbf{x}),$$

then setting $t = 0$ we obtain $f(\mathbf{x}) = h'_1(\mathbf{x},0)g_1(\mathbf{x}) + \cdots + h'_v(\mathbf{x},0)g_v(\mathbf{x}) \in J$ and setting $t = 1$ we obtain $f(\mathbf{x}) = h_1(\mathbf{x},1)f_1(\mathbf{x}) + \cdots + h_u(\mathbf{x},1)f_u(\mathbf{x}) \in I$. □

If we order the variables $t > x_1 > \cdots > x_n$, then in the language of Definition 1.3.1 the proposition says that $I \cap J$ is the first elimination ideal of the ideal in $k[t, x_1, \ldots, x_n]$ given by $\langle tf_1, \ldots, tf_u, (1-t)g_1, \ldots, (1-t)g_v \rangle$. The proposition and the Elimination Theorem (Theorem 1.3.2) together thus yield the algorithm presented in Table 1.5 for computing a generating set for $I \cap J$ from generating sets of I and J. To obtain a basis for the ideal $I : J$ from bases for I and J we will need two more results.

Algorithm for Computing $I \cap J$

Input:

Ideals $I = \langle f_1, \ldots, f_u \rangle$ and $J = \langle g_1, \ldots, g_v \rangle$ in $k[x_1, \ldots, x_n]$

Output:

A Gröbner basis G for $I \cap J$

Procedure:

1. Compute a Gröbner basis G' of

   $\langle tf_1(\mathbf{x}), \ldots, tf_u(\mathbf{x}), (1-t)g_1(\mathbf{x}), \ldots, (1-t)g_v(\mathbf{x}) \rangle$

   in $k[t, x_1, \ldots, x_n]$ with respect to lex with $t > x_1 > \cdots > x_n$.
2. $G = G' \cap k[x_1, \ldots, x_n]$

Table 1.5 Algorithm for Computing $I \cap J$

Proposition 1.3.26. *Suppose I is an ideal in $k[x_1, \ldots, x_n]$ and g is a nonzero element of $k[x_1, \ldots, x_n]$. If $\{h_1, \ldots, h_s\}$ is a basis for $I \cap \langle g \rangle$, then $\{h_1/g, \ldots, h_s/g\}$ is a basis for $I : \langle g \rangle$.*

Proof. First note that for each j, $1 \le j \le s$, because $h_j \in \langle g \rangle$, h_j/g is a polynomial, and because $h_j \in I$, h_j/g is in $I : \langle g \rangle$. For any element f of $I : \langle g \rangle$, $gf \in I \cap \langle g \rangle$, hence by hypothesis $gf = \sum_{j=1}^{s} u_j h_j$ for some choice of $u_j \in k[x_1, \ldots, x_n]$, from which it follows that $f = \sum_{j=1}^{s} u_j(h_j/g)$. □

Proposition 1.3.27. *Suppose that I and $J_1,\dots,J_m$ are ideals in $k[x_1,\dots,x_n]$. Then* $I : (\sum_{s=1}^m J_s) = \cap_{s=1}^m (I : J_s)$.

Proof. Exercise 1.44. □

Using the fact that any ideal $J = \langle g_1,\dots,g_s\rangle$ can be represented as the sum of the principal ideals of its generators, $J = \langle g_1\rangle + \cdots + \langle g_s\rangle$, we obtain from Propositions 1.3.25, 1.3.26, and 1.3.27 the algorithm presented in Table 1.6 for computing generators of the ideal quotient.

Algorithm for Computing $I : J$

Input:

Ideals $I = \langle f_1,\dots,f_u\rangle$ and $J = \langle g_1,\dots,g_v\rangle$ in $k[x_1,\dots,x_n]$

Output:

A basis $\{f_{v1},\dots,f_{vp_v}\}$ for $I : J$

Procedure:

FOR $j = 1,\dots,v$
  Compute $I \cap \langle g_j\rangle = \langle h_{j1},\dots,h_{jm_j}\rangle$
FOR $j = 1,\dots,v$
  Compute $I : \langle g_j\rangle = \langle h_{j1}/g_j,\dots,h_{jm_j}/g_j\rangle$
$K := I : \langle g_1\rangle$
FOR $j = 2,\dots,v$
  Compute $K := K \cap (I : \langle g_j\rangle) = \langle f_{j1},\dots,f_{jp_j}\rangle$

Table 1.6 Algorithm for Computing $I : J$

1.4 Decomposition of Varieties

Consider the ideal $I = \langle x^3y^3, x^2z^2\rangle$. It is obvious that its variety $V = \mathbf{V}(I)$ is the union of two varieties, $V = V_1 \cup V_2$, where V_1 is the plane $x = 0$ and V_2 is the line $\{(x,y,z) : y = 0 \text{ and } z = 0\}$. It is geometrically clear that we cannot further decompose either V_1 or V_2 into a union of two varieties as we have just done for V. Of course, each could be decomposed as the uncountable union of its individual points, but we are thinking in terms of finite unions. With this restriction, our variety V is the union of two irreducible varieties V_1 and V_2.

Definition 1.4.1. A nonempty affine variety $V \subset k^n$ is *irreducible* if $V = V_1 \cup V_2$, for affine varieties V_1 and V_2, only if either $V_1 = V$ or $V_2 = V$.

The ideal $I = \langle x^3y^3, x^2z^2 \rangle$ considered above defines the same variety as any ideal $\langle x^py^q, x^rz^s \rangle$, where $p,q,r,s \in \mathbb{N}$. Among all these ideals the one having the simplest description is the radical of I, $\sqrt{I} = \langle xy, xz \rangle$, and it is easily seen that $\sqrt{I}$ is the intersection of two radical ideals $\langle x \rangle$ and $\langle y,z \rangle$, which in turn correspond to the irreducible components of $\mathbf{V}(I)$. The radical ideals that define irreducible varieties are the so-called prime ideals.

Definition 1.4.2. A proper ideal $I \subsetneq k[x_1,\dots,x_n]$ is a *prime* ideal if $fg \in I$ implies that either $f \in I$ or $g \in I$.

The following property of prime ideals is immediate from the definition.

Proposition 1.4.3. *Every prime ideal is a radical ideal.*

Another interrelation between prime and radical ideals is the following fact.

Proposition 1.4.4. *Suppose $P_1,\dots,P_s$ are prime ideals in $k[x_1,\dots,x_n]$. Then the ideal $I = \cap_{j=1}^s P_j$ is a radical ideal.*

Proof. Assume that for some $p \in \mathbb{N}$, $f^p \in \cap_{j=1}^s P_j$. Then for each j, $1 \le j \le s$, $f^p \in P_j$, a prime ideal, which by Proposition 1.4.3 is a radical ideal, hence $f \in P_j$. Therefore $f \in \cap_{j=1}^s P_j = I$, so that I is a radical ideal. □

We know from Theorem 1.3.15 that in the case $k = \mathbb{C}$ there is a one-to-one correspondence between radical ideals and affine varieties. We now show that there is always a one-to-one correspondence between prime ideals and irreducible varieties.

Theorem 1.4.5. *Let $V \in k^n$ be a nonempty affine variety. Then V is irreducible if and only if $\mathbf{I}(V)$ is a prime ideal.*

Proof. Suppose V is irreducible and that $fg \in \mathbf{I}(V)$. Setting $V_1 = V \cap \mathbf{V}(f)$ and $V_2 = V \cap \mathbf{V}(g)$, $V = V_1 \cup V_2$, hence either $V = V_1$ or $V = V_2$. If $V = V_1$, then $V \subset \mathbf{V}(f)$, therefore $f \in \mathbf{I}(V)$. If $V = V_2$, then $g \in \mathbf{I}(V)$ similarly, so the ideal $\mathbf{I}(V)$ is prime.

Conversely, suppose that $\mathbf{I}(V)$ is prime and that $V = V_1 \cup V_2$. If $V = V_1$, then there is nothing to show. Hence we assume $V \neq V_1$ and must show that $V = V_2$. First we will show that $\mathbf{I}(V) = \mathbf{I}(V_2)$. Indeed, $V_2 \subset V$, so Theorem 1.1.14 implies that $\mathbf{I}(V) \subset \mathbf{I}(V_2)$. For the reverse inclusion we note that by Proposition 1.1.12, $\mathbf{I}(V)$ is strictly contained in $\mathbf{I}(V_1)$, $\mathbf{I}(V) \subsetneq \mathbf{I}(V_1)$, because V_1 is a proper subset of V (possibly the empty set). Therefore there exists a polynomial

$$f \in \mathbf{I}(V_1) \setminus \mathbf{I}(V). \tag{1.48}$$

Choose any polynomial $g \in \mathbf{I}(V_2)$. Then since $f \in \mathbf{I}(V_1)$ and $g \in \mathbf{I}(V_2)$, by the fact that $V = V_1 \cup V_2$ and (1.40), $fg \in \mathbf{I}(V)$. Because $\mathbf{I}(V)$ is prime, either $f \in \mathbf{I}(V)$ or $g \in \mathbf{I}(V)$. But (1.48) forces $g \in \mathbf{I}(V)$, yielding $\mathbf{I}(V_2) \subset \mathbf{I}(V)$, hence $\mathbf{I}(V) = \mathbf{I}(V_2)$. But then by part (2) of Proposition 1.1.12, $V = V_2$, so that V is irreducible. □

We saw above that the variety of any ideal $\langle x^py^q, x^rz^s \rangle$ where $p,q,r,s \in \mathbb{N}$, is a union of a plane and a line. As one would expect, the possibility of decomposing a variety into a finite union of irreducible varieties is a general property:

Theorem 1.4.6. *Let $V \in k^n$ be an affine variety. Then V is a union of a finite number of irreducible varieties.*

Proof. If V is not itself irreducible, then $V = V_L \cup V_R$, where V_L and V_R are affine varieties but neither is V. If they are both irreducible, then the proof is complete. Otherwise, one or both of them decomposes into a union of proper subsets, each one an affine variety, say $V = (V_{LL} \cup V_{LR}) \cup (V_{RL} \cup V_{RR})$. If every one of these four sets is is irreducible, the proof is complete; otherwise, the decomposition continues. In this way a collection of strictly descending chains of affine varieties is created, such as (with the obvious meaning to the notation)

$$V \supsetneqq V_L \supsetneqq V_{LR} \supsetneqq V_{LRL} \supsetneqq \cdots,$$

each of which in turn generates a strictly increasing chain of ideals, such as

$$\mathbf{I}(V) \subsetneqq \mathbf{I}(V_L) \subsetneqq \mathbf{I}(V_{LR}) \subsetneqq \mathbf{I}(V_{LRL}) \subsetneqq \cdots$$

in $k[x_1,\ldots,x_n]$. (The inclusions are proper because, by Theorem 1.1.14, $\mathbf{I}$ is injective.) By Corollary 1.1.7 every such chain of ideals terminates, hence (again by injectivity of $\mathbf{I}$) each chain of affine varieties terminates, and so the decomposition of V terminates with the expression of V as a union of finitely many irreducible affine varieties. □

A decomposition $V = V_1 \cup \cdots \cup V_m$ of the variety $V \subset k^n$ into a finite union of irreducible subvarieties is called a *minimal decomposition* if $V_i \not\subset V_j$ for $i \neq j$.

Theorem 1.4.7. *Every variety $V \subset k^n$ has a minimal decomposition*

$$V = V_1 \cup \cdots \cup V_m, \tag{1.49}$$

and this decomposition is unique up to the order of the V_j in (1.49).

The proof is left as Exercise 1.48 (or see [60]).

Note that for the statement of uniqueness in the theorem above it is important that the minimal decomposition be a union of a finite number of varieties. Otherwise, for instance, a plane can be represented as the union of points or the union of lines, and they, of course, are different decompositions.

We already knew that an intersection of prime ideals is a radical ideal. As a direct corollary of Theorems 1.3.15, 1.4.5, and 1.4.7 and identity (1.40), when $k = \mathbb{C}$ we obtain as a converse the following property of radical ideals.

Theorem 1.4.8. *Every radical ideal $I \subset \mathbb{C}[x_1,\ldots,x_n]$ can be uniquely represented as an intersection of prime ideals, $I = \cap_{j=1}^m P_j$, where $P_r \not\subset P_s$ if $r \neq s$.*

Our goal is to understand the set of solutions of a given system

$$f_1(x_1,\ldots,x_n) = 0, \ \ldots, \ f_s(x_1,\ldots,x_n) = 0 \tag{1.50}$$

of polynomial equations, which, as we know (Proposition 1.1.9), really depends on the ideal $I = \langle f_1, \ldots, f_s \rangle$ rather than on the individual polynomials themselves. The solution set of (1.50) is the variety $V = \mathbf{V}(f_1, \ldots, f_s) = \mathbf{V}(I)$. To solve the system is to obtain a description of V that is as complete and explicit as possible. If V is composed of a finite number c of points of k^n, then as illustrated in Example 1.3.3, we can find them by computing a Gröbner basis of the ideal I with respect to lex and then iteratively finding roots of univariant polynomials. It is known that this procedure will always yield full knowledge of the solution set V (including the case $c = 0$, for then the reduced Gröbner basis is $\{1\}$). If V is not finite, then we turn to the fact that there must exist a unique minimal decomposition (1.49); the best way to describe V is to find this minimal decomposition. Application of $\mathbf{I}$ to (1.49) yields

$$\mathbf{I}(V) = \mathbf{I}(V_1) \cap \cdots \cap \mathbf{I}(V_m), \tag{1.51}$$

where each term on the right is a prime ideal (Theorem 1.4.5). These prime ideals correspond precisely to the irreducible varieties that we seek. However, unlike $\mathbf{I}(V)$, the ideal I with which we have to work need not be radical, and from Proposition 1.4.4 and Theorem 1.4.8 it is apparent that radical ideals, but only radical ideals, can be represented as intersections of prime ideals. A solution is to work over $\mathbb{C}$ and appeal to Proposition 1.3.16: supposing that the prime decomposition of $\sqrt{I}$ is $\cap_{j=1}^m P_j$ and applying the map $\mathbf{V}$ then yields

$$V = \mathbf{V}(I) = \mathbf{V}(\sqrt{I}) = \mathbf{V}(\cap_{j=1}^m P_j) = \cup_{j=1}^m \mathbf{V}(P_j), \tag{1.52}$$

and each term $\mathbf{V}(P_j)$ specifies the irreducible subvariety V_j. A first step in following this procedure is to compute $\sqrt{I}$ from the known ideal I, and there exist algorithms for computing the radical of an arbitrary ideal ([81, 173, 192]). It is more convenient to work directly with I, however; in order to decompose it into an intersection of ideals, we need the following weaker condition on the components of the decomposition.

Definition 1.4.9. An ideal $I \subset k[x_1, \ldots, x_n]$ is called a *primary ideal* if, for any pair $f, g \in k[x_1, \ldots, x_n]$, $fg \in I$ only if either $f \in I$ or $g^p \in I$ for some $p \in \mathbb{N}$.

An ideal I is primary if and only if $\sqrt{I}$ is prime (Exercise 1.49); $\sqrt{I}$ is called the *associated prime ideal of I*.

Definition 1.4.10. A *primary decomposition* of an ideal $I \subset k[x_1, \ldots, x_n]$ is a representation of I as a finite intersection of primary ideals Q_j:

$$I = \cap_{j=1}^m Q_j. \tag{1.53}$$

The decomposition is called a *minimal* primary decomposition if the associated prime ideals $\sqrt{Q_j}$ are all distinct and $\cap_{i \neq j} Q_i \not\subset Q_j$ for any j.

A minimal primary decomposition of a polynomial ideal always exists, but it is not necessarily unique:

Theorem 1.4.11 (Lasker–Noether Decomposition Theorem). *Every ideal I in $k[x_1,\ldots,x_n]$ has a minimal primary decomposition* (1.53)*. All such decompositions have the same number m of primary ideals and the same collection of associated prime ideals.*

The reader can find a proof in [18, 60].

Returning to the problem of finding the variety of the ideal I generated by the polynomials in (1.50), whose primary decomposition is given by (1.53), by (1.39) the prime decomposition of $\sqrt{I}$ is $\sqrt{I} = \cap_{j=1}^{m} \sqrt{Q_j} := \cap_{j=1}^{m} P_j$. Now apply $\mathbf{V}$, which by Theorem 1.3.18(2) yields (1.52), in which each term $\mathbf{V}(P_j)$ specifies the irreducible subvariety V_j. There exist algorithms for computing the primary decomposition and, as mentioned above, the radical of a given ideal ([81, 173, 192]). They are time- and memory-consuming, however, and have yet to be implemented in some general-purpose computer algebra systems. They *can* be found in Maple and in specialized computer systems designed specifically for algebraic computations, like CALI ([85]), Macaulay ([88]), and Singular ([89]). We illustrate the ideas with a pair of examples.

Example 1.4.12. We look at Example 1.3.3 again, this time using Singular in order to obtain a prime decomposition of the ideal $I = \langle f_1,\ldots,f_4\rangle$. Code for carrying out the decomposition is as follows.

```
> LIB "primdec.lib";
> ring r=0,(x,y,z),dp;
> poly f1=y^2+x+z-1;
> poly f2=x^2+2*y*x+2*z*x-4*x-3*y+2*y*z-3*z+3;
> poly f3=z^2+x+y-1;
> poly f4=2*x^3+6*z*x^2-5*x^2-4*x-7*y+4*y*z-7*z+7;
> ideal i=f1,f2,f3,f4;
> primdecGTZ(i);
```

The first command downloads a Singular library that enables computation of primary and prime decompositions. The second command declares that the polynomial ring involved has characteristic zero, that the variables are x, y, and z and in the order $x > y > z$, and that the term order to be used (as specified by the parameter `dp`) is degree lexicographic order. The next four lines specify the polynomials in question; in the following line `ideal` is the declaration that the ideal under investigation is $I = \langle f_1, f_2, f_3, f_4\rangle$. Finally, `primdecGTZ` ([62]) commands the computation of a primary decomposition of I using the Gianni–Trager–Zacharias algorithm ([81]). The output is displayed in Table 1.7 on page 43. In the output the symbol `z2` is Singular's "short" notation for z^2. To have the output presented as `z^2`, switch off the `short` function with the command `short=0`.

The output is a list of pairs of ideals, where each ideal is, of course, specified by a list of generators. In this case there are four pairs. The first ideal Q_j in each pair is a primary ideal in a primary decomposition of I; the second ideal P_j in each pair is the associated prime ideal, that is, the radical, of the first, $P_j = \sqrt{Q_j}$. Singular's output indicates therefore that the ideal I in Example 1.4.12 is the intersection of the four primary ideals

```
[1]:
   [1]:
      _[1]=z2-2z+1
      _[2]=y+z-1
      _[3]=z2+x+y-1
   [2]:
      _[1]=z-1
      _[2]=y
      _[3]=z2+x+y-1
[2]:
   [1]:
      _[1]=z2-2z+1
      _[2]=y-z
      _[3]=z2+x+y-1
   [2]:
      _[1]=z-1
      _[2]=y-1
      _[3]=z2+x+y-1
[3]:
   [1]:
      _[1]=z2
      _[2]=y-z
      _[3]=z2+x+y-1
   [2]:
      _[1]=z
      _[2]=y
      _[3]=z2+x+y-1
[4]:
   [1]:
      _[1]=z2
      _[2]=y+z-1
      _[3]=z2+x+y-1
   [2]:
      _[1]=z
      _[2]=y-1
      _[3]=z2+x+y-1
```

Table 1.7 Singular Output of Example 1.4.12

$$\begin{aligned}
Q_1 &= \langle z^2 - 2z + 1, y + z - 1, z^2 + x + y - 1 \rangle \\
Q_2 &= \langle z^2 - 2z + 1, y - z, z^2 + x + y - 1 \rangle \\
Q_3 &= \langle z^2, y - z, z^2 + x + y - 1 \rangle \\
Q_4 &= \langle z^2, y + z - 1, z^2 + x + y - 1 \rangle .
\end{aligned}$$

We see that for each j, $1 \le j \le 4$, $Q_j \ne P_j = \sqrt{Q_j}$, so that Q_j is not prime. It is also apparent that the varieties $\mathbf{V}(P_j)$ are the four points from (1.27).

Example 1.4.13. The discussion in Section 6.4 on the bifurcation of critical periods from centers will be built around the system of differential equations

$$\begin{aligned} \dot{x} &= \quad x - a_{20}x^3 - a_{11}x^2y - a_{02}xy^2 - a_{-13}y^3 \\ \dot{y} &= -y + b_{02}y^3 + b_{11}xy^2 + b_{20}x^2y + b_{3,-1}x^3 \end{aligned} \tag{1.54}$$

(the negative indices are merely a product of the indexing system that will simplify certain expressions). Associated to (1.54) is an infinite collection of polynomials g_{kk}, $k \in \mathbb{N}$, in the coefficients a_{pq} and b_{qp} of (1.54), called the *focus quantities* of the system. Here we consider the ideal I generated by just the first five focus quantities, $I = \langle g_{11}, \dots, g_{55} \rangle$. They are the polynomials denoted in the Singular code that follows by `g11`, ..., `g55`. To carry out the primary decomposition of I we use the following Singular code (where `a13` stands for a_{-13} and `b31` stands for $b_{3,-1}$):

```
>LIB "primdec.lib";
>ring r=0,(a20,a11,a02,a13,b31,b20,b11,b02),dp;
>poly g11=a11-b11;
>poly g22=a20*a02-b02*b20;
>poly g33=(3*a20^2*a13+8*a20*a13*b20+3*a02^2*b31
          -8*a02*b02*b31-3*a13*b20^2-3*b02^2*b31)/8;
>poly g44=(-9*a20^2*a13*b11+a11*a13*b20^2
          +9*a11*b02^2*b31-a02^2*b11*b31)/16;
>poly g55=(-9*a20^2*a13*b02*b20+a20*a02*a13*b20^2
          +9*a20*a02*b02^2*b31+18*a20*a13^2*b20*b31
          +6*a02^2*a13*b31^2-a02^2*b02*b20*b31
          -18*a02*a13*b02*b31^2
          -6*a13^2*b20^2*b31)/36;
>ideal i = g11,g22,g33,g44,g55;
>primdecSY(i);
```

The only significant difference between this code and that of the previous example is the last line; `primdecSY` commands the computation of a primary decomposition of I using the Shimoyama–Yokoyama algorithm ([173]). The output is displayed in Table 1.8 on page 45.

Thus in this case the ideal I is the intersection of three primary ideals

$$\begin{aligned} Q_1 &= \langle a_{02} - 3b_{02},\, a_{11} - b_{11},\, 3a_{20} - b_{20} \rangle \\ Q_2 &= \langle b_{11},\, 3a_{02} + b_{02},\, a_{11},\, a_{20} + 3b_{20},\, 3a_{-13}b_{3,-1} + 4b_{20}b_{02} \rangle \\ Q_3 &= \langle a_{11} - b_{11},\, a_{20}a_{02} - b_{20}b_{02},\, a_{20}a_{-13}b_{20} - a_{02}b_{3,-1}b_{02}, \\ &\qquad\qquad a_{02}^2 b_{3,-1} - a_{-13}b_{20}^2,\, {a_{20}}^2 a_{-13} - b_{3,-1}{b_{02}}^2 \rangle \end{aligned}$$

which correspond to the first ideals of each pair in the output. In contrast to the first example, in this case the second ideal in each output pair is the same as the first one, which means that $Q_j = \sqrt{Q_j} = P_j$ for $j = 1, 2, 3$, hence is prime (see Exercise 1.49). Thus, by Proposition 1.4.4, I is a radical ideal. The variety $\mathbf{V}(I)$ is the union of three irreducible varieties $\mathbf{V}(P_1)$, $\mathbf{V}(P_2)$, and $\mathbf{V}(P_3)$.

Note that in many cases the simplest output form of the minimal associated prime ideals is provided by Singular's routine `minAssChar`, which computes the minimal associated primes by the characteristic sets method ([198]). For instance, for the

```
[1]:
   [1]:
      _[1]=a02-3*b02
      _[2]=a11-b11
      _[3]=3*a20-b20
   [2]:
      _[1]=a02-3*b02
      _[2]=a11-b11
      _[3]=3*a20-b20
[2]:
   [1]:
      _[1]=b11
      _[2]=3*a02+b02
      _[3]=a11
      _[4]=a20+3*b20
      _[5]=3*a13*b31+4*b20*b02
   [2]:
      _[1]=b11
      _[2]=3*a02+b02
      _[3]=a11
      _[4]=a20+3*b20
      _[5]=3*a13*b31+4*b20*b02
[3]:
   [1]:
      _[1]=a11-b11
      _[2]=a20*a02-b20*b02
      _[3]=a20*a13*b20-a02*b31*b02
      _[4]=a02^2*b31-a13*b20^2
      _[5]=a20^2*a13-b31*b02^2
   [2]:
      _[1]=a11-b11
      _[2]=a20*a02-b20*b02
      _[3]=a20*a13*b20-a02*b31*b02
      _[4]=a02^2*b31-a13*b20^2
      _[5]=a20^2*a13-b31*b02^2
```

Table 1.8 Singular Output of Example 1.4.13

polynomials of Example 1.4.12 the output is given in Table 1.9 on page 46. These output ideals look simpler than those obtained by `primdecGTZ`, but of course they are the same.

Finally, we turn to the problem of "rational implicitizations." Consider the unit circle in the real plane $\mathbb{R}^2$, which is the real affine variety $V = \mathbf{V}(x^2 + y^2 - 1)$. V is represented by the well-known parametrization $(x,y) = (\cos t, \sin t)$. This is not an algebraic parametrization, however, but a transcendental one. An algebraic parametrization of the circle is given by the rational functions

$$x = \frac{1-t^2}{1+t^2}, \quad y = \frac{2t}{1+t^2}, \qquad t \in \mathbb{R}, \tag{1.55}$$

```
[1]:
   _[1]=z
   _[2]=y-1
   _[3]=x
[2]:
   _[1]=z-1
   _[2]=y
   _[3]=x
[3]:
   _[1]=z-1
   _[2]=y-1
   _[3]=x+1
[4]:
   _[1]=z
   _[2]=y
   _[3]=x-1
```

Table 1.9 Singular Output of Example 1.4.12 Using `minAssChar`

which covers all points of the variety V except the point $(-1,0)$. (For a derivation of this parametrization see [60, §4.4].)

Now we consider the opposite problem: given a rational or even a polynomial parametrization of a subset S of k^n, such as (1.55) for the unit circle, try to in essence eliminate the parameters so as to express the set in terms of polynomials in $x_1,\ldots,x_n$. Hence suppose we are given the system of equations

$$x_1 = \frac{f_1(t_1,\ldots,t_m)}{g_1(t_1,\ldots,t_m)}, \ \ldots, \ x_n = \frac{f_n(t_1,\ldots,t_m)}{g_n(t_1,\ldots,t_m)}, \tag{1.56}$$

where $f_j, g_j \in k[t_1,\ldots,t_m]$ for $j = 1,\ldots,n$. Let $W = \mathbf{V}(g_1 \cdots g_n)$. Equations (1.56) define a function

$$F : k^m \setminus W \to k^n$$

by the formula

$$F(t_1,\ldots,t_m) = \left(\frac{f_1(t_1,\ldots,t_m)}{g_1(t_1,\ldots,t_m)}, \ \ldots, \ \frac{f_n(t_1,\ldots,t_m)}{g_n(t_1,\ldots,t_m)}\right). \tag{1.57}$$

The image of $k^m \setminus W$ under F, which we denote by $F(k^m \setminus W)$, is not necessarily an affine variety. For instance, for system (1.55) with $k = \mathbb{R}$, $W = \varnothing$ and $F(\mathbb{R})$ is not the circle $x^2 + y^2 = 1$, but the circle with the point $(-1,0)$ deleted. Consequently, we look for the smallest affine variety that contains $F(k^m \setminus W)$, that is, its Zariski closure $\overline{F(k^m \setminus W)}$. In the case of (1.55), this is the whole unit circle. The problem of finding $\overline{F(k^m \setminus W)}$ is known as the problem of *rational implicitization.* If the right-hand sides of (1.56) are polynomials, then it is the problem of *polynomial implicitization.* The terminology comes from the fact that the collection of polynomials $f_1,\ldots,f_s$ that determine the variety $V(f_1,\ldots,f_s)$ defines it only implicitly. The following theorem gives an algorithm for polynomial implicitization.

Theorem 1.4.14. *Let k be an infinite field, let $f_1,\dots,f_n$ be elements of $k[t_1,\dots,t_m]$, and let $F : k^m \to k^n$ be the function defined by the equations*

$$x_1 = f_1(t_1,\dots,t_m),\ \dots,\ x_n = f_n(t_1,\dots,t_m). \tag{1.58}$$

Form the ideal $I = \langle f_1 - x_1,\dots,f_n - x_n\rangle \subset k[t_1,\dots,t_m,x_1,\dots,x_n]$. Then the smallest variety in k^n that contains $F(k^m)$, the image of k^m under F, is the variety $\mathbf{V}(I_m)$ of the mth elimination ideal $I_m = I \cap k[x_1,\dots,x_n]$.

Proof. We prove the theorem only for the case $k = \mathbb{C}$; for the general case see [60]. The function F is defined by

$$F(t_1,\dots,t_m) = (f_1(t_1,\dots,t_m),\dots,f_n(t_1,\dots,t_m)). \tag{1.59}$$

Its graph is that subset of $\mathbb{C}^{m+n}$ for which each equation in (1.58) holds, hence is precisely the affine variety $V = \mathbf{V}(I)$. The set $F(\mathbb{C}^m) \subset \mathbb{C}^n$ is simply the projection of the graph of the function F onto $\mathbb{C}^n$. That is, $F(\mathbb{C}^m) = \pi_m(V)$, where π_m is the projection from $\mathbb{C}^{n+m}$ to $\mathbb{C}^n$ defined by $\pi_m(t_1,\dots,t_m,x_1,\dots,x_n) = (x_1,\dots,x_n)$. By the Closure Theorem (Theorem 1.3.8) $\mathbf{V}(I_m)$ is the smallest variety that contains the set $\pi_m(V)$. □

To obtain an algorithm for rational implicitization we add an additional variable in order to eliminate the influence of vanishing denominators.

Theorem 1.4.15. *Let k be an infinite field, let $f_1,\dots,f_n$ and $g_1,\dots,g_n$ be elements of $k[t_1,\dots,t_m]$, let $W = \mathbf{V}(g_1\cdots g_n)$, and let $F : k^m \setminus W \to k^n$ be the function defined by equations* (1.57)*. Set $g = g_1\cdots g_n$. Consider the ideal*

$$J = \langle f_1 - g_1x_1,\dots,f_n - g_nx_n, 1 - gy\rangle \subset k[y,t_1,\dots,t_m,x_1,\dots,x_n],$$

and let

$$J_{m+1} = J \cap k[x_1,\dots,x_n] \tag{1.60}$$

be the $(m+1)$st elimination ideal. Then $\mathbf{V}(J_{m+1})$ is the smallest variety in k^n containing $F(k^m \setminus W)$.

For a proof see [60]. We will use this theorem in our derivation of the center variety of complex quadratic systems (Theorem 3.7.1) and later in our investigation of the symmetry ideal of more general families (Section 5.2).

Definition 1.4.16. An affine variety $V \subset k^n$ *admits a rational parametrization* (or *can be parametrized by rational functions*) if there exists a function F of the form (1.57), $F : k^m \setminus W \to k^n$, where $f_j, g_j \in k[t_1,\dots,t_m]$ for $j = 1,\dots,n$ and $W = \mathbf{V}(g_1\cdots g_n)$, such that V is the Zariski closure of $F(k^m \setminus W)$.

Theorem 1.4.17. *Let $f_1,\dots,f_n$ and $g_1,\dots,g_n$ be elements of $\mathbb{C}[t_1,\dots,t_m]$. Suppose I is an ideal $\mathbb{C}[x_1,\dots,x_n]$ that satisfies*

$$\mathbb{C}[x_1,\dots,x_n] \cap \langle 1 - tg, x_1g_1 - f_1,\dots,x_ng_n - f_n\rangle = I, \tag{1.61}$$

where $g = g_1 g_2 \cdots g_n$ and the f_j and g_j are evaluated at $(t_1,\ldots,t_m)$. If the variety $\mathbf{V}(I)$ of I admits a rational parametrization (1.56)*, then I is a prime ideal in $\mathbb{C}[x_1,\ldots,x_n]$.*

Corollary 1.4.18. *If an affine variety $V \subset \mathbb{C}^n$ can be parametrized by rational functions, then it is irreducible.*

Proof. Let (1.56) be the parametrization of a variety V. Then by Theorem 1.4.15 $V = \mathbf{V}(J_{m+1})$ where J_{m+1} is defined by (1.60). However, J_{m+1} is the same as the ideal I defined by (1.61). According to Theorem 1.4.17, the ideal I is prime, hence radical (Proposition 1.4.3). Then using the Strong Hilbert Nullstellensatz (Theorem 1.3.14) in the second step, $\mathbf{I}(V) = \mathbf{I}(\mathbf{V}(I)) = \sqrt{I} = I$ is prime, so by Theorem 1.4.5 the variety $V = \mathbf{V}(I)$ is irreducible. $\square$

The statement of the corollary remains true with $\mathbb{C}$ replaced by any infinite field (for the proof see, for example, [60, Chap. 4]).

Proof of Theorem 1.4.17. It is sufficient to show that the ideal

$$H = \langle 1 - tg, x_1 g_1(t_1,\ldots,t_m) - f_1(t_1,\ldots,t_m),\ldots,x_n g_n(t_1,\ldots,t_m) - f_n(t_1,\ldots,t_m)\rangle$$

is prime in $\mathbb{C}[x_1,\ldots,x_n,t_1,\ldots,t_m,t]$. To do so we exploit the following facts of abstract algebra:

(i) if $\psi : R \to R'$ is a ring homomorphism, then the kernel $\ker(\psi) = \{r : \psi(r) = 0\}$ is an ideal in R (which means that it is a subring of R and for all $r \in R$ and $s \in \ker(\psi)$, $rs \in \ker(\psi)$, which the reader should be able to verify); and

(ii) if $\ker(\psi)$ is a proper subset of R and R' is an integral domain (which means that $ab = 0$ only if either $a = 0$ or $b = 0$), then $\ker(\psi)$ is prime (which is true because $ab \in \ker(\psi)$ means that $0 = \psi(ab) = \psi(a)\psi(b)$, which forces $\psi(a) = 0$ or $\psi(b) = 0$ by the hypothesis on R').

Let $\mathbb{C}(t_1,\ldots,t_m)$ denote the ring of rational functions of m variables with coefficients in $\mathbb{C}$, and consider the ring homomorphism

$$\psi : \mathbb{C}[x_1,\ldots,x_n,t_1,\ldots,t_m,t] \to \mathbb{C}(t_1,\ldots,t_m)$$

defined by

$$t_i \to t_i, \quad x_j \to f_j(t_1,\ldots,t_m)/g_j(t_1,\ldots,t_m), \quad t \to 1/g(t_1,\ldots,t_m),$$

$i = 1,\ldots,m$, $j = 1,\ldots,n$. We will prove that $H = \ker(\psi)$, which is clearly a proper subset of $\mathbb{C}[x_1,\ldots,x_n,t_1,\ldots,t_m,t]$. It is immediate that $H \subset \ker(\psi)$. We will show the other inclusion by induction on the degree in just the variables $x_1,\ldots,x_n,t$ of the polynomial $h \in \mathbb{C}[x_1,\ldots,x_n,t_1,\ldots,t_m,t]$.

Basis step. Suppose that $h \in \ker(\psi)$ and that h is linear in $x_1,\ldots,x_n,t$, so that h may be written as

$$h = \sum_{j=1}^{n} \alpha_j(t_1,\ldots,t_m)x_j + \alpha(t_1,\ldots,t_m)t + \alpha_0(t_1,\ldots,t_m).$$

Then $\alpha_0 tg = -\sum_{j=1}^{n} \alpha_j f_j t\tilde{g}_j - \alpha t$, where $\tilde{g}_j = g/g_j$ and g, f_j, and$\tilde{g}_j$ are all evaluated at $(t_1,\dots,t_m)$. Therefore $h = \sum_{j=1}^{n} \alpha_j(x_j g_j - f_j)t\tilde{g}_j - \alpha t(1-tg) + (1-tg)h$, so that $h \in H$.

Inductive step. Now assume that for all polynomials of degree d in $x_1,\dots,x_n,t$, if $h \in \ker(\psi)$, then $h \in H$, and let $h \in \ker(\psi)$ be of degree $d+1$ in $x_1,\dots,x_n,t$. We can write h as

$$h = \sum_{j=1}^{n} h_j(x_j,x_{j+1},\dots,x_n,t_1,\dots,t_m,t) + \widehat{h}(t_1,\dots,t_m,t) + h_0(t_1,\dots,t_m),$$

where every term of h_j contains x_j and every term of $\widehat{h}$ contains t, which allows us to express h as $h = u + v$ for

$$u = \sum_{j=1}^{n} (h_j/x_j)(x_j g_j - f_j)t\tilde{g}_j - \widehat{h}(1-tg) + (1-tg)h \in H \subset \ker(\psi)$$

and

$$v = t\sum_{j=1}^{n} f_j\tilde{g}_j(h_j/x_j) + \widehat{h} + tgh_0 .$$

Since $h, u \in \ker(\psi)$, $v \in \ker(\psi)$, which clearly implies that $v/t \in \ker(\psi)$. Then by the induction hypothesis, $v/t \in H$, hence $v \in H$, so that $h \in H$ as well. □

Example 1.4.19. As an illustration of the use to which we put these ideas, consider the ideal Q_3 of Example 1.4.13. Of course, we know already that Q_3 is prime, so that $\mathbf{V}(Q_3)$ is irreducible. Imagine, however, that we merely suspect that $\mathbf{V}(Q_3)$ is irreducible and that we wish to establish its irreducibility directly. We attempt to parametrize $\mathbf{V}(Q_3)$. There is no systematic procedure for finding a parametrization; one must proceed on a case-by-case basis. In this instance, the first generator of Q_3 suggests that we introduce the single parameter u by writing $a_{11} = b_{11} = u$. The equation $a_{20}a_{02} - b_{20}b_{02} = 0$ corresponding to the second generator suggests that we will need three more parameters, since we have one equation (albeit nonlinear) in three variables. If we arbitrarily select a_{20} from among a_{20} and a_{02} to play the role of a parameter and similarly select b_{02} as a second parameter, then writing $a_{20} = v$ and $b_{02} = w$, we can satisfy the equation using just one more parameter s if we require $a_{02} = sw$ and $b_{20} = sv$. Inserting the expressions already specified into the third generator of Q_3 yields $a_{20}a_{-13}b_{20} = a_{02}b_{3,-1}b_{02} = s(a_{-13}v^2 - b_{3,-1}w^2)$, which can be made to vanish by the introduction of just one more parameter t if we require $a_{-13} = tw^2$ and $b_{3,-1} = tv^2$. The last two generators then vanish automatically when all the conditions specified so far are applied. Thus the set $S \subset \mathbb{C}^8$ defined by the parametric equations

$$\begin{aligned} a_{11} &= u & a_{02} &= sw \\ b_{11} &= u & b_{02} &= w \\ a_{20} &= v & a_{-13} &= tw^2 \\ b_{20} &= sv & b_{3,-1} &= tv^2 \end{aligned} \tag{1.62}$$

satisfies $S \subset \mathbf{V}(Q_3)$. To show that $\bar{S} = \mathbf{V}(Q_3)$ we apply Theorem 1.4.14. A Gröbner basis of the ideal

$$I = \langle a_{11} - u,\ b_{11} - u,\ a_{20} - v,\ b_{20} - sv, a_{02} - sw,\ b_{02} - w,\ a_{-13} - tw^2, b_{3,-1} - tv^2 \rangle$$

with respect to lex with

$$u > s > w > t > v > a_{11} > b_{11} > a_{20} > b_{20} > a_{02} > b_{02} > a_{-13} > b_{3,-1}$$

is

$$\begin{aligned}
\{&a_{-13}b_{20}^2 - a_{02}^2 b_{3,-1}, \quad a_{02}a_{20} - b_{02}b_{20}, \quad -a_{-13}a_{20}b_{20} + a_{02}b_{02}b_{3,-1},\\
&a_{-13}a_{20}^2 - b_{02}^2 b_{3,-1}, \quad a_{11} - b_{11}, \quad -a_{20} + v, \quad a_{-13} - b_{02}^2 t,\\
&a_{02}b_{3,-1} - a_{20}b_{02}b_{20}t, \quad b_{3,-1} - a_{20}^2 t, \quad b_{02} - w, \quad -b_{3,-1}s + a_{20}b_{20}t,\\
&\qquad a_{-13}s - a_{02}b_{02}t, \quad -a_{02} + b_{02}s, \quad b_{20} - a_{20}s, \quad b_{11} - u\}.
\end{aligned}$$

The first five polynomials listed in the Gröbner basis, the ones that do not depend on any of s, t, u, v, or w, are the generators of Q_3. According to Theorem 1.4.14 the variety $\mathbf{V}(Q_3)$ is the Zariski closure of S. Thus, by Definition 1.4.16, (1.62) defines a polynomial parametrization of $\mathbf{V}(Q_3)$, which by Corollary 1.4.18 is irreducible.

Example 1.4.19 gives us an intuitive idea of the concept of the *dimension* of an irreducible affine variety. The variety $S = \mathbf{V}(Q_3)$ is the Zariski closure of an image of $\mathbb{C}^5$; therefore, it is a five-dimensional variety. If a variety is not irreducible then its dimension is the maximal dimension of its components. For the precise definition of dimension of a variety and methods for computing it see, for example, [18] or [60]. Here we only observe that with reference to Example 1.4.19, the naive supposition that because the variety $\mathbf{V}(Q_3)$ arises from five conditions on eight variables its dimension must be $8 - 5 = 3$ is incorrect.

1.5 Notes and Complements

Gröbner bases were introduced by Bruno Buchberger in 1965 ([26]) in the context of his work on performing algorithmic computations in residue classes of polynomial rings. He named them after his thesis advisor Wolfgang Gröbner, a professor at the University of Innsbruck, who stimulated the research on the subject. The concept of a Gröbner basis has proven to be an extremely useful computational tool and has provided new insights in various subjects of modern algebra. It has also found enormous application in studies of many fundamental problems in various branches of mathematics and engineering. For further reading on Gröbner basis theory, its development, and applications, the reader is referred to [1, 18, 28, 60, 184, 185].

We have written the algorithms in this and succeeding chapters in pseudocode in order to facilitate the reader's writing his own programs in the computer algebra system or programming language of his choice. Each system has its own peculiarities, which the user must take into account when using it. For instance, Singular performs decompositions of polynomial ideals over the field $\mathbb{Q}$ and fields of positive characteristic, but not over $\mathbb{R}$ or $\mathbb{C}$. Thus although the decomposition (1.53) $I = \cap_{j=1}^m Q_j$ and the companion decomposition $\sqrt{I} = \cap_{j=1}^m \sqrt{Q_j} := \cap_{j=1}^m P_j$ that it gives are still valid over $\mathbb{C}$, the components in the corresponding decomposition (1.52) $V = \mathbf{V}(I) = \mathbf{V}(\sqrt{I}) = \mathbf{V}(\cap_{j=1}^m P_j) = \cup_{j=1}^m \mathbf{V}(P_j)$ might not all be irreducible as varieties in $\mathbb{C}^n$.

The decomposition of varieties that arise in actual applications, such as the problems in the qualitative theory of ordinary differential equations that we will study in Chapters 3–6, can lead to computations so vast that they cannot be completed even with rather powerful computers. In such cases it can be helpful to employ modular arithmetic; see [70, 148].

Exercises

1.1 Let $\mathbb{F}_2 = \{0,1\}$, define an addition by $0+0=0$, $0+1=1$, $1+0=1$, $1+1=0$, and define a multiplication by $1\cdot 0=0$, $0\cdot 0=0$, $0\cdot 1=0$, $1\cdot 1=1$. Show that $\mathbb{F}_2$ is a field and that $f = x^2 - x$ is a nonzero element of the ring $\mathbb{F}_2[x]$ that defines the zero function $f : \mathbb{F}_2 \to \mathbb{F}_2$.

1.2 The Factor Theorem states that for $f \in k[x]$ and $c \in k$, $f(c) = 0$ (c is a root of f) if and only if $(x-c)$ divides f. Use this to show that a polynomial f of degree $s > 0$ in $k[x]$ has at most s roots.

1.3 a. Prove Proposition 1.1.4.
b. More generally, prove that if T is any indexing set whatsoever and if for each $t \in T$ the set V_t is an affine variety, then $\cap_{t\in T} V_t$ is an affine variety.
Hint. Use part (b) of Exercise 1.4 and the Hilbert Basis Theorem (Theorem 1.1.6).

1.4 a. Prove that the set $\langle f_1, \ldots, f_s \rangle$ defined by (1.4) is an ideal in $k[x_1, \ldots, x_n]$.
b. More generally, prove that $\langle f : f \in F \rangle$ is an ideal in $k[x_1, \ldots, x_n]$.

1.5 Prove that congruence modulo an ideal I (Definition 1.1.8) is an equivalence relation on $k[x_1, \ldots, x_n]$.

1.6 a. Let $[f]$ denote the equivalence class of f in $k[x_1, \ldots, x_n]/I$. Show that for any f_1 and f_2 in $[f]$ and for any g_1 and g_2 in $[g]$, $(f_1 + g_1) - (f_2 + g_2) \in I$ and $f_1 g_1 - f_2 g_2 \in I$.
b. Use part (a) to show that addition and multiplication in $k[x_1, \ldots, x_n]/I$ can be validly defined by $[f] + [g] = [f + g]$ and $[f][g] = [fg]$.
c. Show that the operations on $k[x_1, \ldots, x_n]/I$ of part (b) define a ring structure on $k[x_1, \ldots, x_n]/I$. Note in particular what plays the role of the zero object.

1.7 Prove that for any subset V of k, not necessarily an affine variety, the set $\mathbf{I}(V)$ specified by Definition 1.1.10 is an ideal of $k[x_1, \ldots, x_n]$.

1.8 Prove that the maps (1.5) and (1.6) are inclusion-reversing.

1.9 The greatest common divisor of polynomials f and g in $k[x]$, denoted $\mathrm{GCD}(f,g)$, is the polynomial h such that (i) h divides both f and g, and (ii) if $p \in k[x]$ divides both f and g, then p divides h. Prove that

a. $\mathrm{GCD}(f,g)$ exists (and is unique except for multiplication by nonzero elements of k);

b. the Euclidean Algorithm, shown in Table 1.10, produces $\mathrm{GCD}(f,g)$.

Hint. The idea of the algorithm is that if R is the remainder upon division of F by G, then the set of common divisors of F and G is precisely the set of common divisors of G and R, so $\mathrm{GCD}(F,G) = \mathrm{GCD}(G,R)$. Note that $\mathrm{GCD}(f,0) = f$.

The Euclidean Algorithm

Input:

Polynomials $f,g \in k[x]$

Output:

$h = \mathrm{GCD}(f,g)$

Procedure:

$h := f,\ s := g$
WHILE $s \neq 0$
DO
  $h \xrightarrow{s} r$
  $h := s$
  $s := r$

Table 1.10 The Euclidean Algorithm

1.10 Suppose $f,g \in k[x]$ and let $d = \mathrm{GCD}(f,g)$. Show that there exist polynomials u_1, $u_2 \in k[x]$ such that $d = u_1 f + u_2 g$. (Compare with Exercise 5.7.)

Hint. The polynomials u_1 and u_2 can be shown to exist and simultaneously computed using the Extended Euclidean Algorithm, which is to write successive passes through the Euclidean Algorithm as

$$\begin{aligned} f &= q_1 g + r_1 \\ g &= q_2 r_1 + r_2 \\ r_1 &= q_3 r_2 + r_3 \\ &\vdots \\ r_{n-3} &= q_{n-1} r_{n-2} + r_{n-1} \\ r_{n-2} &= q_n r_{n-1} + 0 \end{aligned}$$

and retain all the quotients. For $j \geq 2$, each remainder r_j can be expressed in terms of r_1 through r_{j-1}, and r_1 can be expressed in terms of f and g.

1.11 Use Exercise 1.10 to show that in $k[x]$, $\langle f_1, f_2 \rangle = \langle \mathrm{GCD}(f_1, f_2) \rangle$.

1.12 Show that any ideal in $k[x]$ is generated by a single polynomial f.
Hint. The Hilbert Basis Theorem and the idea in Exercise 1.11.

1.13 Without factoring the polynomials involved, find a single generator of the ideal in $\mathbb{R}[x]$ indicated.
a. $I = \langle x^2 - 4,\ x + 3 \rangle$
b. $I = \langle x^3 + 3x^2 - x - 3,\ x^3 + x^2 - 9x - 9 \rangle$
c. $I = \langle 6x^3 - 11x^2 - 39x + 14,\ 4x^3 - 12x^2 - x - 21,\ 10x^3 - 35x^2 + 8x - 28 \rangle$

1.14 Determine whether or not $f \in I$ by finding a single generator of I, and then applying the Division Algorithm.
a. $f = 6x^3 + 12x^2 + 4x + 15, I = \langle x - 5,\ x^2 + x + 1 \rangle$
b. $f = 2x^3 + 5x^2 - 6x - 9$, I the ideal of Exercise 1.13(b)
c. $f = x^4 - 7x^3 + 12x - 13$, I the ideal of Exercise 1.13(b)

1.15 Show that for any term order $>$ on $k[x_1, \ldots, x_n]$, if $\alpha \in \mathbb{N}_0^n$ is not $(0, \ldots, 0)$, then $\alpha > (0, \ldots, 0)$, hence that if $\alpha + \beta = \gamma$, then $\gamma > \alpha$ and $\gamma > \beta$.

1.16 Show that the term orders introduced in Definition 1.2.3 satisfy the properties of Definition 1.2.1.

1.17 Suppose f, g, h, and p are in $k[x_1, \ldots, x_n]$ and $g \neq 0$. Show that if $f \xrightarrow{g} h$, then $(f + p) \xrightarrow{g} (h + p)$.

1.18 a. Suppose $f, g, h, p \in k[x_1, \ldots, x_n]$ and $p \neq 0$. Show that if $g \xrightarrow{p} h$ and f is a monomial, then $fg \xrightarrow{p} fh$.
b. Suppose $f_1, \ldots, f_s, f, g \in k[x_1, \ldots, x_n]$ with $f_j \neq 0$ for $1 \leq j \leq s$, and write $F = \{f_1, \ldots, f_s\}$. Show that if $g \xrightarrow{F} h$ and f is a monomial, then $fg \xrightarrow{F} fh$.

1.19 Suppose $f_1, \ldots, f_s, f, g \in k[x_1, \ldots, x_n]$, $f_s = f + g$, and $g \in \langle f_1, \ldots, f_{s-1} \rangle$. Show that $\langle f_1, \ldots, f_s \rangle = \langle f_1, \ldots, f_{s-1}, f \rangle$

1.20 Suppose $f_1, \ldots, f_s, g \in k[x_1, \ldots, x_n]$, $f_j \neq 0$ for $1 \leq j \leq s$, and set $F = \{f_1, \ldots, f_s\}$. Show that if $g \xrightarrow{F} h$, then $\langle f_1, \ldots, f_s, g \rangle = \langle f_1, \ldots, f_s, h \rangle$.

1.21 Show that for the polynomials of Example 1.2.9 the remainder upon division of f by $\{f_2, f_1\}$ is $-2x + 1$, whether the term order is lex, deglex, or degrevlex.

1.22 Find the remainder upon division of the polynomial $f = x^7 y^2 + x^3 y^2 - y + 1$ by $F = \{xy^2 - x, x - y^3\}$ (in the order listed), first with respect to the term order lex on $\mathbb{R}[x, y]$, then with respect to the term order deglex on $\mathbb{R}[x, y]$.

1.23 In $\mathbb{R}[x, y, z]$ let $g_1 = x + z$, $g_2 = y + z$, and $I = \langle g_1, g_2 \rangle$. Use Buchberger's Criterion (Theorem 1.2.19) as in Example 1.2.20 to show that under lex with $x > y > z$, $G = \{g_1, g_2\}$ is a Gröbner basis of I, but under lex with $x < y < z$ it is not.

1.24 Prove Lemma 1.2.18.
Hint. Let $a_j = \mathrm{LC}(f_j) \neq 0$, $1 \leq j \leq s$. Observe that $\sum_j c_j a_j = 0$ and that $S(f_j, f_\ell) = a_j^{-1} f_j - a_\ell^{-1} f_\ell$, and show that (1.14) can be represented in the form

$$f = c_1a_1S(f_1,f_2) + (c_1a_1 + c_2a_2)S(f_2,f_3) + \cdots + (c_1a_1 + \cdots + c_{s-1}a_{s-1})S(f_{s-1},f_s).$$

1.25 Find the reduced Gröbner basis for the ideal $\langle f_1, f_2 \rangle$ of Example 1.2.20.

1.26 Find an example of a set $S \subset k[x,y]$ such that $\langle \mathrm{LT}(S) \rangle \subsetneq \langle \mathrm{LT}(\langle S \rangle) \rangle$.

1.27 Prove Proposition 1.2.22.
Hint. Use the reasoning applied in the proof that (iv) implies (i) in Theorem 1.2.16.

1.28 Find two Gröbner bases for the ideal $I = \langle y^2 + yx + x^2, y + x, y \rangle$ in the polynomial ring $\mathbb{R}[x,y]$ under the term order lex with $y > x$. Use Proposition 1.2.22 to make two different choices of elements to discard so as to obtain two distinct *minimal* Gröbner bases of I.

1.29 Given a Gröbner basis G of an ideal I, let us say that an element g of G is *reduced for* G if no monomial in g lies in $\langle \mathrm{LT}(G) \setminus \{g\} \rangle$, and then define G to be reduced if each element of G is reduced for G and has leading coefficient 1. Show that this definition of a reduced Gröbner basis agrees with Definition 1.2.26.

1.30 Prove that a line with one point deleted is not a variety.

1.31 Prove Corollary 1.3.7.

1.32 Prove that for any ideal I, $\sqrt{I}$ is an ideal.
Hint. For $f^u, g^v \in I$, $u, v \in \mathbb{N}$, consider the binomial expansion of $(f+g)^{u+v-1}$.

1.33 Let k be a field and I an ideal in $k[x_1, \ldots, x_n]$. Show that $\mathbf{V}(\sqrt{I}) = \mathbf{V}(I)$.

1.34 Construct counterexamples to the analogues of Theorems 1.3.10 and 1.3.14 that arise when $\mathbb{C}$ is replaced by $\mathbb{R}$.

1.35 Let $I = \langle f_1, \ldots, f_s \rangle$ and $J = \langle g_1, \ldots, g_t \rangle$ be ideals in $\mathbb{C}[x_1, \ldots, x_n]$. Show that $\sqrt{I} = \sqrt{J}$ if and only if $f_i \in \sqrt{J}$ for all i, $1 \le i \le s$ and $g_j \in \sqrt{I}$ for all j, $1 \le j \le t$.

1.36 Let $I = \langle f_1, \ldots, f_u \rangle$ and $J = \langle g_1, \ldots, g_v \rangle$ be ideals in $k[x_1, \ldots, x_n]$.
a. Prove that $I \cap J$ and $I + J$ are ideals in $k[x_1, \ldots, x_n]$.
b. Show that $I + J = \langle f_1, \ldots, f_u, g_1, \ldots, g_v \rangle$.

1.37 Suppose I and J are ideals in $k[x_1, \ldots, x_m, y_1, \ldots, y_n]$.
a. Show that $I \cap k[y_1, \ldots, y_n]$ is an ideal in $k[y_1, \ldots, y_n]$.
b. Let $k[y]$ denote $k[y_1, \ldots, y_n]$. Show that $(I \cap k[y]) + (J \cap k[y]) \subset (I + J) \cap k[y]$ but that the reverse inclusion need not hold.

1.38 a. If A is any set and $\{I_a : a \in A\}$ is a collection of ideals indexed by A, show that $\cap_{a \in A} I_a$ is an ideal.
b. Let any subset F of $k[x_1, \ldots, x_n]$ be given. By part (a) of this exercise the set $J := \cap\{I : I \text{ is an ideal and } I \supset F\}$ is an ideal, the "smallest ideal that contains every element of F." Show that $J = \langle f : f \in F \rangle$.

1.39 Prove that for any two ideals I and J, $\sqrt{I \cap J} = \sqrt{I} \cap \sqrt{J}$. Prove that for any two sets A and B in k^n, $\mathbf{I}(A \cup B) = \mathbf{I}(A) \cap \mathbf{I}(B)$.

1.40 Prove Theorem 1.3.18.

1.41 See Exercise 1.7. Show that if S is not an affine variety, then $S \subset \mathbf{V}(\mathbf{I}(S))$ always holds, but the inclusion could be strict.

1.42 Show that if V is a variety, then V is a topologically closed set. Suppose S is a set (not necessarily a variety) and $C\ell(S)$ is its topological closure; show that $S \subset C\ell(S) \subset \bar{S}$. Show by example that just the first, just the second, both, or neither of the two set inclusions could be strict.

1.43 Prove that the set $I : J$ introduced in Definition 1.3.23 is an ideal in $k[x_1, \ldots, x_n]$.

1.44 Prove Proposition 1.3.27.

1.45 [Referenced in Theorem 3.7.1, proof.] Let V be a variety and V_1 a subvariety of V, $V_1 \subsetneqq V$.

a. Since $V_1 \subset V$, by definition $V = V_1 \cup (V \setminus V_1)$. Show that no points outside V are picked up in forming the Zariski closure of $V \setminus V_1$. That is, show that $V_1 \cup \overline{(V \setminus V_1)} \subset V$.

b. Use the result in part (a) to show that if V is irreducible, then $\overline{(V \setminus V_1)} = V$. (Compare this result to Proposition 1.3.20.)

c. Show that the set equality in part (b) can fail if V is not irreducible or if V_1 is not a variety.

1.46 Consider $J = \langle 2a_1 + b_0, a_0 + 2b_1, a_1 b_1 - a_2 b_2 \rangle$ in $\mathbb{C}[a_0, a_1, a_2, b_2, b_1, b_0]$.

a. Use a computer algebra system to verify that J is a radical ideal. (For example, in Maple apply the `IsRadical` command that is part of the `PolynomialIdeals` package.)

b. Use the algorithm of Table 1.6 on page 38 to verify that $J : \langle b_1 b_2 \rangle = J$.

c. Conclude that $\overline{\mathbf{V}(J) \setminus \mathbf{V}(b_1 b_2)} = \mathbf{V}(J)$.

d. Could the conclusion in part (c) be drawn from the result in Exercise 1.45 without computing $J : \langle b_1 b_2 \rangle$? If so, explain why; if not, what additional information is needed?

1.47 In the context of the Exercise 1.46, define a pair of polynomials g and h by $g = a_0 b_0 b_2 - a_2 b_2^2$ and $h = 4 a_0^3 a_2 - a_0^2 b_0^2 - 18 a_0 a_2 b_0 b_2 + 4 b_0^3 b_2 + 27 a_2^2 b_2^2$. Show that $\overline{\mathbf{V}(J) \setminus \mathbf{V}(\langle g \rangle \cap \langle h \rangle)} = \mathbf{V}(J)$.

1.48 Prove Proposition 1.4.7.

1.49 Prove that an ideal I is primary if and only if the ideal $\sqrt{I}$ is prime.

1.50 Show that the minimal associated prime ideals given in Tables 1.7 and 1.9 are the same.

Chapter 2
Stability and Normal Forms

In this chapter our concern is with a system of ordinary differential equations $\dot{\mathbf{x}} = \mathbf{f}(\mathbf{x})$ in $\mathbb{R}^n$ or $\mathbb{C}^n$ in a neighborhood of a point $\mathbf{x}_0$ at which $\mathbf{f}(\mathbf{x}_0) = \mathbf{0}$. Early investigations into the nature of solutions of the system of differential equations in a neighborhood of such a point were made in the late 19th century by A. M. Lyapunov ([114, 115]) and H. Poincaré ([143]). Lyapunov developed two methods for investigating the stability of $\mathbf{x}_0$. The so-called First Method involves transformation of the system to *normal form*; the Second or Direct Method involves the use of what are now termed *Lyapunov functions*. In the first section of this chapter we present several of the principal theorems of Lyapunov's Direct Method. Since smoothness of $\mathbf{f}$ is not necessary for these results, we do not assume it in this section. The second and third sections are devoted to the basics of the theory of normal forms.

2.1 Lyapunov's Second Method

Let Ω be an open subset of $\mathbb{R}^n$, $\mathbf{x}_0$ a point of Ω, and $\mathbf{f} : \Omega \to \mathbb{R}^n$ continuous and such that solutions of initial value problems associated with the autonomous system of differential equations

$$\dot{\mathbf{x}} = \mathbf{f}(\mathbf{x}) \tag{2.1}$$

are unique. For $\mathbf{x}_1 \in \Omega$, we let $\mathbf{x}_1(t)$ denote the unique solution of (2.1) that satisfies $\mathbf{x}(0) = \mathbf{x}_1$, on its maximal interval of existence J_1; this is the *trajectory through* $\mathbf{x}_1$. The point set $\{\mathbf{x}_1(t) : t \in J_1\}$ is the *orbit* of (or through) $\mathbf{x}_1$. If $\mathbf{f}(\mathbf{x}_0) = \mathbf{0}$, then $\mathbf{x}(t) \equiv \mathbf{x}_0$ solves (2.1) uniquely, the orbit through $\mathbf{x}_0$ is just $\{\mathbf{x}_0\}$, and $\mathbf{x}_0$ is termed an *equilibrium* or *rest point* of the system, or a *singularity* or *singular point* (particularly when we view $\mathbf{f}$ as a vector field on Ω; see Remark 3.2.4(b)). Any orbit is topologically a point, a circle (a *closed orbit* or a *cycle*), or a line (see [44] or [140]). The decomposition of Ω, the *phase space* of (2.1), into the union of disjoint orbits is the *phase portrait* of (2.1). In part (a) of the following definition we have incorporated into the definition the result from the theory of differential equations

V.G. Romanovski, D.S. Shafer, *The Center and Cyclicity Problems*,
DOI 10.1007/978-0-8176-4727-8_2,

that if $\mathbf{x}_1(t)$ is confined to a compact set for all nonnegative t in its maximal interval of existence, then that interval must contain the half-line $[0,\infty)$, so that existence of $\mathbf{x}_1(t)$ for all nonnegative t need not be assumed in advance. The same consideration applies to part (b) of the definition.

Definition 2.1.1.
(a) An equilibrium $\mathbf{x}_0$ of (2.1) is *stable* if, for every $\varepsilon > 0$, there exists $\delta > 0$ such that if $\mathbf{x}_1$ satisfies $|\mathbf{x}_1 - \mathbf{x}_0| < \delta$ then $|\mathbf{x}_1(t) - \mathbf{x}_0| < \varepsilon$ for all $t \geq 0$.
(b) An equilibrium $\mathbf{x}_0$ of (2.1) is *asymptotically stable* if it is stable and if there exists $\delta_1 > 0$ such that if $\mathbf{x}_1$ satisfies $|\mathbf{x}_1 - \mathbf{x}_0| < \delta_1$, then $\lim_{t\to\infty} \mathbf{x}_1(t) = \mathbf{x}_0$.
(c) An equilibrium of (2.1) is *unstable* if it is not stable.

Note that it is possible that the trajectory of every point in a neighborhood of an equilibrium $\mathbf{x}_0$ tends to $\mathbf{x}_0$ in forward time, yet in every neighborhood there exist a point whose forward trajectory travels a uniformly large distance away from $\mathbf{x}_0$ before returning to limit on $\mathbf{x}_0$ (Exercise 2.1). This is the reason for the specific requirement in point (b) that $\mathbf{x}_0$ be stable. What we have called *stability* and *asymptotic stability* are sometimes referred to in the literature as *positive stability* and *positive asymptotic stability*, and the equilibrium is then said to be *negatively stable* or *negatively asymptotically stable* for (2.1) if it is respectively positively stable or positively asymptotically stable for $\dot{\mathbf{x}} = -\mathbf{f}(\mathbf{x})$.

The problem of interest in this section is to obtain a means of determining when an equilibrium of system (2.1) is stable, asymptotically stable, or unstable without actually solving or estimating solutions of the system, particularly when linear estimates fail. In a system of differential equations that models a mechanical system, the total energy function typically holds the answer: if total energy strictly decreases along positive trajectories near the equilibrium, then it is asymptotically stable. The concept of Lyapunov function generalizes this idea.

Since without loss of generality we may assume that the equilibrium is located at the origin of $\mathbb{R}^n$, we henceforth assume that Ω is a neighborhood of $\mathbf{0} \in \mathbb{R}^n$ and that $\mathbf{f}(\mathbf{0}) = \mathbf{0}$. To understand the meaning of the quantity $\dot{W}$ in point (b) of the following definition, note that if $\mathbf{x}(t)$ is a trajectory of system (2.1) (corresponding to some initial condition $\mathbf{x}(0) = \mathbf{x}_0$), then the expression $w(t) = W(\mathbf{x}(t))$ defines a differentiable function from a punctured neighborhood of $0 \in \mathbb{R}$ into $\mathbb{R}$, and by the chain rule its derivative is $\mathrm{d}W(\mathbf{x}) \cdot \mathbf{f}(\mathbf{x})$. The expression in (b) can thus be understood as giving the instantaneous rate of change (at $\mathbf{x}$) of the function W along the unique trajectory of (2.1) which is at $\mathbf{x}$ at time zero.

Definition 2.1.2. Let U be an open neighborhood of $\mathbf{0} \in \mathbb{R}^n$ and let $W : U \to \mathbb{R}$ be a continuous function that is C^1 on $U \setminus \{\mathbf{0}\}$.
(a) W is *positive definite* if $W(\mathbf{0}) = 0$ and $W(\mathbf{x}) > 0$ for $\mathbf{x} \neq \mathbf{0}$.
(b) W is a *Lyapunov function* for system (2.1) if it is positive definite and if the function $\dot{W} : U \setminus \{\mathbf{0}\} \to \mathbb{R} : \mathbf{x} \mapsto \mathrm{d}W(\mathbf{x}) \cdot \mathbf{f}(\mathbf{x})$ is nonpositive.
(c) W is a *strict Lyapunov function* for system (2.1) if it is positive definite and if $\dot{W}$ is negative.

Theorem 2.1.3. *Let Ω be an open neighborhood of $\mathbf{0} \in \mathbb{R}^n$, and let $\mathbf{0}$ be an equilibrium for system* (2.1) *on Ω.*

1. *If there exists a Lyapunov function for system* (2.1) *on a neighborhood U of $\mathbf{0}$, then $\mathbf{0}$ is stable.*
2. *If there exists a strict Lyapunov function for system* (2.1) *on a neighborhood U of $\mathbf{0}$, then $\mathbf{0}$ is asymptotically stable.*

Proof. Suppose there exists a Lyapunov function W defined on a neighborhood U of $\mathbf{0}$, and let $\varepsilon > 0$ be given. Decrease ε if necessary so that $\{\mathbf{x} : |\mathbf{x}| \leq \varepsilon\} \subset U$, and let $S = \{\mathbf{x} : |\mathbf{x}| = \varepsilon\}$. By continuity of W, compactness of S, and the fact that W is positive definite, it is clear that $m := \min\{W(\mathbf{x}) : \mathbf{x} \in S\}$ is finite and positive, and that there exists a positive number $\delta < \varepsilon$ such that $M := \max\{W(\mathbf{x}) : |\mathbf{x}| \leq \delta\} < m$. We claim that δ is as required. For fix $\mathbf{x}_1$ such that $0 < |\mathbf{x}_1| \leq \delta$, and as usual let $\mathbf{x}_1(t)$ denote the trajectory through $\mathbf{x}_1$. If, contrary to what we wish to establish, there exists a positive value of t for which $|\mathbf{x}_1(t)| = \varepsilon$, then there exists a smallest such value T. Then for $0 \leq t \leq T$, $\mathbf{x}_1(t)$ is in $U \setminus \{\mathbf{0}\}$, hence $v(t) := W(\mathbf{x}_1(t))$ is defined and smooth, and $v'(t) = \dot{W}(\mathbf{x}_1(t)) \leq 0$, so that $v(0) \geq v(T)$, in contradiction to the fact that $v(0) \leq M < m \leq v(T)$.

For our proof of the second point we recall the notion of the omega limit set $\omega(\mathbf{x})$ of a point $\mathbf{x}$ for which $\mathbf{x}(t)$ is defined for all $t \geq 0$: $\omega(\mathbf{x})$ is the set of all points $\mathbf{y}$ for which there exists a sequence $t_1 < t_2 < \cdots$ of numbers such that $t_n \to \infty$ and $\mathbf{x}(t_n) \to \mathbf{y}$ as $n \to \infty$. Note that $\omega(\mathbf{x})$ is closed and invariant under the flow of system (2.1) (see [44] or [140]).

Suppose there exists a strict Lyapunov function W defined on a neighborhood U of $\mathbf{0}$. Since W is also a Lyapunov function, by point (1) $\mathbf{0}$ is stable. Choose $\varepsilon > 0$ so small that $\{\mathbf{x} : |\mathbf{x}| \leq \varepsilon\} \subset U$, and let δ be such that if $|\mathbf{x}| < \delta$, then $\mathbf{x}(t)$ exists for all $t \geq 0$ and satisfies $|\mathbf{x}(t)| < \varepsilon/2$, hence the omega limit set $\omega(\mathbf{x})$ is defined and is a nonempty, compact, connected set ([44, 140]). Note that if $0 < |\mathbf{x}_1| < \delta$, then for all $t \geq 0$, $W(\mathbf{x}_1(t))$ exists and $\dot{W}(\mathbf{x}_1(t)) < 0$.

Fix any $\mathbf{x}_1$ such that $0 < |\mathbf{x}_1| < \delta$. If $\mathbf{a}, \mathbf{b} \in \omega(\mathbf{x}_1)$, then $W(\mathbf{a}) = W(\mathbf{b})$. Indeed, there exist sequences t_n and s_n, $t_n \to \infty$ and $s_n \geq 0$, such that both $\mathbf{x}_1(t_n) \to \mathbf{a}$ and $\mathbf{x}_1(t_n + s_n) \to \mathbf{b}$. Since W is continuous and strictly decreases on every trajectory in $\Omega \setminus \{\mathbf{0}\}$, $W(\mathbf{a}) \geq W(\mathbf{b})$. Similarly, $W(\mathbf{b}) \geq W(\mathbf{a})$.

Again, if $\mathbf{a} \in \omega(\mathbf{x}_1)$, then $\mathbf{a}$ must be an equilibrium. For if $|\mathbf{a}| \neq \mathbf{0}$ and $\mathbf{f}(\mathbf{a}) \neq \mathbf{0}$, then $\mathbf{a}(t) \in \Omega \setminus \{\mathbf{0}\}$ when it is defined, and for sufficiently small $\tau > 0$ the time-τ image $\mathbf{a}(\tau)$ of $\mathbf{a}$ satisfies $\mathbf{a}(\tau) \neq \mathbf{a}$. But if $\mathbf{a} \in \omega(\mathbf{x}_1)$, then $|\mathbf{a}| \leq \varepsilon/2 < \varepsilon$, so that $W(\mathbf{a}(\tau)) < W(\mathbf{a})$. Yet $\mathbf{a}(\tau) \in \omega(\mathbf{x}_1)$, hence $W(\mathbf{a}(\tau)) = W(\mathbf{a})$, a contradiction.

Finally, $\mathbf{0}$ is the only equilibrium in $\{\mathbf{x} : |\mathbf{x}| < \delta\}$. For given any $\mathbf{x}$ satisfying $0 < |\mathbf{x}| < \delta$, for sufficiently small $\tau > 0$, $W(\mathbf{x}(\tau))$ is defined and $W(\mathbf{x}(\tau)) < W(\mathbf{x})$, hence $\mathbf{x}(\tau) \neq \mathbf{x}$.

In short, $\omega(\mathbf{x}_1) = \{\mathbf{0}\}$, so $\mathbf{x}_1(t) \to \mathbf{0}$ as $t \to \infty$, as required. □

The following slight generalization of part (2) of Theorem 2.1.3 is sometimes useful (see Example 2.1.6, but also Exercise 2.4).

Theorem 2.1.4. *Let Ω be an open neighborhood of $\mathbf{0} \in \mathbb{R}^n$, and let $\mathbf{0}$ be an equilibrium for system* (2.1) *on Ω. Let C be a smooth curve in Ω to which $\mathbf{f}$ is nowhere* *tangent except at* $\mathbf{0}$. *If there exists a Lyapunov function* W *for system* (2.1) *on a neighborhood* U *of* $\mathbf{0}$ *such that* $\dot{W}$ *is negative on* $\Omega \setminus C$, *then* $\mathbf{0}$ *is asymptotically stable.*

Proof. The proof of part (2) of Theorem 2.1.3 goes through unchanged, since it continues to be true that W is strictly decreasing on every trajectory in $\Omega \setminus \{\mathbf{0}\}$. □

An instability theorem analogous to Theorem 2.1.3 is the following.

Theorem 2.1.5. *Let* Ω *be an open neighborhood of* $\mathbf{0} \in \mathbb{R}^n$, *and let* $\mathbf{0}$ *be an equilibrium for system* (2.1) *on* Ω.
1. *If there exists a positive definite function* W *on a neighborhood* U *of* $\mathbf{0}$ *such that* $\dot{W} > 0$ *on* $U \setminus \{\mathbf{0}\}$, *then* $\mathbf{0}$ *is unstable.*
2. *If there exists a function* W *on a neighborhood* U *of* $\mathbf{0}$ *such that* $W(\mathbf{0}) = 0$, $\dot{W}$ *is positive definite, and* W *takes a positive value in every neighborhood of* $\mathbf{x}_0$, *then* $\mathbf{x}_0$ *is unstable.*

Proof. The proof is left as Exercise 2.5. □

Example 2.1.6. Consider a quadratic system on $\mathbb{R}^2$ with an equilibrium at which the linear part has one eigenvalue negative and the other eigenvalue zero. (To say that the system is *quadratic* means that the right-hand sides in (2.1) are polynomials, the maximum of whose degrees is two.) By translating the equilibrium to the origin, performing a linear change of coordinates, and rescaling time, we may place the system in the form

$$\begin{aligned} \dot{x} &= -x + ax^2 + bxy + cy^2 \\ \dot{y} &= \qquad\; dx^2 + exy + fy^2 . \end{aligned}$$

We will use a Lyapunov function to show that the equilibrium is stable if $f = 0$ and $ce < 0$, or if $c = e = f = 0$. To do so, consider any function W of the form

$$W(x,y) = (Ax^2 + By^2)/2. \tag{2.2}$$

Then

$$\dot{W}(x,y) = -Ax^2 + Aax^3 + (Ab + Bd)x^2y + (Ac + Be)xy^2 + Bfy^3 .$$

Choosing $A = |e|$ and $B = |c|$ if $ce < 0$ and $A = B = 1$ if $c = e = 0$, W is positive definite. When $f = 0$, $\dot{W}$ becomes $\dot{W}(x,y) = -x^2(A - Aax - (Ab + Bd)y)$, hence is nonpositive on a neighborhood of the origin, implying stability. We note that any equilibrium of a planar system at which the linear part has exactly one eigenvalue zero is a node, a (topological) saddle, or a saddle-node (see, for example, Theorem 65, §21 of [12]), so that the equilibrium in question must actually be a stable node, hence be asymptotically stable, a fact that the Lyapunov function does not reveal if only Theorem 2.1.3 is used, but one that is shown by Theorem 2.1.4 when $c \neq 0$ (but not when $c = 0$).

Example 2.1.7. Consider a quadratic system on $\mathbb{R}^2$ with an equilibrium at the origin at which the linear part has purely imaginary eigenvalues. By a linear change of coordinates and time rescaling (Exercise 2.7), we may write the system in the form

$$\begin{aligned}\dot{x} &= -y + ax^2 + bxy + cy^2\\ \dot{y} &= x + dx^2 + exy + fy^2.\end{aligned} \tag{2.3}$$

If we look for a Lyapunov function in the form (2.2), then

$$\dot{W}(x,y) = (B-A)xy + Aax^3 + (Ab+Bd)x^2y + (Ac+Be)xy^2 + Bfy^3,$$

which is nonpositive in a neighborhood of the origin if and only if $A = B$ and

$$a = f = 0, \quad b = -d, \quad c = -e \tag{2.4}$$

(in which case $\dot{W} \equiv 0$). Therefore the origin of system (2.3) is stable if the coefficients satisfy (2.4). The sufficient conditions (2.4) that we have found for stability of the origin of system (2.3) in fact are not necessary, however, but arose simply as a result of our specific choice of the form (2.2) of our trial function W. Indeed, consider system (2.3) with $a = f = 1$ and $c = d = e = 0$, that is,

$$\dot{x} = -y + x^2 + bxy, \quad \dot{y} = x + y^2, \tag{2.5}$$

and in place of (2.2) the far more complicated trial function

$$\begin{aligned}W(x,y) = x^2 + y^2 &\\ &+ \frac{2\,(2+b)}{3}x^3 - 2x^2y + 2xy^2 - \frac{4}{3}y^3\\ &- \frac{(26-30b^2-9b^3)}{6\,(4+3b)}x^4 + \frac{4}{3\,(4+3b)}x^3y\\ &- \frac{2\,(1+3b)}{3\,(4+3b)}x^2y^2 + \frac{4\,(1+b)\,(1+3b)}{3\,(4+3b)}xy^3 - \frac{(26+36b+9b^2)}{6\,(4+3b)}y^4,\end{aligned}$$

for which

$$\dot{W}(x,y) = \frac{2(26+30b+9b^2)}{12+9b}x^4 - \frac{4\,(13+13b+3b^2)\,y^4}{12+9b} + o(|x^4+y^4|).$$

When $\frac{-13-\sqrt{13}}{6} < b < \frac{-13+\sqrt{13}}{6}$, W is a strict Lyapunov function, hence for such b the origin is an asymptotically stable singular point of system (2.5).

Example 2.1.7 illustrates that even for a system that is apparently as simple as system (2.3) it is a difficult problem to find in the space of parameters $\{a,b,c,d,e,f\}$ the subsets corresponding to systems with a stable, unstable, or asymptotically stable singularity at the origin. The theorems of this section do not provide a procedure for resolving this problem. We will study this situation in detail in Chapter 3.

2.2 Real Normal Forms

Suppose $\mathbf{x}_0$ is a regular point of system (2.1), that is, a point $\mathbf{x}_0$ at which $\mathbf{f}(\mathbf{x}_0) \neq \mathbf{0}$, and that $\mathbf{f}$ is C^∞ (or real analytic). The Flowbox Theorem (see, for example, §1.7.3 of [44]) states that there is a C^∞ (respectively, real analytic) change of coordinates $\mathbf{x} = \mathbf{H}(\mathbf{y})$ in a neighborhood of $\mathbf{x}_0$ so that with respect to the new coordinates system (2.1) becomes

$$\dot{y}_1 = 1 \qquad \text{and} \qquad \dot{y}_j = 0 \qquad \text{for} \quad 2 \leq j \leq n. \tag{2.6}$$

The Flowbox Theorem confirms the intuitively obvious answers to two questions about regular points, those of *identity* and of *structural stability* or *bifurcation*: regardless of the infinitesimal generator $\mathbf{f}$, the phase portrait of system (2.1) in a neighborhood of a regular point $\mathbf{x}_0$ is topologically equivalent to that of the parallel flow of system (2.6) at any point, and when $\mathbf{f}$ is regarded as an element of any "reasonably" topologized set V of vector fields, there is a neighborhood N of $\mathbf{f}$ in V such that for any $\tilde{\mathbf{f}}$ in N, the flows of $\mathbf{f}$ and $\tilde{\mathbf{f}}$ (or we often say $\mathbf{f}$ and $\tilde{\mathbf{f}}$ themselves) are topologically equivalent in a neighborhood of $\mathbf{x}_0$. (Two phase portraits, or the systems of differential equations or vector fields that generate them, are said to be *topologically equivalent* if there is a homeomorphism carrying the orbits of one onto the orbits of the other, preserving the direction of flow along the orbits, but not necessarily their actual parametrizations, say as solutions of the differential equation.) In short, the phase portrait in a neighborhood of a regular point is known (up to diffeomorphism), and there is no bifurcation under sufficiently small perturbation.

Now suppose that $\mathbf{x}_0$ is an equilibrium of system (2.1). As always we assume, by applying a translation of coordinates if necessary, that the equilibrium is located at the origin and that $\mathbf{f}$ is defined and is sufficiently smooth on some open neighborhood of the origin. The questions of identity and of structural stability or bifurcation have equally simple answers in the case that the real parts of the eigenvalues of the linear part $A := d\mathbf{f}(\mathbf{0})$ of $\mathbf{f}$ are nonzero, in which case the equilibrium is called *hyperbolic*. For the Hartman–Grobman Theorem (see for example [44]) states that, in such a case, in a neighborhood of the origin the local flows $\phi(t, \mathbf{x})$ generated by (2.1) and $\psi(t, \mathbf{x})$ generated by the linear system

$$\dot{\mathbf{x}} = A\mathbf{x} \qquad (A = d\mathbf{f}(\mathbf{0})) \tag{2.7}$$

are topologically conjugate: there is a homeomorphism $\mathbf{H}$ of a neighborhood of the origin onto its image such that $\mathbf{H}(\phi(t, \mathbf{x})) = \psi(t, \mathbf{H}(\mathbf{x}))$. (Here, for each $\mathbf{x}_0$, $\phi(t, \mathbf{x}_0)$ (respectively, $\psi(t, \mathbf{x}_0)$) is the unique solution $\mathbf{x}(t)$ of (2.1) (respectively, of (2.7)) satisfying the initial condition $\mathbf{x}(0) = \mathbf{x}_0$.) Because the homeomorphism $\mathbf{H}$ thus carries trajectories of system (2.1) onto those of system (2.7), preserving their sense (in fact, their parametrization), it is a fortiori a topological equivalence between the full system (2.1) and the linear system (2.7) in a neighborhood of $\mathbf{0}$, the latter of which is explicitly known. Furthermore, if $\mathbf{f}$ is an element of a set V of vector fields that is topologized in such a way that eigenvalues of linear parts depend continuously on the vector fields, then system (2.1) is structurally stable as well.

The real situation of interest then is that of an isolated equilibrium of system (2.1) that is nonhyperbolic. The identity problem is most fully resolved in dimension two. If there is a characteristic direction of approach to the equilibrium, then apart from certain exceptional cases, the topological type of the equilibrium can be found by means of a finite sequence of "blow-ups" and is determined by a finite initial segment of the Taylor series expansion of $\mathbf{f}$ at $\mathbf{0}$ (see [65]). In higher dimensions, results on the topological type of degenerate equilibria are less complete.

Normal form theory enters in when we wish to understand the bifurcations of system (2.1) in a neighborhood of a nonhyperbolic equilibrium. Supposing that we have solved the identity problem (although in actual practice an initial normal form computation may be done in order to simplify the identity problem for the original system), we wish to know what phase portraits are possible in a neighborhood of $\mathbf{0}$, for any vector field in a neighborhood of $\mathbf{f}$ in some particular family V of vector fields, with a given topology. The idea of normal form theory quite simply is to perform a change of coordinates $\mathbf{x} = \mathbf{H}(\mathbf{y})$, or a succession of coordinate transformations, so as to place the original system (2.1) into a form most amenable to study. Typically, this means eliminating as many terms as possible from an initial segment of the power series expansion of $\mathbf{f}$ at the origin.

It is useful to compare this idea to an application of the Hartman–Grobman Theorem. Although the homeomorphism $\mathbf{H}$ guaranteed by the Hartman–Grobman Theorem can rightly be regarded as a change of coordinates in a neighborhood of the origin, in that it is invertible, it would be incorrect to say that system (2.1) has been transformed into system (2.7) by the change of coordinates $\mathbf{H}$, since $\mathbf{H}$ could fail to be smooth. In fact, for some choices of $\mathbf{f}$ in (2.1) it is impossible to choose $\mathbf{H}$ to be smooth. An example for which this happens is instructive and will lay the groundwork for the general approach in the nonhyperbolic case.

Example 2.2.1. Consider the linear system

$$\begin{aligned} \dot{x}_1 &= 2x_1 \\ \dot{x}_2 &= x_2, \end{aligned} \tag{2.8}$$

which has a hyperbolic equilibrium at the origin, and the general quadratic system with the same linear part,

$$\begin{aligned} \dot{x}_1 &= 2x_1 + ax_1^2 + bx_1x_2 + cx_2^2 \\ \dot{x}_2 &= x_2 + a'x_1^2 + b'x_1x_2 + c'x_2^2 . \end{aligned} \tag{2.9}$$

We make a C^2 change of coordinates $\mathbf{x} = \mathbf{H}(\mathbf{y}) = \mathbf{h}^{(0)}(\mathbf{y}) + \mathbf{h}^{(1)}(\mathbf{y}) + \mathbf{h}^{[2]}(\mathbf{y})$, where $\mathbf{h}^{(0)}(\mathbf{y})$ denotes the constant terms, $\mathbf{h}^{(1)}(\mathbf{y})$ the linear terms in y_1 and y_2, and $\mathbf{h}^{[2]}(\mathbf{y})$ *all* the remaining terms. Our goal is to eliminate all the quadratic terms in (2.9). To keep the equilibrium situated at the origin, we choose $\mathbf{h}^{(0)}(\mathbf{y}) = \mathbf{0}$, and to maintain the same linear part, which is already in "simplest" form, namely Jordan normal form, we choose $\mathbf{h}^{(1)}(\mathbf{y}) = \mathbf{y}$. Thus the change of coordinates is

$$\mathbf{x} = \mathbf{y} + \mathbf{h}^{[2]}(\mathbf{y}),$$

whence

$$\dot{\mathbf{x}} = \dot{\mathbf{y}} + \mathbf{dh}^{[2]}(\mathbf{y})\dot{\mathbf{y}} = (\mathbf{Id} + \mathbf{dh}^{[2]}(\mathbf{y}))\dot{\mathbf{y}}, \tag{2.10}$$

where for the n-dimensional vector function $\mathbf{u}$ we denote by $\mathbf{du}$ the Jacobian matrix

$$\begin{pmatrix} \frac{\partial u_1}{\partial y_1} & \cdots & \frac{\partial u_1}{\partial y_n} \\ \vdots & \ddots & \vdots \\ \frac{\partial u_n}{\partial y_1} & \cdots & \frac{\partial u_n}{\partial y_n} \end{pmatrix}.$$

For $\mathbf{y}$ sufficiently close to $\mathbf{0}$ the geometric series $\sum_{k=0}^{\infty}(-\mathbf{dh}^{[2]}(\mathbf{y}))^k$ converges in the real vector space of linear transformations from $\mathbb{R}^n$ to $\mathbb{R}^n$ with the uniform norm $\|\mathbf{T}\| = \max\{|\mathbf{Tx}| : |\mathbf{x}| \leq 1\}$. Therefore, the linear transformation $\mathbf{Id} + \mathbf{dh}^{[2]}(\mathbf{y})$ is invertible and $(\mathbf{Id} + \mathbf{dh}^{[2]}(\mathbf{y}))^{-1} = \mathbf{Id} - \mathbf{dh}^{[2]}(\mathbf{y}) + \cdots$, so that (2.10) yields

$$\dot{\mathbf{y}} = (\mathbf{Id} + \mathbf{dh}^{[2]}(\mathbf{y}))^{-1}\dot{\mathbf{x}} = (\mathbf{Id} - \mathbf{dh}^{[2]}(\mathbf{y}) + \cdots)\dot{\mathbf{x}}. \tag{2.11}$$

Writing $\mathbf{h}^{[2]}$ as

$$\mathbf{h}^{[2]}(\mathbf{y}) = \begin{pmatrix} a_{20}y_1^2 + a_{11}y_1y_2 + a_{02}y_2^2 + \cdots \\ b_{20}y_1^2 + b_{11}y_1y_2 + b_{02}y_2^2 + \cdots \end{pmatrix},$$

we have

$$\mathbf{dh}^{[2]}(\mathbf{y}) = \begin{pmatrix} 2a_{20}y_1 + a_{11}y_2 + \cdots & a_{11}y_1 + 2a_{02}y_2 + \cdots \\ 2b_{20}y_1 + b_{11}y_2 + \cdots & b_{11}y_1 + 2b_{02}y_2 + \cdots \end{pmatrix},$$

so that (2.11) is

$$\begin{aligned} \begin{pmatrix} \dot{y}_1 \\ \dot{y}_2 \end{pmatrix} &= \begin{pmatrix} 1 - 2a_{20}y_1 - a_{11}y_2 - \cdots & -a_{11}y_1 - 2a_{02}y_2 - \cdots \\ -2b_{20}y_1 - b_{11}y_2 - \cdots & 1 - b_{11}y_1 - 2b_{02}y_2 - \cdots \end{pmatrix} \\ &\quad \times \begin{pmatrix} 2(y_1 + a_{20}y_1^2 + a_{11}y_1y_2 + a_{02}y_2^2 + \cdots) \\ \qquad + a(y_1 + \cdots)^2 + b(y_1 + \cdots)(y_2 + \cdots) + c(y_2 + \cdots)^2 \\ y_2 + b_{20}y_1^2 + b_{11}y_1y_2 + b_{02}y_2^2 + \cdots \\ \qquad + a'(y_1 + \cdots)^2 + b'(y_1 + \cdots)(y_2 + \cdots) + c'(y_2 + \cdots)^2 \end{pmatrix} \\ &= \begin{pmatrix} 2y_1 + (a - 2a_{20})y_1^2 + (b - \ a_{11})y_1y_2 + cy_2^2 + \cdots \\ y_2 + (a' - 3b_{20})y_1^2 + (b' - 2b_{11})y_1y_2 + (c' + b_{02})y_2^2 + \cdots \end{pmatrix}. \end{aligned}$$

From this last expression it is clear that suitable choices of the coefficients a_{ij} and b_{ij} exist to eliminate all the quadratic terms in the transformed system except for the y_2^2 term in $\dot{y}_1$. In short, no matter what our choice of the transformation $\mathbf{H}(\mathbf{y})$, if $\mathbf{H}$ is twice differentiable, then system (2.9) can be simplified to

$$\begin{aligned} \dot{y}_1 &= 2y_1 + cy_2^2 \\ \dot{y}_2 &= \ y_2 \end{aligned} \tag{2.12}$$

through order two, but the quadratic term in the first component cannot be removed; (2.12) is the *normal form* for system (2.9) through order two.

We now turn our attention to the general situation in which system (2.1) has an isolated equilibrium at the origin, not necessarily hyperbolic. We will need the following notation and terminology. As in Chapter 1, for $\alpha = (\alpha_1, \ldots, \alpha_n) \in \mathbb{N}_0^n$, $\mathbf{x}^\alpha$ denotes $x_1^{\alpha_1} \cdots x_n^{\alpha_n}$ and $|\alpha| = \alpha_1 + \cdots + \alpha_n$. We let $\mathscr{H}_s$ denote the vector space of functions from $\mathbb{R}^n$ to $\mathbb{R}^n$ each of whose components is a homogeneous polynomial function of degree s; elements of $\mathscr{H}_s$ will be termed vector homogeneous functions. If $\{\mathbf{e}_1, \ldots, \mathbf{e}_n\}$ is the standard basis of $\mathbb{R}^n$, $\mathbf{e}_j = (0, \ldots, 0, \overset{j}{1}, 0, \ldots, 0)^T$, then a basis for $\mathscr{H}_s$ is the collection of vector homogeneous functions

$$\mathbf{v}_{j,\alpha} = \mathbf{x}^\alpha \mathbf{e}_j \tag{2.13}$$

for all j such that $1 \leq j \leq n$ and all α such that $|\alpha| = s$ (this is the product of a monomial and a vector). For example, the first three vectors listed in the basis for $\mathscr{H}_2$ given in Example 2.2.2 below are $\mathbf{v}_{1,(2,0)}$, $\mathbf{v}_{1,(1,1)}$, and $\mathbf{v}_{1,(0,2)}$. Thus $\mathscr{H}_s$ has dimension $N = nC(s+n-1, s)$ (see Exercise 2.6).

Assuming that $\mathbf{f}$ is C^2, we expand $\mathbf{f}$ in a Taylor series so as to write (2.1) as

$$\dot{\mathbf{x}} = A\mathbf{x} + \mathbf{f}^{(2)}(\mathbf{x}) + \mathbf{R}(\mathbf{x}), \tag{2.14}$$

where $\mathbf{f}^{(2)} \in \mathscr{H}_2$ and the remainder satisfies the condition that $|\mathbf{R}(\mathbf{x})|/|\mathbf{x}|^2 \to 0$ as $|\mathbf{x}| \to 0$. We assume that A has been placed in a standard form by a preliminary linear transformation, so that in practical situations the nonlinear terms may be different from what they were in the system as originally encountered.

Applying the reasoning of Example 2.2.1, we make a coordinate transformation of the form

$$\mathbf{x} = \mathbf{H}(\mathbf{y}) = \mathbf{y} + \mathbf{h}^{(2)}(\mathbf{y}), \tag{2.15}$$

where in this case $\mathbf{h}^{(2)} \in \mathscr{H}_2$, that is, each of the n components of $\mathbf{h}^{(2)}(\mathbf{y})$ is a homogeneous quadratic polynomial in $\mathbf{x}$. Since $\mathrm{d}\mathbf{H}(\mathbf{0}) = \mathbf{Id}$ is invertible, the Inverse Function Theorem guarantees that $\mathbf{H}$ has an analytic inverse on a neighborhood of $\mathbf{0}$. Using (2.15) in the right-hand side of (2.14) and inserting that in turn into the analogue of (2.11) (that is, into (2.11) with $\mathbf{h}^{[2]}$ replaced by $\mathbf{h}^{(2)}$) yields

$$\dot{\mathbf{y}} = A\mathbf{y} + A\mathbf{h}^{(2)}(\mathbf{y}) + \mathbf{f}^{(2)}(\mathbf{y}) - \mathbf{dh}^{(2)}(\mathbf{y})A\mathbf{y} + \mathbf{R}_2(\mathbf{y}), \tag{2.16}$$

where the remainder satisfies the condition $|\mathbf{R}_2(\mathbf{y})|/|\mathbf{y}|^2 \to 0$ as $|\mathbf{y}| \to 0$. The quadratic terms can be eliminated from (2.16) if and only if $\mathbf{h}^{(2)}(\mathbf{y})$ can be chosen so that

$$\mathscr{L}\mathbf{h}^{(2)}(\mathbf{y}) = \mathbf{dh}^{(2)}(\mathbf{y})A\mathbf{y} - A\mathbf{h}^{(2)}(\mathbf{y}) = \mathbf{f}^{(2)}(\mathbf{y}), \tag{2.17}$$

where $\mathscr{L}$, the so-called *homological operator*, is the linear operator on $\mathscr{H}_2$ defined by

$$\mathscr{L} : \mathbf{p}(\mathbf{y}) \mapsto \mathbf{dp}(\mathbf{y})A\mathbf{y} - A\mathbf{p}(\mathbf{y}). \tag{2.18}$$

In other words, all quadratic terms can be eliminated from (2.16) if and only if $\mathscr{L}$ maps $\mathscr{H}_2$ *onto* itself. If $\mathscr{L}$ is not onto, then because $\mathscr{H}_2$ is finite-dimensional, it decomposes as a direct sum $\mathscr{H}_2 = \text{Image}(\mathscr{L}) \oplus \mathscr{K}_2$, where Image($\mathscr{L}$) denotes the image of $\mathscr{L}$ in $\mathscr{H}_2$, although the complementary subspace $\mathscr{K}_2$ is not unique. The quadratic terms in (2.16) that can be eliminated by a C^2 change of coordinates are precisely those that lie in Image($\mathscr{L}$). Those that remain have a form dependent on the choice of the complementary subspace $\mathscr{K}_2$.

Example 2.2.2. Let us reconsider the system (2.9) of Example 2.2.1 in this context. The basis (2.13) for $\mathscr{H}_2$ is

$$\left\{ \begin{pmatrix} x_1^2 \\ 0 \end{pmatrix}, \begin{pmatrix} x_1x_2 \\ 0 \end{pmatrix}, \begin{pmatrix} x_2^2 \\ 0 \end{pmatrix}, \begin{pmatrix} 0 \\ x_1^2 \end{pmatrix}, \begin{pmatrix} 0 \\ x_1x_2 \end{pmatrix}, \begin{pmatrix} 0 \\ x_2^2 \end{pmatrix} \right\},$$

which we order by the order in which we have listed the basis elements, and which for ease of exposition we label $\mathbf{u}_1$, $\mathbf{u}_2$, $\mathbf{u}_3$, $\mathbf{u}_4$, $\mathbf{u}_5$, and $\mathbf{u}_6$. A straightforward computation based on the definition (2.18) of the homological operator $\mathscr{L}$ yields

$$\mathscr{L} \begin{pmatrix} y_1^{\alpha_1} y_2^{\alpha_2} \\ 0 \end{pmatrix} = (2\alpha_1 + \alpha_2 - 2) \begin{pmatrix} y_1^{\alpha_1} y_2^{\alpha_2} \\ 0 \end{pmatrix} \tag{2.19a}$$

and

$$\mathscr{L} \begin{pmatrix} 0 \\ y_1^{\alpha_1} y_2^{\alpha_2} \end{pmatrix} = (2\alpha_1 + \alpha_2 - 1) \begin{pmatrix} 0 \\ y_1^{\alpha_1} y_2^{\alpha_2} \end{pmatrix}, \tag{2.19b}$$

so that each basis vector is an eigenvector. The eigenvalues are, in the order of the basis vectors to which they correspond, 2, 1, 0, 3, 2, and 1. Image($\mathscr{L}$) is thus the five-dimensional subspace of $\mathscr{H}_2$ spanned by the basis vectors other than $\mathbf{u}_3$, and a natural complement to Image($\mathscr{L}$) is Span$\{\mathbf{u}_3\}$, corresponding precisely to the quadratic term in (2.12).

Returning to the general situation, beginning with (2.1), written in the form (2.14), we compute the operator $\mathscr{L}$ of (2.18), choose a complement $\mathscr{K}_2$ to Image($\mathscr{L}$) in $\mathscr{H}_2$, and decompose $\mathbf{f}^{(2)}$ as $\mathbf{f}^{(2)} = (\mathbf{f}^{(2)})_0 + \tilde{\mathbf{f}}^{(2)} \in \text{Image}(\mathscr{L}) \oplus \mathscr{K}_2$. Then for any $\mathbf{h}^{(2)} \in \mathscr{H}_2$ satisfying $\mathscr{L}\mathbf{h}^{(2)} = (\mathbf{f}^{(2)})_0$, by (2.16) the change of coordinates (2.15) reduces (2.1) (which is the same as (2.14)) to

$$\dot{\mathbf{y}} = A\mathbf{y} + \tilde{\mathbf{f}}^{(2)}(\mathbf{y}) + \tilde{\mathbf{R}}_2(\mathbf{y}), \tag{2.20}$$

where, to repeat, $\tilde{\mathbf{f}}^{(2)} \in \mathscr{K}_2$, and $|\tilde{\mathbf{R}}_2(\mathbf{y})|/|\mathbf{y}|^2 \to 0$ as $|\mathbf{y}| \to 0$. This is the normal form for (2.1) through order two: the quadratic terms have been simplified as much as possible.

Turning to the cubic terms, and assuming one more degree of differentiability, let us return to $\mathbf{x}$ for the current coordinates and write (2.20) as

$$\dot{\mathbf{x}} = A\mathbf{x} + \tilde{\mathbf{f}}^{(2)}(\mathbf{x}) + \mathbf{f}^{(3)}(\mathbf{x}) + \mathbf{R}(\mathbf{x}), \tag{2.21}$$

where $\mathbf{R}$ denotes a new remainder term, satisfying the condition $|\mathbf{R}(\mathbf{x})|/|\mathbf{x}|^3 \to 0$ as $|\mathbf{x}| \to 0$. A change of coordinates that will leave the constant, linear, and quadratic terms in (2.21) unchanged is one of the form

$$\mathbf{x} = \mathbf{y} + \mathbf{h}^{(3)}(\mathbf{y}), \tag{2.22}$$

where $\mathbf{h}^{(3)} \in \mathscr{H}_3$, that is, each of the n components of $\mathbf{h}^{(3)}(\mathbf{y})$ is a homogeneous cubic polynomial function in $\mathbf{y}$. Using (2.22) in the right-hand side of (2.21) and inserting that in turn into the analogue of (2.11) (that is, into (2.11) with $\mathbf{h}^{[2]}$ replaced by $\mathbf{h}^{(3)}$) yields

$$\dot{\mathbf{y}} = A\mathbf{y} + \tilde{\mathbf{f}}^{(2)}(\mathbf{y}) + A\mathbf{h}^{(2)}(\mathbf{y}) + \mathbf{f}^{(3)}(\mathbf{y}) - d\mathbf{h}^{(3)}(\mathbf{y})A\mathbf{y} + \mathbf{R}_3(\mathbf{y}), \tag{2.23}$$

where the remainder satisfies the condition $|\mathbf{R}_3(\mathbf{y})|/|\mathbf{y}|^2 \to 0$ as $|\mathbf{y}| \to 0$. The cubic terms can be eliminated from (2.23) if and only if $\mathbf{h}^{(3)}(\mathbf{y})$ can be chosen so that

$$\mathscr{L}\mathbf{h}^{(3)}(\mathbf{y}) = d\mathbf{h}^{(3)}(\mathbf{y})A\mathbf{y} - A\mathbf{h}^{(3)}(\mathbf{y}) = \mathbf{f}^{(3)}(\mathbf{y}). \tag{2.24}$$

Comparing (2.17) and (2.24), the reader can see that the condition for the elimination of all cubic terms is exactly the same as that for the elimination of all quadratic terms. The homological operator $\mathscr{L}$ is again defined by (2.18), except that it now operates on the vector space $\mathscr{H}_3$ of functions from $\mathbb{R}^n$ to $\mathbb{R}^n$ all of whose components are homogeneous *cubic* polynomial functions. If $\mathscr{L}$ does not map onto $\mathscr{H}_3$, then exactly as for $\mathscr{H}_2$ when $\mathscr{L}$ does not map onto $\mathscr{H}_2$, $\mathscr{H}_3$ decomposes as a direct sum $\mathscr{H}_3 = \text{Image}(\mathscr{L}) \oplus \mathscr{K}_3$, although again the complementary subspace $\mathscr{K}_3$ is not unique. Once we have chosen a complement $\mathscr{K}_3$, $\mathbf{f}^{(3)}$ decomposes as $\mathbf{f}^{(3)} = (\mathbf{f}^{(3)})_0 + \tilde{\mathbf{f}}^{(3)} \in \text{Image}(\mathscr{L}) \oplus \mathscr{K}_3$. Then for any $\mathbf{h}^{(3)} \in \mathscr{H}_3$ satisfying $\mathscr{L}\mathbf{h}^{(3)} = (\mathbf{f}^{(3)})_0$, by (2.23) the change of coordinates (2.22) reduces (2.1) (which is the same as (2.21)) to

$$\dot{\mathbf{y}} = A\mathbf{y} + \tilde{\mathbf{f}}^{(2)}(\mathbf{y}) + \tilde{\mathbf{f}}^{(3)}(\mathbf{y}) + \tilde{\mathbf{R}}_3(\mathbf{y}), \tag{2.25}$$

where $\tilde{\mathbf{f}}^{(3)} \in \mathscr{K}_3$, and $|\tilde{\mathbf{R}}_3(\mathbf{y})|/|\mathbf{y}|^3 \to 0$ as $|\mathbf{y}| \to 0$. This is the normal form for (2.1) through order three: the quadratic and cubic terms have been simplified as much as possible.

It is apparent that the pattern continues through all orders, as long as $\mathbf{f}$ is sufficiently differentiable. Noting that a composition of transformations of the form $\mathbf{x} = \mathbf{y} + \mathbf{p}(\mathbf{y})$, where each component of $\mathbf{p}(\mathbf{y})$ is a polynomial function, is a transformation of the same form and has an analytic inverse on a neighborhood of $\mathbf{0}$, we have the following theorem.

Theorem 2.2.3. *Let* $\mathbf{f}$ *be defined and* C^r *on a neighborhood of* $\mathbf{0}$ *in* $\mathbb{R}^n$ *and satisfy* $\mathbf{f}(\mathbf{0}) = \mathbf{0}$. *Let* $A = d\mathbf{f}(\mathbf{0})$. *For* $2 \le k \le r$, *let* $\mathscr{H}_k$ *denote the vector space of functions from* $\mathbb{R}^n$ *to* $\mathbb{R}^n$ *all of whose components are homogeneous polynomial functions of degree k, let* $\mathscr{L}$ *denote the linear operator on* $\mathscr{H}_k$ *(the "homological operator") defined by* $\mathscr{L}\mathbf{p}(\mathbf{y}) = d\mathbf{p}(\mathbf{y})A\mathbf{y} - A\mathbf{p}(\mathbf{y})$, *and let* $\mathscr{K}_k$ *be any complement to* Image($\mathscr{L}$) *in* $\mathscr{H}_k$, *so that* $\mathscr{H}_k = \text{Image}(\mathscr{L}) \oplus \mathscr{K}_k$. *Then there is a polynomial change of coordinates* $\mathbf{x} = \mathbf{H}(\mathbf{y}) = \mathbf{y} + \mathbf{p}(\mathbf{y})$ *such that in the new coordinates system* (2.1) *is*

$$\dot{\mathbf{y}} = A\mathbf{y} + \mathbf{f}^{(2)}(\mathbf{y}) + \cdots + \mathbf{f}^{(r)}(\mathbf{y}) + \mathbf{R}(\mathbf{y}), \tag{2.26}$$

where for $2 \le k \le r$, $\mathbf{f}^{(k)} \in \mathscr{H}_k$, *and the remainder* $\mathbf{R}$ *satisfies* $|\mathbf{R}(\mathbf{y})|/|\mathbf{y}|^r \to 0$ *as* $|\mathbf{y}| \to 0$.

Definition 2.2.4. In the context of Theorem 2.2.3, expression (2.26) is a *normal form through order r* for system (2.1).

2.3 Analytic and Formal Normal Forms

In this section we study in detail the homological operator $\mathscr{L}$ and normal forms of the system

$$\dot{\mathbf{x}} = A\mathbf{x} + \mathbf{X}(\mathbf{x}), \tag{2.27}$$

where now $\mathbf{x} \in \mathbb{C}^n$, A is a possibly complex $n \times n$ matrix, and each component $X_k(\mathbf{x})$ of $\mathbf{X}$, $1 \le k \le n$, is a formal or convergent power series, possibly with complex coefficients, that contains no constant or linear terms. Our treatment mainly follows the lines of [19].

To see why this is of importance, even when our primary interest is in real systems, recall that the underlying assumption in the discussion leading up to Theorem 2.2.3, which was explicitly stated in the first sentence following (2.14), was that the linear terms in the right-hand side of (2.1), that is, the $n \times n$ matrix A in (2.14), had already been placed in some standard form by a preliminary linear transformation. To elaborate on this point, typically at the beginning of an investigation of system (2.1), expressed as (2.14), the matrix A has no particularly special form. From linear algebra we know that there exists a nonsingular $n \times n$ matrix S such that the similiarity transformation $SAS^{-1} = J$ produces the Jordan normal form J of A. If we use the matrix S to make the linear coordinate transformation

$$\mathbf{y} = S\mathbf{x} \tag{2.28}$$

of phase space, then in the new coordinates (2.14) becomes

$$\dot{\mathbf{y}} = J\mathbf{y} + S\mathbf{f}^{(2)}(S^{-1}\mathbf{y}) + S\mathbf{R}(S^{-1}\mathbf{y}). \tag{2.29}$$

Although the original system (2.1) or (2.14) is real, the matrices J and S can be complex, hence system (2.29) can be complex as well. Thus even if we are primarily interested in studying real systems, it is nevertheless fruitful to investigate normal forms of complex systems (2.27). Since we are working with systems whose right-hand sides are power series, we will also allow formal rather than convergent series as well.

Whereas previously $\mathscr{H}_s$ denoted the vector space of functions from $\mathbb{R}^n$ to $\mathbb{R}^n$, all of whose components are homogeneous polynomial functions of degree s, we now let $\mathscr{H}_s$ denote the vector space of functions from $\mathbb{C}^n$ to $\mathbb{C}^n$, all of whose components are homogeneous polynomial functions of degree s. The collection $\mathbf{v}_{j,\alpha}$ of (2.13) remains a basis of $\mathscr{H}_s$. For $\alpha = (\alpha_1, \dots, \alpha_n) \in \mathbb{N}_0^n$ and $\kappa = (\kappa_1, \dots, \kappa_n) \in \mathbb{C}^n$ we will let (α, κ) denote the scalar product

$$(\alpha, \kappa) = \sum_{j=1}^{n} \alpha_j \kappa_j .$$

Lemma 2.3.1. *Let A be an $n \times n$ matrix with eigenvalues $\kappa_1, \dots, \kappa_n$, and let $\mathscr{L}$ be the corresponding homological operator on $\mathscr{H}_s$, that is, the linear operator on $\mathscr{H}_s$ defined by*

$$\mathscr{L}\mathbf{p}(\mathbf{y}) = d\mathbf{p}(\mathbf{y})A\mathbf{y} - A\mathbf{p}(\mathbf{y}) . \tag{2.30}$$

Let $\kappa = (\kappa_1, \dots, \kappa_n)$. Then the eigenvalues λ_j, $i = j, \dots, N$, of $\mathscr{L}$ are

$$\lambda_j = (\alpha, \kappa) - \kappa_m ,$$

where m ranges over $\{1, \dots, n\} \subset \mathbb{N}$ and α ranges over $\{\beta \in \mathbb{N}_0^n : |\beta| = s\}$.

Proof. For ease of exposition, just for this paragraph let $\mathscr{T}$ denote the linear transformation of $\mathbb{C}^n$ whose matrix representative with respect to the standard basis $\{\mathbf{e}_1, \dots, \mathbf{e}_n\}$ of $\mathbb{C}^n$ is A. There exists a nonsingular $n \times n$ matrix S such that $J = SAS^{-1}$ is the lower-triangular Jordan form of A (omitted entries are zero),

$$J = \begin{pmatrix} \kappa_1 & & & & & \\ \sigma_2 & \kappa_2 & & & & \\ & \sigma_3 & \kappa_3 & & & \\ & & \sigma_4 & \kappa_4 & & \\ & & & \ddots & \ddots & \\ & & & & \sigma_n & \kappa_n \end{pmatrix} ,$$

where κ_1 through κ_n are the eigenvalues of A (repeated eigenvalues grouped together) and $\sigma_j \in \{0, 1\}$ for $2 \le j \le n$. We will compute the eigenvalues of $\mathscr{L}$ by finding its $N \times N$ matrix representative L with respect to a basis of $\mathscr{H}_s$ corresponding to the new basis of $\mathbb{C}^n$ in which J is the matrix of $\mathscr{T}$. This corresponds to the change of coordinates $\mathbf{x} = S\mathbf{y}$ in $\mathbb{C}^n$.

In Exercise 2.11 the reader is led through a derivation of the fact that with respect to the new coordinates the expression for $\mathscr{L}$ changes to (2.30) with A replaced by J, that is,

$$\mathscr{L}\mathbf{p}(\mathbf{y}) = d\mathbf{p}(\mathbf{y})J\mathbf{y} - J\mathbf{p}(\mathbf{y}) . \tag{2.31}$$

Column (j, α) of L is the coordinate vector of the image under $\mathscr{L}$ of the basis vector $\mathbf{v}_{j,\alpha}$ of (2.13). Computing directly from (2.31), $\mathscr{L}\mathbf{v}_{j,\alpha}(\mathbf{y})$ is (omitted entries are zeros)

$$\mathscr{L}\mathbf{v}_{j,\alpha}(\mathbf{y}) = \begin{pmatrix} \\ \alpha_1 y_1^{\alpha_1-1} y_2^{\alpha_2}\cdots y_n^{\alpha_n} & \cdots & \alpha_n y_1^{\alpha_1} y_2^{\alpha_2}\cdots y_n^{\alpha_n-1} \\ \\ \end{pmatrix} \begin{pmatrix} \kappa_1 y_1 \\ \kappa_2 y_2 + \sigma_2 y_1 \\ \kappa_3 y_3 + \sigma_3 y_2 \\ \vdots \\ \kappa_n y_n + \sigma_n y_{n-1} \end{pmatrix}$$

$$- \begin{pmatrix} \\ \kappa_j y_1^{\alpha_1} y_2^{\alpha_2}\cdots y_n^{\alpha_n} \\ \sigma_{j+1} y_1^{\alpha_1} y_2^{\alpha_2}\cdots y_n^{\alpha_n} \\ \\ \end{pmatrix}$$

$$= \left[((\alpha,\kappa)-\kappa_j))\mathbf{y}^\alpha + \sum_{i=2}^{n} \sigma_i \alpha_i y_1^{\alpha_1}\cdots y_{i-1}^{\alpha_{i-1}+1} y_i^{\alpha_i-1}\cdots y_n^{\alpha_n} \right] \mathbf{e}_j$$
$$+ \sigma_{j+1}\mathbf{y}^\alpha \mathbf{e}_{j+1}$$
$$= ((\alpha,\kappa)-\kappa_j))\mathbf{v}_{j,\alpha}(\mathbf{y}) + \sum_{i=2}^{n} \sigma_i \alpha_i \mathbf{v}_{j,(\alpha_1,\ldots,\alpha_{i-1}+1,\alpha_i-1,\ldots,\alpha_n)}(\mathbf{y})$$
$$+ \sigma_{j+1}\mathbf{v}_{j+1,\alpha}(\mathbf{y}).$$

If the basis of $\mathscr{H}_s$ is ordered so that $\mathbf{v}_{r,\beta}$ precedes $\mathbf{v}_{s,\gamma}$ if and only if the first nonzero entry (reading left to right) in the row vector $(r-s, \beta-\gamma)$ is negative, then the basis vector $\mathbf{v}_{j,\alpha}$ precedes all the remaining vectors in the expression for $\mathscr{L}\mathbf{v}_{j,\alpha}$. This implies that the corresponding $N \times N$ matrix L for $\mathscr{L}$ is lower triangular, and has the numbers $(\alpha,\kappa)-\kappa_m$ ($|\alpha| = s$, $1 \le m \le n$) on the main diagonal. □

The order of the basis of $\mathscr{H}_s$ referred to in the proof of Lemma 2.3.1 is the lexicographic order. See Exercise 2.12.

We say that our original system (2.27) under consideration is *formally equivalent* to a like system

$$\dot{\mathbf{y}} = A\mathbf{y} + \mathbf{Y}(\mathbf{y}) \tag{2.32}$$

if there is a change of variables

$$\mathbf{x} = \mathbf{H}(\mathbf{y}) = \mathbf{y} + \mathbf{h}(\mathbf{y}) \tag{2.33}$$

that transforms (2.27) into (2.32), where the coordinate functions of $\mathbf{Y}$ and $\mathbf{h}$, Y_j and h_j, $j = 1,\ldots,n$, are formal power series. (Of course, in this context it is natural to allow the coordinate functions X_j of $\mathbf{X}$ to be merely formal series as well.) If all Y_j and h_j are convergent power series (and all X_j are as well), then by the Inverse Function Theorem the transformation (2.33) has an analytic inverse on a neighborhood of $\mathbf{0}$ and we say that (2.27) and (2.32) are *analytically equivalent*. (See the paragraph following Corollary 6.1.3 for comments on the convergence of power series of several variables.)

We alert the reader to the fact that, as indicated by the notation introduced in (2.33), $\mathbf{h}$ stands for just the terms of order at least two in the equivalence transformation $\mathbf{H}$ between the two systems.

Lemma 2.3.1 yields the following theorem.

Theorem 2.3.2. *Let $\kappa_1,\ldots,\kappa_n$ be the eigenvalues of the $n \times n$ matrix A in* (2.27) *and* (2.32), *set $\kappa = (\kappa_1,\ldots,\kappa_n)$, and suppose that*

$$(\alpha,\kappa) - \kappa_m \neq 0 \tag{2.34}$$

for all $m \in \{1,\ldots,n\}$ and for all $\alpha \in \mathbb{N}_0^n$ for which $|\alpha| \geq 2$. Then systems (2.27) *and* (2.32) *are formally equivalent for all $\mathbf{X}$ and $\mathbf{Y}$, and the equivalence transformation* (2.33) *is uniquely determined by $\mathbf{X}$ and $\mathbf{Y}$.*

Proof. Differentiating (2.33) with respect to t and applying (2.27) and (2.32) yields the condition

$$d\mathbf{h}(\mathbf{y})A\mathbf{y} - A\mathbf{h}(\mathbf{y}) = \mathbf{X}(\mathbf{y}+\mathbf{h}(\mathbf{y})) - d\mathbf{h}(\mathbf{y})\mathbf{Y}(\mathbf{y}) - \mathbf{Y}(\mathbf{y}) \tag{2.35}$$

that $\mathbf{h}$ must satisfy. We determine $\mathbf{h}$ by a recursive process like that leading up to Theorem 2.2.3.

Decomposing $\mathbf{X}$, $\mathbf{Y}$, and $\mathbf{h}$ as the sum of their homogeneous parts,

$$\mathbf{X} = \sum_{s=2}^{\infty} \mathbf{X}^{(s)}, \qquad \mathbf{Y} = \sum_{s=2}^{\infty} \mathbf{Y}^{(s)}, \qquad \mathbf{h} = \sum_{s=2}^{\infty} \mathbf{h}^{(s)}, \tag{2.36}$$

where $\mathbf{X}^{(s)}, \mathbf{Y}^{(s)}, \mathbf{h}^{(s)} \in \mathcal{H}_s$, (2.35) decomposes into the infinite sequence of equations

$$\mathcal{L}(\mathbf{h}^{(s)}) = \mathbf{g}^{(s)}(\mathbf{h}^{(2)},\ldots,\mathbf{h}^{(s-1)},\mathbf{Y}^{(2)},\ldots,\mathbf{Y}^{(s-1)},\mathbf{X}^{(2)},\ldots,\mathbf{X}^{(s)}) - \mathbf{Y}^{(s)}, \tag{2.37}$$

for $s = 2,3,\ldots$, where $\mathbf{g}^{(s)}$ denotes the function that is obtained after the substitution into $\mathbf{X}(\mathbf{y}+\mathbf{h}(\mathbf{y})) - d\mathbf{h}(\mathbf{y})\mathbf{Y}(\mathbf{y})$ of the expression $\mathbf{y} + \sum_{i=1}^{s} \mathbf{h}^{(i)}$ in the place of $\mathbf{y}+\mathbf{h}(\mathbf{y})$ and the expression $\sum_{i=1}^{s} \mathbf{Y}^{(i)}(\mathbf{y})$ in the place of $\mathbf{Y}(\mathbf{y})$, and maintaining only terms that are of order s. For $s = 2$, the right-hand side of (2.37) is to be understood to stand for $\mathbf{X}^{(2)}(\mathbf{y}) - \mathbf{Y}^{(2)}(\mathbf{y})$, which is known. For $s > 2$, the right-hand side of (2.37) is known if $\mathbf{h}^{(2)},\ldots,\mathbf{h}^{(s-1)}$ have already been computed. By the hypothesis (2.34) and Lemma 2.3.1 the operator $\mathcal{L}$ is invertible. Thus for any $s \geq 2$ there is a unique solution $\mathbf{h}^{(s)}$ to (2.37). Therefore a unique solution $\mathbf{h}(\mathbf{y})$ of (2.35) is determined recursively. □

Choosing $\mathbf{Y} = \mathbf{0}$ in (2.32) yields the following corollary and motivates the definition that follows it.

Corollary 2.3.3. *If condition* (2.34) *holds, then system* (2.27) *is formally equivalent to its linear approximation $\dot{\mathbf{y}} = A\mathbf{y}$. The (possibly formal) coordinate transformation that transforms* (2.27) *into $\dot{\mathbf{y}} = A\mathbf{y}$ is unique.*

Definition 2.3.4. System $\dot{\mathbf{x}} = A\mathbf{x} + \mathbf{X}(\mathbf{x})$ is *linearizable* if there is an analytic normalizing transformation $\mathbf{x} = \mathbf{y} + \mathbf{h}(\mathbf{y})$ that places it in the normal form $\dot{\mathbf{y}} = A\mathbf{y}$.

Both the linear system $\dot{\mathbf{y}} = A\mathbf{y}$ and the linearizing transformation that produces it are referred to as a "linearization" of the system $\dot{\mathbf{x}} = A\mathbf{x} + \mathbf{X}(\mathbf{x})$. We will see later (Corollary 4.2.3) that, at least when A is diagonal and $n = 2$ (the only case of practical interest for us), the existence of a merely formal linearization implies the existence of a convergent linearization.

As we saw in Example 2.2.2, when (2.34) does not hold some equations in (2.37) might not have a solution. This means that in such a case we might not be able to transform system (2.27) into a linear system by even a formal transformation (2.33). The best we are sure to be able to do is to transform (2.27) into a system in which all terms that correspond to pairs (m,α) for which (2.34) holds have been eliminated. However, terms corresponding to those pairs (m,α) for which (2.34) fails might be impossible to eliminate. These troublesome terms have a special name.

Definition 2.3.5. Let $\kappa_1,\ldots,\kappa_n$ be the eigenvalues of the matrix A in (2.27), ordered according to the choice of a Jordan normal form J of A, and let $\kappa = (\kappa_1,\ldots,\kappa_n)$. Suppose $m \in \{1,\ldots,n\}$ and $\alpha \in \mathbb{N}_0^n$, $|\alpha| = \alpha_1 + \cdots + \alpha_n \geq 2$, are such that

$$(\alpha,\kappa) - \kappa_m = 0\,.$$

Then m and α are called a *resonant pair*, the corresponding coefficient $X_m^{(\alpha)}$ of the monomial $\mathbf{x}^\alpha$ in the mth component of $\mathbf{X}$ is called a *resonant coefficient*, and the corresponding term is called a *resonant term* of $\mathbf{X}$. Index and multi-index pairs, terms, and coefficients that are not resonant are called *nonresonant.*

A "normal form" for system (2.27) should be a form that is as simple as possible. The first step in the simplification process is to apply (2.28) to change the linear part A in (2.27) into its Jordan normal form. We will assume that this preliminary step has already been taken, so we begin with (2.27) in the form

$$\dot{\mathbf{x}} = J\mathbf{x} + \mathbf{X}(\mathbf{x}), \tag{2.38}$$

where J is a lower-triangular Jordan matrix. (Note that the following definition is based on this supposition.) The simplest form that we are sure to be able to obtain is one in which all nonresonant terms are zero, so we will take this as the meaning of the term "normal form" from now on.

Definition 2.3.6. A *normal form* for system (2.27) is a system (2.38) in which every nonresonant coefficient is equal to zero. A *normalizing transformation* for system (2.27) is any (possibly formal) change of variables (2.33) that transforms (2.27) into a normal form; it is called *distinguished* if for each resonant pair m and α, the corresponding coefficient $h_m^{(\alpha)}$ is zero, in which case the resulting normal form is likewise termed *distinguished.*

Two remarks about this definition are in order. The first is that it is more restrictive than the definition of normal form through order k for a smooth function, Definition 2.2.4, since it requires that every nonresonant term be eliminated. In Example 2.2.2, for instance, $\text{Span}\{c_1\mathbf{u}_1 + c_2\mathbf{u}_2 + \mathbf{u}_3 + c_4\mathbf{u}_4 + c_5\mathbf{u}_5 + c_6\mathbf{u}_6\}$ for any fixed choice of the constants c_j is an acceptable complement to Image($\mathscr{L}$), hence

$$\begin{aligned} \dot{x} &= 2x + c(c_1x^2 + c_2xy + y^2) \\ \dot{y} &= y + c(c_4x^2 + c_5xy + c_6y^2) \end{aligned} \tag{2.39}$$

is a normal form through order two according to Definition 2.2.4. But equations (2.19) show that the single resonant term is the y^2 term in the first component, so that (2.39) does not give a normal form according to Definition 2.3.6 unless the c_j are all chosen to be zero.

The second remark concerning Definition 2.3.6 is almost the reverse of the first: although a normal form is the simplest form that we are sure to be able to obtain in general, for a particular system it might not be the absolute simplest. In other words, the fact that a coefficient is resonant does not mean that it must be (or remain) nonzero under every normalization: a normalizing transformation that eliminates all the nonresonant terms could very well eliminate some resonant terms as well. For example, the condition in Corollary 2.3.3 is sufficent for linearizability, but it is by no means necessary. In Chapter 4 we will treat the question of the possibility of removing all resonant as well as nonresonant terms under normalization.

Remark 2.3.7. Suppose the system $\dot{\mathbf{x}} = A\mathbf{x} + \mathbf{X}(\mathbf{x})$ is transformed into the normal form $\dot{\mathbf{y}} = A\mathbf{y} + \mathbf{Y}(\mathbf{y})$ by the normalizing transformation $\mathbf{x} = \mathbf{y} + \mathbf{h}(\mathbf{y})$, and let λ be any nonzero number. It is an immediate consequence of Definitions 2.3.5 and 2.3.6 that $\dot{\mathbf{y}} = \lambda A\mathbf{y} + \lambda\mathbf{Y}(\mathbf{y})$ is a normal form for $\dot{\mathbf{x}} = \lambda A\mathbf{x} + \lambda\mathbf{X}(\mathbf{y})$, and inspection of (2.35) shows that the same transformation $\mathbf{x} = \mathbf{y} + \mathbf{h}(\mathbf{y})$ is a normalizing transformation between the scaled systems.

Henceforth we will use the following notation. For any multi-index α, the coefficient of the monomial $\mathbf{x}^\alpha$ in the mth component X_m of $\mathbf{X}$ will be denoted $X_m^{(\alpha)}$. Thus for example in system (2.9), $X_1^{((2,0))} = a$ and $X_2^{((1,1))} = b'$, although when α is given explicitly, by slight abuse of notation we will write simply $X_m^{(\alpha_1,\dots,\alpha_n)}$ instead of $X_m^{((\alpha_1,\dots,\alpha_n))}$. Hence, for example, we write $X_1^{(2,0)} = a$ instead of $X_1^{((2,0))} = a$. We will use the same notational convention for $\mathbf{Y}$ and $\mathbf{h}$.

Every system is at least formally equivalent to a normal form, and as the proof of the following theorem shows, there is some freedom in choosing it, although that freedom disappears if we restrict ourselves to distinguished normalizing transformations. Theorem 2.3.11 has more to say about this.

Theorem 2.3.8. *Any system* (2.38) *is formally equivalent to a normal form (which need not be unique). The normalizing transformation can be chosen to be distinguished.*

Proof. Since the linear part is already in simplest form, we look for a change of coordinates of the form (2.33) that transforms system (2.38) into $\dot{\mathbf{y}} = J\mathbf{y} + \mathbf{Y}(\mathbf{y})$ in which all nonresonant coefficients are zero. Writing $\mathbf{h}(\mathbf{y}) = \sum_{s=2}^{\infty}\mathbf{h}^{(s)}(\mathbf{y})$, for each s the function $\mathbf{h}^{(s)}$ must satisfy equation (2.37), arising from (2.35) by the process described immediately below (2.37). We have shown in the proof of Lemma 2.3.1 that the matrix of the operator $\mathscr{L}$ on the left-hand side of (2.37) is lower triangular with the eigenvalues $(\alpha,\kappa)-\kappa_m$ on the main diagonal. Therefore any coefficient $h_m^{(\alpha)}$ of $\mathbf{h}^{(s)}$ is determined by the equation

$$[(\alpha,\kappa)-\kappa_m]h_m^{(\alpha)} = g_m^{(\alpha)} - Y_m^{(\alpha)}, \tag{2.40}$$

where $g_m^{(\alpha)}$ is a known expression depending on the coefficients of $\mathbf{h}^{(i)}$ satisfying $j<s$. Suppose that for $i=2,3,\ldots,s-1$, the homogeneous terms $h^{(j)}$ and $Y^{(j)}$ have been determined . Then for any $m\in\{1,\ldots,n\}$ and any multi-index α with $|\alpha|=s$, if the pair m and α is nonresonant, that is, if $(\alpha,\kappa)-\kappa_m\neq 0$, then we choose $Y_m^{\alpha}=0$ so that $\mathbf{Y}$ will be a normal form, and choose $h_m^{(\alpha)}$ as uniquely determined by equation (2.40). If $(\alpha,\kappa)-\kappa_m=0$, then we may choose $h_m^{(\alpha)}$ arbitrarily (and in particular, the choice $h_m^{(\alpha)}=0$ every time yields a distinguished transformation), but the resonant coefficient $Y_m^{(\alpha)}$ must be chosen to be $g_m^{(\alpha)}$, $Y_m^{(\alpha)}=g_m^{(\alpha)}$. The process can be started because for $s=2$ the right-hand side of (2.40) is $X_m^{(\alpha)}-Y_m^{(\alpha)}$. Thus formal series for a normal form and a normalizing transformation, distinguished or not, as we decide, are obtained. □

For simplicity, from now on we will assume that the matrix J is diagonal, that is, that $\sigma_k=0$ for $k=2,\ldots,n$. (All applications of normal form theory in this book will be confined to systems that meet this condition.) Then the mth component on the right-hand side of (2.35) is obtaining by expanding $X_m(\mathbf{y}+\mathbf{h}(\mathbf{y}))$ in powers of $\mathbf{y}$; we let $\{X_m(\mathbf{y}+\mathbf{h}(\mathbf{y}))\}^{(\alpha)}$ denote the coefficient of $\mathbf{y}^{(\alpha)}$ in this expansion. Using this fact and Exercise 2.14, the coefficient $g_m^{(\alpha)}$ in (2.40) is given by the expression

$$g_m^{(\alpha)} = \{X_m(\mathbf{y}+\mathbf{h}(\mathbf{y}))\}^{(\alpha)} - \sum_{j=1}^{n} \sum_{\substack{2\le|\beta|\le|\alpha|-1\\ \alpha-\beta+e_j\in\mathbb{N}_0^n}} \beta_j h_m^{(\beta)} Y_j^{(\alpha-\beta+e_j)}, \tag{2.41}$$

where again $\{X_m(\mathbf{y}+\mathbf{h}(\mathbf{y}))\}^{(\alpha)}$ denotes the coefficient of $\mathbf{y}^{\alpha}$ obtained after expanding $X_m(\mathbf{y}+\mathbf{h}(\mathbf{y}))$ in powers of $\mathbf{y}$, and $e_j=(0,\ldots,0,\overset{j}{1},0,\ldots,0)\in\mathbb{N}_0^n$. Note that for $|\alpha|=2$ the sum over β is empty, so that $g_m^{(\alpha)}=\{X_m(\mathbf{y}+\mathbf{h}(\mathbf{y}))\}^{(\alpha)}$, which reduces to $g_m^{(\alpha)}=X_m^{(\alpha)}$, since $\mathbf{X}$ and $\mathbf{h}$ begin with quadratic terms. For $|\alpha|>2$, $|\beta|<|\alpha|$ and $|\alpha-\beta+e_j|<|\alpha|$ ensure that $g_m^{(\alpha)}$ is uniquely determined by (2.41).

The proof of Theorem 2.3.8 and formula (2.41) yield the normalization procedure that is displayed in Table 2.1 on page 75 for system (2.38) in the case that the matrix J is diagonal.

Example 2.3.9. Fix any C^∞ system (2.1) with an equilibrium at $\mathbf{0}$ whose linear part is the same as that of system (2.8):

$$\begin{aligned} \dot{x}_1 &= 2x_1 + ax_1^2 + bx_1x_2 + cx_2^2 + \cdots \\ \dot{x}_2 &= x_2 + a'x_1^2 + b'x_1x_2 + c'x_2^2 + \cdots . \end{aligned} \tag{2.42}$$

The resonant coefficients are determined by the equations

Normal Form Algorithm

Input:

system $\dot{\mathbf{x}} = J\mathbf{x} + \mathbf{X}(\mathbf{x})$,
J diagonal with eigenvalues $\kappa_1, \dots, \kappa_n$
$\kappa = (\kappa_1, \dots, \kappa_n)$
an integer $k > 1$

Output:

a normal form $\dot{\mathbf{y}} = J\mathbf{y} + \mathbf{Y}(\mathbf{y}) + o(|\mathbf{y}|^k)$ up to order k
a distinguished transformation $\mathbf{x} = \mathbf{H}(\mathbf{y}) = \mathbf{y} + \mathbf{h}(\mathbf{y}) + o(|\mathbf{y}|^k)$ up to order k

Procedure:

$\mathbf{h}(\mathbf{y}) := \mathbf{0}$; $\mathbf{Y}(\mathbf{y}) := \mathbf{0}$
FOR $s = 2$ TO $s = k$ DO
  FOR $m = 1$ TO $m = n$ DO
    compute $\mathbf{X}_m(\mathbf{y} + \mathbf{h}(\mathbf{y}))$ through order s
  FOR $|\alpha| = s$ DO
    FOR $m = 1$ TO $m = n$ DO
      compute $g_m^{(\alpha)}$ using (2.41)
      IF
      $(\alpha, \kappa) - \kappa_m \neq 0$
      THEN
      $h_m^{(\alpha)} := \dfrac{g_m^{(\alpha)}}{(\alpha, \kappa) - \kappa_m}$
      $h_m(\mathbf{y}) := h_m(\mathbf{y}) + h_m^{(\alpha)} \mathbf{y}^\alpha$
      ELSE
      $Y_m^{(\alpha)} := g_m^{(\alpha)}$
      $Y_m(\mathbf{y}) := Y_m(\mathbf{y}) + Y_m^{(\alpha)} \mathbf{y}^\alpha$
$\dot{\mathbf{y}} := J\mathbf{y} + \mathbf{Y}(\mathbf{y}) + o(|\mathbf{y}|^k)$
$\mathbf{H} := \mathbf{y} + \mathbf{h}(\mathbf{y}) + o(|\mathbf{y}|^k)$

Table 2.1 Normal Form Algorithm

$$
\begin{aligned}
(\alpha, \kappa) - 2 &= 2\alpha_1 + \alpha_2 - 2 = 0 \\
(\alpha, \kappa) - 1 &= 2\alpha_1 + \alpha_2 - 1 = 0 .
\end{aligned}
$$

When $|\alpha| = 2$, the first equation has the unique solution $(\alpha_1, \alpha_2) = (0, 2)$ and the second equation has no solution; for $|\alpha| \geq 3$, neither equation has a solution. Thus by Definition 2.3.6, for any $k \in \mathbb{N}_0$, the normal form through order k is

$$
\begin{aligned}
\dot{y}_1 &= 2y_1 + Y_1^{(0,2)} y_2^2 + o(|\mathbf{y}|^k) \\
\dot{y}_2 &= \ \ y_2 + o(|\mathbf{y}|^k) .
\end{aligned}
$$

As remarked immediately after equation (2.41), for $|\alpha| = 2$, $g_m^{(\alpha)} = X_m^{(\alpha)}(\mathbf{y})$, so we know that in fact $Y_1^{(0,2)} = c$. If we ignore the question of convergence, then (2.42) is formally equivalent to

$$\begin{aligned}\dot{y}_1 &= 2y_1 + cy_2^2 \\ \dot{y}_2 &= \ \ y_2 .\end{aligned}$$

In Exercise 2.15 the reader is asked to use the Normal Form Algorithm to find the normalizing transformation $\mathbf{H}$ through order two.

Example 2.3.10. Let us change the sign of the coefficient of x_2 in the second equation of system (2.42) and consider the resulting system:

$$\begin{aligned}\dot{x}_1 &= 2x_1 + \ ax_1^2 + \ bx_1x_2 + \ cx_2^2 + \cdots \\ \dot{x}_2 &= -x_2 + a'x_1^2 + b'x_1x_2 + c'x_2^2 + \cdots .\end{aligned} \tag{2.43}$$

The normal form of (2.43) is drastically different from the normal form of (2.42). For now the resonant coefficients are determined by the equations

$$\begin{aligned}(\alpha,\kappa) - 2 &= 2\alpha_1 - \alpha_2 - 2 = 0 \\ (\alpha,\kappa) + 1 &= 2\alpha_1 - \alpha_2 + 1 = 0 .\end{aligned}$$

Solutions of the first equation that correspond to $|\alpha| \geq 2$ are the pairs $(k, 2k-2)$, $k \in \mathbb{N}_0$, $k \geq 2$; solutions of the second equation that correspond to $|\alpha| \geq 2$ are the pairs $(k, 2k+1)$, $k \in \mathbb{N}_0$. By Definition 2.3.6, the normal form of (2.43) is

$$\begin{aligned}\dot{y}_1 &= 2y_1 + y_1 \sum_{k=1}^{\infty} Y_1^{(k+1,2k)} (y_1 y_2^2)^k , \\ \dot{y}_2 &= -y_2 + y_2 \sum_{k=1}^{\infty} Y_2^{(k,2k+1)} (y_1 y_2^2)^k .\end{aligned} \tag{2.44}$$

In Exercise 2.16 the reader is asked to use the Normal Form Algorithm to find the resonant coefficients $Y_1^{(2,2)}$ and $Y_2^{(1,3)}$ and the normalizing transformation $\mathbf{H}$ through order three.

Theorem 2.3.11. *Let system* (2.38)*, with J diagonal, be given. There is a unique normal form that can be obtained from system* (2.38) *by means of a distinguished normalizing transformation, which we call the distinguished normal form for* (2.38)*, and the distinguished normalizing transformation that produces it is unique. The resonant coefficients of the distinguished normal form are given by the formula*

$$Y_m^{(\alpha)} = \{X_m(\mathbf{y} + \mathbf{h}(\mathbf{y}))\}^{(\alpha)} , \tag{2.45}$$

where $\{X_m(\mathbf{y} + \mathbf{h}(\mathbf{y}))\}^{(\alpha)}$ denotes the coefficient of $\mathbf{y}^\alpha$ obtained after expanding $X_m(\mathbf{y} + \mathbf{h}(\mathbf{y}))$ in powers of $\mathbf{y}$.

Proof. In the inductive process in the proof of Theorem 2.3.8, at each step the choice of $h_m^{(\alpha)}$ is already uniquely determined if m and α are a nonresonant pair, while if they are resonant, then $h_m^{(\alpha)}$ must be chosen to be zero so that the transformation will be distinguished. Consider now the choice of $Y_m^{(\alpha)}$ at any step. If m and α are a nonresonant pair then of course $Y_m^{(\alpha)}$ must be chosen to be zero. If m and α are a resonant pair, so that $(\alpha,\kappa)-\kappa_m=0$, then by (2.40) and (2.41)

$$Y_m^{(\alpha)} = g_m^{(\alpha)} = \{X_m(\mathbf{y}+\mathbf{h}(\mathbf{y}))\}^{(\alpha)} - \sum_{j=1}^{n} \sum_{\substack{2\le|\beta|\le|\alpha|-1\\ \alpha-\beta+e_j\in\mathbb{N}_0^n}} \beta_j h_m^{(\beta)} Y_j^{(\alpha-\beta+e_j)}, \tag{2.46}$$

so again $Y_m^{(\alpha)}$ is uniquely determined at this stage.

At the start of the process, when $|\alpha|=2$, as noted immediately below (2.41) the sum in (2.46) is empty; if m and α are a resonant pair, then $Y_m^{(\alpha)}$ is determined as $Y_m^{(\alpha)} = \{X_m(\mathbf{y}+\mathbf{h}(\mathbf{y}))\}^{(\alpha)} = X_m^{(\alpha)}$, since $\mathbf{X}$ and $\mathbf{h}$ begin with quadratic terms. To obtain formula (2.45), consider the sum in (2.46). Any particular coefficient $Y_j^{(\alpha-\beta+e_j)}$ can be nonzero only if j and $\alpha-\beta+e_j$ are a resonant pair, that is, only if $(\alpha-\beta+e_j,\kappa)-\kappa_j=0$, which by linearity and the equality $(e_j,\kappa)=\kappa_j$ holds if and only if $(\alpha-\beta,\kappa)=0$. That is,

$$Y_j^{(\alpha-\beta+e_j)} \neq 0 \quad \text{implies} \quad (\alpha-\beta,\kappa)=0. \tag{2.47}$$

Thus if in the sum j and β are such that $Y_j^{(\alpha-\beta+e_j)} \neq 0$, then

$$(\beta,\kappa)-\kappa_m = (\beta-\alpha,\kappa)+(\alpha,\kappa)-\kappa_m = 0$$

by (2.47) and the assumption that m and α are a resonant pair. Since $\mathbf{h}$ is distinguished this means that $h_m^{(\beta)}=0$. Thus every term in the sum in (2.46) is zero, and the theorem is established. □

We close this section with a theorem that gives a criterion for convergence of normalizing transformations that applies to all the situations of interest in this book. Recall that a series $v(\mathbf{z})=\sum_\alpha v^{(\alpha)}\mathbf{z}^\alpha$ is said to *majorize* a series $u(\mathbf{z})=\sum_\alpha u^{(\alpha)}\mathbf{z}^\alpha$, and $v(\mathbf{z})$ is called a *majorant* of $u(\mathbf{z})$, denoted $u(\mathbf{z}) \prec v(\mathbf{z})$, if $|u^{(\alpha)}|\le v^{(\alpha)}$ for all $\alpha\in\mathbb{N}_0^n$. If a convergent series $v(\mathbf{z})$ majorizes a series $u(\mathbf{z})$, then $u(\mathbf{z})$ converges on some neighborhood of $\mathbf{0}$. By way of notation, for any series $f(\mathbf{z})=\sum_\alpha f^{(\alpha)}\mathbf{z}^\alpha$ we denote by $f^\natural(\mathbf{z})$ its trivial majorant, the series that is obtained by replacing each coefficient of f by its modulus: $f^\natural(\mathbf{z}) := \sum_\alpha |f^{(\alpha)}|\mathbf{z}^\alpha$. Note that in the following lemma $f(\mathbf{x})$ begins with terms of order at least two.

Lemma 2.3.12. *Suppose the series $f(\mathbf{x})=\sum_{\alpha:|\alpha|\ge2} f^{(\alpha)}\mathbf{x}^\alpha$ converges on a neighborhood of $\mathbf{0}\in\mathbb{C}^n$. Then there exist positive real numbers a and b such that*

$$f^\natural(\mathbf{x}) \prec \frac{a\left(\sum_{j=1}^n x_j\right)^2}{1-b\sum_{j=1}^n x_j}. \tag{2.48}$$

Proof. There exist positive real numbers a_0 and b_0 such that $f(\mathbf{x})$ converges on a neighborhood of $M := \{\mathbf{x} : |x_j| \leq b_0 \text{ for } 1 \leq j \leq n\}$, and $|f(\mathbf{x})| \leq a_0$ for $\mathbf{x} \in M$. Then the Cauchy Inequalities state that

$$|f^{(\alpha)}| \leq \frac{a_0}{b_0^{\alpha_1} \cdots b_0^{\alpha_n}} \quad \text{for all} \quad \alpha \in \mathbb{N}_0^n. \tag{2.49}$$

From the identity $\sum_{|\alpha|\geq 0} y_1^{\alpha_1} \cdots y_n^{\alpha_n} = \prod_{j=1}^n \left[\sum_{s=0}^\infty y_j^s\right]$ it follows that

$$\begin{aligned}\sum_{|\alpha|\geq 0} \frac{a_0}{b_0^{\alpha_1}\cdots b_0^{\alpha_n}} x_1^{\alpha_1}\cdots x_n^{\alpha_n} &= a_0 \sum_{|\alpha|\geq 0} \left(\frac{x_1}{b_0}\right)^{\alpha_1} \cdots \left(\frac{x_n}{b_0}\right)^{\alpha_n} = a_0 \prod_{j=1}^n \left[\sum_{s=0}^\infty \left(\frac{x_j}{b_0}\right)^s\right] \\ &= a_0 \prod_{j=1}^n \left(1 - \frac{x_j}{b_0}\right)^{-1}\end{aligned} \tag{2.50}$$

holds on $\mathrm{Int}(M)$. By the definition of majorization, (2.49) and (2.50) yield

$$f^\natural(\mathbf{x}) \prec a_0 \prod_{j=1}^n \left(1-\frac{x_j}{b_0}\right)^{-1}. \tag{2.51}$$

It follows readily from the fact that for all $n, k \in \mathbb{N}$, $(n+k)/(1+k) \leq n$, and the series expansion of $(1+x)^{-n}$ about $0 \in \mathbb{C}$ that for any $n \in \mathbb{N}$,

$$(1+x)^{-n} \prec \frac{1}{1-nx}. \tag{2.52}$$

It is also readily verified that

$$\prod_{j=1}^n \left(1-\frac{x_j}{b_0}\right)^{-1} \prec \left(1 - \frac{1}{b_0}\sum_{j=1}^n x_j\right)^{-n}. \tag{2.53}$$

Thus applying (2.52) with $\mathbf{x}$ replaced by $-\frac{1}{b_0}\sum_{j=1}^n x_j$, and using the fact that for $c \in \mathbb{R}^+$, $1/(1+cu) \prec 1/(1-cu)$, (2.53) yields

$$\prod_{j=1}^n \left(1-\frac{x_j}{b_0}\right)^{-1} \prec \frac{1}{1-\frac{n}{b_0}\sum_{j=1}^n x_j}. \tag{2.54}$$

Combining (2.51) and (2.54) yields

$$f^\natural(\mathbf{x}) \prec \frac{a_0}{1-\frac{n}{b_0}\sum_{j=1}^n x_j}.$$

But since f has no constant or linear terms, the constant and linear terms in the right-hand side may be removed, yielding finally

$$f^{\natural}(\mathbf{x}) \prec a_0 \left[\frac{1}{1-\frac{n}{b_0}\sum_{j=1}^{n} x_j} - 1 - \frac{n}{b_0}\sum_{j=1}^{n} x_j \right],$$

so that (2.48) holds with $a = (n/b_0)^2 a_0$ and $b = n/b_0$. □

Theorem 2.3.13. *Let $\kappa_1, \dots, \kappa_n$ be the eigenvalues of the matrix J in* (2.38) *and set $\kappa = (\kappa_1, \dots, \kappa_n)$. Suppose $\mathbf{X}$ is analytic, that is, that each component X_m is given by a convergent power series, and that for each resonant coefficient $Y_j^{(\alpha)}$ in the distinguished normal form $\mathbf{Y}$ of $\mathbf{X}$, $\alpha \in \mathbb{N}^n$ (that is, every entry in the multi-index α is positive). Suppose further that there exist positive constants d and ε such that the following conditions hold:*
(a) for all $\alpha \in \mathbb{N}_0^n$ and all $m \in \{1, \dots, n\}$ such that $(\alpha, \kappa) - \kappa_m \neq 0$,

$$|(\alpha, \kappa) - \kappa_m| \geq \varepsilon; \tag{2.55}$$

(b) for all α and β in $\mathbb{N}_0^n$ for which $2 \leq |\beta| \leq |\alpha| - 1$, $\alpha - \beta + e_m \in \mathbb{N}_0^n$ for all $m \in \{1, \dots, n\}$, and

$$(\alpha - \beta, \kappa) = 0, \tag{2.56}$$

the following inequality holds:

$$\left| \sum_{j=1}^{n} \beta_j Y_j^{(\alpha-\beta+e_j)} \right| \leq d |(\beta, \kappa)| \sum_{j=1}^{n} \left| Y_j^{(\alpha-\beta+e_j)} \right|. \tag{2.57}$$

Then the distinguished normalizing transformation $\mathbf{x} = \mathbf{H}(\mathbf{y})$ is analytic as well, that is, each component $h_m(\mathbf{y})$ of $\mathbf{h}$ is given by a convergent power series, so that system (2.38) *is analytically equivalent to its normal form.*

Proof. Suppose that a particular pair m and α correspond to a nonresonant term. Then $Y_m^{(\alpha)} = 0$ and by (2.40) and (2.41)

$$\begin{aligned} |h_m^{(\alpha)}| \leq \tfrac{1}{\varepsilon} & \left| \{X_m(\mathbf{y} + \mathbf{h}(\mathbf{y}))\}^{(\alpha)} \right| \\ & + \frac{1}{|(\alpha, \kappa) - \kappa_m|} \left| \sum_{j=1}^{n} \sum_{\substack{2 \leq |\beta| \leq |\alpha|-1 \\ \alpha-\beta+e_j \in \mathbb{N}_0^n}} \beta_j h_m^{(\beta)} Y_j^{(\alpha-\beta+e_j)} \right|. \end{aligned} \tag{2.58}$$

For any nonzero term $Y_j^{(\alpha-\beta+e_j)}$ in (2.58), by (2.47)

$$(\alpha, \kappa) - \kappa_m = (\alpha - \beta, \kappa) + (\beta, \kappa) - \kappa_m = (\beta, \kappa) - \kappa_m,$$

so by hypothesis (a) $|(\beta, \kappa) - \kappa_m| = |(\alpha, \kappa) - \kappa_m| \geq \varepsilon$. We adopt the convention that $Y_j^{(\gamma)} = 0$ if $\gamma \notin \mathbb{N}_0^n$, so that we can reverse the order of summation in (2.58) and apply hypothesis (b). Thus the second term in (2.58) is bounded above by

$$\frac{1}{|(\alpha,\kappa)-\kappa_m|}\sum_{2\le|\beta|\le|\alpha|-1}|h_m^{(\beta)}|\left|\sum_{j=1}^{n}\beta_i Y_j^{(\alpha-\beta+e_j)}\right|$$
$$\le\frac{1}{|(\alpha,\kappa)-\kappa_m|}\sum_{2\le|\beta|\le|\alpha|-1}|h_m^{(\beta)}|d|(\alpha,\kappa)-\kappa_m+\kappa_m|\sum_{j=1}^{n}|Y_j^{(\alpha-\beta+e_j)}|$$
$$\le d\left(1+\frac{|\kappa_m|}{\varepsilon}\right)\sum_{j=1}^{n}\sum_{\substack{2\le|\beta|\le|\alpha|-1\\ \alpha-\beta+e_j\in\mathbb{N}_0^n}}|h_m^{(\beta)}||Y_j^{(\alpha-\beta+e_j)}|.$$

Applying this to (2.58) gives

$$|h_m^{(\alpha)}|\le\left|\frac{1}{\varepsilon}\{X_m(\mathbf{y}+\mathbf{h}(\mathbf{y}))\}^{(\alpha)}\right|+d_0\sum_{j=1}^{n}\sum_{\substack{2\le|\beta|\le|\alpha|-1\\ \alpha-\beta+e_j\in\mathbb{N}_0^n}}|h_m^{(\beta)}||Y_j^{(\alpha-\beta+e_j)}|, \tag{2.59}$$

where $d_0=\max_{1\le m\le n}d(1+|\kappa_m|/\varepsilon)$. Thus

$$h_m^{\natural}(\mathbf{y})\prec\tfrac{1}{\varepsilon}\sum_{|\alpha|\ge2}|\{X_m(\mathbf{y}+\mathbf{h}(\mathbf{y}))\}^{(\alpha)}|\mathbf{y}^{\alpha}+d_0\sum_{|\alpha|\ge2}\sum_{j=1}^{n}\sum_{\substack{2\le|\beta|\le|\alpha|-1\\ \alpha-\beta+e_j\in\mathbb{N}_0^n}}|h_m^{(\beta)}||Y_j^{(\alpha-\beta+e_j)}|\mathbf{y}^{\alpha}. \tag{2.60}$$

Clearly

$$\sum_{|\alpha|\ge2}|\{X_m(\mathbf{y}+\mathbf{h}(\mathbf{y}))\}^{(\alpha)}|\mathbf{y}^{\alpha}\prec\sum_{|\alpha|\ge2}X_m^{\natural}(\mathbf{y}+\mathbf{h}^{\natural}(\mathbf{y})). \tag{2.61}$$

Turning to the term to the right of the plus sign in (2.60), for convenience index elements of $\mathbb{N}^n$ as $\{\gamma_r\}_{r=1}^{\infty}$. Recalling our convention that $Y^{(\gamma)}=0$ if $\gamma\notin\mathbb{N}_0^n$,

$$\sum_{|\alpha|\ge2}\sum_{j=1}^{n}\sum_{\substack{2\le|\beta|\le|\alpha|-1\\ \alpha-\beta+e_j\in\mathbb{N}_0^n}}|h_m^{(\beta)}||Y_j^{(\alpha-\beta+e_j)}|\mathbf{y}^{\alpha}$$
$$=\sum_{r=1}^{\infty}\left[|h_m^{(\beta_r)}|\mathbf{y}^{\beta_r}\sum_{j=1}^{n}\sum_{s=1}^{\infty}|Y_j^{(\alpha_s-\beta_r+e_j)}|\mathbf{y}^{\alpha_s-\beta_r}\right]. \tag{2.62}$$

For any fixed multi-index β_r, consider the sum

$$\sum_{j=1}^{n}\sum_{s=1}^{\infty}|Y_j^{(\alpha_s-\beta_r+e_j)}|\mathbf{y}^{\alpha_s-\beta_r}. \tag{2.63}$$

The number $Y_j^{(\alpha_s-\beta_r+e_j)}$ is nonzero only if j and $\alpha_s-\beta_r+e_j$ form a resonant pair. The same term, times the same power of $\mathbf{y}$, will occur in the sum (2.63) corresponding to the multi-index $\beta_{r'}$ if and only if there exists a multi-index $\alpha_{s'}$ that satisfies $\alpha_{s'}-\beta_{r'}=\alpha_s-\beta_r$, which is true if and only if $\alpha_s-\beta_r+\beta_{r'}\in\mathbb{N}_0^n$. Writing $\alpha_s=(\alpha_s^1,\dots,\alpha_s^n)$ and $\beta_r=(\beta_r^1,\dots,\beta_r^n)$, then by the fact that $\alpha_s-\beta_r+e_j$ is part of a resonant pair and the hypothesis that no entry in the multi-index of a resonant pair is zero, if $k \neq j$, then $\alpha_s^k - \beta_r^k \geq 1$, while $\alpha_s^j - \beta_r^j \geq 0$. Thus $\alpha_s - \beta_r + \beta_{r'} \in \mathbb{N}_0^n$, so we conclude that the expression in (2.63) is the same for all multi-indices $\beta_{r'}$, and may be written

$$\sum_{j=1}^{n} \sum_{(\gamma,j) \text{ resonant}} |Y_j^{(\gamma)}| \mathbf{y}^{\gamma - e_j},$$

hence (2.62) is

$$\begin{aligned}
&\sum_{|\alpha|\geq 2} \sum_{j=1}^{n} \sum_{\substack{2\leq|\beta|\leq|\alpha|-1 \\ \alpha-\beta+e_j\in\mathbb{N}_0^n}} \left|h_m^{(\beta)}\right| \left|Y_j^{(\alpha-\beta+e_j)}\right| \mathbf{y}^\alpha \\
&\prec \left(\sum_{|\beta|\geq 2} \left|h_m^{(\beta)}\right| \mathbf{y}^\beta\right) \left(\sum_{j=1}^{n} \sum_{(\gamma,j) \text{ resonant}} \left|Y_j^{(\gamma)}\right| \mathbf{y}^{\gamma-e_j}\right) \\
&= h_m^\natural(\mathbf{y}) \sum_{j=1}^{n} \sum_{(\gamma,j) \text{ resonant}} y_j^{-1} \left|Y_j^{(\gamma)}\right| \mathbf{y}^\gamma \\
&= h_m^\natural(\mathbf{y}) \sum_{j=1}^{n} y_j^{-1} Y_j^\natural(\mathbf{y}),
\end{aligned} \tag{2.64}$$

which is well-defined even at $\mathbf{y} = \mathbf{0}$ by the hypothesis on resonant pairs. Thus applying (2.61) and (2.64) to (2.60) yields

$$h_m^\natural(\mathbf{y}) \prec \tfrac{1}{\varepsilon} X_m^\natural(\mathbf{y} + \mathbf{h}^\natural(\mathbf{y})) + h_m^\natural(\mathbf{y}) \sum_{j=1}^{n} y_j^{-1} Y_j^\natural(\mathbf{y}). \tag{2.65}$$

Multiplying (2.45) by $\mathbf{y}^\alpha$ and summing it over all α for which $|\alpha| \geq 2$, from (2.61) we obtain $Y_m^\natural(\mathbf{y}) \prec X_m^\natural(\mathbf{y} + \mathbf{h}^\natural(\mathbf{y}))$. Summing this latter expression and expression (2.65) over m, and recalling that for a distinguished transformation $h_m^{(\alpha)} = 0$ for all nonresonant pairs, we obtain the existence of a real constant $c_1 > 0$ such that the following majorizing relation holds:

$$\sum_{m=1}^{n} Y_m^\natural(\mathbf{y}) + \sum_{m=1}^{n} h_m^\natural(\mathbf{y}) \prec c_1 \sum_{m=1}^{n} X_m^\natural(\mathbf{y} + \mathbf{h}^\natural(\mathbf{y})) + c_1 \sum_{m=1}^{n} h_m^\natural(\mathbf{y}) \sum_{j=1}^{n} \frac{Y_j^\natural(\mathbf{y})}{y_j}. \tag{2.66}$$

By Lemma 2.3.12

$$X_m^\natural(\mathbf{y} + \mathbf{h}^\natural(\mathbf{y})) \prec \frac{a\left(\sum_{j=1}^n y_j + \sum_{j=1}^n h_j^\natural(\mathbf{y})\right)^2}{1 - b\left(\sum_{j=1}^n y_j + \sum_{j=1}^n h_j^\natural(\mathbf{y})\right)}, \tag{2.67}$$

so that (2.66) becomes

$$\sum_{m=1}^{n} Y_m^{\natural}(\mathbf{y}) + \sum_{m=1}^{n} h_m^{\natural}(\mathbf{y}) \prec \frac{c_1 a\left(\sum_{j=1}^{n} y_j + \sum_{j=1}^{n} h_j^{\natural}(\mathbf{y})\right)^2}{1 - b\left(\sum_{j=1}^{n} y_j + \sum_{j=1}^{n} h_j^{\natural}(\mathbf{y})\right)} + c_1 \sum_{m=1}^{n} h_m^{\natural}(\mathbf{y}) \sum_{j=1}^{n} y_j^{-1} Y_j^{\natural}(\mathbf{y}). \tag{2.68}$$

To prove the theorem it suffices to show that the series

$$S(\mathbf{y}) = \sum_{m=1}^{n} Y_m^{\natural}(\mathbf{y}) + \sum_{m=1}^{n} h_m^{\natural}(\mathbf{y})$$

converges at some nonzero value of $\mathbf{y}$. We will show convergence at $\mathbf{y} = (\eta, \dots, \eta)$ for some positive value of the real variable η. To do so, we consider the real series $S(\eta, \dots, \eta)$. Since $\mathbf{Y}$ and $\mathbf{h}$ begin with quadratic or higher terms we can write $S(\eta, \dots, \eta) = \eta U(\eta)$ for a real series $U(\eta) = \sum_{k=1}^{\infty} u_k \eta^k$, where the coefficients u_k are nonnegative. By the definition of S and U, $U(\eta)$ clearly satisfies $\sum_{m=1}^{n} h_m^{\natural}(\eta, \dots, \eta) \prec \eta U(\eta)$ and $\sum_{m=1}^{n} Y_m^{\natural}(\eta, \dots, \eta) \prec \eta U(\eta)$, which when applied to (2.68) yield

$$\eta U(\eta) = S(\eta, \dots, \eta) \prec \frac{c_1 a (n\eta + \eta U(\eta))^2}{1 - b(n\eta + \eta U(\eta))} + c_1 \eta U^2(\eta)$$

so that

$$U(\eta) \prec c_1 U^2(\eta) + \frac{c_1 a \eta\, (n + U(\eta))^2}{1 - b\eta\, (n + U(\eta))}. \tag{2.69}$$

Consider the real analytic function F defined on a neighborhood of $(0,0) \in \mathbb{R}^2$ by

$$F(x,y) = c_1 x^2 + \frac{c_1 a y (n+x)^2}{1 - b y (n+x)}.$$

Using the geometric series to expand the second term, it is clear that

$$F(x,y) = F_0(y) + F_1(y)x + F_2(y)x^2 + \cdots,$$

where $F_j(0) = 0$ for $j \neq 2$ and $F_j(y) \geq 0$ if $y \geq 0$. Thus for any sequence of real constants $r_1, r_2, r_3, \dots,$

$$F\left(\sum_{k=1}^{\infty} r_k y^k, y\right) = \delta_1 y + \delta_2(r_1) y^2 + \delta_3(r_1, r_2) y^3 + \cdots, \tag{2.70}$$

where $\delta_1 \geq 0$ and for $k \geq 2$ δ_k is a polynomial in $r_1, r_2, \dots, r_{k-1}$ with nonnegative coefficients. Thus

$$0 \leq a_j \leq b_j \text{ for } j = 1, \dots, k-1 \quad \text{implies} \quad \delta_k(a_1, \dots, a_{k-1}) \leq \delta_k(b_1, \dots, b_{k-1}). \tag{2.71}$$

By the Implicit Function Theorem there is a unique real analytic function $w = w(y)$ defined on a neighborhood of 0 in $\mathbb{R}$ such that $w(0) = 0$ and $x - F(x,y) = 0$ if and only if $x = w(y)$; write $w(y) = \sum_{k=1}^{\infty} w_k y^k$. By (2.70) the coefficients w_k satisfy

$$w_k = \delta_k(w_1, \ldots, w_{k-1}) \quad \text{for} \quad k \geq 2, \tag{2.72}$$

and by (2.69) the coefficients u_k satisfy

$$u_k \leq \delta_k(u_1, \ldots, u_{k-1}) \quad \text{for} \quad k \geq 2. \tag{2.73}$$

A simple computation shows that $u_1 = c_1 a n^2 = w_1$, hence (2.71), (2.72), and (2.73) imply by mathematical induction that $u_k \leq w_k$ for $k \geq 1$. Thus $U(\eta) \prec w(\eta)$, implying the convergence of U on a neighborhood of 0 in $\mathbb{R}$, which implies convergence of S, hence of $\mathbf{h}$ and $\mathbf{Y}$, on a neighborhood of 0 in $\mathbb{C}^n$. □

2.4 Notes and Complements

Our concern in this chapter has been with vector fields in a neighborhood of an equilibrium. In Section 2.1 we presented a few of Lyapunov's classical results on the stability problem. Other theorems of this sort are available in the literature. The reader is referred to the monograph of La Salle ([109]) for generalizations of the classical Lyapunov theory.

The idea of placing system (2.1) in some sort of normal form in preparation for a more general study, or in order to take advantage of a general property of the collections of systems under consideration, has wide applicability. To cite just one example, a class of systems of differential equations that has received much attention is the set of quadratic systems in the plane, which are those of the form

$$\begin{aligned} \dot{x} &= a_{00} + a_{10}x + a_{01}y + a_{20}x^2 + a_{11}xy + a_{02}y^2 \\ \dot{y} &= b_{00} + b_{10}x + b_{01}y + b_{20}x^2 + b_{11}xy + b_{02}y^2 . \end{aligned} \tag{2.74}$$

Using special properties of such systems, it is possible by a sequence of coordinate transformations and time rescalings to place any such system that has a cycle in its phase portrait into the special form

$$\begin{aligned} \dot{x} &= \delta x - y + \ell x^2 + mxy + ny^2 \\ \dot{y} &= x(1 + ax + by) , \end{aligned} \tag{2.75}$$

thereby simplifying the expression, reducing the number of parameters, and making one parameter (δ) into a rotation parameter on the portion of the plane in which a cycle can exist. (See §12 of [202].)

Turning to the problem of computing normal forms, if we wish to know the actual coefficients in the normal form in terms of the *original* coefficients, whether numerical or symbolic, we must keep track of all coefficients exactly at each successive step of the sequence of transformations leading to the normal form. Hand computation can quickly become infeasible. Computer algebra approaches to the actual computation of normal forms is treated in the literature; the reader is referred to [71, 139, 146]. Sample code for the algorithm in Table 2.1 is in the Appendix.

We have noted that unless $\mathscr{L}$ maps $\mathscr{H}_s$ onto itself, the subspace $\mathscr{K}_s$ complementary to Image($\mathscr{L}$) is not unique. It is reasonable to attempt to make a uniform, or at least systematic, choice of the subspace $\mathscr{K}_s$, $s \geq 2$. Such a systematic choice is termed in [139] a normal form *style*, and the reader is directed there for a full discussion.

For more exhaustive treatments of the theory of normal forms the reader can consult, for example, the references [15, 16, 19, 24, 25, 115, 139, 142, 180, 181].

Exercises

2.1 Show that the trajectory of every point in a neighborhood of the equilibrium $\mathbf{x}_0 = (1,0)$ of the system

$$\begin{aligned} \dot{x} &= x - (x+y)\sqrt{x^2+y^2} + xy \\ \dot{y} &= y + (x-y)\sqrt{x^2+y^2} - x^2 \end{aligned}$$

on $\mathbb{R}^2$ tends to $\mathbf{x}_0$ in forward time, yet in every neighborhood of $\mathbf{x}_0$ there exists a point whose forward trajectory travels distance at least 1 away from $\mathbf{x}_0$ before returning to limit on $\mathbf{x}_0$.
Hint. Change to polar coordinates.

2.2 Consider the general second-order linear homogeneous ordinary differential equation in one dependent variable in standard form,

$$\ddot{x} + f(x)\dot{x} + g(x) = 0, \tag{2.76}$$

generally called a *Liénard equation*.

a. Show that if there exists $\varepsilon > 0$ such that $xg(x) > 0$ whenever $0 < |x| < \varepsilon$, then the function $W(x,y) = \frac{1}{2}y^2 + \int_0^x g(u)\,du$ is positive definite on a neighborhood of $\mathbf{0} \in \mathbb{R}^2$.
b. Assuming the truth of the condition on $g(x)$ in part (a), use the function W to show that the equilibrium of the system

$$\begin{aligned} \dot{x}_1 &= x_2 \\ \dot{x}_2 &= -g(x_1) - f(x_1)x_2, \end{aligned}$$

which is equivalent to (2.76), is stable if $f(x) \equiv 0$ and is asymptotically stable if $f(x) > 0$ for $0 < |x| < \varepsilon$.

2.3 Transform system (2.76) into the equivalent Liénard form

$$\begin{aligned} \dot{x}_1 &= x_2 - F(x_1) \\ \dot{x}_2 &= -g(x_1), \end{aligned} \tag{2.77}$$

where $F(x) = \int_0^x f(u)du$. Formulate conditions in terms of $g(x)$ and $F(x)$ so that the equilibrium of (2.77) is stable, asymptotically stable, or unstable.

2.4 Construct a counterexample to Theorem 2.1.4 if tangency to $C \setminus \{\mathbf{0}\}$ is allowed, even if at only countably many points.

2.5 Prove Theorem 2.1.5.

2.6 Show that the dimension of the vector space $\mathscr{H}_s$ of functions from $\mathbb{R}^n$ to $\mathbb{R}^n$ (or from $\mathbb{C}^n$ to $\mathbb{C}^n$), all of whose components are homogeneous polynomial functions of degree s, is $nC(s+n-1,s) = n(s+n-1)!/(s!(n-1)!)$.
Hint. Think in terms of distributing p identical objects into q different boxes, which is the same as selecting with repetition p objects from q types of objects.

2.7 Show that if the eigenvalues of the matrix $\left(\begin{smallmatrix} a & b \\ c & d \end{smallmatrix}\right)$ are $\pm i\beta$ ($\beta \in \mathbb{R}$), then by a linear transformation the system

$$\begin{aligned} \dot{x} &= ax + by \\ \dot{y} &= cx + dy \end{aligned}$$

can be brought to the form

$$\begin{aligned} \dot{x} &= \quad \beta y \\ \dot{y} &= -\beta x. \end{aligned}$$

2.8 For system (2.1) on $\mathbb{R}^2$, suppose $\mathbf{f}(\mathbf{0}) = \mathbf{0}$ and $A = \mathbf{df}(\mathbf{0})$ has exactly one zero eigenvalue. Show that for each $k \in \mathbb{N}$, $k \geq 2$, each vector in the ordered basis of $\mathscr{H}_k$ of Example 2.2.2 is an eigenvector of $\mathscr{L}$, and that $\mathscr{L}$ has zero as an eigenvalue of multiplicity two, with corresponding eigenvectors $\begin{pmatrix} xy^{k-1} \\ 0 \end{pmatrix}$ and $\begin{pmatrix} 0 \\ y^k \end{pmatrix}$, where x and y denote the usual coordinates on $\mathbb{R}^2$. Write down the normal form for (2.1) through order k. (Assume from the outset that A has been placed in Jordan normal form, that is, diagonalized.)

2.9 For system (2.1) on $\mathbb{R}^2$, suppose $\mathbf{f}(\mathbf{0}) = \mathbf{0}$ and $A = \mathbf{df}(\mathbf{0})$ has both eigenvalues zero but is not itself the zero transformation, hence has upper-triangular Jordan normal form $\left(\begin{smallmatrix} 0 & 1 \\ 0 & 0 \end{smallmatrix}\right)$. (This form for the linear part is traditional in this context.)

a. Show that the matrix of $\mathscr{L} : \mathscr{H}_2 \to \mathscr{H}_2$ with respect to the ordered basis of Example 2.2.2 is

$$L = \begin{pmatrix} 0 & 0 & 0 & -1 & 0 & 0 \\ 2 & 0 & 0 & 0 & -1 & 0 \\ 0 & 1 & 0 & 0 & 0 & -1 \\ 0 & 0 & 0 & 0 & 0 & 0 \\ 0 & 0 & 0 & 2 & 0 & 0 \\ 0 & 0 & 0 & 0 & 1 & 0 \end{pmatrix}.$$

b. Show that the rank of L (the dimension of its column space) is four, hence that $\dim \operatorname{Image}(\mathscr{L}) = 4$.

c. Show that none of the basis vectors $\begin{pmatrix} x^2 \\ 0 \end{pmatrix}$, $\left(\begin{smallmatrix} 0 \\ xy \end{smallmatrix}\right)$, and $\begin{pmatrix} 0 \\ x^2 \end{pmatrix}$ lies in $\operatorname{Image}(\mathscr{L})$.
Hint. Work in coordinates, for example replacing the second vector in the list by its coordinate vector, the standard basis vector e_5 of $\mathbb{R}^6$.

d. Explain why two possible normal forms of $\mathbf{f}$ through order two are

$$\text{(i)}\quad \begin{aligned}\dot{x} &= y + O(3)\\ \dot{y} &= ax^2 + bxy + O(3)\end{aligned} \qquad \text{and} \qquad \text{(ii)}\quad \begin{aligned}\dot{x} &= y + ax^2 + O(3)\\ \dot{y} &= bx^2 + O(3).\end{aligned}$$

Remark: This is the "Bogdanov–Takens Singularity." Form (i) in part (d) is the normal form of Bogdanov ([20]); form (ii) is that of Takens ([187]).

2.10 In the same situation as that of Exercise 2.9, show that for all $k \in \mathbb{N}$, $k \geq 3$, $\dim(\operatorname{Image}(\mathscr{L})) = \dim(\mathscr{H}_k) - 2$ and that $\mathscr{K}_k = \operatorname{Span}\left\{\binom{x^k}{0}, \binom{0}{x^k}\right\}$ is one choice of $\mathscr{K}_k$. Hence a normal form for (2.1) through order r is

$$\begin{aligned}\dot{x} &= y + a_2x^2 + a_3x^3 + \cdots + a_rx^r + O(r+1)\\ \dot{y} &= \phantom{y +{}} b_2x^2 + b_3x^3 + \cdots + b_rx^r + O(r+1).\end{aligned}$$

2.11 Show that if S is a nonsingular $n \times n$ matrix such that $J = SAS^{-1}$, then under the coordinate change $\mathbf{x} = S\mathbf{y}$, expression (2.30) for $\mathscr{L}$ is transformed into (2.31), as follows.

a. Writing $\mathscr{S} : \mathbb{C}^n \to \mathbb{C}^n : \mathbf{x} \mapsto \mathbf{y} = S\mathbf{x}$ and for $\mathbf{h} \in \mathscr{H}_s$ letting $\mathbf{u} = \mathscr{S}^{-1} \circ \mathbf{h} \circ \mathscr{S}$, use the fact that $d\mathscr{S}(\mathbf{y})z = Sz$ to show that $\mathscr{L}\mathbf{h}(\mathbf{x}) = S d\mathbf{u}(\mathbf{y})J\mathbf{y} - AS\mathbf{u}(\mathbf{y})$.

b. Use the fact that $\mathscr{L}\mathbf{u} = \mathscr{S}^{-1} \circ \mathscr{L}\mathbf{h} \circ \mathscr{S}$ and the result of part (a) to obtain (2.31).

2.12 a. Order the basis vectors of Example 2.2.2 according to the lexicographic order of the proof of Lemma 2.3.1.

b. Rework part (a) of Exercise 2.9 using lexicographic order. (Merely use the result of Exercise 2.9 without additional computation.)

c. By inspection of the diagonal entries of the matrix in part (b), determine the eigenvalues of $\mathscr{L}$.

d. Verify by direct computation of the numbers $(\kappa, \alpha) - \kappa_j$ that the eigenvalues of $\mathscr{L}$ are all zero.

2.13 [Referenced in Section 3.2.] Let $\kappa_1, \ldots, \kappa_n$ be the eigenvalues of the matrix A in display (2.27).

a. Suppose $n = 2$. Show that if $\kappa_1 \neq 0$, then a necessary condition that there be a resonant term in either component of $\mathbf{X}$ is that κ_2/κ_1 be a rational number. Similarly if $\kappa_2 \neq 0$. (If the ratio of the eigenvalues is p/q, $\operatorname{GCD}(p,q) = 1$, then the resonance is called a $p:q$ resonance.)

b. Show that the analogous statement is not true when $n \geq 3$.

2.14 Derive the rightmost expression in (2.41) in the following three steps.

a. Use the expansions of $\mathbf{h}$ and $\mathbf{Y}$ analogous to (2.36) to determine that the vector homogeneous function of degree s in $d\mathbf{h}(\mathbf{y})\mathbf{Y}(\mathbf{y})$ is a sum of $s-2$ products of the form $d\mathbf{h}^{(k)}(\mathbf{y})\mathbf{Y}^{(\ell)}(\mathbf{y})$.

b. Show that for each $r \in \{2, \ldots, s-1\}$, the mth component of the corresponding product in part (a) is

$$\sum_{j=1}^{n}\left(\left[\sum_{|\beta|=r}\beta_j h_m^{(\beta)} y_1^{\beta_1}\cdots y_j^{\beta_j-1}\cdots y_n^{\beta_n}\right]\left[\sum_{|\gamma|=s-r+1} Y_j^{(\gamma)}\mathbf{y}^{\gamma}\right]\right).$$

c. For any α such that $|\alpha| = s$, use the expression in part (b) to find the coefficient of $\mathbf{y}^{\alpha}$ in the mth component of $d\mathbf{h}(\mathbf{y})\mathbf{Y}(\mathbf{y})$, thereby obtaining the negative of the rightmost expression in (2.41).

2.15 Apply the Normal Form Algorithm displayed in Table 2.1 on page 75 to show that the normalizing transformation $\mathbf{H}$ through order two in Example 2.2.2 is

$$\mathbf{H}\begin{pmatrix} y_1 \\ y_2 \end{pmatrix} = \begin{pmatrix} y_1 \\ y_2 \end{pmatrix} + \begin{pmatrix} \frac{1}{2}ay_1^2 + by_1y_2 \\ \frac{1}{3}a'y_1^2 + \frac{1}{2}b'y_1y_2 + c'y_2^2 \end{pmatrix} + \cdots.$$

2.16 Apply the Normal Form Algorithm displayed in Table 2.1 on page 75 to compute the coefficients $Y_1^{(2,2)}$ and $Y_2^{(1,3)}$ of the normal form (2.44) of Example 2.3.10 and the normalizing transformation $\mathbf{H}(\mathbf{y})$ up to order three.

2.17 In system (2.38) let J be a diagonal 2×2 matrix with eigenvalues $\kappa_1 = p$ and $\kappa_2 = -q$, where $p, q \in \mathbb{N}$ and $\mathrm{GCD}(p,q) = 1$. Show that the normal form of system (2.38) is

$$\begin{aligned} \dot{y}_1 &= py_1 + y_1Y_1(w) \\ \dot{y}_2 &= -qy_2 + y_2Y_2(w), \end{aligned}$$

where Y_1 and Y_2 are formal power series without constant terms and $w = y_1^q y_2^p$.

2.18 For system (2.1) on $\mathbb{R}^n$, suppose that $\mathbf{f}(\mathbf{0}) = \mathbf{0}$ and that the linear part $A = d\mathbf{f}(\mathbf{0})$ of f at $\mathbf{0}$ is diagonalizable. For $k \in \mathbb{N}$, $k \geq 2$, choose as a basis of $\mathscr{H}_k$ the analogue of the basis of Example 2.2.2, consisting of functions of the form

$$\mathbf{h}_{\alpha,j} : \mathbb{R}^n \to \mathbb{R}^n : \begin{pmatrix} x_1 \\ x_2 \\ \vdots \\ x_n \end{pmatrix} \mapsto x_1^{\alpha_1} x_2^{\alpha_2} \cdots x_n^{\alpha_n} \mathbf{e}_j,$$

where $\alpha = (\alpha_1, \ldots, \alpha_n) \in \mathbb{N}_0^n$, $\alpha_1 + \alpha_2 + \cdots + \alpha_n = k$, $j \in \{1, 2, \ldots, n\}$, and for each such j, $\mathbf{e}_j$ is the jth standard basis vector of $\mathbb{R}^n$, as always. Show by direct computation using the definition (2.18) of $\mathscr{L}$ that each basis vector $\mathbf{h}_{\alpha,j}$ is an eigenvector of $\mathscr{L}$, with corresponding eigenvalue $\lambda_j - \sum_{i=1}^{n} \alpha_i \lambda_i$. (Note how Example 2.2.2 and Exercise 2.8 fit into this framework but Exercises 2.9 and 2.10 do not.) When $\lambda_j \neq 0$ for all j the equilibrium is hyperbolic, and the non-resonance conditions

$$\lambda_j - \sum_{k=1}^{n} \alpha_k \lambda_k \neq 0 \quad \text{for all } 1 \leq j \leq n, \quad \text{for all } \alpha \in \mathbb{N}_0^n$$

are necessary conditions for the existence of a smooth linearization. Results on the existence of smooth linearizations are given in [183] (for the C^∞ case) and in [180] (for the analytic case).

Chapter 3
The Center Problem

Consider a real planar system of differential equations $\dot{\mathbf{u}} = \mathbf{f}(\mathbf{u})$, defined and analytic on a neighborhood of $\mathbf{0}$, for which $\mathbf{f}(\mathbf{0}) = \mathbf{0}$ and the eigenvalues of the linear part of $\mathbf{f}$ at $\mathbf{0}$ are $\alpha \pm i\beta$ with $\beta \neq 0$. If the system is actually linear, then a straightforward geometric analysis (see, for example, [44], [95], or [110]) shows that when $\alpha \neq 0$ the trajectory of every point spirals towards or away from $\mathbf{0}$ (see Definition 3.1.1: $\mathbf{0}$ is a *focus*), but when $\alpha = 0$, the trajectory of every point except $\mathbf{0}$ is a cycle, that is, lies in an oval (see Definition 3.1.1: $\mathbf{0}$ is a *center*). When the system is nonlinear, then in the first case ($\alpha \neq 0$) trajectories in a sufficiently small neighborhood of the origin follow the behavior of the linear system determined by the linear part of $\mathbf{f}$ at $\mathbf{0}$: they spiral towards or away from the origin in accordance with the trajectories of the linear system. The second case is different: the linear approximation does not necessarily determine the geometric behavior of the trajectories of the nonlinear system in a neighborhood of the origin. This phenomenon is illustrated by the system

$$\begin{aligned} \dot{u} &= -v - u(u^2+v^2) \\ \dot{v} &= \quad u - v(u^2+v^2)\,. \end{aligned} \tag{3.1}$$

In polar coordinates system (3.1) is $\dot{r} = -r^3$, $\dot{\varphi} = 1$. Thus whereas the origin is a center for the corresponding linear system, every trajectory of (3.1) spirals towards the origin, which is thus a stable focus. On the other hand, one can just as easily construct examples in which the addition of higher-order terms does not destroy the center. We see then that in the case of a singular point for which the eigenvalues of the linear part are purely imaginary the topological type of the point is not determined by the linear approximation, and a special investigation is needed. Here we face the fascinating problem in the qualitative theory of differential equations known as *the problem of distinguishing between a center and a focus*, or just *the center problem* for short, which is the subject of this chapter. Since there always exists a nonsingular linear transformation that transforms a system whose linear part has eigenvalues $\alpha \pm i\beta$ into the form (3.4) below, and a time rescaling $\tau = \beta t$ will eliminate β, the objects of study will be systems of the form

V.G. Romanovski, D.S. Shafer, *The Center and Cyclicity Problems*,
DOI 10.1007/978-0-8176-4727-8_3,

$$\begin{aligned} \dot{u} &= -v + U(u,v) \\ \dot{v} &= u + V(u,v), \end{aligned} \tag{3.2}$$

where U and V are convergent real series starting with quadratic terms. We reiterate that every real analytic planar system of differential equations with a nonzero linear part at the origin that has purely imaginary eigenvalues can be placed in this form by a translation, followed by a nonsingular linear change of coordinates, followed by a time rescaling.

Our principal concern, however, is not so much with this or that particular system of the form (3.2) as it is with *families* of such systems, and in particular with families of polynomial systems, such as the family of all quadratic systems (see (2.74)) of the form (3.2) (which is precisely the collection of quadratic systems that can have a center at the origin, after the transformations described in the previous sentence). Conditions that characterize when a member of such a family has a center at the origin are given by the vanishing of polynomials in the coefficients of the members of the family, hence they yield a variety in the space of coefficients. Because the eigenvalues of the linear part at the origin of every system in question are complex, and because complex varieties are more amenable to study than real varieties, it is natural to complexify the family. This leads to the study of families of systems that are of the form

$$\begin{aligned} \dot{x} = \widetilde{P}(x,y) &= i\Big(x - \sum_{(p,q)\in S} a_{pq}x^{p+1}y^{q}\Big) \\ \dot{y} = \widetilde{Q}(x,y) &= -i\Big(y - \sum_{(p,q)\in S} b_{qp}x^{q}y^{p+1}\Big), \end{aligned} \tag{3.3}$$

where the variables x and y are complex, the coefficients of $\widetilde{P}$ and $\widetilde{Q}$ are complex, where $S \subset (\{-1\} \cup \mathbb{N}_0) \times \mathbb{N}_0$ is a finite set, every element (p,q) of which satisfies $p+q \geq 1$, and where $b_{pq} = \bar{a}_{qp}$ for all $(p,q) \in S$. The somewhat unusual indexing is chosen to simplify expressions that will arise later. In this chapter we develop an approach to solving the center problem for such a family.

An overview of the approach, and of this chapter, is the following. We begin in Section 3.1 with a local analysis of any system for which the eigenvalues of the real part are complex, then turn in Section 3.2 to the idea of complexifying the system and examining the resulting normal form. The geometry of a center suggests that there should be a kind of potential function Ψ defined on a neighborhood of the origin into $\mathbb{R}$ whose level sets are the closed orbits about the center; such a function is a so-called first integral of the motion and is shown in Section 3.2 to be intimately connected with the existence of a center. Working from characterizations of centers thus obtained, in Section 3.3 we generalize the concept of a center on $\mathbb{R}^2$ to systems of the form (3.3) on $\mathbb{C}^2$. We then derive, for a *family* of systems of the form (3.3), the *focus quantities*, a collection of polynomials $\{g_{kk} : k \in \mathbb{N}\}$ in the coefficients of family (3.3) whose simultaneous vanishing picks out the coefficients of the systems that have a center at the origin. The variety so identified in the space of coefficients is the *center variety* $V_{\mathscr{C}}$. Since by the Hilbert Basis Theorem (Theorem 1.1.6) every polynomial ideal is generated by finitely many polynomials, there must exist a $K \in \mathbb{N}$ such that the ideal $\mathscr{B} = \langle g_{11}, g_{22}, \ldots \rangle$ of the center variety $V_{\mathscr{C}}$ satisfies $\mathscr{B} = \mathscr{B}_K = \langle g_{11}, \ldots, g_{KK} \rangle$. Thus the center problem will be solved if we find a number K such that $\mathscr{B} = \mathscr{B}_K$. Actually, finding such a number K is too strong a demand: we are interested only in the *variety* of the ideal, not the ideal itself, so by Proposition 1.3.16 it is sufficient to find a number K such that $\sqrt{\mathscr{B}} = \sqrt{\mathscr{B}_K}$. To do so for a specific family (3.3) we compute the first few focus quantities. An efficient computational algorithm is developed in Section 3.4. We compute the g_{kk} until $g_{J+1,J+1} \in \sqrt{\langle g_{11}, \ldots, g_{J,J} \rangle}$, indicating that perhaps $K = J$; we might compute the next several focus quantities, confirming that $g_{J+s,J+s} \in \sqrt{\mathscr{B}_J}$ for small $s \in \mathbb{N}$ and thereby strengthening our conviction that $K = J$. Since

$$\mathbf{V}(g_{11}) \supset \mathbf{V}(g_{11}, g_{22}) \supset \mathbf{V}(g_{11}, g_{22}, g_{33}) \supset \cdots \supset V_{\mathscr{C}}$$

the vanishing of the J polynomials $g_{11}, \ldots, g_{JJ}$ on the coefficients of a particular system in (3.3) is a necessary condition that the system have a center at the origin. We will have shown that $K = J$, and thus have solved the center problem, if we can show that the vanishing of these J polynomials on the coefficients of a given system is a sufficient condition that the system have a center at the origin. To do so we need techniques for determining that a system of a given form has a center, and we present two of the most important methods for doing so in the context of our problem in Sections 3.5 and 3.6. The whole approach is illustrated in Section 3.7, in which we apply it to find the center variety of the full set of quadratic systems, as well as a family of cubic systems. As a complement to this approach, in the final section we study an important class of real planar systems, the Liénard systems.

Since throughout this chapter we will be dealing with both real systems and their complexifications, we will denote real variables by u and v and complex variables by x and y, except in Section 3.8 on Liénard systems, where only real systems are in view and we employ the customary notation. For formal and convergent power series we will consistently follow the notational convention introduced in the paragraph following Remark 2.3.7. In variables subscripted by ordered pairs of integers, a comma is occasionally introduced for clarity, but carries no other meaning, so that for example we write g_{kk} in (3.56) but g_{k_1,k_2} in (3.55). Finally, it will be convenient to have the notation $\mathbb{N}_{-n} = \{-n, \ldots, -1, 0\} \cup \mathbb{N}$.

3.1 The Poincaré First Return Map and the Lyapunov Numbers

Throughout this and the following section the object of study is a real analytic system $\dot{\mathbf{u}} = \mathbf{f}(\mathbf{u})$ on a neighborhood of $\mathbf{0}$ in $\mathbb{R}^2$, where $\mathbf{f}(\mathbf{0}) = \mathbf{0}$ and the eigenvalues of the linear part of $\mathbf{f}$ at $\mathbf{0}$ are $\alpha \pm i\beta$ with $\beta \neq 0$. By a nonsingular linear coordinate change, such a system can be written in the form

$$\begin{aligned} \dot{u} &= \alpha u - \beta v + P(u,v) \\ \dot{v} &= \beta u + \alpha v + Q(u,v), \end{aligned} \tag{3.4}$$

where $P(u,v) = \sum_{k=2}^{\infty} P^{(k)}(u,v)$ and $Q(u,v) = \sum_{k=2}^{\infty} Q^{(k)}(u,v)$, and $P^{(k)}(u,v)$ and $Q^{(k)}(u,v)$ (if nonzero) are homogeneous polynomials of degree k. In Chapter 6 we will need information about the function $\mathscr{P}(r)$ of Definition 3.1.3 in the case that $\alpha \neq 0$, so we will not specialize to the situation $\alpha = 0$ until the next section. The following definition, which applies equally well to systems that are only C^1, is central. For the second part recall that the omega limit set of a point $\mathbf{a}$ for which $\mathbf{a}(t)$ is defined for all $t \geq 0$ is the set of all points $\mathbf{p}$ for which there exists a sequence $t_1 < t_2 < \cdots$ of numbers such that $t_n \to \infty$ and $\mathbf{a}(t_n) \to \mathbf{p}$ as $n \to \infty$.

Definition 3.1.1. Let $\dot{\mathbf{u}} = \mathbf{f}(\mathbf{u})$ be a real analytic system of differential equations on a neighborhood of $\mathbf{0}$ in $\mathbb{R}^2$ for which $\mathbf{f}(\mathbf{0}) = \mathbf{0}$.

(a) The singularity at $\mathbf{0}$ is a *center* if there exists a neighborhood Ω of $\mathbf{0}$ such that the trajectory of any point in $\Omega \setminus \{\mathbf{0}\}$ is a simple closed curve γ that contains $\mathbf{0}$ in its interior (the bounded component of the complement $\mathbb{R}^2 \setminus \{\gamma\}$ of γ). The *period annulus* of the center is the set $\Omega_M \setminus \{\mathbf{0}\}$, where Ω_M is the largest such neighborhood Ω, with respect to set inclusion.

(b) The singularity at $\mathbf{0}$ is a *stable focus* if there exists a neighborhood Ω of $\mathbf{0}$ such that $\mathbf{0}$ is the omega limit set of every point in Ω and for every trajectory $\mathbf{u}(t)$ in $\Omega \setminus \{\mathbf{0}\}$ a continuous determination of the angular polar coordinate $\varphi(t)$ along $\mathbf{u}(t)$ tends to ∞ or to $-\infty$ as t increases without bound. The singularity is an *unstable focus* if it is a stable focus for the time-reversed system $\dot{\mathbf{u}} = -\mathbf{f}(\mathbf{u})$. The singularity is a *focus* if it is either a stable focus or an unstable focus.

For smooth systems the behavior of trajectories near a focus can be somewhat bizarre (even in the C^∞ case), but the theory presented in this section shows that for analytic systems the angular polar coordinate of any trajectory sufficently near zero changes monotonically, so that every trajectory properly spirals onto or away from $\mathbf{0}$ in future time.

In polar coordinates $x = r\cos\varphi$, $y = r\sin\varphi$, system (3.4) becomes

$$\begin{aligned}
\dot{r} &= \alpha r + P(r\cos\varphi, r\sin\varphi)\cos\varphi + Q(r\cos\varphi, r\sin\varphi)\sin\varphi \\
&= \alpha r + r^2\left[P^{(2)}(\cos\varphi,\sin\varphi)\cos\varphi + Q^{(2)}(\cos\varphi,\sin\varphi)\sin\varphi + \cdots\right] \\
\dot{\varphi} &= \beta - r^{-1}[P(r\cos\varphi, r\sin\varphi)\sin\varphi - Q(r\cos\varphi, r\sin\varphi)\cos\varphi] \\
&= \beta - r\left[P^{(2)}(\cos\varphi,\sin\varphi)\sin\varphi - Q^{(2)}(\cos\varphi,\sin\varphi)\cos\varphi + \cdots\right].
\end{aligned} \tag{3.5}$$

It is clear that for $|r|$ sufficiently small, if $\beta > 0$ then the polar angle φ increases as t increases, while if $\beta < 0$ then the angle decreases as t increases.

It is convenient to consider, in place of system (3.5), the equation of its trajectories

$$\frac{dr}{d\varphi} = \frac{\alpha r + r^2 F(r,\sin\varphi,\cos\varphi)}{\beta + rG(r,\sin\varphi,\cos\varphi)} = R(r,\varphi). \tag{3.6}$$

The function $R(r,\varphi)$ is a 2π-periodic function of φ and is analytic for all φ and for $|r| < r^*$, for some sufficiently small r^*. The fact that the origin is a singularity for (3.4) corresponds to the fact that $R(0,\varphi) \equiv 0$, so that $r = 0$ is a solution of (3.6). We can expand $R(r,\varphi)$ in a power series in r:

$$\frac{dr}{d\varphi} = R(r,\varphi) = rR_1(\varphi) + r^2R_2(\varphi) + \cdots = \frac{\alpha}{\beta}r + \cdots, \tag{3.7}$$

where $R_k(\varphi)$ are 2π-periodic functions of φ. The series is convergent for all φ and for all sufficiently small r.

Denote by $r = f(\varphi, \varphi_0, r_0)$ the solution of system (3.7) with initial conditions $r = r_0$ and $\varphi = \varphi_0$. The function $f(\varphi, \varphi_0, r_0)$ is an analytic function of all three variables φ, φ_0, and r_0 and has the property that

$$f(\varphi, \varphi_0, 0) \equiv 0 \tag{3.8}$$

(because $r = 0$ is a solution of (3.7)). Equation (3.8) and continuous dependence of solutions on parameters yield the following proposition.

Proposition 3.1.2. *Every trajectory of system* (3.4) *in a sufficiently small neighborhood of the origin crosses every ray $\varphi = c$, $0 \le c < 2\pi$.*

The proposition implies that in order to investigate all trajectories in a sufficiently small neighborhood of the origin it is sufficient to consider all trajectories passing through a segment $\Sigma = \{(u,v) : v = 0, 0 \le u \le r^*\}$ for r^* sufficiently small, that is, all solutions $r = f(\varphi, 0, r_0)$. We can expand $f(\varphi, 0, r_0)$ in a power series in r_0,

$$r = f(\varphi, 0, r_0) = w_1(\varphi)r_0 + w_2(\varphi)r_0^2 + \cdots, \tag{3.9}$$

which is convergent for all $0 \le \varphi \le 2\pi$ and for $|r_0| < r^*$. This function is a solution of (3.7), hence

$$\begin{aligned}
&w_1' r_0 + w_2' r_0^2 + \cdots \\
&\equiv R_1(\varphi)(w_1(\varphi)r_0 + w_2(\varphi)r_0^2 + \cdots) + R_2(\varphi)(w_1(\varphi)r_0 + w_2(\varphi)r_0^2 + \cdots)^2 + \cdots,
\end{aligned}$$

where the primes denote differentiation with respect to φ. Equating the coefficients of like powers of r_0 in this identity, we obtain recurrence differential equations for the functions $w_j(\varphi)$:

$$\begin{aligned}
w_1' &= R_1(\varphi)w_1, \\
w_2' &= R_1(\varphi)w_2 + R_2(\varphi)w_1^2, \\
w_3' &= R_1(\varphi)w_3 + 2R_2(\varphi)w_1w_2 + R_3(\varphi)w_1^3, \\
&\vdots
\end{aligned} \tag{3.10}$$

The initial condition $r = f(0, 0, r_0) = r_0$ yields

$$w_1(0) = 1, \qquad w_j(0) = 0 \quad \text{for} \quad j > 1. \tag{3.11}$$

Using these conditions, we can consequently find the functions $w_j(\varphi)$ by integrating equations (3.10). In particular,

$$w_1(\varphi) = e^{\frac{\alpha}{\beta}\varphi}. \tag{3.12}$$

Setting $\varphi = 2\pi$ in the solution $r = f(\varphi, 0, r_0)$ we obtain the value $r = f(2\pi, 0, r_0)$, corresponding to the point of Σ where the trajectory $r = f(\varphi, 0, r_0)$ first intersects Σ again.

Definition 3.1.3. Fix a system of the form (3.4).
(a) The function

$$\mathscr{R}(r_0) = f(2\pi, 0, r_0) = \tilde{\eta}_1 r_0 + \eta_2 r_0^2 + \eta_3 r_0^3 + \cdots \tag{3.13}$$

(defined for $|r_0| < r^*$), where $\tilde{\eta}_1 = w_1(2\pi)$ and $\eta_j = w_j(2\pi)$ for $j \geq 2$, is called the *Poincaré first return map* or just the *return map*.
(b) The function

$$\mathscr{P}(r_0) = \mathscr{R}(r_0) - r_0 = \eta_1 r_0 + \eta_2 r_0^2 + \eta_3 r_0^3 + \cdots \tag{3.14}$$

is called the *difference function*.
(c) The coefficient η_j, $j \in \mathbb{N}$, is called the jth *Lyapunov number*.

Note in particular that by (3.12) the first Lyapunov number η_1 has the value $\eta_1 = \tilde{\eta}_1 - 1 = e^{2\pi\alpha/\beta} - 1$. Zeros of (3.14) correspond to *cycles* (closed orbits, that is, orbits that are ovals) of system (3.4); *isolated* zeros correspond to *limit cycles* (isolated closed orbits).

Proposition 3.1.4. *The first nonzero coefficient of the expansion* (3.14) *is the coefficient of an odd power of* r_0.

The proof is left as Exercise 3.3.

The Lyapunov numbers completely determine the behavior of the trajectories of system (3.4) near the origin:

Theorem 3.1.5. *System* (3.4) *has a center at the origin if and only if all the Lyapunov numbers are zero. Moreover, if* $\eta_1 \neq 0$, *or if for some* $k \in \mathbb{N}$

$$\eta_1 = \eta_2 = \cdots = \eta_{2k} = 0, \ \eta_{2k+1} \neq 0, \tag{3.15}$$

then all trajectories in a neighborhood of the origin are spirals and the origin is a focus, which is stable if $\eta_1 < 0$ *or* (3.15) *holds with* $\eta_{2k+1} < 0$ *and is unstable if* $\eta_1 > 0$ *or* (3.15) *holds with* $\eta_{2k+1} > 0$.

Proof. If all the Lyapunov numbers vanish, then by (3.14) the difference function $\mathscr{P}$ is identically zero, so every trajectory in a neighborhood of the origin closes.

If (3.15) holds (including the case $\eta_1 \neq 0$), then

$$\mathscr{P}(r_0) = \eta_{2k+1} r_0^{2k+1} + \sum_{m=2k+2}^{\infty} \alpha_m r_0^m = r_0^{2k+1} \left[\eta_{2k+1} + \widetilde{\mathscr{P}}(r_0) \right], \tag{3.16}$$

where $\widetilde{\mathscr{P}}$ is analytic in a neighborhood of the origin and $\widetilde{\mathscr{P}}(0) = 0$. Suppose that $\eta_{2k+1} < 0$. By (3.16), $r_0 = 0$ is the only zero of the equation $\mathscr{P}(r_0) = 0$ in an open interval about 0. Thus there exists $\bar{r}_0 > 0$ such that

$$\mathscr{P}(r_0) < 0 \qquad \text{for} \qquad 0 < r_0 < \bar{r}_0 . \tag{3.17}$$

Then the trajectory passing through the point (expressed in polar coordinates) $P(0, r_0)$ $(0 < r_0 < \bar{r})$ reaches points $P(0, r_0^{(1)})$, $P(0, r_0^{(2)})$, ... as φ increases through 2π, 4π, ... Here (since $\mathscr{P}$ is the difference function) $r_0^{(1)} = \mathscr{P}(r_0) + r_0$ and $r_0^{(n+1)} = \mathscr{P}(r_0^{(n)}) + r_0^{(n)}$ for $n \geq 1$, and $r_0 > r_0^{(1)} > \cdots > r_0^{(n)} > \cdots > 0$. The sequence $\{r_0^{(n)}\}$ has a limit $\hat{r} \geq 0$, which must in fact be zero, since continuity of $\mathscr{P}(r_0)$ and the computation

$$\lim_{n\to\infty} \mathscr{P}(r_0^{(n)}) = \lim_{n\to\infty} \left(r_0^{(n+1)} - r_0^{(n)} \right) = 0$$

imply that $\mathscr{P}(\hat{r}) = 0$, so $\hat{r} = 0$ follows from (3.17). Therefore, all trajectories in a sufficiently small neighborhood of the origin are spirals and tend to the origin as $\varphi \to +\infty$. Similarly, if $\eta_{2k+1} > 0$, then the trajectories are also all spirals but tend to the origin as $\varphi \to -\infty$. □

Remark 3.1.6. When (3.15) holds for $k > 0$ the focus at the origin is called a *fine focus of order k*. The terminology reflects the fact that in the early literature a hyperbolic singularity was termed "coarse" since it is typically structurally stable, so that a nonhyperbolic singularity is "fine."

Remark 3.1.7. A singularity that is known to be either a node (Definition 3.1.1(b) but $\varphi(t)$ has a finite limit), a focus, or a center is termed an *antisaddle*. Thus Theorem 3.1.5 implies that the origin is an antisaddle of focus or center type for any analytic system of the form (3.4).

Remark. Our conclusions depend in an essential way on our assumption that the functions P and Q on the right-hand side of (3.4) are analytic. If either P or Q is not an analytic function, then the return map need not be analytic, and the function $\mathscr{P}(r_0)$ can have infinitely many isolated zeros that accumulate at zero. When this happens a neighborhood of the origin contains a countably infinite number of limit cycles separated by regions in which all trajectories are spirals or Reeb components of the foliation of a punctured neighborhood of the origin by orbits. Such a singular point is sometimes called a *center-focus*. A concrete example is given by the system $\dot{u} = -v + uh(u^2 + v^2)$, $\dot{v} = u + vh(u^2 + v^2)$, where the function $h : \mathbb{R} \to \mathbb{R}$ is the infinitely flat function defined by $h(0) = 0$ and $h(w) = e^{-w^{-2}} \sin w^{-1}$ for $w \neq 0$. A more exotic example illustrating the range of possibilities for C^∞ (but flat) systems, in which there does not even exist a topological structure at the singularity, is constructed in [171], or consult the appendix of [13].

Remark. If we integrate system (3.10) with initial conditions other than (3.11), then we get a new set of functions $\widehat{w}_j(\varphi)$ and a new set of Lyapunov numbers $\widehat{\eta}_j$. However, the first nonzero Lyapunov number, which determines the behavior of trajectories in a sufficiently small neighborhood of the origin, is the same in each case. That is, if

$$\eta_1 = \cdots = \eta_j = 0,\ \eta_{j+1} \neq 0 \quad \text{and} \quad \widehat{\eta}_1 = \cdots = \widehat{\eta}_k = 0,\ \widehat{\eta}_{k+1} \neq 0,$$

then $k = j$ and $\widehat{\eta}_{k+1} = \eta_{j+1}$.

3.2 Complexification of Real Systems, Normal Forms, and the Center Problem

Since we are investigating the nature of solutions of a real analytic system in the case that the eigenvalues of the linear part at a singularity are complex, hence when the system has the canonical form (3.4), it is natural to attempt to associate to it a two-dimensional complex system that can be profitably studied to gain information about the original real system. There are different ways to do so, some more useful than others. The most convenient and commonly used technique is to begin by considering the real plane (u, v) as the complex line

$$x_1 = u + iv. \tag{3.18}$$

Differentiating (3.18) and applying (3.4), we find that system (3.4) is equivalent to the single complex differential equation

$$\dot{x}_1 = (\alpha + i\beta)x_1 + X_1(x_1, \bar{x}_1), \tag{3.19}$$

where $X_1 = P + iQ$, and P and Q are evaluated at $((x_1 + \bar{x}_1)/2, (x_1 - \bar{x}_1)/(2i))$. At this point we have merely expressed our real system using the notation of complex variables. To obtain a *system* of complex equations in a natural way, we now adjoin to equation (3.19) its complex conjugate, to obtain the pair of equations

$$\begin{aligned} \dot{x}_1 &= (\alpha + i\beta)x_1 + X_1(x_1, \bar{x}_1) \\ \dot{\bar{x}}_1 &= (\alpha - i\beta)\bar{x}_1 + \overline{X_1(x_1, \bar{x}_1)}, \end{aligned} \tag{3.20}$$

where, as the notation indicates, we have taken the complex conjugate of both the coefficients and the variables in $X_1(x_1, \bar{x}_1)$. Thus, for example, from the real system

$$\begin{aligned} \dot{u} &= 2u - 3v + 4u^2 - 8uv \\ \dot{v} &= 3u + 2v + 16u^2 + 12uv \end{aligned}$$

we obtain first

$$\begin{aligned}\dot{x}_1 &= (2+3i)x_1 + 4\left(\frac{x_1+\bar{x}_1}{2}\right)^2 - 8\left(\frac{x_1+\bar{x}_1}{2}\right)\left(\frac{x_1-\bar{x}_1}{2i}\right) \\ &\qquad + i\left[16\left(\frac{x_1+\bar{x}_1}{2}\right)^2 - 12\left(\frac{x_1-\bar{x}_1}{2i}\right)^2\right] \\ &= (2+3i)x_1 + (1+3i)x_1^2 + (2+14i)x_1\bar{x}_1 + (1-i)\bar{x}_1^2,\end{aligned}$$

to which we adjoin

$$\dot{\bar{x}}_1 = (2-3i)\bar{x}_1 + (1+i)x_1^2 + (2-14i)x_1\bar{x}_1 + (1-3i)\bar{x}_1^2.$$

Finally, if we replace $\bar{x}_1$ everywhere by x_2 and regard it as a new complex variable that is independent of x_1, then from (3.20) we obtain a full-fledged system of analytic differential equations on $\mathbb{C}^2$:

$$\begin{aligned}\dot{x}_1 &= (\alpha+i\beta)x_1 + X_1(x_1,x_2) \\ \dot{x}_2 &= (\alpha-i\beta)x_2 + X_2(x_1,x_2)\end{aligned}, \qquad X_2(x_1,\bar{x}_1) = \overline{X_1(x_1,\bar{x}_1)}. \tag{3.21}$$

System (3.21) on $\mathbb{C}^2$ is the *complexification* of the real system (3.4) on $\mathbb{R}^2$. The complex line $\Pi := \{(x_1,x_2) : x_2 = \bar{x}_1\}$ is invariant for system (3.21); viewing Π as a two-dimensional hyperplane in $\mathbb{R}^4$, the flow on Π is precisely the original flow of system (3.4) on $\mathbb{R}^2$ (Exercise 3.6). In this sense the phase portrait of the real system has been embedded in an invariant set in the phase portrait of a complex one. An important point is that (3.21) is a member of the family

$$\begin{aligned}\dot{x}_1 &= \lambda_1 x_1 + X_1(x_1,x_2) \\ \dot{x}_2 &= \lambda_2 x_2 + X_2(x_1,x_2)\end{aligned} \tag{3.22}$$

of systems of analytic differential equations on $\mathbb{C}^2$ with diagonal linear part, in which the eigenvalues need not be complex conjugates, and the higher-order terms can be unrelated. See Exercise 3.7. Two of the first few results in this section will be stated and proved for elements of this more general collection of systems.

In the case of (3.21), since we are assuming that $\beta \neq 0$, the eigenvalues λ_1, λ_2 of the linear part satisfy $\lambda_1\lambda_2 \neq 0$ and

$$\frac{\lambda_2}{\lambda_1} = \left(\frac{\alpha^2-\beta^2}{\alpha^2+\beta^2}\right) - i\left(\frac{2\alpha\beta}{\alpha^2+\beta^2}\right),$$

hence by Exercise 2.13 a necessary condition for the existence of resonant terms is that $\alpha\beta = 0$, hence that $\alpha = 0$. Thus when $\alpha \neq 0$ the hypothesis of Corollary 2.3.3 holds and system (3.21) is formally equivalent to its normal form

$$\begin{aligned}\dot{y}_1 &= (\alpha+i\beta)y_1 \\ \dot{y}_2 &= (\alpha-i\beta)y_2,\end{aligned} \tag{3.23}$$

a decoupled system in which each component spirals onto or away from $\mathbf{0} \in \mathbb{C}$, reflecting the fact that the underlying real system has a focus at the origin of $\mathbb{R}^2$ in this case.

We wish to know the normal form for system (3.21) and conditions guaranteeing that the normalizing transformation (2.33) that changes it into its normal form is convergent, rather than merely formal. We will usually restrict to distinguished normalizing transformations, hence by Theorem 2.3.11 (which applies because the linear part in (3.22) is diagonal) can speak of *the* normal form and *the* distinguished normalizing transformation that produces it. However, we will allow more general normalizing transformations when it is not much more trouble to do so.

Proposition 3.2.1. *Fix an analytic real system* (3.4) *for which* $\beta \neq 0$ *(our standing assumption throughout this chapter).*

1. *The normal form of the complexification* (3.21) *of system* (3.4) *that is produced by the unique distinguished transformation is the complexification of a real system, and its coefficients satisfy* $Y_2^{(j,k)} = \overline{Y_1^{(k,j)}}$ *for all* $(j,k) \in \mathbb{N}_0^2$, $j+k \geq 2$. *Moreover, the distinguished normalizing transformation satisfies* $h_2^{(j,k)} = \overline{h_1^{(k,j)}}$ *for all* $(j,k) \in \mathbb{N}_0^2$, $j+k \geq 2$.
2. *More generally, if* $\mathbf{x} = \mathbf{y} + \mathbf{h}(\mathbf{y})$ *is any normalizing transformation of the complexification* (3.21) *of* (3.4), *chosen so as to satisfy the condition* $h_2^{(j,k)} = \overline{h_1^{(k,j)}}$ *for every resonant pair* $(1,(k,j))$ *(which is always possible), then the corresponding normal form is the complexification of a real system, its coefficients satisfy* $Y_2^{(j,k)} = \overline{Y_1^{(k,j)}}$ *for all* $(j,k) \in \mathbb{N}_0^2$, $j+k \geq 2$, *and the normalizing transformation satisfies* $h_2^{(j,k)} = \overline{h_1^{(k,j)}}$ *for all* $(j,k) \in \mathbb{N}_0^2$, $j+k \geq 2$.

Proof. If eigenvalues of the linear part of the real system have nonzero real part, then as we just saw the normal form is (3.23), which is the complexification of the linearization of (3.4), $\dot{u} = \alpha u - \beta v$, $\dot{v} = \beta u + \alpha v$. The unique normalizing transformation $\mathbf{x} = \mathbf{y} + \mathbf{h}(\mathbf{y})$ is vacuously distinguished. The fact that its coefficients satisfy $h_2^{(j,k)} = \overline{h_1^{(k,j)}}$ for all $(j,k) \in \mathbb{N}_0^2$, $j+k \geq 2$, follows from the arguments that will be given in the proof of the proposition in the case that $\alpha = 0$.

Thus suppose that the eigenvalues are purely imaginary, so that in the notation of Section 2.3, $\kappa = (\kappa_1, \kappa_2) = (\beta i, -\beta i)$. We write $X_1(x_1,x_2) = \Sigma X_1^{(j,k)} x_1^j x_2^k$ and $X_2(x_1,x_2) = \Sigma X_2^{(j,k)} x_1^j x_2^k$, following the notational convention established in the same section. Let Y_1 and Y_2 denote the higher-order terms in the normal form of (3.21) and use the same notation $Y_1^{(j,k)}$ and $Y_2^{(j,k)}$ for their respective coefficients. By Exercise 3.7, the hypothesis is that $X_2^{(j,k)} = \overline{X_1^{(k,j)}}$ for all $(j,k) \in \mathbb{N}_0^2$, $j+k \geq 2$. An easy computation using the definition of resonance and the fact that $(\kappa_1, \kappa_2) = (\beta i, -\beta i)$ shows that for $(j,k) \in \mathbb{N}_0^2$, $j+k \geq 2$, 1 and (k,j) form a resonant pair if and only if 2 and (j,k) form a resonant pair; this shows that the choice described in statement (2) of the proposition is always possible. Since statement (2) implies statement (1), the proposition will be fully proved if we can establish that for any transformation $\mathbf{x} = \mathbf{y} + \mathbf{h}(\mathbf{y})$ that normalizes (3.21), the identities

$$Y_2^{(j,k)} = \overline{Y_1^{(k,j)}} \tag{3.24a}$$

$$h_2^{(j,k)} = \overline{h_1^{(k,j)}} \tag{3.24b}$$

hold for all pairs $(j,k) \in \mathbb{N}_0^2$, $j+k \geq 2$, provided (3.24b) is required when 1 and (k,j) form a resonant pair (but $h_1^{(k,j)}$ and $h_2^{(j,k)}$ are otherwise arbitrary).

The coefficients $Y_m^{(j,k)}$ and $h_m^{(j,k)}$ are computed using (2.40) and (2.41). The proof will be by induction on $s = j+k$. Recall that for $m = 1$, 2 and a multi-index α, $\{X_m(\mathbf{y}+\mathbf{h}(\mathbf{y}))\}^{(\alpha)}$ denotes the coefficient of $\mathbf{y}^{(\alpha)}$ when $X_m(\mathbf{y}+\mathbf{h}(\mathbf{y}))$ is expanded in powers of $\mathbf{y}$. To carry out the proof we will simultaneously prove by induction that

$$\{X_2(\mathbf{y}+\mathbf{h}(\mathbf{y}))\}^{(j,k)} = \overline{\{X_1(\mathbf{y}+\mathbf{h}(\mathbf{y}))\}^{(k,j)}}\,. \tag{3.25}$$

Basis step. A simple set of computations using Definition 2.3.5 shows that for $s = 2$ there are no resonant terms, so $Y_m^{(j,k)} = 0$ and $g_m^{(j,k)} = X_m^{(j,k)}$ (comment after (2.41)). We leave it to the reader to compute the coefficients $h_m^{(j,k)}$ and verify the truth of (3.24b) in this case. Since $\mathbf{X}$ and $\mathbf{h}$ begin with quadratic terms, the truth of (3.25) for $s = 2$ is immediate.

Inductive step. Suppose (3.24) and (3.25) hold for $j+k \leq s$. (It is understood that we make the choice that (3.24b) holds when 1 and (k,j) form a resonant pair.) First we establish the truth of (3.25) for $j+k = s+1$. Let $\overset{(s+1)}{\sim}$ denote agreement of series through order $s+1$. Then

$$\begin{aligned}
&\sum_{j+k\geq 2} \{X_1(\mathbf{y}+\mathbf{h}(\mathbf{y}))\}^{(j,k)} y_1^j y_2^k \\
&= X_1(y_1 + h_1(y_1,y_2), y_2 + h_2(y_1,y_2)) \\
&= \sum_{r+t\geq 2} X_1^{(r,t)} \big(y_1 + h_1(\mathbf{y})\big)^r \big(y_2 + h_2(\mathbf{y})\big)^t \\
&\overset{(s+1)}{\sim} \sum_{2\leq r+t\leq s+1} X_1^{(r,t)} \Big(y_1 + \sum_{2\leq|\alpha|\leq s} h_1^{(\alpha)} y_1^{\alpha_1} y_2^{\alpha_2}\Big)^r \Big(y_2 + \sum_{2\leq|\alpha|\leq s} h_2^{(\alpha)} y_1^{\alpha_1} y_2^{\alpha_2}\Big)^t .
\end{aligned}$$

We apply to each term in the sum the involution $y_1 \leftrightarrow y_2$ and obtain

$$\sum_{2\leq r+t\leq s+1} X_1^{(r,t)} \Big(y_2 + \sum_{2\leq|\alpha|\leq s} h_1^{(\alpha)} y_2^{\alpha_1} y_1^{\alpha_2}\Big)^r \Big(y_1 + \sum_{2\leq|\alpha|\leq s} h_2^{(\alpha)} y_2^{\alpha_1} y_1^{\alpha_2}\Big)^t .$$

Now we conjugate the coefficients, obtaining

$$\sum_{2\leq r+t\leq s+1} \bar{X}_1^{(r,t)} \Big(y_2 + \sum_{2\leq|\alpha|\leq s} \bar{h}_1^{(\alpha)} y_2^{\alpha_1} y_1^{\alpha_2}\Big)^r \Big(y_1 + \sum_{2\leq|\alpha|\leq s} \bar{h}_2^{(\alpha)} y_2^{\alpha_1} y_1^{\alpha_2}\Big)^t$$

which, applying the hypothesis on X_1 and X_2 and the induction hypothesis, is

$$
\begin{aligned}
&= \sum_{2\le r+t\le s+1} X_2^{(t,r)}\Big(y_2+\sum_{2\le|\alpha|\le s} h_2^{(\alpha_2,\alpha_1)} y_2^{\alpha_1}y_1^{\alpha_2}\Big)^r\Big(y_1+\sum_{2\le|\alpha|\le s} h_1^{(\alpha_2,\alpha_1)} y_2^{\alpha_1}y_1^{\alpha_2}\Big)^t \\
&\overset{(s+1)}{\sim} \sum_{r+t\ge 2} X_2^{(t,r)}(y_1+h_1(y_1,y_2))^t(y_2+h_2(y_1,y_2))^r \\
&= X_2(y_1+h_1(y_1,y_2),y_2+h_2(y_1,y_2)) \\
&= \sum_{j+k\ge 2}\{X_2(\mathbf{y}+\mathbf{h}(\mathbf{y}))\}^{(j,k)} y_1^j y_2^k,
\end{aligned}
$$

so that (3.25) holds for $j+k=s+1$.

Turning our attention to $g_m^{(j,k)}$, from (2.41) we compute the value of $\overline{g_1^{(k,j)}}$ as

$$
\overline{\{X_1(\mathbf{y}+\mathbf{h}(\mathbf{y}))\}^{(k,j)}}-\sum\beta_1\overline{h_1^{(\beta_1,\beta_2)}}\,\overline{Y_1^{(k-\beta_1+1,j-\beta_2)}}-\sum\beta_2\overline{h_1^{(\beta_1,\beta_2)}}\,\overline{Y_1^{(k-\beta_1,j-\beta_2+1)}},
$$

which by (3.25) and the induction hypothesis is

$$
\{X_2(\mathbf{y}+\mathbf{h}(\mathbf{y}))\}^{(j,k)}-\sum\beta_1 h_2^{(\beta_2,\beta_1)}Y_2^{(j-\beta_2,k-\beta_1+1)}-\sum\beta_2 h_2^{(\beta_2,\beta_1)}Y_2^{(j-\beta_2+1,k-\beta_1)},
$$

which is $g_2^{(j,k)}$, so that

$$
g_2^{(j,k)}=\overline{g_1^{(k,j)}}. \tag{3.26}
$$

Since 1 and (k,j) form a resonant pair if and only if 2 and (j,k) form a resonant pair, $Y_1^{(k,j)}\neq 0$ if and only if $Y_2^{(j,k)}\neq 0$ and we may freely choose $h_1^{(k,j)}$ if and only if we may freely choose $h_2^{(j,k)}$. Thus for any (j,k) with $j+k=s+1$, by (2.40) we see that when they are nonzero,

$$
\overline{Y_1^{(k,j)}}=\overline{g_1^{(k,j)}}=g_2^{(j,k)}=Y_2^{(j,k)}
$$

by (3.26), so that (3.24a) holds, and (again by (2.40)) when its value is forced

$$
h_m^{(j,k)}=\frac{g_m^{(j,k)}}{((j,k),(\kappa_1,\kappa_2))-\kappa_m},
$$

so that by (3.26) and the fact that $(\kappa_1,\kappa_2)=(i\beta,-i\beta)$

$$
\begin{aligned}
\overline{h_1^{(k,j)}} &= \overline{\left[\frac{g_1^{(k,j)}}{((k,j),(i\beta,-i\beta))-i\beta}\right]}=\overline{\left[\frac{g_1^{(k,j)}}{i\beta(k-j-1)}\right]} \\
&= \frac{\overline{g_1^{(k,j)}}}{i\beta(j-k+1)}=\frac{g_2^{(j,k)}}{((j,k),(i\beta,-i\beta))+i\beta}=h_2^{(j,k)},
\end{aligned}
$$

so that (3.24b) holds. Thus (3.24) is true for the pair (j,k). □

It is apparent that if a real system is complexified and that system is transformed into normal form, then the real system that the normal form represents is a transformation of the original real system.

The first two results that we state apply to more general systems than those that arise as complexifications of real systems. The condition that λ_1 and λ_2 be rationally related is connected to the condition for resonance described in Exercise 2.13.

Proposition 3.2.2. *Suppose that in the system* (3.22) $\lambda_1/\lambda_2 = -p/q$ *for* $p,q \in \mathbb{N}$ *with* $\mathrm{GCD}(p,q) = 1$.

1. *The normal form for* (3.22) *produced by any normalizing transformation* (2.33) *(not necessarily satisfying the condition of Proposition 3.2.1(2), hence in particular not necessarily distinguished) is of the form (using the notational convention introduced just before Theorem 2.3.8)*

$$\begin{aligned}\dot{y}_1 &= \lambda_1 y_1 + y_1 \sum_{j=1}^{\infty} Y_1^{(jq+1,jp)} (y_1^q y_2^p)^j = \lambda_1 y_1 + y_1 Y_1(y_1^q y_2^p)\\ \dot{y}_2 &= \lambda_2 y_2 + y_2 \sum_{j=1}^{\infty} Y_2^{(jq,jp+1)} (y_1^q y_2^p)^j = \lambda_2 y_2 + y_2 Y_2(y_1^q y_2^p)\,.\end{aligned} \tag{3.27}$$

2. *Let* $Y_1(w)$ *and* $Y_2(w)$ *be the functions of a single complex variable* w *defined by* (3.27). *If the normalizing transformation that produced* (3.27) *is distinguished, and if* $qY_1(w) + pY_2(w) \equiv 0$, *then the normalizing transformation is convergent.*

Proof. That the normal form for (3.22) has the form (3.27) follows immediately from Definitions 2.3.5 and 2.3.6 and a simple computation that shows that the resonant pairs are $(1,(jq+1,jp))$ and $(2,(jq,jp+1))$ for $j \in \mathbb{N}$.

To establish point (2) we use Theorem 2.3.13. Let $\lambda = (\lambda_1, \lambda_2)$. To obtain a uniform lower bound for all nonzero $|(\alpha,\lambda) - \lambda_m|$, $m = 1,2$, define $E_m : \mathbb{R}^2 \to \mathbb{R}$ for $m = 1,2$ by

$$\begin{aligned}E_1(x,y) &= |((x,y),(\lambda_1,\lambda_2)) - \lambda_1| = \tfrac{1}{q}|\lambda_2|\,|px - qy - p|\\ E_2(x,y) &= |((x,y),(\lambda_1,\lambda_2)) - \lambda_2| = \tfrac{1}{q}|\lambda_2|\,|px - qy + q|.\end{aligned}$$

Then for both $m = 1$ and $m = 2$, $E_m(x+q, y+p) = E_m(x,y)$, so the minimum nonzero value of E_m on $\mathbb{N}_0 \times \mathbb{N}_0$ is the same as the minimum nonzero value of E_m on $\{1,2,\ldots,q+1\} \times \{0,1,\ldots,p\}$, which is some positive constant ε_m. Thus (2.55) holds with $\varepsilon = \min\{\varepsilon_1, \varepsilon_2\}$.

Since p, $q \geq 1$, if $qY_1(w) + pY_2(w) \equiv 0$ we have that, for any multi-indices β, $\gamma \in \mathbb{N}_0^2$ for which $|\beta| \geq 2$ and $|\gamma| \geq 2$,

$$\begin{aligned}\left|\beta_1 Y_1^{(\gamma)} + \beta_2 Y_2^{(\gamma)}\right| &= \left|\left(\beta_1 - \frac{q}{p}\beta_2\right) Y_1^{(\gamma)}\right| = \frac{1}{|\lambda_1|}|(\beta,\lambda)|\left|Y_1^{(\gamma)}\right|\\ &\leq \frac{1}{|\lambda_1|}|(\beta,\lambda)|\left(\left|Y_1^{(\gamma)}\right| + \frac{q}{p}\left|Y_1^{(\gamma)}\right|\right) = \frac{1}{|\lambda_1|}|(\beta,\lambda)|\left(\left|Y_1^{(\gamma)}\right| + \left|Y_2^{(\gamma)}\right|\right).\end{aligned}$$

We conclude that condition (2.57) holds with $d = 1/|\lambda_1|$. □

For any normalization (3.27) of a system (3.22) that satisfies the hypotheses of Proposition 3.2.2 we let

$$G(w) = \sum_{k=1}^{\infty} G_{2k+1} w^k, \qquad H(w) = \sum_{k=1}^{\infty} H_{2k+1} w^k \tag{3.28a}$$

be the functions of the complex variable w defined by

$$G = qY_1 + pY_2, \qquad H = qY_1 - pY_2. \tag{3.28b}$$

The significance of the function G is shown by the next two theorems. The significance of H will be revealed in the next chapter. But first we need an important definition. It is motivated by the geometric considerations described in the introduction to this chapter, namely that we should be able to realize the ovals that surround a center as level curves of a smooth function. That our intuition is correct will be borne out by Theorems 3.2.9 and 3.2.10.

Definition 3.2.3. A *first integral* on an open set Ω in $\mathbb{R}^n$ or $\mathbb{C}^n$ of a smooth or analytic system of differential equations

$$\dot{x}_1 = f_1(\mathbf{x}), \ldots, \dot{x}_n = f_n(\mathbf{x}) \tag{3.29}$$

defined everywhere on Ω is a nonconstant differentiable function $\Psi : \Omega \to \mathbb{C}$ that is constant on trajectories (that is, for any solution $\mathbf{x}(t)$ of (3.29) in Ω the function $\psi(t) = \Psi(\mathbf{x}(t))$ is constant). A *formal first integral* is a formal power series in $\mathbf{x}$, not all of whose coefficients are zero, which under term-by-term differentiation satisfies $\frac{d}{dt}[\Psi(\mathbf{x}(t))] \equiv 0$ in Ω .

Remark 3.2.4. (a) If Ψ is a first integral or formal first integral for system (3.29) on Ω, if $F : \mathbb{C} \to \mathbb{C}$ is any nonconstant differentiable function, and if λ is any constant, then $\Phi = F \circ \Psi$ is a first integral or formal first integral for the system $\dot{x}_1 = \lambda f_1(\mathbf{x}), \ldots, \dot{x}_n = \lambda f_n(\mathbf{x})$ on Ω.
(b) If

$$\mathscr{X}(\mathbf{x}) = f_1(\mathbf{x}) \frac{\partial}{\partial x_1} + \cdots + f_n(\mathbf{x}) \frac{\partial}{\partial x_n} \tag{3.30}$$

is the smooth or analytic vector field on Ω associated to system (3.29), then a nonconstant differentiable function (or formal powers series) Ψ on Ω is a first integral (or formal first integral) for (3.29) if and only if the function $\mathscr{X}\Psi$ vanishes throughout Ω:

$$\mathscr{X}\Psi = f_1 \frac{\partial \Psi}{\partial x_1} + \cdots + f_n \frac{\partial \Psi}{\partial x_n} \equiv 0 \quad \text{on} \quad \Omega. \tag{3.31}$$

(c) Our concern is only with a neighborhood of the origin, so by "existence of a first integral" we will always mean "existence on a neighborhood of the origin."

We now show that the existence of a formal first integral is enough to guarantee that $G(w) = qY_1(w) + pY_2(w) \equiv 0$, hence by Proposition 3.2.2(2) that the normalizing transformation that transforms (3.22) to (3.27), if distinguished, is convergent. Note that if Ψ is any formal first integral of a system (3.22) that meets the conditions of Proposition 3.2.2, does not have a constant term, and begins with terms of order no higher than $p+q$, then Ψ must have the form $\Psi(x_1,x_2) = x_1^q x_2^p + \cdots$ (Exercise 3.8). The assertion made here and in the following theorem about the form of the function $\Psi(x_1,x_2)$ means only that up to terms of degree $p+q$ it has the form indicated, not that it is a function of the product $x_1^q x_2^p$ alone.

Theorem 3.2.5. *Suppose that in system* (3.22) $\lambda_1/\lambda_2 = -p/q$ *for* $p,q \in \mathbb{N}$ *with* $\mathrm{GCD}(p,q) = 1$. *Let* G *be the function defined by* (3.28b), *computed from some normal form of* (3.22).

1. *If system* (3.22) *has a formal first integral of the form* $\Psi = x_1^q x_2^p + \cdots$, *then* $G \equiv 0$. *Thus if the normalizing transformation in question is distinguished, it is convergent.*
2. *Conversely, if* $G \equiv 0$ *then system* (3.22) *has a formal first integral of the form* $\Psi(x_1,x_2) = x_1^q x_2^p + \cdots$, *which is analytic when the normalizing transformation is distinguished.*

Proof. By Remarks 2.3.7 and 3.2.4(a), we may assume that $\lambda_1 = p$ and $\lambda_2 = -q$. Suppose system (3.22) has a formal first integral of the form $\Psi(x_1,x_2) = x_1^q x_2^p + \cdots$. If $\mathbf{H}$ is the normalizing transformation that converts (3.22) into its normal form (3.27), then $F = \Psi \circ \mathbf{H}$ is a formal first integral for the normal form, hence

$$\frac{\partial F}{\partial y_1}(y_1,y_2)\left[py_1 + y_1Y_1(y_1^q y_2^p)\right] + \frac{\partial F}{\partial y_2}(y_1,y_2)\left[-qy_2 + y_2Y_2(y_1^q y_2^p)\right] \equiv 0, \tag{3.32}$$

which we rearrange as

$$\begin{aligned} py_1\frac{\partial F}{\partial y_1}(y_1,y_2) &- qy_2\frac{\partial F}{\partial y_2}(y_1,y_2) \\ &= -y_1\frac{\partial F}{\partial y_1}(y_1,y_2)Y_1(y_1^q y_2^p) - y_2\frac{\partial F}{\partial y_2}(y_1,y_2)Y_2(y_1^q y_2^p). \end{aligned} \tag{3.33}$$

Recalling from (2.33) the form of $\mathbf{H}$ and writing F according to our usual convention, $F(y_1,y_2)$ has the form

$$F(y_1,y_2) = \sum_{(\alpha_1,\alpha_2)} F^{(\alpha_1,\alpha_2)} y_1^{\alpha_1} y_2^{\alpha_2} = y_1^q y_2^p + \cdots. \tag{3.34}$$

A simple computation on the left-hand side of (3.33), and insertion of (3.27) into the right, yields

$$\sum_{(\alpha_1,\alpha_2)} (\alpha_1 p - \alpha_2 q) F^{(\alpha_1,\alpha_2)} y_1^{\alpha_1} y_2^{\alpha_2}$$
$$= -\left[\sum_{(\alpha_1,\alpha_2)} \alpha_1 F^{(\alpha_1,\alpha_2)} y_1^{\alpha_1} y_2^{\alpha_2}\right]\left[\sum_{j=1}^{\infty} Y_1^{(jq+1,jp)} (y_1^q y_2^p)^j\right] \qquad (3.35)$$
$$- \left[\sum_{(\alpha_1,\alpha_2)} \alpha_2 F^{(\alpha_1,\alpha_2)} y_1^{\alpha_1} y_2^{\alpha_2}\right]\left[\sum_{j=1}^{\infty} Y_2^{(jq,jp+1)} (y_1^q y_2^p)^j\right].$$

We claim that $F(y_1,y_2)$ is a function of $y_1^q y_2^p$ alone, so that it may be written $F(y_1,y_2) = f(y_1^q y_2^p) = f_1 y_1^q y_2^p + f_2(y_1^q y_2^p)^2 + \cdots$. The claim is precisely the statement that for any term $F^{(\alpha_1,\alpha_2)} y_1^{\alpha_1} y_2^{\alpha_2}$ of F,

$$p\alpha_1 - q\alpha_2 \neq 0 \quad \text{implies} \quad F^{(\alpha_1,\alpha_2)} = 0. \qquad (3.36)$$

Equation (3.34) shows that (3.36) holds for $|(\alpha_1,\alpha_2)| = \alpha_1 + \alpha_2 \leq p+q$. This implies that the right-hand side of (3.35) has the form $c_2(y_1^q y_2^p)^2 + \cdots$ for some c_2, hence by (3.35) implication (3.36) holds for $\alpha_1 + \alpha_2 \leq 2(p+q)$. But if that is true, then it must be the case that the right-hand side of (3.35) has the form $c_2(y_1^q y_2^p)^2 + c_3(y_1^q y_2^p)^3 + \cdots$ for some c_3, hence by (3.35) implication (3.36) must hold for $\alpha_1 + \alpha_2 \leq 3(p+q)$. Clearly by mathematical induction (3.36) must hold in general, establishing the claim.

But if $F(y_1,y_2) = f(y_1^q y_2^p)$, then

$$y_1 \frac{\partial F}{\partial y_1}(y_1,y_2) = q y_1^q y_2^p f'(y_1^q y_2^p) \quad \text{and} \quad y_2 \frac{\partial F}{\partial y_2}(y_1,y_2) = p y_1^q y_2^p f'(y_1^q y_2^p),$$

so that, letting $w = y_1^q y_2^p$, (3.33) becomes

$$0 \equiv -qwf'(w)Y_1(w) - pwf'(w)Y_2(w).$$

But because F is a formal first integral it is not a constant, so we immediately obtain $qY_1(w) + pY_2(w) \equiv 0$, which in conjunction with part (2) of Proposition 3.2.2 proves part (1).

Direct calculations show that if $G \equiv 0$, then $\widehat{\Psi}(y_1,y_2) = y_1^q y_2^p$ is a first integral of (3.27). The coordinate transformation that places (3.22) in normal form has the form given in (2.33), hence has an inverse of the form $\mathbf{y} = \mathbf{x} + \widehat{\mathbf{h}}(\mathbf{x})$. Therefore system (3.22) admits a formal first integral of the form $\Psi(x_1,x_2) = x_1^q x_2^p + \cdots$. By part (2) of Proposition 3.2.2, if the transformation to the normal form (3.27) is distinguished, then it is convergent, hence so is $\Psi(x_1,x_2)$. □

Corollary 3.2.6. *An analytic system* (3.22) *possesses a formal first integral of the form* $\Psi(x_1,x_2) = x_1^q x_2^p + \cdots$ *only if it possesses an analytic first integral of that form.*

Proof. If there exists a formal first integral of the form $\Psi(x_1,x_2) = x_1^q x_2^p + \cdots$, then part (1) of the theorem implies that for any normalizing transformation the corresponding function G vanishes identically. In particular this is true when the normalizing transformation is distinguished. But then by part (2) of the theorem there exists an analytic first integral of the same form. □

Now we specialize to the case that arises from our original problem, the study of a real system with a singularity at which the eigenvalues of the linear part are purely imaginary, $\alpha \pm i\beta = \pm i\omega$, $\omega \in \mathbb{R} \setminus \{0\}$. By a translation and an invertible linear transformation to produce (3.4) and the process described at the beginning of this section, we obtain the complexification

$$\begin{aligned} \dot{x}_1 &= \quad i\omega x_1 + X_1(x_1, x_2) \\ \dot{x}_2 &= -i\omega x_2 + X_2(x_1, x_2) \end{aligned}, \qquad X_2(x_1, \bar{x}_1) = \overline{X_1(x_1, \bar{x}_1)}. \tag{3.37}$$

By Proposition 3.2.2 with $p = q = 1$, any normal form of (3.37) has the form

$$\begin{aligned} \dot{y}_1 &= \quad i\omega y_1 + y_1 Y_1(y_1 y_2) \\ \dot{y}_2 &= -i\omega y_2 + y_2 Y_2(y_1 y_2). \end{aligned} \tag{3.38}$$

Theorem 3.2.7. *Let* (3.38) *be a normal form of* (3.37) *that is produced by some normalizing transformation, and let G be the function computed according to* (3.28b).
1. *If $G \equiv 0$, then the original real system* (3.4) *generating* (3.37) *has a center at the origin.*
2. *Conversely, if $G = G_{2k+1}(y_1 \bar{y}_1)^k + \cdots$, $G_{2k+1} \neq 0$, and if the normalizing transformation satisfies the condition of Proposition 3.2.1(2), then the origin is a focus for* (3.4)*, which is stable if $G_{2k+1} < 0$ and unstable if $G_{2k+1} > 0$.*

Proof. Let G be the function computed according to (3.28b) with respect to some normalization (3.38) of (3.37), and suppose that $G \equiv 0$. By part (2) of Theorem 3.2.5, system (3.37) has a formal first integral of the form $\Psi(x_1, x_2) = x_1 x_2 + \cdots$, hence by Corollary 3.2.6 it has an analytic first integral of the form $\widehat{\Psi}(x_1, x_2) = x_1 x_2 + \cdots$. But then system (3.4) has an analytic first integral of the form $\Phi(u, v) = u^2 + v^2 + \cdots$ (Exercise 3.9), which implies that the origin is a center for (3.4) (Exercise 3.10).

Now suppose that the normalizing transformation satisfies the condition of part (2) of Proposition 3.2.1 and that $G(y_1, y_2) = G_{2k+1}(y_1 y_2)^k + \cdots$, $G_{2k+1} \neq 0$. We can avoid questions of convergence by breaking off the series that defines $\mathbf{h}$ in the normalizing transformation (2.33) with terms of order $2k+1$, thereby computing just an initial segment of the normal form that is nevertheless sufficiently long.

By Proposition 3.2.1(2) and Exercise 3.7, in the normal form (3.38) the coefficients satsify $Y_2^{(j,j+1)} = \overline{Y_1^{(j+1,j)}}$ for all j, so the hypothesis on G implies that $Y_1^{(j+1,j)}$ is purely imaginary for $1 \leq j \leq k-1$. Hence there are real numbers $b_1, \ldots, b_{k-1}$ so that the normal form (3.38) is

$$\begin{aligned} \dot{y}_1 &= \quad i\omega y_1 + i(b_1 y_1 y_2 + \cdots + b_{k-1}(y_1 y_2)^{k-1}) y_1 + y_1 Y_1^{(k+1,k)}(y_1 y_2)^k + \cdots \\ \dot{y}_2 &= -i\omega y_2 - i(b_1 y_1 y_2 + \cdots + b_{k-1}(y_1 y_2)^{k-1}) y_2 + y_1 \overline{Y_1^{(k,k+1)}}(y_1 y_2)^k + \cdots. \end{aligned} \tag{3.39}$$

We appeal to Proposition 3.2.1 once more, reverting to the real system represented by system (3.39) by replacing every occurrence of y_2 in (3.39) by $\bar{y}_1$, thereby obtaining two equations that describe the underlying real system, after some transformation of coordinates, in complex notation. Making the substitution in (3.39), we use both equations to compute the derivatives of r and φ in complex polar coordinates $y_1 = re^{i\varphi}$, obtaining

$$\begin{aligned} \dot{r} &= \tfrac{1}{2r}(\dot{y}_1\bar{y}_1 + y_1\dot{\bar{y}}_1) = \tfrac{1}{2}G_{2k+1}r^{2k+1} + o(r^{2k+1}) \\ \dot{\varphi} &= \tfrac{i}{2r^2}(y_1\dot{\bar{y}}_1 - \dot{y}_1\bar{y}_1) = \omega + b_1 r^2 + o(r^2). \end{aligned} \tag{3.40}$$

The equation of the trajectories of system (3.40) is

$$\frac{dr}{d\varphi} = \frac{G_{2k+1}r^{2k} + o(r^{2k})}{2\omega + o(r)}. \tag{3.41}$$

This equation has the form of (3.6) and, just as for (3.6), we conclude that the orbits of (3.40) are spirals, hence that the origin is a focus. By the first equation of (3.40) the focus is stable if $G_{2k+1} < 0$ and unstable if $G_{2k+1} > 0$. □

Remark 3.2.8. The observation made in the proof that $Y_2^{(j,j+1)} = \overline{Y_1^{(j+1,j)}}$ for all $j \in \mathbb{N}$ implies that when $G \equiv 0$ all the coefficients in Y_1 and Y_2 have real part zero, which in turn implies that the normal form (3.38) of (3.37) may be written

$$\begin{aligned} \dot{y}_1 &= \quad i\omega y_1 + \tfrac{1}{2}y_1 H(y_1, y_2) \\ \dot{y}_2 &= -i\omega y_2 - \tfrac{1}{2}y_2 H(y_1, y_2), \end{aligned}$$

where H is the function defined by (3.28), and in this case has only purely imaginary coefficients as well.

We close this section with two theorems that give characterizations of real planar systems of the form (3.2) that have a center at the origin. The first one was first proved by Poincaré ([143]) in the case that U and V are polynomials. Lyapunov ([114]) generalized it to the case that U and V are real analytic functions. The assertion about the form of the function $\Psi(u,v)$ in the first theorem means only that up to quadratic terms does it have the form indicated, not that it is a function of the product sum u^2+v^2 alone. In the Russian mathematical literature a first integral of this form for system (3.2), and the corresponding first integral of the form $\Psi(x_1,x_2) = x_1x_2 + \cdots$ for the complexification of (3.2) (see the first line of the proof), are sometimes referred to as "Lyapunov first integrals."

Theorem 3.2.9 (Poincaré–Lyapunov Theorem). *System* (3.2) *on* $\mathbb{R}^2$ *has a center at the origin if and only if there exists a formal first integral that has the form* $\Psi(u,v) = u^2 + v^2 + \cdots$.

Proof. By Exercise 3.9, the real system (3.2) has a formal first integral of the form $\Psi(u,v) = u^2+v^2+\cdots$ if and only if its complexification (3.37) (with $\omega = 1$) has a formal first integral of the form $\Phi(x_1,x_2) = x_1x_2 + \cdots$.

If system (3.2) has a formal first integral of the form $\Psi(u,v) = u^2 + v^2 + \cdots$, then by Theorem 3.2.5(1) $G \equiv 0$, hence by Theorem 3.2.7 system (3.2) has a center at the origin.

Conversely, if the origin is a center for (3.2), then the function G defined by (3.28) is identically zero (since otherwise by Theorem 3.2.7 the origin is a focus). Therefore, by Theorem 3.2.5(2), its complexification, system (3.37), possesses an analytic local first integral of the form $\Phi(x_1,x_2) = x_1x_2 + \cdots$, hence system (3.2) admits a first integral of the form $\Psi(u,v) = u^2 + v^2 + \cdots$. □

Remark. The analogue of the Poincaré–Lyapunov Theorem fails when the linear part at the origin vanishes identically. That is, it is possible for a system of polynomial differential equations on $\mathbb{R}^2$ to have a center at the origin, but there be no formal first integral in any neighborhood of the origin. See Exercise 3.11. However, there must exist a C^∞ first integral ([137]).

The last theorem of this section states that system (3.2) has a center at the origin if and only if it can be placed in a special form by an analytic coordinate change. This is not the normal form in the sense of Definition 2.3.6, since the complex eigenvalues of the linear part of (3.2) automatically mean that its normal form is complex, since the Jordan normal form of the linear part is complex. Nevertheless, going over to the complexification is a useful step in establishing the result.

Theorem 3.2.10. *The origin is a center for system* (3.2) *if and only if there is a transformation of the form*

$$\begin{aligned} u &= \xi + \phi(\xi,\eta) \\ v &= \eta + \psi(\xi,\eta) \end{aligned} \tag{3.42}$$

that is real analytic in a neighborhood of the origin and transforms (3.2) *into the system*

$$\begin{aligned} \dot{\xi} &= -\eta F(\xi^2+\eta^2) \\ \dot{\eta} &= \quad \xi F(\xi^2+\eta^2), \end{aligned} \tag{3.43}$$

where $F(z)$ is real analytic in a neighborhood of the origin and $F(0) = 1$.

Proof. By Proposition 3.2.2, the normal form of the complexification of (3.2) is

$$\begin{aligned} \dot{y}_1 &= \quad iy_1 + y_1Y_1(y_1y_2) \\ \dot{y}_2 &= -iy_2 + y_2Y_2(y_1y_2). \end{aligned} \tag{3.44}$$

By Proposition 3.2.1, this is the complexification of a real system, which we can recover by making the substitutions $y_1 = \xi + i\eta$ and $y_2 = \xi - i\eta$ in (3.44) and applying them to $\dot{\xi} = \frac{1}{2}(\dot{y}_1 + \dot{y}_2)$ and $\dot{\eta} = \frac{1}{2i}(\dot{y}_1 - \dot{y}_2)$. Direct computation gives

$$\begin{aligned} \dot{\xi} &= \xi\Phi_2(\xi^2+\eta^2) - \eta\Phi_1(\xi^2+\eta^2) \\ \dot{\eta} &= \xi\Phi_1(\xi^2+\eta^2) + \eta\Phi_2(\xi^2+\eta^2), \end{aligned} \tag{3.45}$$

where

$$\Phi_1(w) = 1 - \tfrac{i}{2}(Y_1(w) - Y_2(w)) \quad \text{and} \quad \Phi_2(w) = \tfrac{1}{2}(Y_1(w) + Y_2(w)),$$

so that $\Phi_1(w)$ and $\Phi_2(w)$ are formal series. The process by which it was derived means that system (3.45) is a transformation of the original system (3.2), and from the fact that $\Phi_1(0) = 1$ and $\Phi_2(0) = 0$ we conclude that it has the form (3.42).

Suppose (3.2) has a center at the origin. Then by Theorem 3.2.7, $Y_1 + Y_2 \equiv 0$, so that $\Phi_2 \equiv 0$. Hence every real system with a center can be brought into the form (3.43) by a substitution (3.42), where the series $\phi(\xi,\eta)$, $\psi(\xi,\eta)$, and $F(\xi^2+\eta^2)$ are convergent for small ξ and η.

Conversely, suppose that for (3.2) there is a substitution (3.42) such that in the new coordinates the system has the form (3.43). System (3.43) has the first integral $\xi^2+\eta^2$. Therefore system (3.2) possesses a first integral of the form $u^2+v^2+\cdots$ and by Theorem 3.2.9 has a center at the origin. □

Although, as we mentioned earlier, system (3.43) is not a normal form in the sense of Definition 2.3.6, it is still a real normal form in the sense of Section 2.2.

3.3 The Center Variety

Reviewing the results of the two previous sections we see that it is possible to give a number of statements that are characterizations of systems of the form (3.2) that have a center at the origin.

Theorem 3.3.1. *The following statements about system* (3.2), *when the right-hand sides are real analytic functions of real variables u and v, are equivalent:*
(a) system (3.2) *has a center at the origin;*
(b) the Lyapunov numbers η_{2k+1} *are all zero [Proposition 3.1.4 and Theorem 3.1.5];*
(c) system (3.2) *has a formal first integral of the form* $\Psi(u,v) = u^2+v^2+\cdots$ *[Theorem 3.2.9];*
(d) the complexification

$$\begin{aligned} \dot{x}_1 &= ix_1 + X_1(x_1,x_2) \\ \dot{x}_2 &= -ix_2 + X_2(x_1,x_2) \end{aligned}, \qquad X_2(x_1,\bar{x}_1) = \overline{X_1(x_1,\bar{x}_1)} \tag{3.46}$$

of system (3.2) *has a formal first integral of the form* $\Psi(x_1,x_2) = x_1x_2+\cdots$ *[Theorems 3.2.5 and 3.2.7];*
(e) the coefficients G_{2k+1} *of the function* $G = Y_1 + Y_2$ *computed from the normal form*

$$\begin{aligned} \dot{y}_1 &= i\omega y_1 + y_1 Y_1(y_1y_2) \\ \dot{y}_2 &= -i\omega y_2 + y_2 Y_2(y_1y_2) \end{aligned} \tag{3.47}$$

of the complexification (3.46) *of* (3.2) *(produced by a normalizing transformation that satisfies the condition of Proposition 3.2.1(2)) are all zero [Theorem 3.2.7];*

(f) *in the normal form* (3.47) *of the complexification* (3.46) *of system* (3.2) *(produced by a normalizing transformation that satisfies the condition of Proposition 3.2.1(2)) the real parts of the functions on the right-hand side are all zero [Theorem 3.2.7 and Remark 3.2.8];*

(g) *there exists a real analytic transformation in a neighborhood of the origin of the form* $u = \xi + \phi(\xi, \eta)$, $v = \eta + \psi(\xi, \eta)$ *that transforms system* (3.2) *into the form* $\dot{\xi} = -\eta F(\xi^2 + \eta^2)$, $\dot{\eta} = \xi F(\xi^2 + \eta^2)$, *for some real analytic function* F *satisfying* $F(0) = 1$ *[Theorem 3.2.10].*

Following Dulac ([64]), we use statement (d) of the theorem to extend the concept of a center to certain complex systems. See also Definition 3.3.7 below.

Definition 3.3.2. Consider the system

$$\begin{aligned} \dot{x}_1 &= \ \ ix_1 + X_1(x_1, x_2) \\ \dot{x}_2 &= -ix_2 + X_2(x_1, x_2), \end{aligned} \tag{3.48}$$

where x_1 and x_2 are complex variables and X_1 and X_2 are complex series without constant or linear terms that are convergent in a neighborhood of the origin. System (3.48) is said to have a *center* at the origin if it has a formal first integral of the form

$$\Psi(x_1, x_2) = x_1 x_2 + \sum_{j+k \geq 3} w_{jk} x_1^j x_2^k. \tag{3.49}$$

If in (3.48) X_1 and X_2 satisfy the condition $X_2(x_1, \bar{x}_1) = \overline{X_1(x_1, \bar{x}_1)}$, then (3.48) is the complexification of a real system that has a center at the origin, and the real system can be recovered from either equation by replacing x_2 by $\bar{x}_1$. It is worth emphasizing that the basis of the definition is that a real system of the form (3.2) has a center at the origin if and only if its complexification has a center at $(0,0) \in \mathbb{C}^2$.

Thus far we have dealt primarily with analytic systems of differential equations in which the coefficients were fixed. We now come to the central object of study in this book, planar polynomial systems having a singularity at the origin at which the eigenvalues of the linear part are purely imaginary, and whose coefficients depend on parameters. Up to a time rescaling any such system can be written in the form (3.2), where U and V are now polynomial functions, hence by (3.21) its complexification can be written in the form

$$\begin{aligned} \dot{x} = \widetilde{P}(x, y) &= \ \ i\Bigl(x - \sum_{(p,q) \in S} a_{pq} x^{p+1} y^q\Bigr) \\ \dot{y} = \widetilde{Q}(x, y) &= -i\Bigl(y - \sum_{(p,q) \in S} b_{qp} x^q y^{p+1}\Bigr), \end{aligned} \tag{3.50}$$

where the coefficients of $\widetilde{P}$ and $\widetilde{Q}$ are complex, where $S \subset \mathbb{N}_{-1} \times \mathbb{N}_0$ is a finite set ($\mathbb{N}_{-1} := \{-1\} \cup \mathbb{N}$), every element (p, q) of which satisfies $p + q \geq 1$, and where $b_{qp} = \bar{a}_{pq}$ for all $(p, q) \in S$. This is system (3.3) mentioned in the opening of this chapter; we will always refer to it using that number. The somewhat unusual indexing is to simplify expressions that will arise later. Similarly, although system (3.3) is a system of the form (3.48), we will find that it is more convenient to completely factor out the i than it is to use the form (3.48) (see Exercise 3.14). We have not scaled out the multiplicative factor of i so as to avoid complex time (but see Proposition 3.3.9).

The complexification of any *individual* system of the form (3.2) can be written in the notation of (3.3) by choosing the set S and the individual coefficients a_{pq} and b_{qp} suitably. However, (3.3) is intended to indicate a parametrized family of systems under consideration. The set S picks out the allowable nonzero coefficients and thereby specifies the family. Thus $S = \{(1,0),(0,1),(-1,2)\}$ corresponds to the collection of complex systems with linear part $\operatorname{diag}(i,-i)$ and with arbitrary quadratic nonlinearities, while $S = \{(1,0),(-1,2)\}$ corresponds to all systems with the same linear part but with quadratic nonlinearities that do not contain a monomial xy. Understood this way, (3.3) consists of only those families of systems that are closed under the involution of the dependent variables and parameters $(x,y,a_{pq},b_{qp},t) \mapsto (y,x,b_{qp},a_{pq},-t)$. Thus while not completely general, (3.3) includes all systems that are complexifications of real systems, and is adapted to our purposes.

We will allow in our considerations the full set of systems of the form (3.3) without the requirement that $b_{qp} = \bar{a}_{pq}$. Thus throughout this book we take $\mathbb{C}^{2\ell}$ as the parameter space of (3.3), where ℓ is the cardinality of the set S, and will denote it by $E(a,b)$. $E(a) = E(a,\bar{a})$ will denote the parameter space of

$$\dot{x} = \widetilde{P}(x,\bar{x}) = i\left(x - \sum_{(p,q)\in S} a_{pq}x^{p+1}\bar{x}^{q}\right), \tag{3.51}$$

which we call the real polynomial system *in complex form.* That is, (3.51) is just (3.19) for system (3.2). To shorten the notation we will let $\mathbb{C}[a,b]$ denote the polynomial ring in the variables a_{pq}, b_{qp}, $(p,q) \in S$. So, for example, if we want the nonlinear terms in (3.3) to be precisely the collection of homogeneous quadratic polynomials we take $S = \{(1,0),(0,1),(-1,2)\}$, and $E(a,b) = \mathbb{C}^6$. When the context makes it clear that we are considering systems of the form (3.3), for economy of expression we will speak of "the system (a,b)" when we mean the system of the form (3.3) with the choice (a,b) of parameter values.

To determine if a system of the form (3.3) has a center at the origin, by Definition 3.3.2 we must look for a formal first integral of the form

$$\Psi(x,y) = xy + \sum_{j+k\geq 3} v_{j-1,k-1}x^j y^k, \tag{3.52}$$

where $j,k \in \mathbb{N}_0$ and the indexing is chosen to simplify the formulas that we will obtain in the next section. In this context the condition (3.31) of Remark 3.2.4 that the function $\Psi(x,y)$ be a formal first integral is the identity

$$\mathscr{X}\Psi = \frac{\partial \Psi}{\partial x}\widetilde{P}(x,y) + \frac{\partial \Psi}{\partial y}\widetilde{Q}(x,y) \equiv 0, \tag{3.53}$$

which by (3.52) is

$$\begin{aligned} &i\Big(y + \sum_{j+k\geq 3} j v_{j-1,k-1} x^{j-1} y^k\Big)\Big(x - \sum_{(p,q)\in S} a_{pq} x^{p+1} y^q\Big) \\ &\quad + i\Big(x + \sum_{j+k\geq 3} k v_{k-1,k-1} x^j y^{k-1}\Big)\Big(-y + \sum_{(p,q)\in S} b_{qp} x^q y^{p+1}\Big) \equiv 0. \end{aligned} \tag{3.54}$$

In agreement with formula (3.52) we set $v_{00} = 1$ and $v_{1,-1} = v_{-1,1} = 0$ (so that v_{00} is the coefficient of xy in $\Psi(x,y)$). We also set $a_{jk} = b_{kj} = 0$ for $(j,k) \notin S$. With these conventions, for k_1, $k_2 \in \mathbb{N}_{-1}$, the coefficient g_{k_1,k_2} of $x^{k_1+1}y^{k_2+1}$ in (3.54) is zero for $k_1 + k_2 \leq 0$ and for $k_1 + k_2 \geq 1$ is

$$g_{k_1,k_2} = i\left[(k_1 - k_2)v_{k_1,k_2} - \sum_{\substack{s_1+s_2=0\\ s_1,s_2\geq -1}}^{k_1+k_2-1} (s_1+1)a_{k_1-s_1,k_2-s_2} - (s_2+1)b_{k_1-s_1,k_2-s_2}\right] v_{s_1,s_2}. \tag{3.55}$$

If we think in terms of starting with a system (3.3) and trying to build a formal first integral Ψ in a step-by-step process, at the first stage finding all v_{k_1,k_2} for which $k_1+k_2 = 1$, at the second all v_{k_1,k_2} for which $k_1+k_2 = 2$, and so on, then for a pair k_1 and k_2, if $k_1 \neq k_2$, and if all coefficients $v_{j_1 j_2}$ are already known for $j_1 + j_2 < k_1+k_2$, then $v_{k_1 k_2}$ is uniquely determined by (3.55) and the condition that g_{k_1,k_2} be zero, and the process is successful at this step. By our specification of $v_{-1,1}$, v_{00}, and $v_{1,-1}$ the procedure can be started. But at every second stage (in fact, at every even value of k_1+k_2), there is the one pair k_1 and k_2 such that $k_1 = k_2 = k > 0$, for which (3.55) becomes

$$g_{kk} = -i\left[\sum_{\substack{s_1+s_2=0\\ s_1,s_2\geq -1}}^{2k-1} \left[(s_1+1)a_{k-s_1,k-s_2} - (s_2+1)b_{k-s_1,k-s_2}\right] v_{s_1,s_2}\right], \tag{3.56}$$

so the process of constructing a formal first integral Ψ succeeds at this step only if the expression on the right-hand side of (3.56) is zero. The value of v_{kk} is not determined by equation (3.55) and may be assigned arbitrarily.

It is evident from (3.55) that for all indices k_1 and k_2 in $\{-1\}\cup\mathbb{N}_0$, $v_{k_1k_2}$ is a polynomial function of the coefficients of (3.3), that is, is an element of the set that we have denoted $\mathbb{C}[a,b]$, hence by (3.56) so are the expressions g_{kk} for all k.

The polynomial g_{11} is unique, but for $k \geq 2$ the polynomial g_{kk} depends on the arbitrary choices made for v_{jj} for $1 \leq j < k$. So while it is clear that if for system (a^*,b^*), $g_{kk}(a^*,b^*) = 0$ for all $k \in \mathbb{N}$, then there is a center at the origin, since the process of constructing the formal first integral Ψ succeeds at every step, the truth of the converse is not immediately apparent. For even if for some $k \geq 2$ we obtained $g_{kk}(a^*,b^*) \neq 0$, it is conceivable that if we had made different choices for the polynomials v_{jj} for $1 \leq j < k$, we might have gotten $g_{kk}(a^*,b^*) = 0$. We will show below (Theorem 3.3.5) that in fact whether or not g_{kk} vanishes at any particular $(a^*,b^*) \in E(a,b)$ is independent of the choices of the v_{kk}. Thus the polynomial g_{kk} may be thought of as the kth "obstacle" to the existence of a first integral (3.52): if at a point (a^*,b^*) of our parameter space $E(a,b)$ $g_{kk}(a^*,b^*) \neq 0$, then the construction process fails at that step, no formal first integral of the form (3.52) exists for the corresponding system (3.3), and by Definition 3.3.2 that system does not have a center at the origin. Only if *all* the polynomials g_{kk} vanish, $g_{kk}(a^*,b^*) = 0$ for all $k > 0$, does the corresponding system (3.3) have a formal first integral of the form (3.52), hence have a center at the origin of $\mathbb{C}^2$. Although it is not generally true that a first integral of the form (3.52) exists, the construction process always yields a series of the form (3.52) for which $\mathscr{X}\Psi = \Psi_x\widetilde{P} + \Psi_y\widetilde{Q}$ reduces to

$$\mathscr{X}\Psi = g_{11}(xy)^2 + g_{22}(xy)^3 + g_{33}(xy)^4 + \cdots. \tag{3.57}$$

Definition 3.3.3. Fix a set S. The polynomial g_{kk} defined by (3.56) is called the *kth focus quantity* for the singularity at the origin of system (3.3). The ideal of focus quantities, $\mathscr{B} = \langle g_{11}, g_{22}, \ldots, g_{jj}, \ldots \rangle \subset \mathbb{C}[a,b]$, is called the *Bautin ideal*, and the affine variety $V_{\mathscr{C}} = \mathbf{V}(\mathscr{B})$ is called the *center variety* for the singularity at the origin of system (3.3), or more simply, of system (3.3). $\mathscr{B}_k$ will denote the ideal generated by the first k focus quantities, $\mathscr{B}_k = \langle g_{11}, g_{22}, \ldots, g_{kk} \rangle$.

Even though the focus quantities g_{kk} are not unique, Theorem 3.3.5(2) shows that the center variety is well-defined. First, though, we'll illustrate the ideas so far with a simple but important example.

Example 3.3.4. Let us consider the set of all systems of the form (3.3) that have quadratic nonlinearities, so that ordered from greatest to least under degree lexicographic order S is the ordered set $S = \{(1,0),(0,1),(-1,2)\}$, and (3.3) reads

$$\begin{aligned} \dot{x} &= \ \ i\left(x - a_{10}x^2 - a_{01}xy - a_{-12}y^2\right) \\ \dot{y} &= -i\left(y - b_{2,-1}x^2 - b_{10}xy - b_{01}y^2\right). \end{aligned} \tag{3.58}$$

We will use (3.55) and (3.56) to compute v_{k_1,k_2} through $k_1 + k_2 = 2$.
Stage $0: k_1 + k_2 = 0: (k_1,k_2) \in \{(-1,1),(0,0),(1,-1)\}$
By definition, $v_{-11} = 0$, $v_{00} = 1$, and $v_{1,-1} = 0$.
Stage $1: k_1 + k_2 = 1: (k_1,k_2) \in \{(-1,2),(0,1),(1,0),(2,-1)\}$
In (3.55) $s_1 + s_2$ runs from 0 to 0, so the sum is over the terms (s_1,s_2) in the index set $\{(-1,1),(0,0),(1,-1)\}$. Inserting the values of v_{s_1,s_2} from the previous stage, (3.55) reduces to $g_{k_1,k_2} = i[(k_1 - k_2)v_{k_1,k_2} - a_{k_1,k_2} + b_{k_1,k_2}]$. Setting g_{k_1,k_2} equal to zero for each choice of (k_1,k_2) yields

$$v_{-12} = -\tfrac{1}{3}a_{-12}, \qquad v_{01} = -a_{01} + b_{01}, \qquad v_{10} = a_{10} - b_{10}, \qquad v_{2,-1} = -\tfrac{1}{3}b_{2,-1},$$

where we have applied the convention that $a_{pq} = b_{qp} = 0$ if $(p,q) \notin S$.

Stage 2 : $k_1+k_2=2$: $(k_1,k_2)\in\{(-1,3),(0,2),(1,1),(2,0),(3,-1)\}$
In (3.55) s_1+s_2 runs from 0 to 1, so the sum is over the terms (s_1,s_2) in the index set $\{(-1,1),(0,0),(1,-1);(-1,2),(0,1),(1,0),(2,-1)\}$, and (3.55) is

$$\begin{aligned}g_{k_1,k_2}=i[&(k_1-k_2)v_{k_1,k_2}+2b_{k_1+1,k_2-1}v_{-11}-(a_{k_1,k_2}-b_{k_1,k_2})v_{00}-2a_{k_1-1,k_2+1}v_{1,-1}\\&+3b_{k_1+1,k_2-2}v_{-12}-(a_{k_1,k_2-1}-2b_{k_1,k_2-1})v_{01}\\&-(2a_{k_1-1,k_2}-b_{k_1-1,k_2})v_{10}-3a_{k_1-2,k_2+1}v_{2,-1}].\end{aligned}$$

For the first choice $(k_1,k_2)=(-1,3)$ this reads

$$\begin{aligned}g_{-13}&=i[-4v_{-13}+2b_{02}v_{-11}\\&\quad-(a_{-13}-b_{-13})v_{00}-2a_{-24}v_{1,-1}+3b_{01}v_{-12}-(a_{-12}-2b_{-12})v_{01}\\&\quad-(2a_{-23}-b_{-23})v_{10}-3a_{-34}v_{2,-1}]\\&=i[-4v_{-1,3}+3b_{01}v_{-12}-a_{-12}v_{01}],\end{aligned}$$

where we have applied the convention that $a_{pq}=b_{qp}=0$ if $(p,q)\notin S$. Setting $g_{-1,3}$ equal to zero and inserting the known values of v_{ij} from the previous two stages yields $v_{-13}=-\frac{1}{2}a_{-12}b_{01}+\frac{1}{4}a_{-12}a_{01}$. Applying the same procedure for all the remaining choices of (k_1,k_2) except $(k_1,k_2)=(1,1)$ yields

$$\begin{aligned}v_{02}&=\tfrac{1}{2}a_{-12}b_{10}+\tfrac{1}{2}a_{01}^2-\tfrac{3}{2}a_{01}b_{01}+b_{01}^2-a_{10}a_{-12},\\v_{20}&=\tfrac{1}{2}a_{01}b_{2,-1}-b_{01}b_{2,-1}+a_{10}^2-\tfrac{3}{2}a_{10}b_{10}+\tfrac{1}{2}b_{10}^2,\\v_{3,-1}&=-\tfrac{1}{2}a_{10}b_{2,-1}+\tfrac{1}{4}b_{10}b_{2,-1}.\end{aligned}$$

When $(k_1,k_2)=(1,1)$, (3.55) becomes

$$\begin{aligned}g_{11}&=i[0\cdot v_{11}+2b_{20}v_{-11}-(a_{11}-b_{11})v_{00}-2a_{02}v_{1,-1}\\&\quad+3b_{2,-1}v_{-12}-(a_{10}-2b_{10})v_{01}-(2a_{01}-b_{01})v_{10}-3a_{-12}v_{2,-1}]\\&=-i[a_{10}a_{01}-b_{10}b_{01}].\end{aligned}$$

This is the first focus quantity, which must be zero in order for an element of family (3.58) to have a center at the origin.

As promised, we now show that for fixed $k\in\mathbb{N}$, the variety $\mathbf{V}(g_{kk})$ is the same for all choices (when $k\geq 2$) of the polynomials v_{jj}, $j<k$, which determine g_{kk}, and thus that the center variety $V_{\mathscr{C}}$ is well-defined.

Theorem 3.3.5. *Fix a set S and consider family* (3.3).

1. *Let Ψ be a formal series of the form* (3.52) *and let $g_{11}(a,b),g_{22}(a,b),\dots$ be polynomials satisfying* (3.57) *with respect to system* (3.3). *Then system (a^*,b^*) has a center at the origin if and only if $g_{kk}(a^*,b^*)=0$ for all $k\in\mathbb{N}$.*
2. *Let Ψ and g_{kk} be as in (1) and suppose there exist another function Ψ' of the form* (3.52) *and polynomials $g'_{11}(a,b),g'_{22}(a,b),\dots$ that satisfy* (3.57) *with respect to family* (3.3). *Then $V_{\mathscr{C}}=V'_{\mathscr{C}}$, where $V_{\mathscr{C}}=\mathbf{V}(g_{11}(a,b),g_{22}(a,b),\dots)$ and where $V'_{\mathscr{C}}=\mathbf{V}(g'_{11}(a,b),g'_{22}(a,b),\dots)$.*

Proof. 1) Suppose that family (3.3) is as in the statement of the theorem. Let Ψ be a formal series of the form (3.52) and let $\{g_{kk}(a,b) : k \in \mathbb{N}\}$ be polynomials in (a,b) that satisfy (3.57).

If, for $(a^*,b^*) \in E(a,b)$, $g_{kk}(a^*,b^*) = 0$ for all $k \in \mathbb{N}$, then Ψ is a formal first integral for the corresponding family in (3.3), so by Definition 3.3.2 the system has a center at the origin of $\mathbb{C}^2$.

To prove the converse, we first make the following observations. Suppose that there exist a $k \in \mathbb{N}$ and a choice (a^*,b^*) of the parameters such that $g_{jj}(a^*,b^*) = 0$ for $1 \le j \le k-1$ but $g_{kk}(a^*,b^*) \neq 0$. Let $\mathbf{H}(x_1,y_1)$ be the distinguished normalizing transformation (2.33), producing the distinguished normal form (3.27), here written $\dot{x}_1 = ix_1 + x_1X(x_1y_1)$, $\dot{y}_1 = -iy_1 + y_1Y(x_1y_1)$, and consider the function $F = \Psi \circ \mathbf{H}$. By construction

$$\begin{aligned}
\Big[ix_1 + x_1X(x_1y_1)\Big]\frac{\partial F}{\partial x_1}(x_1,y_1) + \Big[-iy_1 + y_1Y(x_1y_1)\Big]\frac{\partial F}{\partial y_1}(x_1,y_1) \\
= g_{kk}(a^*,b^*)[x_1 + h_1(x_1,y_1)]^{k+1}[y_1 + h_2(x_1,y_1)]^{k+1} + \cdots \\
= g_{kk}(a^*,b^*)x_1^{k+1}y_1^{k+1} + \cdots .
\end{aligned} \tag{3.59}$$

Through order $2k+1$ this is almost precisely equation (3.32) (which was derived on the admissible simplifying assumption that $\lambda_1 = p$ and $\lambda_2 = q$), so that if we repeat practically verbatim the argument that follows (3.32), we obtain identity (3.36), with $p=q=1$, through order $2k+2$. (We are able to obtain the conclusion about the form of F with regard to the terms of order $2k+2$ from the fact that at the first step at which the terms on the right-hand side of (3.59) must be taken into account, they have the form a constant times $(x_1y_1)^{k+1}$.) Therefore

$$F(x_1,y_1) = f_1 \cdot (x_1y_1) + \cdots + f_{k+1} \cdot (x_1y_1)^{k+1} + U(x_1,y_1) = f(x_1y_1) + U(x_1,y_1),$$

where $f_1 = 1$ and $U(x_1,y_1)$ begins with terms of order at least $2k+3$. Thus

$$x_1\frac{\partial F}{\partial x_1} = x_1y_1f'(x_1y_1) + \alpha(x_1,y_1) \qquad \text{and} \qquad y_1\frac{\partial F}{\partial y_1} = x_1y_1f'(x_1y_1) + \beta(x_1,y_1),$$

where $\alpha(x_1,y_1)$ and $\beta(x_1,y_1)$ begin with terms of order at least $2k+3$, and so the left-hand side of (3.59) is

$$\begin{aligned}
i[\alpha(x_1,y_1) - \beta(x_1,y_1)] + (X(x_1y_1) + Y(x_1y_1))x_1y_1 f'(x_1y_1) \\
+ X(x_1y_1)\,\alpha(x_1,y_1) + Y(x_1y_1)\,\beta(x_1,y_1).
\end{aligned}$$

Hence if we subtract

$$i[\alpha(x_1,y_1) - \beta(x_1,y_1)] + X(x_1y_1)\,\alpha(x_1,y_1) + Y(x_1y_1)\,\beta(x_1,y_1), .$$

which begins with terms of order at least $2k+3$, from each side of (3.59), we obtain

$$G(x_1y_1)x_1y_1 f'(x_1y_1) = g_{kk}(a^*,b^*)(x_1y_1)^k + \cdots, \tag{3.60}$$

where G is the function of (3.28b). Thus supposing, contrary to what we wish to show, that system (3.3) for the choice $(a,b) = (a^*,b^*)$ has a center at the origin of $\mathbb{C}^2$, so that it admits a first integral $\Phi(x,y) = xy + \cdots$. Then by part (1) of Theorem 3.2.5, the function G vanishes identically, hence the left-hand side of (3.60) is identically zero, whereas the right-hand side is not, a contradiction.

2) If $V_{\mathscr{C}} \neq V'_{\mathscr{C}}$, then there exists (a^*,b^*) that belongs to one of the varieties $V_{\mathscr{C}}$ and $V'_{\mathscr{C}}$ but not to the other, say $(a^*,b^*) \in V_{\mathscr{C}}$ but $(a^*,b^*) \notin V'_{\mathscr{C}}$. The inclusion $(a^*,b^*) \in V_{\mathscr{C}}$ means that the system corresponding to (a^*,b^*) has a center at the origin. Therefore by part (1) $g'_{kk}(a^*,b^*) = 0$ for all $k \in \mathbb{N}$. This contradicts our assumption that $(a^*,b^*) \notin V'_{\mathscr{C}}$. □

Remark 3.3.6. Note that it is a consequence of (3.60) that if, for a particular $(a^*,b^*) \in E(A,b)$, $g_{kk}(a^*,b^*)$ is the first nonzero focus quantity, then the first nonzero coefficient of $G(a^*,b^*)$ is $G_{2k+1}(a^*,b^*)$ and $G_{2k+1}(a^*,b^*) = g_{kk}(a^*,b^*)$.

Thus points of $V_{\mathscr{C}}$ correspond precisely to systems in family (3.3) that have a center at the origin of $\mathbb{C}^2$, in the sense that there exists a first integral of the form (3.52). If $(a,b) \in V_{\mathscr{C}}$ and $a_{pq} = \bar{b}_{qp}$ for all $(p,q) \in S$, which we will denote $b = \bar{a}$, then such a point corresponds to a system that is the complexification of the real system expressed in complex coordinates as (3.51), which then has a topological center at the origin. More generally, we can consider the intersection of the center variety $V_{\mathscr{C}}$ with the set $\Pi \stackrel{\text{def}}{=} \{(a,b) : b = \bar{a}\}$ whose elements correspond to complexifications of real systems; we call this the *real center variety* $V_{\mathscr{C}}^R \stackrel{\text{def}}{=} V_{\mathscr{C}} \cap \Pi$. To the set Π there corresponds a family of real systems of ordinary differential equations on $\mathbb{R}^2$ expressed in complex form as (3.51) or in real form as (3.2), and for this family there is a space E^R of real parameters. Within E^R there is a variety V corresponding to systems that have a center at the origin. By Theorem 3.3.1 points of V are in one-to-one correspondence with points of $V_{\mathscr{C}}^R$.

By Theorems 3.1.5, 3.2.7, and 3.3.5, in order to find either $V_{\mathscr{C}}$ or $V_{\mathscr{C}}^R$, one can compute either the Lyapunov numbers η_{2k+1}, the coefficients G_{2k+1} of the function G defined by (3.28), or the focus quantities g_{kk}, all of which are polynomial functions of the parameters. (Recall that the functions U and V in (3.2) are now assumed to be polynomials. It is apparent from the discussion in the proof of Theorem 2.3.8 that the coefficients of the normalizing transformation $\mathbf{x} = \mathbf{H}(\mathbf{y}) = \mathbf{y} + \mathbf{h}(\mathbf{y})$ and of the resulting normal form are polynomial functions of the parameters (a,b) with coefficients in the field of Gaussian rationals, $\mathbb{Q}(i) = \{a + ib : a, b \in \mathbb{Q}\}$, provided that, when m and α form a resonant pair, we choose $h_m^{(\alpha)}$ to be a polynomial in (a,b) with coefficients in $\mathbb{Q}(i)$. Thus $G_{2k+1} \in \mathbb{Q}[a,b]$. That η_{2k+1} is a polynomial in (a,b) is Proposition 6.2.2.) They can be markedly different from each other, but they all pick out the same varieties $V_{\mathscr{C}}$ and $V_{\mathscr{C}}^R$. From the point of view of applications the most interesting systems are the real systems. The trouble is, of course, that the field $\mathbb{R}$ is not algebraically closed, making it far more difficult to study real varieties than complex varieties. This is why we will primarily investigate the center problem for complex systems (3.3). If the real case is the one of interest, then it is a straightforward matter to pass over to the real variety $V_{\mathscr{C}}^R$ by means of the substitution $b = \bar{a}$, and from $V_{\mathscr{C}}$ obtain $V_{\mathscr{C}}^R$.

It must be noted that in the literature all of the polynomials η_{2k+1}, G_{2k+1}, and g_{kk} are called "focus quantities" (or numbers) or "Lyapunov quantities" (or numbers, or constants), but we reserve the term "focus quantities" for the polynomials g_{kk}. When they are viewed as polynomials in the coefficients of a family of systems, we refer to the η_{2k+1} as "Lyapunov quantites." We underscore the fact that once an indexing set S is fixed, the polynomial g_{kk} is determined for any $k \in \mathbb{N}$, up to choices made at earlier stages when $k > 1$. The polynomial g_{kk} belongs to the family (3.3), not to any particular system in the family. If a system is specified, by specifying the parameter string (a^*, b^*), then what is also called a kth focus quantity is obtained for that system, namely, the element of $\mathbb{C}$ that is obtained by evaluating g_{kk} at (a^*, b^*). In this sense the term "the kth focus quantity" has two meanings: first, as the polynomial $g_{kk} \in \mathbb{C}[a,b]$, and second as the number $g_{kk}(a^*, b^*)$ obtained when g_{kk} is regarded as a function from $E(a,b)$ to $\mathbb{C}$ and is evaluated at the specific string (a^*, b^*). The same is true of η_{2k+1}.

Although a system of the form (3.22) can never in any sense correspond to a real system with a center if $\lambda_2 \neq \bar{\lambda}_1$, yet when λ_1 and λ_2 are rationally related, say $\lambda_1/\lambda_2 = -p/q$, the condition $G \equiv 0$ on its normal form for the existence of a first integral of the form $\Psi(x_1, x_2) = x_1^q x_2^p + \cdots$ (Theorem 3.2.5) is precisely the condition (with $p = q = 1$ and expressed in terms of the normal form of its complexification) that a real system have a center (Theorem 3.3.1(e)). Thus it is reasonable to extend Definition 3.3.2 of a center of a system on $\mathbb{C}^2$ as follows. We restrict to $\lambda_1 = p$ and $\lambda_2 = -q$ to avoid complex time that might be introduced by a time rescaling. See also Exercise 3.8.

Definition 3.3.7. Consider the system

$$\begin{aligned} \dot{x}_1 &= px_1 + X_1(x_1, x_2) \\ \dot{x}_2 &= -qx_2 + X_2(x_1, x_2), \end{aligned} \tag{3.61}$$

where x_1 and x_2 are complex variables, X_1 and X_2 are complex series without constant or linear terms and are convergent in a neighborhood of the origin, and p and q are elements of $\mathbb{N}$ satisfying $\mathrm{GCD}(p,q) = 1$. System (3.61) is said to have a $p:-q$ *resonant center* at the origin if it has a formal first integral of the form

$$\Psi(x_1, x_2) = x_1^q x_2^p + \sum_{j+k > p+q} w_{jk} x_1^j x_2^k. \tag{3.62}$$

By Theorem 3.2.5, the origin is a $p:-q$ resonant center for (3.61) if and only if there exists a *convergent* first integral of the form (3.62). (Compare Corollary 3.2.6.) Another immediate corollary of Theorem 3.2.5 is the following characterization of a $p:-q$ resonant center.

Proposition 3.3.8. *The origin is a $p:-q$ resonant center of system* (3.61) *if and only if the function G computed from its normal form (defined by equation* (3.28b)) *vanishes identically.*

When (3.61) is a polynomial system, we can write it in the form

$$
\begin{aligned}
\dot{x} &= px - \sum_{(j,k)\in S} a_{jk}x^{j+1}y^k \\
\dot{y} &= -qy + \sum_{(j,k)\in S} b_{kj}x^k y^{j+1}.
\end{aligned}
\tag{3.63}
$$

Repeating the same procedure used for family (3.3), we now look for a first integral in the form

$$
\Psi(x,y) = x^q y^p + \sum_{\substack{j+k>p+q\\ j,k\in\mathbb{N}_0}} v_{j-q,k-p}x^j y^k, \tag{3.64}
$$

where the indexing has been chosen in direct analogy with that used in (3.52). Condition (3.53) now reads

$$
\begin{aligned}
&\Big(qx^{q-1}y^p + \sum_{j+k>p+q} jv_{j-q,k-p}x^{j-1}y^k\Big)\Big(px - \sum_{(m,n)\in S} a_{mn}x^{m+1}y^n\Big) \\
&+\Big(px^q y^{p-1} + \sum_{j+k>p+q} kv_{j-q,k-p}x^j y^{k-1}\Big)\Big(-qy + \sum_{(m,n)\in S} b_{nm}x^n y^{m+1}\Big) \equiv 0.
\end{aligned}
\tag{3.65}
$$

We augment the set of coefficients in (3.52) with the collection

$$
J = \{v_{-q+s,q-s} : s = 0,\ldots,p+q\},
$$

where, in agreement with formula (3.64), we set $v_{00} = 1$ and $v_{mn} = 0$ for all other elements of J, so that elements of J are the coefficients of the terms of degree $p+q$ in $\Psi(x,y)$. With the convention $a_{mn} = b_{nm} = 0$ for $(m,n) \notin S$, for $(k_1,k_2) \in \mathbb{N}_{-q}\times\mathbb{N}_{-p}$ the coefficient g_{k_1,k_2} of $x^{k_1+q}y^{k_2+p}$ in (3.65) is zero for $k_1+k_2 \le 0$ and for $k_1+k_2 \ge 1$ is given by

$$
\begin{aligned}
&g_{k_1,k_2} \\
&= (pk_1 - qk_2)v_{k_1,k_2} - \sum_{\substack{s_1+s_2=0\\ s_1\ge -q,\ s_2\ge -p}}^{k_1+k_2-1} \big[(s_1+q)a_{k_1-s_1,k_2-s_2} - (s_2+p)b_{k_1-s_1,k_2-s_2}\big]\,v_{s_1,s_2}.
\end{aligned}
\tag{3.66}
$$

The focus quantities are now

$$
g_{kq,kp} = -\sum_{\substack{s_1+s_2=0\\ s_1\ge -q, s_2\ge -p}}^{kq+kp-1} \big[(s_1+q)a_{k_1-s_1,k_2-s_2} - (s_2+p)b_{k_1-s_1,k_2-s_2}\big]\,v_{s_1,s_2}, \tag{3.67}
$$

so that we always obtain a series of the form (3.64) for which $\mathscr{X}\Psi$ reduces to

$$
\mathscr{X}\Psi = g_{q,p}(x^q y^p)^2 + g_{2q,2p}(x^q y^p)^3 + g_{3q,3p}(x^q y^p)^4 + \cdots. \tag{3.68}
$$

The *center variety* for family (3.63) is defined in analogy with Definition 3.3.3. With appropriate changes, the proof of the analogue of Theorem 3.3.5 goes through as before to show that the center variety for family (3.63) is well-defined. Moreover, we have the following result.

Proposition 3.3.9. *The center variety of system* (3.63) *with* $p = q = 1$ *coincides with the center variety of system* (3.3).

Proof. When $p = q = 1$ the computation of the condition that (3.62) be a formal first integral of (3.63) yields (3.54) but without the multiplicative factor of i. □

Based on this proposition we are free to consider the system

$$\begin{aligned} \dot{x} = \widetilde{P}(x,y) &= \quad x - \sum_{(p,q)\in S} a_{pq} x^{p+1} y^q \\ \dot{y} = \widetilde{Q}(x,y) &= -y + \sum_{(p,q)\in S} b_{qp} x^q y^{p+1} \end{aligned} \tag{3.69}$$

instead of system (3.3). As the reader has by now noticed, the expression for the focus quantities may or may not contain a multiplicative factor of i, depending on how the problem was set up. Since we are concerned only with the variety the focus quantities determine, hence only with their zeros, the presence or absence of such a nonzero factor is unimportant.

There are different points of view as to what exactly constitutes the center problem. Under one approach, to resolve the center problem for a family of polynomial systems means to find the center variety of the family. More precisely, to solve the center problem for the family (3.3) or the family (3.63) of polynomial systems, for a fixed set S, means to find the center variety and its irreducible decomposition $\mathbf{V}(\mathscr{B}) = V_1 \cup \cdots \cup V_k$. In the following sections we will present some methods that enable one to resolve the center problem in this sense for many particular families of polynomial systems.

In a wider sense the center problem is the problem of finding an algorithm that will guarantee complete identification of the center variety of *any* family of polynomial systems in a finite number of steps. It is unknown if the methods that we will present are sufficient for the construction of such an algorithm. Understood in this sense, the center problem seems to be still far from resolution.

3.4 Focus Quantities and Their Properties

Calculating any of the polynomials η_{2k+1}, G_{2k+1}, or g_{kk} is a difficult computational problem because the number of terms in these polynomials grows so fast. The focus quantities g_{kk} are the easiest to calculate, but even they are difficult to compute for large k if we use only formulas (3.55) and (3.56). In this section we identify structure in the focus quantities g_{kk} for systems of the form (3.3). This structure is the basis for an efficient algorithm for computing the focus quantities.

We begin by directing the reader's attention back to Example 3.3.4. Consider $v_{02} = \frac{1}{2}a_{-12}b_{10} + \frac{1}{2}a_{01}^2 - \frac{3}{2}a_{01}b_{01} + b_{01}^2 - a_{10}a_{-12}$, and note that for any monomial that appears, the sum of the product of the index of each term (as an element of $\mathbb{N}_{-1} \times \mathbb{N}_0$) with its exponent is the index of v_{02}:

$$\begin{aligned} a_{-12}b_{10} &: 1\cdot(-1,2)+1\cdot(1,0)=(0,2) \\ a_{01}^2 &: 2\cdot(0,1)=(0,2) \\ a_{01}b_{01} &: 1\cdot(0,1)+1\cdot(0,1)=(0,2) \\ b_{01}^2 &: 2\cdot(0,1)=(0,2) \\ a_{10}a_{-12} &: 1\cdot(1,0)+1\cdot(-1,2)=(0,2) \end{aligned} \tag{3.70}$$

The reader can verify that the same is true for every monomial in every coefficient v_{jk} computed in Example 3.3.4.

To express this fact in general we introduce the following notation. We order the index set S in equation (3.3) in some manner, say by degree lexicographic order from least to greatest, and write the ordered set S as $S = \{(p_1,q_1),\dots,(p_\ell,q_\ell)\}$. Consistently with this we then order the parameters as $(a_{p_1,q_1},\dots,a_{p_\ell,q_\ell},b_{q_\ell,p_\ell},\dots,b_{q_1,p_1})$ so that any monomial appearing in v_{ij} has the form $a_{p_1,q_1}^{\nu_1}\cdots a_{p_\ell,q_\ell}^{\nu_\ell}b_{q_\ell,p_\ell}^{\nu_{\ell+1}}\cdots b_{q_1,p_1}^{\nu_{2\ell}}$ for some $\nu=(\nu_1,\dots,\nu_{2\ell})$. To simplify the notation, for $\nu \in \mathbb{N}_0^{2\ell}$ we write

$$[\nu] \stackrel{\text{def}}{=} a_{p_1,q_1}^{\nu_1}\cdots a_{p_\ell,q_\ell}^{\nu_\ell}b_{q_\ell,p_\ell}^{\nu_{\ell+1}}\cdots b_{q_1,p_1}^{\nu_{2\ell}}.$$

We will write just $\mathbb{C}[a,b]$ in place of $\mathbb{C}[a_{p_1,q_1},\dots,a_{p_\ell,q_\ell},b_{q_\ell,p_\ell},\dots,b_{q_1,p_1}]$, and for $f \in \mathbb{C}[a,b]$ write $f=\sum_{\nu\in\mathrm{Supp}(f)} f^{(\nu)}[\nu]$, where $\mathrm{Supp}(f)$ denotes those $\nu \in \mathbb{N}_0^{2\ell}$ such that the coefficient of $[\nu]$ in the polynomial f is nonzero.

Once the ℓ-element set S has been specified and ordered, we let $L : \mathbb{N}_0^{2\ell} \to \mathbb{Z}^2$ be the linear map defined by

$$\begin{aligned} L(\nu) &= (L_1(\nu),L_2(\nu)) \\ &= \nu_1(p_1,q_1)+\cdots+\nu_\ell(p_\ell,q_\ell)+\nu_{\ell+1}(q_\ell,p_\ell)+\cdots+\nu_{2\ell}(q_1,p_1) \\ &= (p_1\nu_1+\cdots+p_\ell\nu_\ell+q_\ell\nu_{\ell+1}+\cdots+q_1\nu_{2\ell}, \\ &\qquad q_1\nu_1+\cdots+q_\ell\nu_\ell+p_\ell\nu_{\ell+1}+\cdots+p_1\nu_{2\ell}), \end{aligned} \tag{3.71}$$

which is just the formal expression for the sums in (3.70). The fact that we have observed about the v_{jk} is that for each monomial $[\nu]$ appearing in v_{jk}, $L(\nu)=(j,k)$. This is the basis for the following definition.

Definition 3.4.1. For $(j,k) \in \mathbb{N}_{-1}\times\mathbb{N}_{-1}$, a polynomial $f=\sum_{\nu\in\mathrm{Supp}(f)} f^{(\nu)}[\nu]$ in $\mathbb{C}[a,b]$ is a *(j,k)-polynomial* if, for every $\nu\in\mathrm{Supp}(f)$, $L(\nu)=(j,k)$.

Theorem 3.4.2. *Let family* (3.3) *(family* (3.50)*) be given. There exists a formal series $\Psi(x,y)$ of the form* (3.52) *(note that $j-1$ and $k-1$ there correspond to j and k in points (2) and (3) below) and polynomials g_{11}, g_{22}, … in $\mathbb{C}[a,b]$ such that*
1. equation (3.57) *holds,*
2. for every pair $(j,k)\in\mathbb{N}_{-1}^2$, $j+k\geq 0$, $v_{jk}\in\mathbb{Q}[a,b]$, and v_{jk} is a (j,k)-polynomial,
3. for every $k\geq 1$, $v_{kk}=0$, and
4. for every $k\geq 1$, $ig_{kk}\in\mathbb{Q}[a,b]$ $(i=\sqrt{-1})$, and g_{kk} is a (k,k)-polynomial.

Proof. The discussion leading from equation (3.52) to equation (3.57) shows that if we define $v_{1,-1} = 0$, $v_{00} = 1$, and $v_{-11} = 0$, then for $k_1, k_2 \geq -1$, if v_{k_1,k_2} are defined recursively by

$$v_{k_1,k_2} = \begin{cases} \frac{1}{k_1-k_2} \sum\limits_{\substack{s_1+s_2=0 \\ s_1,s_2\geq -1}}^{k_1+k_2-1} [(s_1+1)a_{k_1-s_1,k_2-s_2} - (s_2+1)b_{k_1-s_1,k_2-s_2}]v_{s_1,s_2} & \text{if } k_1 \neq k_2 \\ 0 & \text{if } k_1 = k_2, \end{cases} \tag{3.72}$$

where the recursion is on $k_1 + k_2$ (that is, find all v_{k_1,k_2} for which $k_1 + k_2 = 1$, then find all v_{k_1,k_2} for which $k_1 + k_2 = 2$, and so on), and once all v_{k_1,k_2} are known for $k_1 + k_2 \leq 2k - 1$, g_{kk} is defined by

$$g_{kk} = -i \left[\sum_{\substack{s_1+s_2=0 \\ s_1,s_2\geq -1}}^{k_1+k_2-1} [(s_1+1)a_{k_1-s_1,k_2-s_2} - (s_2+1)b_{k_1-s_1,k_2-s_2}]v_{s_1,s_2} \right], \tag{3.73}$$

then for every pair (j,k), $v_{jk} \in \mathbb{Q}[a,b]$, for every k, $ig_{kk} \in \mathbb{Q}[a,b]$, and equation (3.57) holds. (An assumption that should be recalled is that in (3.72), $a_{k_1-s_1,k_2-s_2}$ and $b_{k_1-s_1,k_2-s_2}$ are replaced by zero when $(k_1 - s_1, k_2 - s_2) \notin S$.) By our definition of v_{kk} in (3.72), (3) holds. To show that v_{jk} is a (j,k)-polynomial we proceed by induction on $j+k$.

Basis step. For $k_1 + k_2 = 0$, there are three polynomials, v_{-11}, v_{00}, and $v_{1,-1}$. Since $\operatorname{Supp}(v_{-11}) = \operatorname{Supp}(v_{1,-1}) = \varnothing$, the condition in Definition 3.4.1 is vacuous for $v_{-1,1}$ and $v_{1,-1}$. For v_{00}, $\operatorname{Supp}(v_{00}) = (0,\ldots,0)$, and $L(0,\ldots,0) = (0,0)$.

Inductive step. Suppose that v_{jk} is a (j,k)-polynomial for all (j,k) satisfying $j+k \leq m$, and that $k_1 + k_2 = m+1$. Consider a term $v_{s_1,s_2}a_{k_1-s_1,k_2-s_2}$ in the sum in (3.72). If $(k_1 - s_1, k_2 - s_2) \notin S$, then $a_{k_1-s_1,k_2-s_2} = 0$ by convention, and the term does not appear. If $(k_1 - s_1, k_2 - s_2) = (p_c, q_c) \in S$, then

$$v_{s_1,s_2}a_{k_1-s_1,k_2-s_2} = \left(\sum_{\nu\in\operatorname{Supp}(v_{s_1,s_2})} v_{s_1,s_2}^{(\nu)}[\nu] \right) [\mu] = \sum_{\nu\in\operatorname{Supp}(v_{s_1,s_2})} v_{s_1,s_2}^{(\nu)}[\nu+\mu], \tag{3.74}$$

where $\mu = (0,\ldots,0,1,0,\ldots,0)$, with the 1 in the cth position, counting from the left. Clearly $L(\mu) = (k_1 - s_1, k_2 - s_2)$, hence by the inductive hypothesis and additivity of L, every term in (3.74) satisfies

$$L(\nu+\mu) = L(\nu) + L(\mu) = (s_1,s_2) + (k_1 - s_1, k_2 - s_2) = (k_1,k_2).$$

Similarly, for any term $v_{s_1,s_2}b_{k_1-s_1,k_2-s_2}$ such that $(k_1 - s_1, k_2 - s_2) = (q_c, p_c)$, that is, such that $(k_2 - s_2, k_1 - s_1) = (p_c, q_c) \in S$, we obtain an expression just like (3.74), except that now the 1 in μ is in the cth position counting from the *right*, and we easily compute that every term in this expression satisfies $L(\nu+\mu) = (k_1,k_2)$. Thus point (2) is fully established.

It is clear that the same arguments show that g_{kk} is a (k,k)-polynomial, completing the proof of point (4). □

The import of Theorem 3.4.2 is that any monomial $[\nu]$ can appear in at most one of the polynomials v_{k_1,k_2} (or g_{k_1,k_2} if $k_1-k_2=0$), and we find it by computing $L(\nu)$. For example, in the context of Example 3.3.4, $S=\{(-1,2),(0,1),(1,0)\}$, for the monomial $a_{01}^2 b_{10}^3 b_{2,-1} = a_{-12}^0 a_{01}^2 a_{10}^0 b_{01}^0 b_{10}^3 b_{2,-1}^1$ $\nu=(0,2,0,0,3,1)$, and

$$L(\nu)=0\cdot(-1,2)+2\cdot(0,1)+0\cdot(1,0)+0\cdot(0,1)+3\cdot(1,0)+1\cdot(2,-1)=(5,1),$$

so that $a_{01}^2 b_{10}^3 b_{2,-1}$ can appear in v_{51} but in no other polynomial v_{k_1,k_2}. From the opposite point of view, since in the situation of this example $L_1(\nu)+L_2(\nu)=|\nu|$ for all $\nu\in\mathbb{N}_0^6$, given $(k_1,k_2)\in\mathbb{N}_0^2$, any solution ν of $(L_1(\nu),L_2(\nu))=(k_1,k_2)$ must satisfy $|\nu|=L_1(\nu)+L_2(\nu)=k_1+k_2$. Hence we find all monomials that can appear in v_{k_1,k_2} by checking $L(\nu)=(k_1,k_2)$ for all $\nu\in\mathbb{N}_0^6$ with $|\nu|=k_1+k_2$. Thus, for example, for v_{-13}, we check $L(\nu)=(-1,3)$ for $\nu\in\mathbb{N}_0^6$ satisfying $|\nu|=2$. The set to check is

$$\{(2,0,0,0,0,0),(0,2,0,0,0,0),\ldots,(0,0,0,0,0,2);\\ (1,1,0,0,0,0),(1,0,1,0,0,0),\ldots,(0,0,0,0,1,1);\ldots,(0,0,0,0,1,1)\}$$

consisting of 21 sextuples. A short computation shows that

$$(L_1(\nu),L_2(\nu))=(-\nu_1+\nu_3+\nu_5+2\nu_6,2\nu_1+\nu_2+\nu_4-\nu_6)=(-1,3)$$

only for $\nu=(1,1,0,0,0,0)$ and $\nu=(1,0,0,1,0,0)$, so that $a_{-12}a_{01}$ and $a_{-12}b_{01}$, respectively, appear in v_{-13}. That is, $v_{-13}=v_{-13}^{(1,1,0,0,0,0)}a_{-12}a_{01}+v_{-13}^{(1,0,0,1,0,0)}a_{-12}b_{01}$ for some unknown constants $v_{-13}^{(1,1,0,0,0,0)}$ and $v_{-13}^{(1,0,0,1,0,0)}$. The next theorem shows how to compute the coefficients $v_{k_1,k_2}^{(\nu)}$. The following definition and lemma will be needed.

Definition 3.4.3. Let $f=\sum_{\nu\in\mathrm{Supp}(f)} f^{(\nu)} a_{p_1,q_1}^{\nu_1}\cdots a_{p_\ell,q_\ell}^{\nu_\ell} b_{q_\ell,p_\ell}^{\nu_{\ell+1}}\cdots b_{q_1,p_1}^{\nu_{2\ell}}\in\mathbb{C}[a,b]$. The *conjugate* $\widehat{f}$ of f is the polynomial obtained from f by the involution

$$f^{(\nu)}\to\bar{f}^{(\nu)} \qquad a_{ij}\to b_{ji} \qquad b_{ji}\to a_{ij};$$

that is, $\widehat{f}=\sum_{\nu\in\mathrm{Supp}(f)}\bar{f}^{(\nu)} a_{p_1,q_1}^{\nu_{2\ell}}\cdots a_{p_\ell,q_\ell}^{\nu_{\ell+1}} b_{q_\ell,p_\ell}^{\nu_\ell}\cdots b_{q_1,p_1}^{\nu_1}\in\mathbb{C}[a,b]$.

Since $[\nu]=a_{p_1,q_1}^{\nu_1}\cdots a_{p_\ell,q_\ell}^{\nu_\ell} b_{q_\ell,p_\ell}^{\nu_{\ell+1}}\cdots b_{q_1,p_1}^{\nu_{2\ell}}$, $\widehat{[\nu]}=a_{p_1,q_1}^{\nu_{2\ell}}\cdots a_{p_\ell,q_\ell}^{\nu_{\ell+1}} b_{q_\ell,p_\ell}^{\nu_\ell}\cdots b_{q_1,p_1}^{\nu_1}$, so that

$$[\widehat{(\nu_1,\ldots\nu_{2\ell})}]=[(\nu_{2\ell},\ldots,\nu_1)]. \tag{3.75}$$

For this reason we will also write, for $\nu=(\nu_1,\ldots,\nu_{2\ell})$, $\widehat{\nu}=(\nu_{2\ell},\ldots,\nu_1)$.

Let a family (3.3) (family (3.50)) for some set S of indices be fixed, and for any $\nu\in\mathbb{N}_0^{2\ell}$ define $V(\nu)\in\mathbb{Q}$ recursively, with respect to $|\nu|=\nu_1+\cdots+\nu_{2\ell}$, as follows:

$$V((0,\dots,0)) = 1; \tag{3.76a}$$

for $\nu \neq (0,\dots,0)$,

$$V(\nu) = 0 \qquad \text{if} \qquad L_1(\nu) = L_2(\nu), \tag{3.76b}$$

and when $L_1(\nu) \neq L_2(\nu)$,

$$\begin{aligned} V(\nu) = &\frac{1}{L_1(\nu) - L_2(\nu)} \\ &\times \Bigg[\sum_{j=1}^{\ell} \widetilde{V}(\nu_1,\dots,\nu_j - 1,\dots,\nu_{2\ell})(L_1(\nu_1,\dots,\nu_j - 1,\dots,\nu_{2\ell}) + 1) \\ &- \sum_{j=\ell+1}^{2\ell} \widetilde{V}(\nu_1,\dots,\nu_j - 1,\dots,\nu_{2\ell})(L_2(\nu_1,\dots,\nu_j - 1,\dots,\nu_{2\ell}) + 1) \Bigg], \end{aligned} \tag{3.76c}$$

where

$$\widetilde{V}(\eta) = \begin{cases} V(\eta) & \text{if } \eta \in \mathbb{N}_0^{2\ell} \\ 0 & \text{if } \eta \in \mathbb{N}_{-1}^{2\ell} \setminus \mathbb{N}_0^{2\ell}. \end{cases}$$

Lemma 3.4.4. *Suppose $\nu \in \mathbb{N}_0^{2\ell}$ is such that either $L_1(\nu) < -1$ or $L_2(\nu) < -1$. Then $V(\nu) = 0$.*

Proof. The proof is by induction on $|\nu|$.

Basis step. A quick computation shows that if $|\nu| = 0$ or $|\nu| = 1$, then both $L_1(\nu) \geq -1$ and $L_2(\nu) \geq -1$, so the basis step is $|\nu| = 2$. This happens in two ways:

(α) $\nu = (0,\dots,0,\overset{j}{2},0,\dots,0)$,

(β) $\nu = (0,\dots,0,\overset{j}{1},0,\dots,0,\overset{k}{1},0,\dots,0)$.

In case (α), since for $(p,q) \in S$, $q \geq 0$, $L_1(\nu) \leq -2$ or $L_2(\nu) \leq -2$ holds if and only if:

(i) when $1 \leq j \leq \ell$: $(p_j,q_j) = (-1,q_j)$, and

(ii) when $\ell + 1 \leq j \leq 2\ell$: $(q_{2\ell-j+1}, p_{2\ell-j+1}) = (q_{2\ell-j+1}, -1)$.

In subcase (i),

$$\begin{aligned} V(\nu) &= \frac{1}{-2-2q_j} V(0,\dots,0,\overset{j}{1},0,\dots,0)\,(L_1(0,\dots,0,\overset{j}{1},0,\dots,0)+1) \\ &= \frac{1}{-2-2q_j} V(0,\dots,0,\overset{j}{1},0,\dots,0)\,(-1+1) \\ &= 0. \end{aligned}$$

Subcase (ii) is the same, except for a minus sign.

In case (β), $L_1(\nu) \leq -2$ or $L_2(\nu) \leq -2$ holds if and only if:

(i) when $1 \leq j < k \leq \ell$: $(p_j,q_j) = (-1,q_j)$ and $(p_k,q_k) = (-1,q_k)$, and

(ii) when $\ell + 1 \leq j < k \leq 2\ell$: $(q_{2\ell-j+1}, p_{2\ell-j+1}) = (q_{2\ell-j+1}, -1)$ and $(q_{2\ell-k+1}, p_{2\ell-k+1}) = (q_{2\ell-k+1}, -1)$.

In subcase (i),

$$
\begin{aligned}
V(\nu) &= \frac{1}{-2-q_j-q_k}\Big[V(0,\dots,0,\overset{k}{1},0,\dots,0)\,(L_1(0,\dots,0,\overset{k}{1},0,\dots,0)+1) \\
&\qquad + V(0,\dots,0,\overset{j}{1},0,\dots,0)\,(L_1(0,\dots,0,\overset{j}{1},0,\dots,0)+1)\Big] \\
&= \frac{1}{-2-q_j-q_k}\Big[V(0,\dots,0,\overset{k}{1},0,\dots,0)\,(-1+1) \\
&\qquad + V(0,\dots,0,\overset{j}{1},0,\dots,0)\,(-1+1)\Big] = 0 .
\end{aligned}
$$

Subcase (ii) is the same, except for a minus sign.

Inductive step. Assume the lemma holds for all ν satisfying $|\nu| \le m$ and let ν be such that $|\nu| = m+1$, $L_1(\nu) < -1$ or $L_2(\nu) < -1$, and $L_1(\nu) \neq L_2(\nu)$.

Suppose first that $L_1(\nu) < -2$. For any term in either sum in (3.76c), the argument μ of $\widetilde{V}$ satisfies $|\mu| \le m$, and if μ arises in a term in the first sum,

$$L_1(\mu) = L_1(\nu_1,\dots,\nu_j - 1,\dots,\nu_{2\ell}) = L_1(\nu) - p_j < -1 ,$$

while if μ arises in a term in the second sum,

$$L_1(\mu) = L_1(\nu_1,\dots,\nu_j - 1,\dots,\nu_{2\ell}) = L_1(\nu) - q_{2\ell-j+1} < -2 ,$$

so that in either case, if $\widetilde{V}(\mu)$ is not automatically zero, then by the induction hypothesis $\widetilde{V}(\mu) = V(\mu) = 0$. The proof when $L_2(\nu) < -2$ is similar.

Now suppose $L_1(\nu) = -2$. For any term in either sum in (3.76c), the argument μ of $\widetilde{V}$ satisfies $|\mu| \le m$. For any term in the first sum in (3.76c), we have that $L_1(\mu) = L_1(\nu) - p_j = -2 - p_j$. If $p_j \neq -1$, then $L_1(\mu) \le -2$ and by the induction hypothesis $\widetilde{V}(\mu) = 0$. If $p_j = -1$, then $L_1(\mu) + 1 = -1 + 1 = 0$. In either case, the term is zero. Again, for any term in the second sum in (3.76c), we have that $L_1(\mu) = L_1(\nu) - q_{2\ell-j+1} = -2 - q_{2\ell-j+1} \le -2$, so by the induction hypothesis $\widetilde{V}(\mu) = 0$.

The proof when $L_2(\nu) = -2$ is similar. □

Here is the theorem that is the basis for an efficient computational algorithm for obtaining the focus quantities for family (3.3).

Theorem 3.4.5. *For a family of systems of the form* (3.3) *(family* (3.50)*), let Ψ be the formal series of the form* (3.52) *and let $\{g_{kk} : k \in \mathbb{N}\}$ be the polynomials in $\mathbb{C}[a,b]$ given by* (3.72) *and* (3.73)*, which satisfy the conditions of Theorem 3.4.2. Then*

1. for $\nu \in \operatorname{Supp}(v_{k_1,k_2})$, the coefficient $v^{(\nu)}_{k_1,k_2}$ of $[\nu]$ in v_{k_1,k_2} is $V(\nu)$,

2. for $\nu \in \operatorname{Supp}(g_{kk})$, the coefficient $g^{(\nu)}_{kk}$ of $[\nu]$ in g_{kk} is

$$
\begin{aligned}
g_{kk}^{(\nu)} &= -i\Bigg[\sum_{j=1}^{\ell} \widetilde{V}(\nu_1,\dots,\nu_j-1,\dots,\nu_{2\ell})(L_1(\nu_1,\dots,\nu_j-1,\dots,\nu_{2\ell})+1) \\
&\quad - \sum_{j=\ell+1}^{2\ell} \widetilde{V}(\nu_1,\dots,\nu_j-1,\dots,\nu_{2\ell})(L_2(\nu_1,\dots,\nu_j-1,\dots,\nu_{2\ell})+1)\Bigg],
\end{aligned} \tag{3.77}
$$

and

3. *the following identities hold:*

$$V(\widehat{\nu}) = V(\nu) \quad \textit{and} \quad g_{kk}^{(\widehat{\nu})} = -g_{kk}^{(\nu)} \qquad \textit{for all} \quad \nu \in \mathbb{N}_0^{2\ell} \tag{3.78a}$$

$$V(\nu) = g_{kk}^{(\nu)} = 0 \quad \textit{if} \quad \widehat{\nu} = \nu \neq (0,\dots,0). \tag{3.78b}$$

Proof. The proof of part (1) is by induction on k_1+k_2.

Basis step. For $k_1+k_2=0$, there are three polynomials, v_{-11}, v_{00}, and $v_{1,-1}$. $\operatorname{Supp}(v_{00}) = (0,\dots,0)$, and

$$v_{00} = 1\cdot a^0_{p_1,q_1}\cdots a^0_{p_\ell,q_\ell} b^0_{q_\ell,p_\ell}\cdots b^0_{q_1,p_1} = V((0,\dots,0))\cdot a^0_{p_1,q_1}\cdots a^0_{p_\ell,q_\ell} b^0_{q_\ell,p_\ell}\cdots b^0_{q_1,p_1},$$

as required. Since $\operatorname{Supp}(v_{-11}) = \operatorname{Supp}(v_{1,-1}) = \varnothing$, statement (1) holds vacuously for them.

Inductive step. Suppose statement (1) holds for v_{k_1,k_2} for $k_1+k_2 \le m$, and let k_1 and k_2 be such that $k_1+k_2 = m+1$. If $k_1 = k_2$, then $v_{k_1,k_2} = 0$ by Theorem 3.4.2(3). By Theorem 3.4.2(2), for any $\nu \in \operatorname{Supp}(v_{k_1,k_2})$, $L(\nu) = (k_1,k_2)$, so $L_1(\nu) = L_2(\nu)$ and $V(\nu) = 0$, as required. If $k_1 \neq k_2$, then by (3.72)

$$
\begin{aligned}
v_{k_1,k_2} &= \frac{1}{k_1-k_2} \sum_{\substack{s_1+s_2=0\\ s_1,s_2\ge -1}}^{k_1+k_2-1} [(s_1+1)a_{k_1-s_1,k_2-s_2} - (s_2+1)b_{k_1-s_1,k_2-s_2}]v_{s_1,s_2} \\
&= \frac{1}{k_1-k_2} \sum_{\substack{s_1+s_2=0\\ s_1,s_2\ge -1}}^{k_1+k_2-1} \Bigg[(s_1+1)\Bigg(\sum_{\mu\in \operatorname{Supp}(v_{s_1,s_2})} v^{(\mu)}_{s_1,s_2}[\mu]a_{k_1-s_1,k_2-s_2}\Bigg) \\
&\qquad - (s_2+1)\Bigg(\sum_{\mu\in \operatorname{Supp}(v_{s_1,s_2})} v^{(\mu)}_{s_1,s_2}[\mu]b_{k_1-s_1,k_2-s_2}]\Bigg)\Bigg] \\
&= \frac{1}{k_1-k_2} \sum_{\substack{s_1+s_2=0\\ s_1,s_2\ge -1}}^{k_1+k_2-1} \Bigg[(s_1+1) \sum_{\mu\in \operatorname{Supp}(v_{s_1,s_2})} v^{(\mu)}_{s_1,s_2} a^{\mu_1}_{p_1,q_1}\cdots a^{\mu_c+1}_{p_c,q_c}\cdots b^{\mu_{2\ell}}_{q_1,p_1} \\
&\qquad - (s_2+1) \sum_{\mu\in \operatorname{Supp}(v_{s_1,s_2})} v^{(\mu)}_{s_1,s_2} a^{\mu_1}_{p_1,q_1}\cdots b^{\mu_{2\ell-d+1}+1}_{q_d,p_d}\cdots b^{\mu_{2\ell}}_{q_1,p_1}\Bigg],
\end{aligned} \tag{3.79}
$$

where $(p_c, q_c) = (k_1 - s_1, k_2 - s_2)$ (provided $(k_1 - s_1, k_2 - s_2) \in S$, else by convention the product is zero) and $(q_d, p_d) = (k_1 - s_1, k_2 - s_2)$ (provided $(k_2 - s_2, k_1 - s_1) \in S$, else by convention the product is zero).

Fix $\nu \in \mathbb{N}_0^{2\ell}$ for which $L(\nu) = (k_1, k_2)$. We wish to find the coefficient $v_{k_1,k_2}^{(\nu)}$ of $[\nu]$ in v_{k_1,k_2}. For a fixed $j \in \{1, \ldots, \ell\}$, we first ask which pairs (s_1, s_2) are such that $(p_c, q_c) = (k_1 - s_1, k_2 - s_2) = (p_j, q_j)$. There is at most one such pair: $s_1 = k_1 - p_j$ and $s_2 = k_2 - q_j$; it exists if and only if $k_1 - p_j \geq -1$ and $k_2 - q_j \geq -1$. For that pair, we then ask which $\mu \in \mathbb{N}_0^{2\ell}$ are such that $(\mu_1, \ldots, \mu_j + 1, \ldots, \mu_{2\ell}) = (\nu_1, \ldots, \nu_{2\ell})$. There is at most one such multi-index: $(\mu_1, \mu_2, \ldots, \mu_{2\ell}) = (\nu_1, \ldots, \nu_j - 1, \ldots, \nu_{2\ell})$; it exists if and only if $\nu_j \geq 1$. For this μ,

$$L(\mu) = \nu_1(p_1, q_1) + \cdots + \nu_{2\ell}(q_1, p_1) - (p_j, q_j) = (k_1 - p_j, k_2 - q_j) = (s_1, s_2),$$

although $\mu \notin \operatorname{Supp}(v_{s_1,s_2})$ is possible. Applying the same considerations to the cases $(q_d, p_d) = (q_{2\ell-j+1}, p_{2\ell-j+1})$ for $j = \ell+1, \ldots, 2\ell$, we see that for any term $v_{k_1,k_2}^{(\nu)}[\nu]$ appearing in v_{k_1,k_2} there is at most one term on the right-hand side of (3.79) for which the value of c is 1, at most one for which the value of c is 2, and so on through $c = \ell$, and similarly at most one term for which the value of d is ℓ, at most one for which the value of d is $\ell - 1$, and so on through $d = 1$. Thus the coefficient of $[\nu]$ in (3.72) is (recalling that v_{k_1,k_2} is a (k_1, k_2)-polynomial so that $(k_1, k_2) = (L_1(\nu), L_2(\nu))$, and similarly for v_{s_1,s_2}):

$$\begin{aligned} v_{k_1,k_2}^{(\nu)} &= \frac{1}{L_1(\nu) - L_2(\nu)} \\ &\times \Bigg[\sum_{j=1}^{\ell}{}' (L_1(\nu_1, \ldots, \nu_j - 1, \ldots, \nu_{2\ell}) + 1) v_{k_1-p_j,k_2-q_j}^{(\nu_1,\ldots,\nu_j-1,\ldots,\nu_{2\ell})} \\ &\quad - \sum_{j=\ell+1}^{2\ell}{}' (L_2(\nu_1, \ldots, \nu_j - 1, \ldots, \nu_{2\ell}) + 1) v_{k_1-q_{2\ell-j+1},k_2-p_{2\ell-j+1}}^{(\nu_1,\ldots,\nu_j-1,\ldots,\nu_{2\ell})} \Bigg], \end{aligned} \tag{3.80}$$

where the prime on the first summation symbol indicates that if (i) $\nu_j - 1 < 0$, or if (ii) $k_1 - p_j < -1$, or if (iii) $k_2 - q_j < -1$, or if (iv) on the contrary $\nu_j - 1 \geq 0$, $k_1 - p_j \geq -1$, and $k_2 - q_j \geq -1$, but $(\nu_1, \ldots, \nu_j - 1, \ldots, \nu_{2\ell}) \notin \operatorname{Supp}(v_{k_1-p_j,k_2-q_j})$, then the corresponding term does not appear in the sum, and the prime on the second summation symbol has a similar meaning.

If in either sum j is such that $\nu_j - 1 < 0$, then since the corresponding term does not appear, if we replace $v_{k_1-p_j,k_2-q_j}^{(\nu_1,\ldots,\nu_j-1,\ldots,\nu_{2\ell})}$ by $\widetilde{V}(\nu_1, \ldots, \nu_j - 1, \ldots, \nu_{2\ell})$ the sum is unchanged, since the latter is zero in this situation.

In the first sum, suppose j is such that $\nu_j - 1 \geq 0$. If both $k_1 - p_j \geq -1$ and $k_2 - q_j \geq -1$, then there are two subcases. If the corresponding term appears in the sum, then because $|\nu_1 + \cdots + (\nu_j - 1) + \cdots + \nu_{2\ell}| \leq m$, the induction hypothesis applies and $v_{k_1-p_j,k_2-q_j}^{(\nu_1,\ldots,\nu_j-1,\ldots,\nu_{2\ell})} = V(\nu_1, \ldots, \nu_j - 1, \ldots, \nu_{2\ell})$. Since in this situation $\widetilde{V}(\nu_1, \ldots, \nu_j - 1, \ldots, \nu_{2\ell}) = V(\nu_1, \ldots, \nu_j - 1, \ldots, \nu_{2\ell})$, in the corresponding term we may replace $v_{k_1-p_j,k_2-q_j}^{(\nu_1,\ldots,\nu_j-1,\ldots,\nu_{2\ell})}$ by $\widetilde{V}(\nu_1,\ldots,\nu_j-1,\ldots,\nu_{2\ell})$ and the sum is unchanged. The second subcase is that in which the corresponding term does not appear, meaning that $(\nu_1,\ldots,\nu_j-1,\ldots,\nu_{2\ell}) \notin \mathrm{Supp}(v_{k_1-p_j,k_2-q_j})$. But again the induction hypothesis applies, and now yields $V(\nu_1,\ldots,\nu_j-1,\ldots,\nu_{2\ell}) = 0$, so again the sum is unchanged by the same replacement.

Finally, suppose that in the first sum j is such that $\nu_j - 1 \geq 0$ but that either $k_1 - p_j < -1$ or $k_2 - q_j < -1$, so the corresponding term is not present in the sum. Then because $L(\nu_1,\ldots,\nu_j-1,\ldots,\nu_{2\ell}) = (k_1-p_j, k_2-q_j)$, Lemma 3.4.4 applies, by which we can make the same replacement as above, and thus the first sum in (3.80) is the same as the first sum in (3.76c). The second sum in (3.80) is treated similarly. This proves point (1). The same argument as in the inductive step with only slight modification here and there gives point (2).

Turning to point (3), first note that it follows directly from the definitions of L and $\widehat{\nu}$ that for all $\nu \in \mathbb{N}_0^{2\ell}$,

$$L_1(\widehat{\nu}) = L_2(\nu) \qquad \text{and} \qquad L_2(\widehat{\nu}) = L_1(\nu), \tag{3.81}$$

and that if $\widehat{\nu} = \nu$ then $L_1(\nu) = L_2(\nu)$. This latter statement means that if $\widehat{\nu} = \nu$ and $\nu \neq (0,\ldots,0)$ then by definition of V, $V(\nu) = 0$. Of course if $\nu = (0,\ldots,0)$ then $\widehat{\nu} = \nu$, and by definition $V((0,\ldots,0)) = 1$. In sum, for all $\nu \in \mathbb{N}_0^{2\ell}$ such that $\widehat{\nu} = \nu \neq (0,\ldots,0)$, $V(\nu) = 0$, the assertion about V in (3.78b). This implies that $\widetilde{V}(\nu) = 0$ for all $\nu \in \mathbb{N}_0^{2\ell}$ such that $\widehat{\nu} = \nu \neq (0,\ldots,0)$. We prove that $V(\widehat{\nu}) = V(\nu)$ for $\widehat{\nu} \neq \nu$ by induction on $|\nu|$.

Basis step. The equality $|\nu| = 0$ holds only if $\nu = (0,\ldots,0)$, hence only if $\widehat{\nu} = \nu$, so the statement is vacuously true.

Inductive step. Suppose that if $|\nu| \leq m$ and $\widehat{\nu} \neq \nu$ then $V(\widehat{\nu}) = V(\nu)$, hence $\widetilde{V}(\widehat{\nu}) = \widetilde{V}(\nu)$. Thus, more succinctly, $V(\widehat{\nu}) = V(\nu)$ (and $\widetilde{V}(\widehat{\nu}) = \widetilde{V}(\nu)$) if $|\nu| \leq m$. Fix $\nu \in \mathbb{N}_0^{2\ell}$ for which $|\nu| = m+1$ and $\widehat{\nu} \neq \nu$. In general (whether or not $\widehat{\nu} = \nu$), $L_1(\nu) = L_2(\nu)$ if and only if $L_1(\widehat{\nu}) = L_2(\widehat{\nu})$. Thus if $L_1(\nu) = L_2(\nu)$ then it is also true that $L_1(\widehat{\nu}) = L_2(\widehat{\nu})$ and by definition both $V(\nu) = 0$ and $V(\widehat{\nu}) = 0$, so that $V(\widehat{\nu}) = V(\nu)$.

If $L_1(\nu) \neq L_2(\nu)$,

$$\begin{aligned}
V(\widehat{\nu}) = & \frac{1}{L_1(\widehat{\nu}) - L_2(\widehat{\nu})} \\
& \times \big[\widetilde{V}(\nu_{2\ell}-1, \nu_{2\ell-1},\ldots,\nu_1)(L_1(\nu_{2\ell}-1,\nu_{2\ell-1},\ldots,\nu_1)+1) \\
& + \widetilde{V}(\nu_{2\ell}, \nu_{2\ell-1}-1,\ldots,\nu_1)(L_1(\nu_{2\ell},\nu_{2\ell-1}-1,\ldots,\nu_1)+1) \\
& + \cdots \\
& + \widetilde{V}(\nu_{2\ell},\ldots,\nu_{\ell+1}-1,\nu_\ell,\ldots,\nu_1)(L_1(\nu_{2\ell},\ldots,\nu_{\ell+1}-1,\nu_\ell,\ldots,\nu_1)+1) \\
& - \widetilde{V}(\nu_{2\ell},\ldots,\nu_{\ell+1},\nu_\ell-1,\ldots,\nu_1)(L_2(\nu_{2\ell},\ldots,\nu_{\ell+1},\nu_\ell-1,\ldots,\nu_1)+1) \\
& - \cdots \\
& - \widetilde{V}(\nu_{2\ell},\ldots,\nu_2,\nu_1-1)(L_2(\nu_{2\ell},\ldots,\nu_2,\nu_1-1)+1)\big],
\end{aligned}$$

which, by the induction hypothesis and (3.81), is equal to

$$
\begin{aligned}
&\frac{1}{L_2(\nu)-L_1(\nu)} \\
&\times \big[\widetilde{V}(\nu_1,\ldots,\nu_{2\ell-1},\nu_{2\ell}-1)(L_2(\nu_1,\ldots,\nu_{2\ell-1},\nu_{2\ell}-1)+1) \\
&\quad +\widetilde{V}(\nu_1,\ldots,\nu_{2\ell-1}-1,\nu_{2\ell})(L_2(\nu_1,\ldots,\nu_{2\ell-1}-1,\nu_{2\ell})+1) \\
&\quad +\cdots \\
&\quad +\widetilde{V}(\nu_1,\ldots,\nu_\ell,\nu_{\ell+1}-1,\ldots,\nu_{2\ell})(L_2(\nu_1,\ldots,\nu_\ell,\nu_{\ell+1}-1,\ldots,\nu_{2\ell})+1) \\
&\quad -\widetilde{V}(\nu_1,\ldots,\nu_\ell-1,\nu_{\ell+1},\ldots,\nu_{2\ell})(L_1(\nu_1,\ldots,\nu_\ell-1,\nu_{\ell+1},\ldots,\nu_{2\ell})+1) \\
&\quad -\cdots \\
&\quad -\widetilde{V}(\nu_1-1,\nu_2,\ldots,\nu_{2\ell})(L_1(\nu_1-1,\nu_2,\ldots,\nu_{2\ell})+1)\big] = V(\nu).
\end{aligned}
$$

Thus $V(\widehat{\nu}) = V(\nu)$ if $\widehat{\nu} \neq \nu$, for $|\nu| = m+1$, hence for all $\nu \in \mathbb{N}_0^{2\ell}$.

The identity $V(\widehat{\nu}) = V(\nu)$ for all $\nu \in \mathbb{N}_0^{2\ell}$ and formula (3.77) immediately imply that $g^{(\widehat{\nu})} = -g^{(\nu)}$ for all $\nu \in \mathbb{N}_0^{2\ell}$. Thus (3.78a) is proved, and it implies the assertion about $g_{kk}^{(\nu)}$ in (3.78b). The truth of the assertion about V in (3.78b) has already been noted. □

Theorem 3.4.2(4) and equation (3.78a) show that the focus quantities have the following structure (Exercise 3.16; see also Exercise 4.3), identifiable in (3.133) and (3.134) (but not present in (3.135) because of a reduction modulo $\langle g_{11}, g_{22}\rangle$).

Corollary 3.4.6. *The focus quantities have the form*

$$
g_{kk} = \tfrac{1}{2} \sum_{\{\nu : L(\nu) = (k,k)\}} g_{kk}^{(\nu)}([\nu] - [\widehat{\nu}]) . \tag{3.82}
$$

Remark 3.4.7. Noting that by (3.77) the coefficient $g_{kk}^{(\nu)}$ of $[\nu]$ in g_{kk} is purely imaginary, it is an easy consequence of (3.82) that if system (3.3) (system (3.50)) is the complexification of a real system, that is, if $\bar{b} = a$, then for every fixed $a \in \mathbb{C}^\ell$, g_{kk} is a real number.

The results of this section yield the Focus Quantity Algorithm for computation of the focus quantities for family (3.3) given in Table 3.1 on page 128, where in the last two lines we have abused the notation slightly. The formula for the maximum value of $|\nu|$ such that ν can contribute to g_{kk} (the quantity M in the algorithm) is established in Exercise 3.15; in that formula, $\lfloor r \rfloor$ denotes the greatest integer less than or equal to r. The code for an implementation of this algorithm in Mathematica is given in the Appendix.

Focus Quantity Algorithm

Input:

$K \in \mathbb{N}$
Ordered set $S = \{(p_1, q_1), \dots, (p_\ell, q_\ell)\} \subset (\{-1\} \times \mathbb{N}_0)^2$
satisfying $p_j + q_j \geq 1$, $1 \leq j \leq \ell$

Output:

Focus quantities g_{kk}, $1 \leq k \leq K$, for family (3.3)

Procedure:

$w := \min\{p_1 + q_1, \dots, p_\ell + q_\ell\}$
$M := \lfloor \frac{2K}{w} \rfloor$
$g_{11} := 0; \; \dots, \; g_{KK} := 0; \; V(0, \dots, 0) := 1;$
FOR $m = 1$ TO M DO
  FOR $\nu \in \mathbb{N}_0^{2\ell}$ such that $|\nu| = m$ DO
    Compute $L(\nu)$ using (3.71)
    Compute $V(\nu)$ using (3.76)
    IF
    $L_1(\nu) = L_2(\nu)$
    THEN
    Compute $g_{L(\nu)}^{(\nu)}$ using (3.77)
    $g_{L(\nu)} := g_{L(\nu)} + g_{L(\nu)}^{(\nu)}[\nu]$

Table 3.1 The Focus Quantity Algorithm

3.5 Hamiltonian and Reversible Systems

As we showed in the opening paragraphs of this chapter, it is important that we have means of determining when a system satisfying certain conditions has a center at the origin. In this and the next section we describe several such techniques. This section is devoted to a discussion of two classes of systems that have the property that, in the real case, any antisaddle at the origin must be a center. Because they possess this property, we will find these two classes naturally appearing as components of the center varieties that we study in Section 3.7.

The first class with which we deal is the set of Hamiltonian systems. A system $\dot{x} = \widetilde{P}(x,y)$, $\dot{y} = \widetilde{Q}(x,y)$ on $\mathbb{C}^2$ is said to be a *Hamiltonian system* if there is a function $H : \mathbb{C}^2 \to \mathbb{C}$, called the *Hamiltonian* of the system, such that $\widetilde{P} = -H_y$ and $\widetilde{Q} = H_x$. It is immediately apparent that $\mathscr{X} H \equiv 0$, so that H is a first integral of the system. Thus an antisaddle of any real Hamiltonian system, and the singularity at the origin of any complex Hamiltonian system of the form (3.3) (= (3.50)) or (3.61), is known to be a center. Hamiltonian systems are easy to detect, as the following proposition shows.

Proposition 3.5.1. *The collection of Hamiltonian systems in a family of the form* (3.3) *are precisely those systems whose coefficients satisfy the following condition:*

$$\begin{aligned}&\text{for all } (p,q)\in S \text{ for which } p\geq 0:\\&\quad \text{if } (q,p)\in S, \text{ then } (p+1)a_{pq}=(q+1)b_{pq};\\&\quad \text{if } (q,p)\notin S, \text{ then } a_{pq}=b_{qp}=0.\end{aligned} \tag{3.83}$$

Proof. If there exists a function $H:\mathbb{C}^2\to\mathbb{C}$ such that $\widetilde{P}=H_y$ and $\widetilde{Q}=-H_x$, then $\widetilde{P}_x=-\widetilde{Q}_y$, which from (3.3) is the condition

$$\sum_{\substack{(p,q)\in S\\ p\geq 0}}(p+1)a_{pq}x^py^q=\sum_{\substack{(p,q)\in S\\ p\geq 0}}(p+1)b_{qp}x^qy^p\,. \tag{3.84}$$

If $(r,s)\in\mathbb{N}_0^2$ is such that both $(r,s)\in S$ and $(s,r)\in S$, then the monomial x^ry^s appears in the left-hand side of equation (3.84) with coefficient $(r+1)a_{rs}$ and in the right-hand side with coefficient $(s+1)b_{rs}$, so that $(r+1)a_{rs}=(s+1)b_{rs}$ holds. But if $(r,s)\in\mathbb{N}_0^2$ is such that $(r,s)\in S$ but $(s,r)\notin S$, then the monomial x^ry^s appears in the left-hand side of equation (3.84) with coefficient $(r+1)a_{rs}$ (and $r\geq 0$) but not in the right-hand side, so that $a_{rs}=0$, and the monomial x^sy^r appears in the right-hand side of equation (3.84) with coefficient $(r+1)b_{sr}$ (and $r\geq 0$) but not in the left-hand side, so that $b_{sr}=0$. Thus condition (3.83) holds.

Conversely, if condition (3.83) holds, then because the system is polynomial, $\widetilde{P}(x,y)$ can be integrated with respect to y and $-\widetilde{Q}(x,y)$ can be integrated with respect to x consistently to produce a polynomial H that is a Hamiltonian for system (3.3). □

By Proposition 3.5.1, the set of Hamiltonian systems in family (3.3) corresponds precisely to the variety of the ideal

$$\begin{aligned}I_{\text{Ham}}\stackrel{\text{def}}{=}&\ \langle (p+1)a_{pq}-(q+1)b_{pq}: p\geq 0 \text{ and } (p,q)\in S \text{ and } (q,p)\in S\rangle\\&\cap\langle a_{pq},b_{qp}: p\geq 0 \text{ and } (p,q)\in S \text{ and } (q,p)\notin S\rangle\,.\end{aligned} \tag{3.85}$$

If there are r_1 generators in the first ideal on the right in (3.85) and $2r_2$ in the second, then a polynomial parametrization of I_{Ham} is given by the mapping $F:\mathbb{C}^{2\ell-(r_1+2r_2)}\to\mathbb{C}^{2\ell}$, where r_1 component functions have the form $b_{pq}=t_j$ and another corresponding r_1 component functions have the form $a_{pq}=(\frac{q+1}{p+1})t_j$; where r_2 component functions have the form $a_{pq}=0$ and r_2 component functions have the form $b_{qp}=0$; and where the remaining $2\ell-(r_1+2r_2)$ component functions, corresponding to a_{pq} and b_{qp} not appearing in any generator of I_{Ham}, have the form $a_{pq}=t_j$ or $b_{qp}=t_j$. By Corollary 1.4.18, $\mathbf{V}(I_{\text{Ham}})$ is irreducible.

The second class of systems that we discuss in this section are those having a symmetry that in the real case forces any singularity of focus or center type to be in fact a center. When direction of flow is taken into account, there are two types of symmetry of a real system with respect to a line L: *mirror symmetry*, meaning that when the phase portrait is reflected in the line L it is unchanged; and *time-reversible symmetry* (standard terminology, although *reversed-time symmetry* might be more accurate), meaning that when the phase portrait is reflected in the line L and then the sense of every orbit is reversed (corresponding to a reversal of time), the original phase portrait is obtained. When L is the u-axis, the former situation is exemplified by the canonical linear saddle $\dot{u} = -u$, $\dot{v} = v$, and the latter situation is exemplified by the canonical linear center $\dot{u} = -v$, $\dot{v} = u$. The canonical saddle has four lines of symmetry in all, two exhibiting mirror symmetry and two exhibiting time-reversible symmetry; every line through the origin is a line of time-reversible symmetry for the center. In general, a real system $\dot{u} = \widetilde{U}(u,v)$, $\dot{v} = \widetilde{V}(u,v)$ possesses a time-reversible symmetry with respect to the u-axis if and only if

$$\widetilde{U}(u,-v) = -\widetilde{U}(u,v) \quad \text{and} \quad \widetilde{V}(u,-v) = \widetilde{V}(u,v); \tag{3.86a}$$

it possesses a mirror symmetry with respect to the u-axis if and only if

$$\widetilde{U}(u,-v) = \widetilde{U}(u,v) \quad \text{and} \quad \widetilde{V}(u,-v) = -\widetilde{V}(u,v) \tag{3.86b}$$

(Exercise 3.20). Time-reversible symmetry is of interest in connection with the center-focus problem because for any antisaddle on L, presence of the symmetry precludes the possibility that the singularity be a focus, hence forces it to be of center type. For example, consider the real system

$$\dot{u} = -v - vM(u,v^2), \qquad \dot{v} = u + N(u,v^2), \tag{3.87}$$

where M is an analytic function without constant term and N is an analytic function whose series expansion at $(0,0)$ starts with terms of order at least two. We know by Remark 3.1.7 that the origin is a either a focus or a center. The fact that system (3.87) satisfies condition (3.86a) implies that all orbits near the origin close, so that the origin is in fact a center. Mirror symmetry will not be of interest to us in the present chapter since it is incompatible with the existence of either a focus or a center on the line L. We will take it up again in the last section of Chapter 5.

Definition 3.5.2. A real system $\dot{u} = \widetilde{U}(u,v)$, $\dot{v} = \widetilde{V}(u,v)$ is *time-reversible* if its phase portrait is invariant under reflection with respect to a line and a change in the direction of time (reversal of the sense of every trajectory).

We say that a line L is an *axis of symmetry* of system (3.88) if as point sets (ignoring the sense of the parametrization by time t) the orbits of the system are symmetric with respect to L. Henceforth in this section, except for Definition 3.5.4, which applies to general analytic systems (as did Definition 3.5.2), we restrict to polynomial systems.

The point of view that we take throughout this book is that it is generally easier and much more fruitful to study the center problem for a complex system than it is to do so for a real one. Thus our first objective in this part of this section is to generalize the concept of reversibility to complex systems. As usual, we complexify the (u,v)-plane by making use of the substitution $x = u + iv$. Then the real system

$$\dot{u} = \widetilde{U}(u,v), \qquad \dot{v} = \widetilde{V}(u,v) \tag{3.88}$$

is transformed into

$$\dot{x} = \widetilde{P}(x,\bar{x}), \tag{3.89}$$

where $\widetilde{P}(x,\bar{x}) = \widetilde{U}(\frac{1}{2}(x+\bar{x}), \frac{1}{2i}(x-\bar{x})) + i\widetilde{V}(\frac{1}{2}(x+\bar{x}), \frac{1}{2i}(x-\bar{x}))$. Both mirror symmetry and time-reversible symmetry with respect to the u-axis are captured by a simple condition on the vector a of coefficients of the polynomial $\widetilde{P}$: system (3.88) has one of these two types of symmetry with respect to the u-axis if and only if

$$\widetilde{U}(u,-v) + i\widetilde{V}(u,-v) \equiv \pm(\widetilde{U}(u,v) - i\widetilde{V}(u,v)),$$

which in terms of $\widetilde{P}$ is precisely the condition $\widetilde{P}(\bar{x},x) = \pm\overline{\widetilde{P}(x,\bar{x})}$, as can easily be seen by writing it out in detail, and this latter condition is equivalent to $\bar{a} = \pm a$. The upper sign everywhere corresponds to mirror symmetry; the lower sign everywhere corresponds to time-reversible symmetry. Thus we have established the following fact.

Lemma 3.5.3. *Let a denote the vector of coefficients of the polynomial $\widetilde{P}(x,\bar{x})$ in* (3.89).

1. *System* (3.88) *exhibits mirror symmetry (respectively, time-reversible symmetry) with respect to the u-axis if and only if $a = \bar{a}$ (respectively, $a = -\bar{a}$), that is, if and only if all coefficients are real (respectively, all are purely imaginary).*
2. *If $a = \pm\bar{a}$, then the u-axis is an axis of symmetry for system* (3.88).

The condition in part (2) is only sufficient because the u-axis can be an axis of symmetry in the absence of both mirror and time-reversible symmetry; see Section 5.3.

By the lemma the u-axis is an axis of symmetry for (3.89) if

$$\widetilde{P}(\bar{x},x) = -\overline{\widetilde{P}(x,\bar{x})}, \tag{3.90a}$$

corresponding to (3.86a), time-reversible symmetry, or if

$$\widetilde{P}(\bar{x},x) = \overline{\widetilde{P}(x,\bar{x})}, \tag{3.90b}$$

corresponding to (3.86b), mirror symmetry. If (3.90a) is satisfied, then under the involution

$$x \to \bar{x}, \qquad \bar{x} \to x, \tag{3.91}$$

(3.89) is transformed into its negative,

$$\dot{x} = -\widetilde{P}(x,\bar{x}), \qquad \dot{\bar{x}} = -\overline{\widetilde{P}(x,\bar{x})}. \tag{3.92}$$

So if (3.89) is obtained from (3.88) and the transformation (3.91) when applied to (3.89) and its complex conjugate yields the system (3.92), then the real system (3.88) is time-reversible, hence has a center at the origin. If the line of reflection is not the u-axis but a distinct line L passing through the origin, then we can apply the rotation $x_1 = e^{-i\varphi}x$ through an appropriate angle φ to make L the u-axis. In the new coordinates we have

$$\dot{x}_1 = e^{-i\varphi}\widetilde{P}(e^{i\varphi}x_1, e^{-i\varphi}\bar{x}_1) := \widetilde{P}_1(x_1,\bar{x}_1).$$

According to Lemma 3.5.3, this system is time-reversible with respect to the line $\operatorname{Im} x_1 = 0$ if the analogue $\widetilde{P}_1(\bar{x}_1, x_1) = -\overline{\widetilde{P}_1(x_1,\bar{x}_1)}$ of (3.90a) holds, which by a straightforward computation is

$$e^{i\varphi}\overline{\widetilde{P}(e^{i\varphi}x_1, e^{-i\varphi}\bar{x}_1)} = -e^{-i\varphi}\widetilde{P}(e^{i\varphi}\bar{x}_1, e^{-i\varphi}x_1).$$

Hence (3.89) is time-reversible precisely when there exists a $\varphi \in \mathbb{R}$ such that

$$e^{2i\varphi}\overline{\widetilde{P}(x,\bar{x})} = -\widetilde{P}(e^{2i\varphi}\bar{x}, e^{-2i\varphi}x). \tag{3.93}$$

Recall from the beginning of Section 3.2 that the complexification of system (3.88) is the system that is obtained by adjoining to (3.89) its complex conjugate. An examination of (3.93) and its conjugate suggests the following natural generalization of time-reversibility to systems on $\mathbb{C}^2$. For ease of comparison with the mathematical literature we have stated it in two equivalent forms.

Definition 3.5.4. A system $\dot{x} = \widetilde{P}(x,y)$, $\dot{y} = \widetilde{Q}(x,y)$ on $\mathbb{C}^2$ is *time-reversible* if there exists $\gamma \in \mathbb{C}\setminus\{0\}$ such that

$$\widetilde{P}(x,y) = -\gamma\widetilde{Q}(\gamma y, \gamma^{-1}x). \tag{3.94}$$

Equivalently, letting $\mathbf{z} = (x,y) \in \mathbb{C}^2$, the system

$$\frac{d\mathbf{z}}{dt} = F(\mathbf{z})$$

is *time-reversible* if there exists a linear transformation $T(x,y) = (\gamma y, \gamma^{-1}x)$, for some $\gamma \in \mathbb{C}\setminus\{0\}$, such that

$$\frac{d(T\mathbf{z})}{dt} = -F(T\mathbf{z}).$$

The equivalence of the two forms of the definition follows from the fact, readily verified by direct computation, that a complex system is time-reversible according to the first statement in Definition 3.5.4 if and only if under the change of coordinates

$$x_1 = \gamma y,\ y_1 = \gamma^{-1}x \tag{3.95}$$

and reversal of time the form of the system of differential equations is unaltered.

We note in particular that a system of the form (3.3) is time-reversible if and only if the parameter (a,b) satisfies

$$b_{qp} = \gamma^{p-q}a_{pq} \qquad \text{for all} \qquad (p,q) \in S \tag{3.96}$$

for some fixed $\gamma \in \mathbb{C} \setminus \{0\}$. Observe also that if we complexify the real system (3.88) written in the complex form (3.89) by adjoining to the latter the equation $\dot{\bar{x}} = \widetilde{Q}(x,\bar{x}) = \overline{\widetilde{P}(x,\bar{x})}$, then setting $\gamma = e^{2i\varphi}$ in (3.94) and replacing y by $\bar{x}$ we recover (3.93), as anticipated.

As we have seen above, a real polynomial system that has a singularity that must be either a center or focus and that is time-reversible with respect to a line passing through this point must have a center at the point. We will now show that the analogous fact for complex systems of the form (3.3) is also true.

Theorem 3.5.5. *Every time-reversible system of the form* (3.3) *(not necessarily satisfying $b_{qp} = \bar{a}_{pq}$) has a center at the origin.*

Proof. Suppose the system of the form (3.3) under consideration corresponds to parameter value (a,b), so that condition (3.96) holds for some $\gamma \in \mathbb{C} \setminus \{0\}$. We will compute the value of the focus quantities g_{kk} at (a,b) using (3.82). Hence fix $k \in \mathbb{N}$. Since g_{kk} is a (k,k)-polynomial (Theorem 3.4.2), each $\nu \in \operatorname{Supp}(g_{kk})$ satisfies $L(\nu) = (k,k)$, which by (3.71) is

$$\begin{aligned} \nu_1 p_1 + \cdots + \nu_\ell p_\ell + \nu_{\ell+1} q_\ell + \cdots + \nu_{2\ell} q_1 &= k \\ \nu_1 q_1 + \cdots + \nu_\ell q_\ell + \nu_{\ell+1} p_\ell + \cdots + \nu_{2\ell} p_1 &= k\,. \end{aligned} \tag{3.97}$$

For any such ν, evaluating $[\nu]$ and $\widehat{[\nu]}$ at (a,b) and applying condition (3.96) yields the relation $\widehat{[\nu]} = \gamma^w [\nu]$, where

$$w = \nu_{2\ell}(q_1 - p_1) + \cdots + \nu_{\ell+1}(q_\ell - p_\ell) + \nu_\ell(p_\ell - q_\ell) + \cdots + \nu_1(p_1 - q_1)\,.$$

But (3.97) implies that w is zero, so that by (3.82) $g_{kk}(a,b) = 0$. Since this holds for all k, system (a,b) has a center at the origin. □

The set of time-reversible systems is not generally itself a variety, for although condition (3.96) is a polynomial condition, the value of γ will vary from system to system, so that the full set of time-reversible systems is not picked out by one uniform set of polynomial conditions. It can be made into a variety by forming its Zariski closure, but the proof of Theorem 3.5.5 suggests the following idea for defining a readily identifiable variety that contains it, which turns out to be a useful approach, and will be shown in Section 5.2 to be equivalent. For a fixed index set S determining a family (3.3) and the corresponding mapping L given by (3.71), define

$$\mathscr{M} = \{\nu \in \mathbb{N}_0^{2\ell} : L(\nu) = (j,j) \text{ for some } j \in \mathbb{N}_0\} \tag{3.98}$$

and let I_{sym} be the ideal defined by

$$I_{\text{sym}} \stackrel{\text{def}}{=} \langle [\nu] - [\hat{\nu}] : \nu \in \mathscr{M} \rangle \subset \mathbb{C}[a,b],$$

which is termed the *symmetry ideal* or the *Sibirsky ideal* for family (3.3). It is almost immediate that any time-reversible system in family (3.3) corresponds to an element of $\mathbf{V}(I_{\mathrm{sym}})$: for $\nu \in \mathscr{M}$ means that (3.97) holds, and if system (a,b) is reversible then $[\widehat{\nu}] = \gamma^w[\nu]$ for w as in the proof of Theorem 3.5.5, which by (3.97) is zero, hence $([\nu]-[\widehat{\nu}])|_{(a,b)} = 0$.

(The reader may have wondered why, in the definition of $\mathscr{M}$, the requirement on j is that it be in $\mathbb{N}_0$, when $j \in \mathbb{N}$ would have served. The reason is that the same set $\mathscr{M}$ will be encountered again in Chapter 5, where we will want it to have the structure of a monoid under addition, hence we need it to have an identity element.)

By definition of the Bautin ideal $\mathscr{B}$ (Definition 3.3.3) and equation (3.82) clearly $\mathscr{B} \subset I_{\mathrm{sym}}$, hence $\mathbf{V}(I_{\mathrm{sym}}) \subset \mathbf{V}(\mathscr{B})$, the center variety of system (3.3).

Definition 3.5.6. The variety $\mathbf{V}(I_{\mathrm{sym}})$ is termed the *symmetry* or *Sibirsky subvariety* of the center variety.

Although the parameters a and b of any time-reversible system (a,b) of the form (3.3) satisfy $(a,b) \in \mathbf{V}(I_{\mathrm{sym}})$, the converse is false, as the following example shows.

Example 3.5.7. We will find the symmetry ideal I_{sym} for the family

$$\begin{aligned} \dot{x} &= \quad i(x - a_{10}x^2 - a_{01}xy) \\ \dot{y} &= -i(y - b_{10}xy - b_{01}y^2)\,, \end{aligned} \tag{3.99}$$

family (3.3) when $S = \{(p_1,q_1),(p_2,q_2)\} = \{(1,0),(0,1)\}$, so $\ell = 2$. We must first identify the set $\mathscr{M} = \{\nu \in \mathbb{N}_0^4 : L(\nu) = (k,k) \text{ for some } k \in \mathbb{N}_0\}$. Writing $\nu = (\nu_1,\nu_2,\nu_3,\nu_4)$, $[\nu] = a_{10}^{\nu_1}a_{01}^{\nu_2}b_{10}^{\nu_3}b_{01}^{\nu_4} \in \mathbb{C}[a_{10},a_{01},b_{10},b_{01}]$, $\widehat{[\nu]} = a_{10}^{\nu_4}a_{01}^{\nu_3}b_{10}^{\nu_2}b_{01}^{\nu_1}$, and $L(\nu) = \nu_1(1,0)+\nu_2(0,1)+\nu_3(1,0)+\nu_4(0,1) = (\nu_1+\nu_3,\nu_2+\nu_4)$.

For $k=1$, $(\nu_1,\nu_3),(\nu_2,\nu_4) \in \{(1,0),(0,1)\}$, so there are four 4-tuples ν satisfying $L(\nu) = (1,1)$, and we compute:

$$\begin{aligned} \nu = (1,1,0,0) &: [\nu] - \widehat{[\nu]} = a_{10}a_{01} - b_{10}b_{01}, \\ \nu = (1,0,0,1) &: [\nu] - \widehat{[\nu]} \equiv 0, \\ \nu = (0,1,1,0) &: [\nu] - \widehat{[\nu]} \equiv 0, \\ \nu = (0,0,1,1) &: [\nu] - \widehat{[\nu]} = b_{10}b_{01} - a_{10}a_{01}. \end{aligned}$$

For $k=2$, $(\nu_1,\nu_3),(\nu_2,\nu_4) \in \{(2,0),(1,1),(0,2)\}$, so there are nine 4-tuples ν satisfying $L(\nu) = (2,2)$, and we compute:

$$\begin{aligned} \nu = (2,2,0,0) &: [\nu] - \widehat{[\nu]} = a_{10}^2a_{01}^2 - b_{10}^2b_{01}^2 = (a_{10}a_{01}-b_{10}b_{01})(a_{10}a_{01}+b_{10}b_{01}), \\ \nu = (2,1,0,1) &: [\nu] - \widehat{[\nu]} = a_{10}^2a_{01}b_{01} - a_{10}b_{10}b_{01}^2 = a_{10}b_{01}(a_{10}a_{01}-b_{10}b_{01}), \\ \nu = (2,0,0,2) &: [\nu] - \widehat{[\nu]} \equiv 0, \\ \nu = (1,2,1,0) &: [\nu] - \widehat{[\nu]} = a_{10}a_{01}^2b_{10} - a_{01}b_{10}^2b_{01} = a_{01}b_{10}(a_{10}a_{01}-b_{10}b_{01}). \end{aligned}$$

and so on for the remaining five 4-tuples.

The pattern seen so far holds for all k (Exercise 3.21):

$$I_{\mathrm{sym}} = \langle [\nu] - \widehat{[\nu]} : \nu \in \mathscr{M} \rangle = \langle a_{10}a_{01} - b_{10}b_{01} \rangle .$$

Thus $\mathbf{V}(I_{\mathrm{sym}}) = \{(a_{10}, a_{01}, b_{10}, b_{01}) : a_{10}a_{01} - b_{10}b_{01} = 0\}$.

In particular, $\mathbf{V}(I_{\mathrm{sym}})$ includes $(a_{10}, a_{01}, b_{10}, b_{01}) = (1,0,1,0)$, corresponding to the system $\dot{x} = i(x - x^2)$, $\dot{y} = -i(y - xy)$. But this system is not time-reversible, since for $(p,q) = (0,1)$, (3.94) reads $b_{10} = \gamma^{-1} a_{01}$, which here evaluates to $1 = 0$.

Thus to repeat, while every time-reversible system belongs to the symmetry subvariety of the center variety, the converse fails, as Example 3.5.7 shows. We will nevertheless use the terminology "symmetry" subvariety because the restriction of $\mathbf{V}(I_{\mathrm{sym}})$ to the real center variety $V_{\mathscr{C}}^{R}$ gives precisely the time-reversible real systems with a center and because the name is suggestive. The correct relationship between the set of time-reversible systems in family (3.3) and the symmetry subvariety is described by the following theorem.

Theorem 3.5.8. *Let $\mathscr{R} \subset E(a,b)$ be the set of all time-reversible systems in family* (3.3). *Then*
1. $\mathscr{R} \subset \mathbf{V}(I_{\mathrm{sym}})$ and
2. $\mathbf{V}(I_{\mathrm{sym}}) \setminus \mathscr{R}$
$= \{(a,b) \in \mathbf{V}(I_{\mathrm{sym}}) :$ *there exists* $(p,q) \in S$ *with* $a_{pq}b_{qp} = 0$ *but* $a_{pq} + b_{qp} \neq 0\}$.

We have already seen the proof of part (1). We will present the proof of part (2) in Chapter 5, where we also give a simple and effective algorithm for computing a finite set of generators for the ideal I_{sym} (Table 5.1) and establish the following important property that it possesses.

Theorem 3.5.9. *The ideal I_{sym} is prime in $\mathbb{C}[a,b]$.*

This theorem immediately implies that the variety $\mathbf{V}(I_{\mathrm{sym}})$ is irreducible. In fact, for all polynomial systems investigated up to this point, $\mathbf{V}(I_{\mathrm{sym}})$ is a component, that is, is a proper irreducible subvariety, of the center variety. We conjecture that this is always the case, that is, that for any polynomial system of the form (3.3), $\mathbf{V}(I_{\mathrm{sym}})$ is a component of the center variety.

Using the algorithm mentioned just after Theorem 3.5.8 one can compute the ideal I_{sym} for the general cubic system of the form (3.69):

$$\begin{aligned} \dot{x} &= x - a_{10}x^2 - a_{01}xy - a_{-12}y^2 - a_{20}x^3 - a_{11}x^2y - a_{02}xy^2 - a_{-13}y^3 \\ \dot{y} &= -y + b_{2,-1}x^2 + b_{10}xy + b_{01}y^2 + b_{3,-1}x^3 + b_{20}x^2y + b_{11}xy^2 + b_{02}y^3 \end{aligned} \tag{3.100}$$

(where x, y, a_{ij}, and b_{ij} are in $\mathbb{C}$), which by Proposition 3.3.9 has the same center variety as the general cubic system of the form (3.3).

Theorem 3.5.10. *The ideal I_{sym} of system* (3.100) *is generated by the polynomials listed in Table 3.2 on page 136. Thus the symmetry subvariety $\mathbf{V}(I_{\mathrm{sym}})$ of the center variety of system* (3.100) *is the set of common zeros of these polynomials.*

The proof of Theorem 3.5.10 will also be given in Chapter 5 (page 236), where we will also show that the pattern exhibited in Table 3.2, that every generator has the form $[\mu] - [\hat{\mu}]$ for some $\mu \in \mathbb{N}_0^{2\ell}$, holds in general (Theorem 5.2.5).

$a_{11} - b_{11}$	$a_{01}b_{02}b_{2,-1} - a_{-12}b_{10}a_{20}$
$a_{01}a_{02}b_{2,-1} - a_{-12}b_{10}b_{20}$	$a_{10}^4 a_{-13} - b_{3,-1}b_{01}^4$
$a_{10}a_{-12}b_{20} - b_{01}b_{2,-1}a_{02}$	$a_{10}a_{-12}b_{10}^2 - a_{01}^2 b_{2,-1}b_{01}$
$a_{20}a_{02} - b_{20}b_{02}$	$a_{10}^2 a_{-12}b_{10} - a_{01}b_{2,-1}b_{01}^2$
$a_{10}b_{02}b_{10} - a_{01}a_{20}b_{01}$	$a_{01}^3 b_{2,-1} - a_{-12}b_{10}^3$
$a_{10}a_{02}b_{10} - a_{01}b_{20}b_{01}$	$a_{10}^3 a_{-12} - b_{2,-1}b_{01}^3$
$a_{10}a_{-13}b_{2,-1} - a_{-12}b_{3,-1}b_{01}$	$a_{20}a_{-13}b_{20} - a_{02}b_{3,-1}b_{02}$
$a_{10}^2 b_{02} - a_{20}b_{01}^2$	$a_{02}^2 b_{3,-1} - a_{-13}b_{20}^2$
$a_{01}a_{-12}b_{3,-1} - a_{-13}b_{2,-1}b_{10}$	$a_{01}^2 b_{20} - a_{02}b_{10}^2$
$a_{20}^2 a_{-13} - b_{3,-1}b_{02}^2$	$a_{10}a_{-13}b_{20}b_{10} - a_{01}a_{02}b_{3,-1}b_{01}$
$a_{10}a_{20}a_{-13}b_{10} - a_{01}b_{3,-1}b_{02}b_{01}$	$a_{10}b_{02}^2 b_{2,-1} - a_{-12}a_{20}^2 b_{01}$
$a_{10}^2 a_{02} - b_{20}b_{01}^2$	$a_{10}a_{02}b_{02}b_{2,-1} - a_{-12}a_{20}b_{20}b_{01}$
$a_{10}a_{02}^2 b_{2,-1} - a_{-12}b_{20}^2 b_{01}$	$a_{01}^2 b_{3,-1}b_{02} - a_{20}a_{-13}b_{10}^2$
$a_{01}^2 a_{20} - b_{02}b_{10}^2$	$a_{01}a_{-12}b_{20}^2 - a_{02}^2 b_{2,-1}b_{10}$
$a_{10}^2 a_{-13}b_{20} - a_{02}b_{3,-1}b_{01}^2$	$a_{-12}a_{20}a_{10} - b_{02}b_{2,-1}b_{01}$
$a_{01}a_{-12}a_{20}b_{20} - a_{02}b_{02}b_{2,-1}b_{10}$	$a_{01}^2 a_{02}b_{3,-1} - a_{-13}b_{20}b_{10}^2$
$a_{10}^2 a_{20}a_{-13} - b_{3,-1}b_{02}b_{01}^2$	$a_{01}a_{-12}a_{20}^2 - b_{02}^2 b_{2,-1}b_{10}$
$a_{10}a_{-13}b_{10}^3 - a_{01}^3 b_{3,-1}b_{01}$	$a_{10}^2 a_{-13}b_{10}^2 - a_{01}^2 b_{3,-1}b_{01}^2$
$a_{10}^3 a_{-13}b_{10} - a_{01}b_{3,-1}b_{01}^3$	$a_{01}^4 b_{3,-1} - a_{-13}b_{10}^4$
$a_{01}a_{-13}b_{20}b_{2,-1} - a_{-12}a_{02}b_{3,-1}b_{10}$	$a_{01}a_{20}a_{-13}b_{2,-1} - a_{-12}b_{3,-1}b_{02}b_{10}$
$a_{10}a_{-12}b_{3,-1}b_{02} - a_{20}a_{-13}b_{2,-1}b_{01}$	$a_{10}a_{-12}a_{02}b_{3,-1} - a_{-13}b_{20}b_{2,-1}b_{01}$
$a_{-12}^2 b_{3,-1}b_{20} - a_{02}a_{-13}b_{2,-1}^2$	$a_{-12}^2 a_{20}b_{3,-1} - a_{-13}b_{02}b_{2,-1}^2$
$a_{10}a_{-12}^2 b_{3,-1}b_{10} - a_{01}a_{-13}b_{2,-1}^2 b_{01}$	$a_{01}^2 a_{-13}b_{2,-1}^2 - a_{-12}^2 b_{3,-1}b_{10}^2$
$a_{-12}^2 b_{20}^3 - a_{02}^3 b_{2,-1}^2$	$a_{-12}^2 a_{20}b_{20}^2 - a_{02}^2 b_{02}b_{2,-1}^2$
$a_{-12}^2 a_{20}^2 b_{20} - a_{02}b_{02}^2 b_{2,-1}^2$	$a_{10}^2 a_{-12}^2 b_{3,-1} - a_{-13}b_{2,-1}^2 b_{01}^2$
$a_{-12}^2 a_{20}^3 - b_{02}^3 b_{2,-1}^2$	$a_{-12}^2 b_{3,-1}^2 b_{02} - a_{20}a_{-13}^2 b_{2,-1}^2$
$a_{-12}^4 b_{3,-1}^3 - a_{-13}^3 b_{2,-1}^4$	$a_{-12}^2 a_{02}b_{3,-1}^2 - a_{-13}^2 b_{20}b_{2,-1}^2$
$a_{10}a_{01} - b_{10}b_{01}$	$a_{01}a_{-13}^2 b_{2,-1}^3 - a_{-12}^3 b_{3,-1}^2 b_{10}$
$a_{10}a_{-12}^3 b_{3,-1}^2 - a_{-13}^2 b_{2,-1}^3 b_{01}$	

Table 3.2 Generators of I_{sym} for System (3.100)

3.6 Darboux Integrals and Integrating Factors

We now present the method of Darboux integration for proving the existence of first integrals and integrating factors for polynomial systems of differential equations on $\mathbb{C}^2$. We thus consider systems

$$\dot{x} = \widetilde{P}(x,y), \qquad \dot{y} = \widetilde{Q}(x,y), \tag{3.101}$$

where x, $y \in \mathbb{C}$, $\widetilde{P}$ and $\widetilde{Q}$ are polynomials without constant terms that have no non-constant common factor, and $m = \max(\deg(\widetilde{P}), \deg(\widetilde{Q}))$. Let $\mathscr{X}$ denote the corresponding vector field as defined in Remark 3.2.4. Suppose system (3.101) has a first integral H on a neighborhood of the origin. If, as is not infrequently the case, H has the form $\prod f_j^{\alpha_j}$, where for each j, $\alpha_j \in \mathbb{C}$ and $f_j \in \mathbb{C}[x,y]$ (and we may assume that f_j is irreducible and that f_j and f_k are relatively prime if $j \neq k$), then f_j divides

$\mathscr{X} f_j$ for each j (Exercise 3.22). But the fact that $\mathscr{X} f_j = K_j f_j$ for some polynomial $K_j \in \mathbb{C}[x,y]$ implies that the variety $\mathbf{V}(f_j)$ of f_j is an invariant curve for (3.101), since it is then the case that

$$\left(\frac{\partial f_j}{\partial x}\widetilde{P} + \frac{\partial f_j}{\partial y}\widetilde{Q}\right)\bigg|_{\mathbf{V}(f_j)} = \left(\operatorname{grad} f_j, (\widetilde{P},\widetilde{Q})\right)\bigg|_{\mathbf{V}(f_j)} = \mathscr{X} f_j|_{\mathbf{V}(f_j)} \equiv 0. \tag{3.102}$$

Starting with system (3.101), this suggests that in the search for a first integral one look for a first integral that is a product of powers of polynomials whose zero sets are invariant curves in the phase portrait of (3.101). This discussion is the motivation for the definitions and results in this section.

Definition 3.6.1. A nonconstant polynomial $f(x,y) \in \mathbb{C}[x,y]$ is called an *algebraic partial integral* of system (3.101) if there exists a polynomial $K(x,y) \in \mathbb{C}[x,y]$ such that

$$\mathscr{X} f = \frac{\partial f}{\partial x}\widetilde{P} + \frac{\partial f}{\partial y}\widetilde{Q} = Kf. \tag{3.103}$$

The polynomial K is termed a *cofactor* of f; it has degree at most $m-1$. (See Exercise 3.23.)

Since the vector field $\mathscr{X}$ associated to (3.101) is a derivation (see (4.58)), the following facts are apparent:

1. if f is an algebraic partial integral for (3.101) with cofactor K, then any constant multiple of f is also an algebraic partial integral for (3.101) with cofactor K;
2. if f_1 and f_2 are algebraic partial integrals for (3.101) with cofactors K_1 and K_2, then $f_1 f_2$ is an algebraic partial integral for (3.101) with cofactor $K_1 + K_2$.

We know that if f is an algebraic partial integral, then $\mathbf{V}(f)$ is an algebraic invariant curve. The converse also holds:

Proposition 3.6.2. *Fix $f \in \mathbb{C}[x,y]$. $\mathbf{V}(f)$ is an algebraic invariant curve of system* (3.101) *if and only if f is an algebraic partial integral of system* (3.101).

Proof. Just the "only if" part of the proposition requires proof. Write $f = f_1^{\alpha_1} \cdots f_s^{\alpha_s}$, where, for each j, f_j is an irreducible polynomial. The equality $\mathscr{X} f|_{\mathbf{V}(f)} = 0$ implies that

$$\mathscr{X} f_j|_{\mathbf{V}(f_j)} = 0 \tag{3.104}$$

for all j. Thus $\mathbf{V}(f_j) \subset \mathbf{V}(\mathscr{X} f_j)$; applying Proposition 1.1.12 gives the inclusion $\mathscr{X} f_j \in \mathbf{I}(\mathbf{V}(\mathscr{X} f_j)) \subset \mathbf{I}(\mathbf{V}(f_j))$. Since f_j is irreducible, $\langle f_j \rangle$ is prime, hence radical (Proposition 1.4.3), so by Theorem 1.3.15, $\mathbf{I}(\mathbf{V}(f_j)) = \langle f_j \rangle$ and we conclude that $\mathscr{X} f_j \in \langle f_j \rangle$ for all j. Therefore $\mathscr{X} f_j = K_j f_j$, for some $K_j \in \mathbb{C}[x,y]$, so that every polynomial f_j is an algebraic partial integral of (3.101). As noted just above the statement of the proposition, if g and h are two algebraic partial integrals with cofactors K_g and K_h, then gh is also an algebraic partial integral, with cofactor $K_g + K_h$. Therefore $f = f_1^{\alpha_1} \cdots f_s^{\alpha_s}$ is an algebraic partial integral of (3.101). □

Remark. In the literature a function that meets the condition of Definition 3.6.1 is frequently termed an *algebraic invariant curve*, in keeping with the characterization given in the proposition.

Definition 3.6.3. Suppose that the curves defined by $f_1 = 0, \ldots, f_s = 0$ are algebraic invariant curves of system (3.101), and that $\alpha_j \in \mathbb{C}$ for $1 \leq j \leq s$. A first integral of system (3.101) of the form

$$H = f_1^{\alpha_1} \cdots f_s^{\alpha_s} \tag{3.105}$$

is called a *Darboux first integral* of system (3.101).

The existence of a Darboux first integral can be highly restrictive. For example, if α_j is real and rational for all j, then every trajectory of (3.101) lies in an algebraic curve. (See Exercise 3.25.) If sufficiently many algebraic invariant curves can be found, then they can be used to construct a Darboux first integral, as the following theorem shows.

Theorem 3.6.4 (Darboux). *Suppose system* (3.101) *has q (distinct) algebraic invariant curves $f_j(x,y) = 0$, $1 \leq j \leq q$, where for each j, f_j is irreducible over $\mathbb{C}^2$, and that $q > (m^2 + m)/2$. Then system* (3.101) *admits a Darboux first integral.*

Proof. Let $\mathbb{C}_{m-1}[x,y]$ denote the complex vector space of polynomials of degree at most $m-1$. A homogeneous polynomial of degree p has $p+1$ terms, so $\mathbb{C}_{m-1}[x,y]$ has dimension $m + (m-1) + \cdots + 1 = m(m+1)/2$. By Proposition 3.6.2, there exist polynomials $K_1, \ldots, K_q$ such that $K_j \in \mathbb{C}_{m-1}[x,y]$ and $\mathscr{X} f_j = K_j f_j$ for $1 \leq j \leq q$. Thus the number of vectors K_j is greater than the dimension of $\mathbb{C}_{m-1}[x,y]$, hence the collection $\{K_1, \ldots, K_q\}$ is linearly dependent, so that there exist constants α_j, not all zero, such that

$$\sum_{j=1}^{q} \alpha_j K_j = \sum_{j=1}^{q} \alpha_j \frac{\mathscr{X} f_j}{f_j} = 0,$$

the zero polynomial. Defining H for these polynomials f_j and constants α_j by (3.105) with $s = q$, this yields

$$\mathscr{X} H = H \sum_{j=1}^{q} \alpha_j \frac{\mathscr{X} f_j}{f_j} = H \sum_{j=1}^{q} \alpha_j K_j = 0,$$

the zero function, meaning that H is a first integral of (3.101) if it is not constant. Since the algebraic curves $f_j = 0$ are irreducible and not all the constants α_j are zero, the function H is indeed not constant (Exercise 3.26). □

Corollary 3.6.5. *If system* (3.101) *has at least $q = (m^2 + m)/2$ algebraic invariant curves $f_j(x,y) = 0$, each of which is irreducible over $\mathbb{C}^2$ and does not pass through the origin (that is, $f_j(0,0) \neq 0$), then it admits a Darboux first integral* (3.105).

Proof. In this case all the cofactors are of the form $K_j = ax + by + \cdots$. Thus they are contained in a vector subspace of $\mathbb{C}_{m-1}[x,y]$ of dimension $(m^2 + m - 2)/2$. □

The situation in which a system of the form (3.101) has more independent algebraic partial integrals than $\dim \mathbb{C}_{m-1}[x,y]$ occurs very seldom (but examples are known; see Section 3.7). Sometimes, though, it is possible to find a Darboux first integral using a smaller number of invariant algebraic curves. Indeed, the proof of

Theorem 3.6.4 shows that if $f_1,\ldots,f_s$ are any number of distinct irreducible algebraic partial integrals of system (3.101), as long as a nontrivial linear combination of the cofactors K_j is zero, $\sum_{j=1}^{s}\alpha_j K_j = 0$, $H = f_1^{\alpha_1}\cdots f_s^{\alpha_s}$ will be a first integral of system (3.101). See also Theorem 3.6.8 below.

If a first integral of system (3.101) cannot be found, then we turn our attention to the possible existence of an integrating factor. Classically, an *integrating factor* of the equation

$$M(x,y)dx + N(x,y)dy = 0 \tag{3.106}$$

for differentiable functions M and N on an open set Ω is a differentiable function $\mu(x,y)$ on Ω such that $\mu(x,y)M(x,y)\,dx + \mu(x,y)N(x,y)\,dy = 0$ is an exact differential, which is the case if and only if

$$\frac{\partial(\mu M)}{\partial y} - \frac{\partial(\mu N)}{\partial x} \equiv 0\,.$$

For the following definition, the functions $\widetilde{P}$ and $\widetilde{Q}$ need only be differentiable. Denote by $\operatorname{div}\mathscr{X}$ the divergence of the vector field $\mathscr{X}$, $\operatorname{div}\mathscr{X} = \widetilde{P}_x + \widetilde{Q}_y$.

Definition 3.6.6. An *integrating factor* on an open set Ω for system (3.101) is a differentiable function $\mu(x,y)$ on Ω such that

$$\mathscr{X}\mu = -\mu \operatorname{div}\mathscr{X} \tag{3.107}$$

holds throughout on Ω. An integrating factor on Ω of the form

$$\mu = f_1^{\beta_1}\cdots f_s^{\beta_s}\,, \tag{3.108}$$

where f_j is an algebraic partial integral for (3.101) on Ω for $1 \le j \le s$, is called a *Darboux integrating factor* on Ω.

This definition is consistent with the classical definition of an integrating factor of equation (3.106). Indeed, the trajectories of system (3.101) satisfy the equation

$$\widetilde{Q}(x,y)\,dx - \widetilde{P}(x,y)\,dy = 0\,,$$

so that μ is an integrating factor in the classical sense if and only if

$$\frac{\partial(\mu\widetilde{Q})}{\partial y} - \frac{\partial(-\mu\widetilde{P})}{\partial x} \equiv 0\,,$$

hence if and only if $\mu_x\widetilde{P} + \mu_y\widetilde{Q} + \mu(\widetilde{P}_x + \widetilde{Q}_y) \equiv 0$, which is precisely (3.107). Of course, the importance of μ is that when (3.101) is rescaled by μ the resulting system on Ω, which has the same orbits in Ω as (3.101) except where $\mu = 0$, has the form $\dot{x} = -H_y$, $\dot{y} = H_x$, from which the first integral H on Ω can be found by integration (although H could be multivalued if Ω is not simply connected).

The same computation as in the proof of Theorem 3.6.4 shows that a function μ of the form (3.108) is a Darboux integrating factor if and only if

$$\mathscr{X}\mu + \mu \operatorname{div}\mathscr{X} = \mu\left(\sum_{j=1}^{s}\beta_j K_j + \operatorname{div}\mathscr{X}\right) \equiv 0,$$

where K_j is the cofactor of f_j for $j = 1,\dots,s$. Thus if we have s algebraic invariant curves $f_1,\dots,f_s$ and are able to find s constants β_j such that

$$\sum_{j=1}^{s}\beta_j K_j + \operatorname{div}\mathscr{X} \equiv 0, \tag{3.109}$$

then $\mu = f_1^{\beta_1}\cdots f_s^{\beta_s}$ is an integrating factor of (3.101).

It is possible to extend the concept of Darboux integrability to a wider class of functions that includes limits of first integrals of the form (3.105). We can proceed in the following way. Suppose that for every value of a parameter ε near zero both $f = 0$ and $f + \varepsilon g = 0$ are invariant curves for (3.101), with respective cofactors K_f and $K_{f+\varepsilon g}$. Then

$$\begin{aligned}K_{f+\varepsilon g} &= \frac{\mathscr{X}(f+\varepsilon g)}{f+\varepsilon g} = \frac{\mathscr{X}f}{f+\varepsilon g} + \varepsilon\frac{\mathscr{X}g}{f+\varepsilon g}\\ &= K_f\left(1-\varepsilon\frac{g}{f}+O(\varepsilon^2)\right) + \varepsilon\frac{\mathscr{X}g}{f} + O(\varepsilon^2) = K_f + \varepsilon\frac{\mathscr{X}g - gK_f}{f} + O(\varepsilon^2).\end{aligned}$$

Since $K_{f+\varepsilon g}$ is a polynomial of degree at most $m-1$,

$$K' \stackrel{\text{def}}{=} \frac{\mathscr{X}g - gK_f}{f}$$

is a polynomial of at most degree $m-1$. Thus

$$K_{f+\varepsilon g} = K_f + \varepsilon K' + O(\varepsilon^2)$$

and

$$\mathscr{X}\left(\left[\frac{f+\varepsilon g}{f}\right]^{1/\varepsilon}\right) = \left[\frac{f+\varepsilon g}{f}\right]^{1/\varepsilon}\left(\frac{K_{f+\varepsilon g}}{\varepsilon} - \frac{K_f}{\varepsilon}\right) = \left[\frac{f+\varepsilon g}{f}\right]^{1/\varepsilon}(K' + O(\varepsilon)).$$

As ε tends to zero, $((f+\varepsilon g)/f)^{1/\varepsilon}$ tends to $h \stackrel{\text{def}}{=} e^{g/f}$, which clearly satisfies

$$\mathscr{X}h = K'h. \tag{3.110}$$

Thus the function h satisfies the same equation as an algebraic invariant curve, namely equation (3.103), and has a polynomial cofactor K' of degree at most $m-1$.

Definition 3.6.7. A (possibly multivalued) function of the form

$$e^{g/f}\prod f_j^{\alpha_j},$$

where f, g, and all the f_j are polynomials, is called a *Darboux function*. A function of the form $h = e^{g/f}$ satisfying (3.110), where K' is a polynomial of degree at most $m-1$, is called an *exponential factor*. As before, K' is termed the *cofactor* of h.

Sometimes an exponential factor is called a "degenerate algebraic invariant curve," to emphasize its origin from the coalescence of algebraic invariant curves. The name above is preferable since $e^{g/f}$ is neither algebraic nor a curve. It is easy to check that if $h = e^{g/f}$ is an exponential factor, then $f = 0$ is an algebraic invariant curve, and g satisfies an equation

$$\mathscr{X} g = gK_f + fK_h,$$

where K_f is the cofactor of f and K_h is the cofactor of h. Since the product of two exponential factors is again an exponential factor, it is no loss of generality that the exponential factor in Definition 3.6.7 is unique.

The theory of Darboux integrability presented above for algebraic invariant curves goes through essentially unchanged when exponential factors are allowed in addition to algebraic curves. In particular, the existence of at least $m(m+1)/2+1$ invariant algebraic curves or exponential factors yields the existence of Darboux first integrals, and the existence of at least $m(m+1)/2$ invariant algebraic curves or exponential factors implies the existence of a Darboux integrating factor. The following theorem provides a Darboux first integral when the number of algebraic invariant curves and exponential factors is small.

Theorem 3.6.8. *Suppose system* (3.101) *has q (distinct) irreducible algebraic partial integrals f_j with corresponding cofactors K_j, $1 \le j \le q$, and has r (distinct) exponential factors* $\exp(g_j/h_j)$ *with corresponding cofactors L_j, $1 \le j \le r$, for which there exist q complex constants α_j, $1 \le j \le q$, and r complex constants β_j, $1 \le j \le r$, not all zero, such that $\sum_{j=1}^{q} \alpha_j K_j + \sum_{j=1}^{r} \beta_j L_j \equiv 0$. Then system* (3.101) *admits a first integral of the form $H = f_1^{\alpha_1} \cdots f_q^{\alpha_q} (\exp(g_1/h_1))^{\beta_1} \cdots (\exp(g_1/h_1))^{\beta_1}$.*

Proof. Simply differentiate the expression for H and apply the same reasoning as in the proof of Theorem 3.6.4. □

Darboux's method is one of the most efficient tools for studying the center problem for polynomial systems (3.101). In particular, if we are able to construct a Darboux first integral (3.105) or a Darboux integrating factor (3.108) with algebraic curves $f_j = 0$ that do not pass through the origin, then we are sure to have a first integral that is analytic in a neighborhood of the origin. If for system (3.48) (respectively, (3.61)) the first integral has the form (3.49) (respectively, (3.62)), or can be modified in accordance with Remark 3.2.4 to produce a first integral on a neighborhood of the origin that does, then the system has a center at the origin.

In the case that at least one of the invariant curves that are used to construct a first integral or an integrating factor passes through the origin, the first integral that we obtain need not exist on a neighborhood of the origin. In certain situations this poses no real difficulty. Suppose, for instance, that a first integral H has the form $H = f/g^p$

for $p \in \mathbb{N}$ and algebraic partial integrals f and g with $g(0,0)=0$ but $f(0,0)\neq 0$. Then $G=g^p/f$ is certainly a first integral on a neighborhood of $(0,0)$ in $\mathbb{C}^2\setminus \mathbf{V}(g)$. Since $\mathbf{V}(g)$ is an algebraic invariant curve and G takes the constant value 0 on $\mathbf{V}(g)$, G is constant on all orbits in a neighborhood of $(0,0)$, including those confined to $\mathbf{V}(g)$, hence G is a first integral on a neighborhood of $(0,0)$. Similar situations are examined in Exercises 3.28–3.30.

Under certain circumstances it is sufficient to have an integrating factor that is defined only in a punctured neighborhood of the origin in order to distinguish between a center and a focus for a *real* polynomial system of the form (3.2), that is, of the form

$$\dot u = -v + U(u,v) = \widetilde U(u,v), \qquad \dot v = u + V(u,v) = \widetilde V(u,v), \tag{3.111}$$

where $\max(\deg \widetilde U, \deg \widetilde V) = m$. The following theorem is an example of this sort of result. As always, we let $\mathscr{X}$ denote the vector field associated with the system of differential equations, here $\mathscr{X}(u,v) = \widetilde U(u,v)\frac{\partial}{\partial u} + \widetilde V(u,v)\frac{\partial}{\partial v}$.

Theorem 3.6.9. *Suppose u^2+v^2 is an algebraic partial integral of the real system* (3.111) *and that for some negative real number α, $(u^2+v^2)^{\alpha/2}$ is an integrating factor for* (3.111) *on a punctured neighborhood of the origin. Consider the polynomial* $\operatorname{div}\mathscr{X}$, *and decompose it as a sum of homogeneous polynomials*

$$\operatorname{div}\mathscr{X}(u,v) = \sum_{j=1}^{m-1} d_j(u,v). \tag{3.112}$$

Then the origin is a center if either
(i) $\int_0^{2\pi} \operatorname{div}\mathscr{X}(r\cos\varphi, r\sin\varphi)\,d\varphi \equiv 0$ *on a neighborhood of* 0 *in* $\mathbb{R}$, *or*
(ii) $\int_0^{2\pi} d_j(\cos\varphi,\sin\varphi)\,d\varphi = 0$ *for all j such that* $j < -\alpha - 1$.
In particular, the origin is a center if $\alpha \geq -2$.

Before presenting a proof of this theorem, we will consider two specific examples. More examples of applying the Darboux method are presented in Section 3.7.

Example 3.6.10. Consider the family of cubic systems of the form (3.111) given by

$$\begin{aligned} \dot u &= -v + B(u^2-v^2) + 2Auv + 2Du^3 - (4E-C)u^2v - 2Duv^2 + Cv^3 \\ \dot v &= u - A(u^2-v^2) + 2Buv - Cu^3 + 2Du^2v - (4E+C)uv^2 - 2Dv^3, \end{aligned} \tag{3.113}$$

where A, B, C, D, and E are real constants. To search for invariant lines, we use the method of undetermined coefficients: we insert $f(u,v) = f_{00} + f_{10}u + f_{01}v$ and, since $m=3$, $K(u,v) = K_{00} + K_{10}u + K_{01}v + K_{20}u^2 + K_{11}uv + K_{02}v^2$ into (3.103), written as $\mathscr{X}f - Kf \equiv 0$, and collect terms. This yields a cubic polynomial equation

$$f_{00}K_{00} + (K_{00}f_{10} + K_{10}f_{00} - f_{01})u + (K_{00}f_{01} + K_{01}f_{00} + f_{10})v + \cdots \equiv 0. \tag{3.114}$$

To make the constant term zero we arbitrarily choose $f_{00}=0$. With that choice made, if, in order to make the coefficient of either linear term zero, we choose either

$f_{10} = 0$ or $f_{01} = 0$, then $f_{00} = f_{10} = f_{01} = 0$ is forced, which is of no interest, so instead we eliminate the linear terms by choosing $K_{00} = f_{01}/f_{10}$ and $K_{00} = -f_{10}/f_{01}$, which in turn forces $f_{10}^2 + f_{01}^2 = 0$. Although system (3.113) is real, we allow complex coefficients in this computation, so that we proceed by choosing $f_{01} = if_{10}$ or $f_{01} = -if_{10}$. These choices ultimately lead to the algebraic partial integrals

$$f_1(u,v) = u + iv \qquad \text{and} \qquad f_2(u,v) = u - iv$$

with corresponding cofactors

$$K_1(u,v) = \quad i + (B - iA)u + (A + iB)v + (2D - iC)u^2 - 4Euv - (2D + iC)v^2$$

and

$$K_2(u,v) = -i + (B + iA)u + (A - iB)v + (2D + iC)u^2 - 4Euv - (2D - iC)v^2.$$

This example illustrates the fact that a real system can have complex algebraic partial integrals, although what happens here is true in general: they always occur in complex conjugate pairs (Exercise 3.27).

Then $f(u,v) = f_1(u,v)f_2(u,v) = u^2 + v^2$ must be a real algebraic partial integral with cofactor $K(u,v) = K_1(u,v) + K_2(u,v) = 2(Bu + Av + 2Du^2 - 4Euv - 2Dv^2)$. We will not pursue the other possible ways of forcing the truth of (3.114).

The equation

$$\begin{aligned}(u - A(u^2 - v^2) + 2Buv - Cu^3 + 2Du^2v - (4E + C)uv^2 - 2Dv^3)\,du \\ - (-v + B(u^2 - v^2) + 2Auv + 2Du^3 - (4E - C)u^2v - 2Duv^2 + Cv^3)\,dv = 0\end{aligned}$$

corresponding to (3.113) is exact if and only if $A = B = D = E = 0$. We seek to use the algebraic partial integral $u^2 + v^2$ to construct an integrating factor by looking for a constant β such that (3.109) holds. Substitution of f and K into (3.109) gives

$$2\beta(Bu + Av + 2Du^2 - 4Euv - 2Dv^2) + 4(Bu + Av + 2Du^2 - 4Euv - 2Dv^2) \equiv 0,$$

hence condition (3.109) holds if $\beta = -2$. Thus $\mu(u,v) = f^{-2} = (u^2 + v^2)^{-2}$ is an integrating factor for (3.113) on the set $\Omega = \mathbb{R}^2 \setminus \{(0,0)\}$. By integrating $-H_v = \mu\widetilde{P}$ and $H_u = \mu\widetilde{Q}$, we obtain a first integral

$$H(u,v) = C\log(u^2 + v^2) + \frac{1 - 2Au + 2Bv + 4Duv + 4Eu^2}{u^2 + v^2}$$

on the set Ω. The first two hypotheses of Theorem 3.6.9 are satisfied (with $\alpha = -4$), and a simple computation shows that condition (i) is met, so that system (3.113) has a center at the origin.

Example 3.6.11. In practice, it is typically the case that when a Darboux first integral can be found for system (3.111) on a punctured neighborhood of the origin, then the origin is a center, but this is not always the case, as the following example shows. For real constants A, B, and C, consider the family of cubic systems

$$\begin{aligned}\dot{u} &= -v + Au^3 + Bu^2v + Cv^3\\ \dot{v} &= \quad u - Cu^3 + Au^2v + (B-2C)uv^2.\end{aligned} \tag{3.115}$$

Exactly the same procedure as in the previous example leads to the algebraic partial integral $f(u,v) = u^2 + v^2$ with cofactor $K(u,v) = 2Au^2 + 2(B-C)uv$, which can be used as above to find the integrating factor $\mu(u,v) = f^{-2} = (u^2+v^2)^{-2}$ on the set $\Omega = \mathbb{R}^2 \setminus \{(0,0)\}$. Further calculation gives the first integral

$$\begin{aligned}H(u,v) &= C\log(u^2+v^2) + \frac{1+(C-B)u^2+Auv}{u^2+v^2} + A\arctan\frac{v}{u}\\ &= C\log(u^2+v^2) + \frac{1+(C-B)u^2+Auv}{u^2+v^2} + A\left(\tfrac{\pi}{2} - \arctan\frac{u}{v}\right)\end{aligned}$$

on Ω. The first two hypotheses of Theorem 3.6.9 are satisfied (with $\alpha = -4$). Since

$$\operatorname{div}\mathscr{X} = 4Au^2 + 4(B-C)uv$$

is homogeneous of degree two, the condition $\int_0^{2\pi} d_j(r\cos\varphi, r\sin\varphi)\,d\varphi = 0$ for $j < 3$ is restrictive; Proposition 3.6.9 guarantees that system (3.115) has a center at the origin provided

$$\int_1^{2\pi} 4A\cos^2\varphi + 4(B-C)\cos\varphi\sin\varphi\,d\varphi = 4A\pi = 0,$$

that is, if $A = 0$.

Whether the origin is a center or a focus when $A \neq 0$ can be determined using the theory of Lyapunov stability from Section 2.1 since, by Remark 3.1.7, we know that the origin is a focus or a center. (See Exercise 3.31 for an alternate approach.) The function $W(u,v) = u^2 + v^2 + (B-C)u^4 - Au^3v + (B-C)u^2v^2 - Auv^3$ is positive definite on a neighborhood of the origin, and differentiation with respect to t gives $\dot{W}(u,v) = (u^2+v^2)(A(u^2+v^2)+\cdots)$, where the omitted terms are of order three or greater. Thus $\dot{W} < 0$ on a punctured neighborhood of the origin if $A < 0$ and $\dot{W} > 0$ on a punctured neighborhood of the origin if $A > 0$. Hence by Theorems 2.1.3(2) and 2.1.5(1) and the fact that the origin is known to be a focus or a center, the origin is a stable focus if $A < 0$ and an unstable focus if $A > 0$.

In order to prove Theorem 3.6.9 we will need an additional result, which we state and prove first.

Theorem 3.6.12. *Suppose that for the real system* (3.111) *there exist a punctured neighborhood Ω of the origin in $\mathbb{R}^2$ and a function $B : \Omega \to [0,\infty) \subset \mathbb{R}$ that is continuously differentiable, $B \not\equiv 0$ on any punctured neighborhood of the origin but*

$$\operatorname{div} B\mathscr{X} = \frac{\partial}{\partial u}(B\widetilde{U}) + \frac{\partial}{\partial v}(B\widetilde{V}) \equiv 0 \text{ on } \Omega, \tag{3.116}$$

and for which there exists a finite positive number M such that

$$\left| \int_0^{2\pi} (BU\cos\varphi + BV\sin\varphi)\Big|_{\substack{u=r\cos\varphi \\ v=r\sin\varphi}} d\varphi \right| < M \tag{3.117}$$

on some punctured neighborhood of $r = 0$ in $\mathbb{R}$. Then system (3.111) *has a center at the origin.*

Proof. There are a ray ρ at the origin and a sequence of points p_j in ρ, $j \in \mathbb{N}$, such that $B(p_j) > 0$ and $p_j \to (0,0)$ as $j \to \infty$. Since the linear part of system (3.111) is invariant under a rotation of the coordinate system about the origin, we may assume that ρ is the positive u-axis. By Remark 3.1.7, the origin is a focus or a center for system (3.111). Moreover the Poincaré first return map $\mathscr{R}(r)$ is defined for $r > 0$ sufficiently small. Suppose, contrary to what we wish to show, that the origin is a focus. Reversing the flow if necessary so that the origin is a sink (that is, attracts every point in a sufficiently small neighborhood of itself), this means that if we choose any point $(u,v) = (r_1,0)$ in ρ with $r_1 > 0$ sufficiently small, a sequence r_j, $j \in \mathbb{N}$, is generated, satisfying $0 < r_{j+1} = \mathscr{R}(r_j) < r_j$ and $r_j \to 0$ as $j \to \infty$. The corresponding points in $\mathbb{R}^2$ have (u,v)-coordinates $(r_j,0)$. For j sufficiently large, $\int_{r_{j+1}}^{r_j} u + V(u,0)\,du > 0$. If the flow was not reversed in order to form the sequence r_j, choose J such that $B(u,0)$ has a nonzero value in the interval $r_{J+1} \le u \le r_J$ and consequently

$$\int_{r_{J+1}}^{r_J} B(u,0)[u + V(u,0)]\,du > 0. \tag{3.118}$$

If the flow was reversed for the construction, choose J so that $B(u,0)$ has a nonzero value in the interval $r_J \le u \le r_{J-1}$ and consequently

$$\int_{r_J}^{r_{J-1}} B(u,0)[u + V(u,0)]\,du > 0. \tag{3.119}$$

Now let Γ be the positively oriented simple closed curve composed of the arc γ of the trajectory of system (3.111) (not reversed) from $(r_J,0)$ to its next intersection (either $(r_{J+1},0)$ or $(r_{J-1},0)$) with ρ, followed by the segment λ oriented from that point back to $(r_J,0)$. We conclude from (3.118) or (3.119), whichever applies, that

$$\int_\lambda B(u,0)[u + V(u,0)]\,du \neq 0. \tag{3.120}$$

Let C_r denote the negatively oriented circle of radius r centered at the origin and, for $r > 0$ sufficiently small, let $\mathscr{U}$ denote the region bounded by Γ and C_r. Then by (3.116) and Green's Theorem,

$$0 = \iint_{\mathscr{U}} \operatorname{div} B\mathscr{X}\,dA = \int_\gamma (-B\widetilde{V}du + B\widetilde{U}dv) \pm \int_\lambda B\widetilde{V}du + \int_{\mathscr{C}_r} (B\widetilde{V}du - B\widetilde{U}dv), \tag{3.121}$$

where the immaterial ambiguity in the sign of the second term arises from the question of whether or not the flow was reversed in forming the sequence r_j.

The first term in (3.121) is

$$\int_0^T (-B\widetilde{V}\widetilde{U} + B\widetilde{U}\widetilde{V})(u(t),v(t))dt = 0,$$

where T is the time taken to describe γ. The third term is, up to sign,

$$r\int_0^{2\pi} (BV\sin\varphi + BU\cos\varphi)\Big|_{\substack{u=r\cos\varphi\\ v=r\sin\varphi}} d\varphi,$$

which by (3.117) tends to zero as r tends to zero. But by (3.120) the second term in (3.121) is a fixed nonzero constant, yielding a contradiction. □

We are now in a position to prove Theorem 3.6.9.

Proof of Theorem 3.6.9. By (3.103) and a simple computation, the condition that $u^2+v^2=0$ be an invariant curve is that there exist a polynomial $K(u,v)$ satisfying

$$uU(u,v)+vV(u,v) = \tfrac{1}{2}(u^2+v^2)K(u,v) \tag{3.122}$$

or

$$U(r\cos\varphi, r\sin\varphi)\cos\varphi + V(r\cos\varphi, r\sin\varphi)\sin\varphi = \tfrac{1}{2}rK(r\cos\varphi, r\sin\varphi). \tag{3.123}$$

The condition that $(u^2+v^2)^{\alpha/2} = r^\alpha$ be an integrating factor is, by (3.107) and straightforward manipulations,

$$\alpha(uU(u,v)+vV(u,v)) = -(u^2+v^2)\left(\frac{\partial U}{\partial u}+\frac{\partial V}{\partial v}\right)(u,v), \tag{3.124}$$

which, when combined with (3.122), is

$$\tfrac{1}{2}\alpha K(u,v) = -\left(\frac{\partial U}{\partial u}+\frac{\partial V}{\partial v}\right)(u,v). \tag{3.125}$$

We now apply Theorem 3.6.12. When $B(u,v) = (u^2+v^2)^{\alpha/2} = r^\alpha$ the left-hand side of (3.116) may be computed as

$$(u^2+v^2)^{\frac{\alpha}{2}-1}\left[\alpha(uU(u,v)+vV(u,v)) + (u^2+v^2)\left(\frac{\partial U}{\partial u}+\frac{\partial V}{\partial v}\right)(u,v)\right],$$

which by (3.124) vanishes identically; the first integral in (3.117) is

$$\int_0^{2\pi} r^{\alpha-1}[uU(u,v)+vV(u,v)]\Big|_{\substack{u=r\cos\varphi\\ v=r\sin\varphi}} d\varphi,$$

which, again by (3.122), is

$$\frac{1}{2}\int_0^{2\pi} r^{\alpha+1} K(r\cos\varphi, r\sin\varphi)\,d\varphi\,,$$

so condition (3.117) holds if

$$\int_0^{2\pi} K(r\cos\varphi, r\sin\varphi)\,d\varphi = O(r^{-\alpha-1})\,, \tag{3.126}$$

in which case by Theorem 3.6.12 (noting that $B(u,v) > 0$ on $\mathbb{R}^2 \setminus \{(0,0)\}$) the origin is a center for system (3.111). We observe in particular that if we write

$$K = \sum_{j=1}^{m-1} K_j, \tag{3.127}$$

where each K_j is a homogeneous polynomial of degree j, then (3.126) holds (and system (3.111) has a center) if $\int_0^{2\pi} K_j(\cos\varphi, \sin\varphi)\,d\varphi = 0$ for $j < -\alpha - 1$. This is automatic for $\alpha \geq -2$.

If now condition (i) in Theorem 3.6.9 holds, then replacing $K(r\cos\varphi, r\sin\varphi)$ in (3.126) using (3.125) we see that (3.126) holds, so there is a center at the origin.

If condition (ii) in Theorem 3.6.9 holds, then (3.125) shows that the homogeneous components in the decomposition (3.112) are $d_j(u,v) = -K_j(u,v)/\alpha$, where K_j is as in (3.127). The theorem then follows from the remarks immediately following (3.127). □

3.7 Applications: Quadratic Systems and a Family of Cubic Systems

To find the center variety of a given family of polynomial systems of the form (3.3) we use the following approach, outlined in the introduction to this chapter. We compute the first focus quantity that is different from zero, say g_{KK}, and set $G = \{g_{KK}\}$. We then compute the next focus quantity $g_{K+1,K+1}$, reduce it modulo g_{KK} (Definition 1.2.15), and, using the Radical Membership Test, check if the reduced polynomial $g'_{K+1,K+1}$ belongs to $\sqrt{\langle g_{K,K}\rangle}$. If not, then we add it to G, so that now $G = \{g_{KK}, g'_{K+1,K+1}\}$. We then compute $g_{K+2,K+2}$, reduce it modulo $\langle G\rangle$ (where $\langle G\rangle$ denotes the ideal generated by the polynomials in G), and check whether the reduced polynomial $g'_{K+2,K+2}$ is in $\sqrt{\langle G\rangle}$. (Of course, the set G need not be a Gröbner basis of $\langle G\rangle$.) If not, we adjoin it to G, and so continue until we reach the smallest value of s such that $G = \{g_{K,K}, g'_{K+1,K+1}, \ldots, g'_{K+s,K+s}\}$ and $g'_{K+s+1,K+s+1} \in \sqrt{\langle G\rangle}$. At this point we expect that

$$V_{\mathscr{C}} = \mathbf{V}(\mathscr{B}) = \mathbf{V}(\langle G\rangle), \tag{3.128}$$

where $\mathscr{B}$ is the Bautin ideal and $V_{\mathscr{C}}$ is the center variety of the family of systems under consideration. To increase our confidence, we might compute the next few focus quantities and verify that they also lie in $\sqrt{\langle G\rangle}$. Certainly $\mathbf{V}(\mathscr{B}) \subset \mathbf{V}(\langle G\rangle)$; we must establish the reverse inclusion so as to prove (3.128). To this end, we next find the irreducible decomposition of $\mathbf{V}(\langle G\rangle)$,

$$\mathbf{V}(\langle G\rangle) = V_1 \cup \cdots \cup V_q\,,$$

and then for every component V_s of the decomposition use the methods presented in the previous sections, augmented as necessary by other techniques, to prove that all systems from that component have a center at the origin.

In this section we will apply this theory to find the center variety of two families of systems, all systems of the form (3.3) with quadratic nonlinearities,

$$\begin{aligned} \dot{x} &= i\left(x - a_{10}x^2 - a_{01}xy - a_{-12}y^2\right) \\ \dot{y} &= -i\left(y - b_{2,-1}x^2 - b_{10}xy - b_{01}y^2\right), \end{aligned} \tag{3.129}$$

and the restricted family of cubic systems of the form (3.3) given by

$$\begin{aligned} \dot{x} &= i\left(x - a_{10}x^2 - a_{01}xy - a_{-13}y^3\right) \\ \dot{y} &= -i\left(y - b_{10}xy - b_{01}y^2 - b_{3,-1}x^3\right). \end{aligned} \tag{3.130}$$

According to Proposition 3.3.9, the center variety of each system is the same as the center variety of the respective system

$$\begin{aligned} \dot{x} &= x - a_{10}x^2 - a_{01}xy - a_{-12}y^2 \\ \dot{y} &= -y + b_{2,-1}x^2 + b_{10}xy + b_{01}y^2 \end{aligned} \tag{3.131}$$

and

$$\begin{aligned} \dot{x} &= x - a_{10}x^2 - a_{01}xy - a_{-13}y^3 \\ \dot{y} &= -y + b_{10}xy + b_{01}y^2 + b_{3,-1}x^3\,. \end{aligned} \tag{3.132}$$

This particular cubic family was selected both because it is amenable to study and because the Darboux theory of integrability is inadequate to treat it completely, so that it gives us an opportunity to illustrate yet another technique for finding a first integral. It also provides an example illustrating the following important remark.

Remark. The problem of finding an initial string of focus quantities $g_{11},\dots,g_{KK}$ such that $\mathbf{V}(\mathscr{B}_K) = \mathbf{V}(\mathscr{B})$ and the problem of finding an initial string of focus quantities $g_{11},\dots,g_{JJ}$ such that $\mathscr{B}_J = \mathscr{B}$ are not the same problem, and need not have the same answer. The first equality tells us only that the ideals $\mathscr{B}_K$ and $\mathscr{B}$ have the same radical, not that they are the same ideal. We will see in Section 6.3 that for the general quadratic family (3.129) both $\mathbf{V}(\mathscr{B}_3) = \mathbf{V}(\mathscr{B})$ and $\mathscr{B}_3 = \mathscr{B}$ hold true, but that for family (3.130) $\mathbf{V}(\mathscr{B}_5) = \mathbf{V}(\mathscr{B})$ but $\mathscr{B}_5 \subsetneq \mathscr{B}$. In the same section we will see that these two families also show that $\mathscr{B}_K$ for the least K such that $\mathbf{V}(\mathscr{B}_K) = \mathbf{V}(\mathscr{B})$ may or may not be radical (true for family (3.129) but false for (3.130)).

Theorem 3.7.1. *The center variety of families* (3.129) *and* (3.131) *is the variety of the ideal $\mathscr{B}_3$ generated by the first three focus quantities,, and is composed of the following four irreducible components:*
1. $V_1 = \mathbf{V}(J_1)$, *where* $J_1 = \langle 2a_{10} - b_{10},\ 2b_{01} - a_{01}\rangle$;
2. $V_2 = \mathbf{V}(J_2)$, *where* $J_2 = \langle a_{01},\ b_{10}\rangle$;
3. $V_3 = \mathbf{V}(J_3)$, *where* $J_3 = \langle 2a_{01} + b_{01},\ a_{10} + 2b_{10},\ a_{01}b_{10} - a_{-12}b_{2,-1}\rangle$;
4. $V_4 = \mathbf{V}(J_4)$, *where* $J_4 = \langle f_1, f_2, f_3, f_4, f_5\rangle$, *where*
 (a) $f_1 = a_{01}^3 b_{2,-1} - a_{-12}b_{10}^3$,
 (b) $f_2 = a_{10}a_{01} - b_{01}b_{10}$,
 (c) $f_3 = a_{10}^3 a_{-12} - b_{2,-1}b_{01}^3$,
 (d) $f_4 = a_{10}a_{-12}b_{10}^2 - a_{01}^2 b_{2,-1}b_{01}$, *and*
 (e) $f_5 = a_{10}^2 a_{-12}b_{10} - a_{01}b_{2,-1}b_{01}^2$.
Moreover, $V_1 = \mathbf{V}(I_{\text{Ham}})$, $V_4 = \mathbf{V}(I_{\text{sym}})$, *$V_2$ is the Zariski closure of those systems having three invariant lines, and V_3 is the Zariski closure of those systems having an invariant conic and an invariant cubic.*

Proof. Following the general approach outlined above, by means of the algorithm of Section 3.4, we compute the first three focus quantities for family (3.129) and reduce g_{22} modulo g_{11} and g_{33} modulo $\{g_{11}, g_{22}\}$. Actually, before performing the reduction of g_{33}, we compute a Gröbner basis for $\langle g_{11}, g_{22}\rangle$ (with respect to lex and with the ordering $a_{10} > a_{01} > a_{-12} > b_{2,-1} > b_{10} > b_{01}$) and reduce it modulo this basis. Our abbreviated terminology for this procedure is that we "reduce g_{33} modulo $\langle g_{11}, g_{22}\rangle$" (see Definition 1.2.15). Maintaining the notation g_{kk} for the reduced quantities, the result is

$$g_{11} = -i(a_{10}a_{01} - b_{01}b_{10}) \tag{3.133}$$

$$\begin{aligned} g_{22} = -i(&a_{10}a_{-12}b_{10}^2 - b_{01}b_{2,-1}a_{01}^2 - \tfrac{2}{3}(a_{-12}b_{10}^3 - b_{2,-1}a_{01}^3) \\ &- \tfrac{2}{3}(a_{01}b_{01}^2 b_{2,-1} - b_{10}a_{10}^2 a_{-12})) \end{aligned} \tag{3.134}$$

$$\begin{aligned} g_{33} = i\tfrac{5}{8}(&-a_{01}a_{-12}b_{10}^4 + 2a_{-12}b_{01}b_{10}^4 + a_{01}^4 b_{10}b_{2,-1} \\ &- 2a_{01}^3 b_{01}b_{10}b_{2,-1} - 2a_{10}a_{-12}^2 b_{10}^2 b_{2,-1} + a_{-12}^2 b_{10}^3 b_{2,-1} \\ &- a_{01}^3 a_{-12} b_{2,-1}^2 + 2a_{01}^2 a_{-12} b_{01} b_{2,-1}^2). \end{aligned} \tag{3.135}$$

Note that g_{33} is not of the form (3.82) because of the reduction.

The reader is encouraged to verify (Exercise 3.32) using the Radical Membership Test that

$$g_{22} \notin \sqrt{\langle g_{11}\rangle}, \quad g_{33} \notin \sqrt{\langle g_{11}, g_{22}\rangle}, \quad g_{44}, g_{55}, g_{66} \in \sqrt{\langle g_{11}, g_{22}, g_{33}\rangle}. \tag{3.136}$$

Thus we expect that

$$\mathbf{V}(\langle g_{11}, g_{22}, g_{33}\rangle) = \mathbf{V}(\mathscr{B}_3) = \mathbf{V}(\mathscr{B}). \tag{3.137}$$

To verify that this is the case, we will find the irreducible decomposition of $\mathbf{V}(\mathscr{B}_3)$ and then check that an arbitrary element of each component has a center at the origin, thus confirming that $\mathbf{V}(\mathscr{B}_3) \subset \mathbf{V}(\mathscr{B})$ and thereby establishing (3.137).

In general, finding the irreducible decomposition of a variety is a difficult computational problem, which relies on rather laborious algorithms for the primary decomposition of polynomial ideals. Although, as we have mentioned in Chapter 1, there are at present implementations of such algorithms in some specialized computer algebra systems (for example, CALI, Macaulay, and Singular), for system (3.129) we can find the irreducible decomposition of $\mathbf{V}(\mathscr{B}_3)$ using only a general-purpose computer algebra system such as Maple or Mathematica. Moreover, we choose to do so here in order to further illustrate some of the concepts that have been developed. We begin by computing the Hamiltonian and symmetry ideals using equation (3.83) and the algorithm of Section 5.2. This gives $I_{\text{Ham}} = J_1$ and $I_{\text{sym}} = J_4$; we know that $V_1 = \mathbf{V}(I_{\text{Ham}})$ and $V_4 = \mathbf{V}(I_{\text{sym}})$ are irreducible; they are, therefore, candidates for components of $\mathbf{V}(\mathscr{B}_3)$.

To proceed further, we parametrize $\mathbf{V}(g_{11})$ as

$$(a_{-12}, a_{01}, a_{10}, b_{01}, b_{10}, b_{2,-1}) = (a_{-12}, a_{01}, s b_{10}, s a_{01}, b_{10}, b_{2,-1}).$$

For $j \in \{1,2,3\}$, define $\widetilde{g}_{jj} \in \mathbb{C}[a_{-12}, a_{01}, b_{10}, b_{2,-1}, s]$ by

$$\widetilde{g}_{jj}(a_{-12}, a_{01}, b_{10}, b_{2,-1}, s) = g_{jj}(a_{-12}, a_{01}, s b_{10}, s a_{01}, b_{10}, b_{2,-1}).$$

For every $(a_{-12}, a_{01}, b_{10}, b_{2,-1}, s) \in \mathbb{C}^5$,

$$\widetilde{g}_{11}(a_{-12}, a_{01}, b_{10}, b_{2,-1}, s) := g_{11}(a_{-12}, a_{01}, s b_{10}, s a_{01}, b_{10}, b_{2,-1}) = 0;$$

$\widetilde{g}_{jj}(a_{-12}, a_{01}, b_{10}, b_{2,-1}, s) = 0$ if and only if $g_{jj}(a_{-12}, a_{01}, s b_{10}, s a_{01}, b_{10}, b_{2,-1}) = 0$ for $j = 2,3$, although some points of $\mathbf{V}(g_{11})$ are missed by the parametrization, and will have to be considered separately (points (iv) and (v) below). The system $\widetilde{g}_{22} = \widetilde{g}_{33} = 0$ has the same solution set, that is, defines the same variety, as the system $h_1 = h_2 = 0$, where $\{h_1, h_2\}$ is a Gröbner basis of $\langle \widetilde{g}_{22}, \widetilde{g}_{33} \rangle$. Such a basis with respect to lex with $a_{-12} > a_{01} > b_{10} > b_{2,-1} > s$ is

$$\{(2s-1)(s+2)f_1, (2s-1)(a_{-12}b_{2,-1} - b_{10}a_{01})f_1\},$$

where f_1 is the first polynomial on the list of generators of $J_4 = I_{\text{sym}}$. (Strictly speaking, we should have $\widetilde{f_1}$, defined analogously to the $\widetilde{g}_{jj}$, in place of f_1.) We solve this system.

(i) One solution is $s = \frac{1}{2}$: $(a_{-12}, a_{01}, b_{10}/2, a_{01}/2, b_{10}, b_{2,-1})$ lies in $\mathbf{V}(g_{11}, g_{22}, g_{33})$ for all $(a_{-12}, a_{01}, b_{10}, b_{2,-1})$. All these sextuples lie in $\mathbf{V}(2a_{10} - b_{10}, 2b_{01} - a_{01})$, and they are precisely this variety. Thus $(a,b) \in \mathbf{V}(2a_{10} - b_{10}, 2b_{01} - a_{01})$ implies $(a,b) \in \mathbf{V}(\mathscr{B}_3)$. This is $\mathbf{V}(J_1)$.

(ii) A second solution is $f_1 = a_{01}^3 b_{2,-1} - a_{-12} b_{10}^3 = 0$. This means that if a sextuple $(a_{-12}, a_{01}, s b_{10}, s a_{01}, b_{10}, b_{2,-1})$ (which lies in $\mathbf{V}(g_{11})$) satisfies the condition $a_{01}^3 b_{2,-1} - a_{-12} b_{10}^3 = 0$, then it is in $\mathbf{V}(g_{22}, g_{33})$, hence in $\mathbf{V}(g_{11}, g_{22}, g_{33}) = \mathbf{V}(\mathscr{B}_3)$. We would like to say that any element of $\mathbf{V}(g_{11}) \cap \mathbf{V}(a_{01}^3 b_{2,-1} - a_{-12} b_{10}^3)$ lies in $\mathbf{V}(\mathscr{B}_3)$, but we have not shown that, because there are points in $\mathbf{V}(g_{11})$ that have been left out of consideration, the points missed by our parametrization: those for

which $a_{01} = 0$ but $b_{01} \neq 0$ and those for which $b_{10} = 0$ but $a_{10} \neq 0$. In the former case $g_{11} = 0$ only if $b_{10} = 0$, and we quickly check that if $a_{01} = b_{10} = 0$, then $g_{22} = g_{33} = 0$. Similarly, if $b_{10} = 0$ but $a_{10} \neq 0$, $g_{11} = 0$ forces $a_{01} = 0$, so we again have that $a_{01} = b_{10} = 0$, so $g_{22} = g_{33} = 0$. Thus we conclude that $(a,b) \in \mathbf{V}(g_{11}, a_{01}^3 b_{2,-1} - a_{-12} b_{10}^3) = \mathbf{V}(a_{10}a_{01} - b_{10}b_{01}, a_{01}^3 b_{2,-1} - a_{-12}b_{10}^3)$ implies $(a,b) \in \mathbf{V}(\mathscr{B}_3)$. We set $J_5 = \langle a_{10}a_{01} - b_{10}b_{01}, a_{01}^3 b_{2,-1} - a_{-12}b_{10}^3 \rangle$ and set $V_5 = \mathbf{V}(J_5)$.
(iii) A third solution is $s = -2$ and $a_{-12}b_{2,-1} - b_{10}a_{01} = 0$, which means that any sextuple $(a_{-12}, a_{01}, -2b_{10}, -2a_{01}, b_{10}, b_{2,-1})$ satisfying $a_{-12}b_{2,-1} - b_{10}a_{01} = 0$ lies in $\mathbf{V}(\mathscr{B}_3)$. These points form $\mathbf{V}(a_{-12}b_{2,-1} - b_{10}a_{01}, a_{10} + 2b_{10}, 2a_{01} + b_{01}) = V_3$, so $V_3 \subset \mathbf{V}(\mathscr{B}_3)$.

Points of $\mathbf{V}(g_{11})$ that are not covered by our parametrization are those points for which
(iv) $b_{10} = 0$ but $a_{10} \neq 0$, hence $a_{01} = 0$; and
(v) $a_{01} = 0$ but $b_{01} \neq 0$, hence $b_{10} = 0$.
In either case, $a_{01} = b_{10} = 0$, so all these points lie in $\mathbf{V}(a_{01}, b_{10})$, and we have already seen that $a_{01} = b_{10} = 0$ implies that $g_{22} = g_{33} = 0$, so we conclude that $(a,b) \in \mathbf{V}(a_{01}, b_{10}) = V_2$ implies $(a,b) \in \mathbf{V}(\mathscr{B}_3)$, so $V_2 \subset \mathbf{V}(\mathscr{B}_3)$.

Our computations have shown that $\mathbf{V}(\mathscr{B}_3) = V_1 \cup V_2 \cup V_3 \cup V_5$. Since the variety $V_4 = \mathbf{V}(I_{\text{sym}})$, which is irreducible, does not appear, but $V_4 \subset V_{\mathscr{C}} \subset \mathbf{V}(\mathscr{B}_3)$, we include it as well, writing

$$\mathbf{V}(\mathscr{B}_3) = V_1 \cup V_2 \cup V_3 \cup V_4 \cup V_5.$$

Since $J_5 = \langle f_1, f_2 \rangle \subset I_{\text{sym}}$, $V_4 = \mathbf{V}(I_{\text{sym}}) \subset \mathbf{V}(J_5) = V_5$, so we suspect that V_5 is the union of V_4 and another subvariety on our list, and recognize that $V_2 \subset V_5$ as well. We can discard V_5 provided $V_2 \cup V_4$ contains it. To check this, we compute (using, say, the algorithm in Table 1.6 in Section 1.3) $J_5 : J_4$ and apply Theorem 1.3.24 to obtain

$$\begin{aligned}\mathbf{V}(J_5) \setminus \mathbf{V}(J_4) &\subset \overline{\mathbf{V}(J_5) \setminus \mathbf{V}(J_4)} \\ &\subset \mathbf{V}(J_5 : J_4) = \mathbf{V}(b_{10}^3,\ a_{01}^2,\ a_{01}b_{10}^2,\ a_{01}^2 b_{10},\ a_{10}a_{01} - b_{10}b_{01}) \\ &= \mathbf{V}(a_{01},\ b_{10}) = \mathbf{V}(J_2),\end{aligned}$$

where the next-to-last equality is by inspection. Thus $\mathbf{V}(J_5) \subset \mathbf{V}(J_2) \cup \mathbf{V}(J_4)$, so V_5 is superfluous, and we remove it from our list of subvarieties composing $\mathbf{V}(\mathscr{B}_3)$.

Turning to the question of the irreducibility of V_1, V_2, V_3, and V_4, we already know that V_1 and V_4 are irreducible. Of course, the irreducibility of both V_1 and V_2 is clear "geometrically" because each is the intersection of a pair of hyperplanes in $\mathbb{C}^6$. Alternatively, each has a polynomial parametrization, namely,

$$(a_{-12}, a_{01}, a_{10}, b_{01}, b_{10}, b_{2,-1}) = (r, 2t, s, t, 2s, u)$$

and

$$(a_{-12}, a_{01}, a_{10}, b_{01}, b_{10}, b_{2,-1}) = (r, 0, s, t, 0, u),$$

respectively, hence by Corollary 1.4.18 is irreducible. We note here a fact that will be needed later, that by Theorem 1.4.17 the ideal J_2 is prime, hence by Proposition 1.4.3 is radical. To show that V_3 is irreducible, we look for a rational parametrization similarly. It is apparent that $S := V_3 \setminus \{a_{-12} = 0\}$ is precisely the image of the map

$$(a_{-12}, a_{01}, a_{10}, b_{01}, b_{10}, b_{2,-1}) = F(r,s,t) = (r, s, -2t, -2s, t, st/r)$$

from $\mathbb{C}^3 \setminus \{(r,s,t) : r \neq 0\}$ into $\mathbb{C}^6$, so that irreducibility of V_3 will follow from Definition 1.4.16 and Corollary 1.4.18 if we can establish that V_3 is the Zariski closure of $V_3 \setminus \mathbf{V}(a_{-12})$. To do so we apply Theorem 1.4.15, which in this context states that the Zariski closure of S is the fourth elimination ideal of the ideal

$$J = \langle r - a_{-12}, s - a_{01}, -2t - a_{10}, -2s - b_{01}, t - b_{10}, st - rb_{2,-1}, 1 - ry \rangle$$

in the ring $\mathbb{C}[r,s,t,y,a_{-12},a_{01},a_{10},b_{01},b_{10},b_{2,-1}]$. A Gröbner basis of J with respect to lex with $r > s > t > y > a_{-12} > a_{01} > a_{10} > b_{01} > b_{10} > b_{2,-1}$ is

$$\{2a_{01} + b_{01}, a_{10} + 2b_{10}, 2a_{-12}b_{2,-1} + b_{01}b_{10}, yb_{10} - 1, t - b_{10}, 2s + b_{01}, r - a_{-12}\}.$$

The generators of the fourth elimination ideal are the basis elements that do not contain r, s, t, or y: $g_1 = 2a_{01} + b_{01}$, $g_2 = a_{10} + 2b_{10}$, and $g_3 = 2a_{-12}b_{2,-1} + b_{01}b_{10}$. These are not the generators of J_3 as given in the statement of the theorem, but the two ideals might still be the same. To see if they are, we must compute a reduced Gröbner basis of J_3 with respect to lex with our usual ordering of the variables (see Theorem 1.2.27). When we do so, we obtain $\{g_1, g_2, g_3\}$ and conclude that J_3 is the fourth elimination ideal of J, as required. As we did for J_2, we note that by Theorem 1.4.17 the ideal J_3 is prime, hence, by Proposition 1.4.3, it is radical.

As an aside we note that we could have approached the problem of the irreducibility of V_1, V_2, and V_3 by proving, without reference to the varieties involved, that the ideals J_1, J_2, and J_3 are prime, hence radical, and appealing to Theorems 1.3.14 and 1.4.5, just as we concluded irreducibility of V_4 from the fact that $J_4 = I_{\text{sym}}$ is prime (Theorem 3.5.9). Direct proofs that J_1, J_2, and J_3 are prime are given in the proof of Theorem 6.3.3.

At this point we have shown that the unique minimal decomposition of $\mathbf{V}(\mathscr{B}_3)$ guaranteed by Theorem 1.4.7 to exist is

$$\mathbf{V}(\mathscr{B}_3) = V_1 \cup V_2 \cup V_3 \cup V_4.$$

It remains to show that every system from V_k, $k = 1,2,3,4$, has a center at the origin. For $V_1 = \mathbf{V}(I_{\text{Ham}})$ and $V_4 = \mathbf{V}(I_{\text{sym}})$ this is automatic. For the remaining two components of $\mathbf{V}(\mathscr{B}_3)$, we look for Darboux first integrals.

Systems from V_2 have the form

$$\dot{x} = x - a_{10}x^2 - a_{-12}y^2, \qquad \dot{y} = -y + b_{2,-1}x^2 + b_{01}y^2. \tag{3.138}$$

If we look for an invariant line of (3.138) that does not pass through the origin, say the zero set of the function $f(x,y) = 1 + rx + sy$, then the cofactor K is a first-degree polynomial, say $K = K_0 + K_1 x + K_2 y$, and satisfies the equation $\mathscr{X} f = K f$. Since

$$\mathscr{X} f - K f = -K_0 + (r - K_1 - K_0 r)x - (s + K_2 + K_0 s)y + \cdots,$$

K is forced to have the form $K(x,y) = rx - sy$, and a computation of the three remaining coefficients in $\mathscr{X} f - K f$ indicates that the zero set of f is an invariant line if and only if

$$r^2 + a_{10} r - b_{2,-1} s = 0 \tag{3.139}$$

and

$$s^2 - a_{-12} r + b_{01} s = 0. \tag{3.140}$$

Suppose $b_{2,-1} \neq 0$. We solve (3.139) for s and insert the resulting expression into (3.140) to obtain

$$r\left[r^3 + 2a_{10} r^2 + (a_{10}^2 + b_{2,-1} b_{01}) r + (a_{10} b_{2,-1} b_{01} - a_{-12} b_{2,-1}^2)\right] = 0.$$

Let g denote the constant term in the cubic and h the discriminant of the cubic, which is a homogeneous polynomial of degree four in a_{10}, a_{-12}, $b_{2,-1}$, and b_{01}. Off $\mathbf{V}(h)$ the cubic has three distinct roots ([191]), call them r_1, r_2, and r_3, and off $\mathbf{V}(g)$ none is zero. Let s_1, s_2, and s_3 denote the corresponding values of s calculated from (3.139), $s_j = (r_j^2 + a_{10} r_j)/b_{2,-1}$. The identity $\sum_{j=1}^{3} \alpha_j K_j \equiv 0$ is the pair of linear equations

$$r_1 \alpha_1 + r_2 \alpha_2 + r_3 \alpha_3 = 0 \tag{3.141}$$

$$s_1 \alpha_1 + s_2 \alpha_2 + s_3 \alpha_3 = 0 \tag{3.142}$$

in the three unknowns α_1, α_2, and α_3, hence by Theorem 3.6.8 there exists a Darboux first integral $H = (1 + r_1 x - s_1 y)^{\alpha_1} (1 + r_2 x - s_2 y)^{\alpha_2} (1 + r_3 x - s_3 y)^{\alpha_3}$. We must show that the exponents in H can be chosen so that H has the form $H(x,y) = xy + \cdots$. Conditions (3.141) and (3.142) imply that $H_x(0,0)$ and $H_y(0,0)$ are both zero. Using them to simplify the second partial derivatives of H, we obtain the additional conditions

$$r_1^2 \alpha_1 + r_2^2 \alpha_2 + r_3^2 \alpha_3 = 0 \tag{3.143}$$

$$s_1^2 \alpha_1 + s_2^2 \alpha_2 + s_3^2 \alpha_3 = 0 \tag{3.144}$$

arising from $H_{xx}(0,0) = 0$ and $H_{yy}(0,0) = 0$ and

$$r_1 s_1 \alpha_1 + r_2 s_2 \alpha_2 + r_3 s_3 \alpha_3 \neq 0$$

arising from $H_{xy}(0,0) \neq 0$, which by (3.139) and (3.141) simplifies to

$$r_1^3 \alpha_1 + r_2^3 \alpha_2 + r_3^3 \alpha_3 \neq 0. \tag{3.145}$$

Given nonzero r_1, r_2, and r_3, for any choice of α_1, α_2, and α_3 meeting conditions (3.141) and (3.143), conditions (3.142) and (3.144) are met automatically: for the relationship $s_j = (r_j^2 + a_{10} r_j)/b_{2,-1}$ for $j = 1,2,3$ means that (3.142) is a linear combination of (3.141) and (3.143), while the truth of the first three yields (3.144) because $s_j^2 = a_{-12} r_j - b_{01} s_j$. Since

$$\det \begin{pmatrix} r_1 & r_2 & r_3 \\ r_1^2 & r_2^2 & r_3^2 \\ r_1^3 & r_2^3 & r_3^3 \end{pmatrix} = -r_1 r_2 r_3 (r_1 - r_2)(r_1 - r_3)(r_2 - r_3)$$

is nonzero the only choice of α_1, α_2, and α_3 that satisfies (3.141) and (3.143) but violates (3.145) is $\alpha_1 = \alpha_2 = \alpha_3 = 0$. Thus for any other choice of α_1, α_2, and α_3, $\Psi = (H-1)/(r_1^3 \alpha_1 + r_2^3 \alpha_2 + r_3^3 \alpha_3)$ is a first integral of the required form.

Thus every element of $V_2 \setminus (\mathbf{V}(g) \cup \mathbf{V}(h) \cup \mathbf{V}(b_{2,-1})) = V_2 \setminus \mathbf{V}(ghb_{2,-1})$ has a center at the origin. Since V_2 is irreducible and $V_2 \setminus \mathbf{V}(ghb_{2,-1})$ is clearly a proper subset of V_2 we conclude by Exercise 1.45 and Proposition 1.3.20 that every element of V_2 has a center at the origin.

Now consider the variety V_3. A search for invariant lines turns up nothing, so we look for a second-degree algebraic partial integral $f_1(x,y)$ that does not pass through the origin and its first-degree cofactor K_1. Solving the system of polynomial equations that arises by equating coefficients of like powers in the defining identity $\mathscr{X} f_1 = K_1 f_1$ yields without complications

$$f_1 = 1 + 2b_{10}x + 2a_{01}y - a_{01}b_{2,-1}x^2 + 2a_{01}b_{10}xy - \frac{a_{01}b_{10}^2}{b_{2,-1}}y^2$$

and its cofactor $K_1 = 2\,(b_{10}x - a_{01}y)$. One algebraic partial integral is inadequate, however, so we look for an invariant cubic curve. The same process gives us

$$\begin{aligned} f_2 = (2b_{10}b_{2,-1}^2)^{-1}\Big[& 2b_{10}b_{2,-1}^2 + 6b_{10}^2 b_{2,-1}^2 x + 6a_{01}b_{10}b_{2,-1}^2 y \\ & + 3b_{10}b_{2,-1}^2(b_{10}^2 - a_{01}b_{2,-1})x^2 + 3b_{2,-1}(2a_{01}b_{10}^2 b_{2,-1} - b_{10}^4 - a_{01}^2 b_{2,-1}^2)xy \\ & + 3a_{01}b_{10}b_{2,-1}(a_{01}b_{2,-1} - b_{10}^2)y^2 \\ & + a_{01}b_{2,-1}^3(a_{01}b_{2,-1} - b_{10}^2)x^3 + 3a_{01}b_{10}b_{2,-1}^2(b_{10}^2 - a_{01}b_{2,-1})x^2 y \\ & + 3a_{01}b_{10}^2 b_{2,-1}(a_{01}b_{2,-1} - b_{10}^2)xy^2 + a_{01}b_{10}^3(b_{10}^2 - a_{01}b_{2,-1})y^3\Big] \end{aligned}$$

and its cofactor $K_2 = 3\,(b_{10}x - a_{01}y)$. From a comparison of the cofactors K_1 and K_2 it is immediately apparent that a nontrivial linear combination $\alpha_1 K_1 + \alpha_2 K_2$ that is zero is $\alpha_1 = -3$, $\alpha_2 = 2$, hence by Theorem 3.6.8 the system has the Darboux first integral $H = f_1^{-3} f_2^2$, provided $b_{10}b_{2,-1} \neq 0$. Set $g = 2a_{01}b_{10}^2 b_{2,-1} + b_{10}^4 + a_{01}^2 b_{2,-1}^2$. Then $H(x,y) = 1 - 3g(b_{2,-1}b_{10})^{-1}xy + \cdots$ so that $\Psi = b_{2,-1}b_{10}(-H+1)/(3g)$ is a first integral of the form $\Psi(x,y) = xy + \cdots$, and every system in $V_3 \setminus \mathbf{V}(b_{2,-1}b_{10}g)$ has a center at the origin. Since V_3 is irreducible and $V_3 \setminus \mathbf{V}(b_{2,-1}b_{10}g)$ is clearly a proper subset of V_3, we conclude by Exercise 1.45 and Proposition 1.3.20 that every element of V_3 has a center at the origin. (See Exercise 3.33 for an alternative way to treat the situation $b_{2,-1}b_{10} = 0$. Note also that we could have rescaled f_2 by

$2b_{10}b_{2,-1}^2$ and still had an algebraic partial integral with the same cofactor, eliminating the problem with vanishing of b_{10}. Rescaling f_1 by $b_{2,-1}$ does not eliminate the problem of $b_{2,-1}$ vanishing, however, since the resulting invariant curve then passes through the origin.) □

A computation shows that for systems in family (3.131) the symmetry subvariety is the Zariski closure of the set of time-reversible systems. Thus the theorem identifies one component of the center variety as the set of Hamiltonian systems, one component as the smallest variety containing the time-reversible systems, and the remaining two components as the smallest varieties that contain systems that possess Darboux first integrals, in one case because the underlying system has three invariant lines, and in the other because it contains an invariant conic and an invariant cubic. The operation of taking the Zariski closure is essential. Although there occur in the literature statements that identify what we have called V_2 as "the set of quadratic systems having three invariant lines," in fact not every element of V_2 does so. For example, the system $\dot{x} = x - y^2$, $\dot{y} = -y + y^2$, corresponding to $(a_{-12}, a_{01}, a_{10}, b_{01}, b_{10}, b_{2,-1}) = (1,0,0,1,0,0) \in V_2$, has exactly two invariant lines, real or complex; the system $\dot{x} = x + x^2 - y^2$, $\dot{y} = -y + x^2 - y^2$, corresponding to $(1,0,-1,-1,0,1) \in V_2$, has exactly one (Exercise 3.34.)

Theorem 3.7.1 shows the utility of finding the minimal decomposition of $\mathbf{V}(\mathscr{B}_k)$: although systems in V_2 and V_3 typically possess Darboux first integrals, the form of the integral for systems from V_2 differs from that for systems from V_3; trying to deal with these two families as a whole, and in particular trying to prove the existence of a Darboux first integral for the whole family, would almost certainly be impossible. In the case of family (3.130), even when attention is restricted to some individual component of the center variety it does not seem possible to produce first integrals solely by Darboux's method of invariant algebraic curves, and we resort to a different approach.

Theorem 3.7.2. *The center variety of families* (3.130) *and* (3.132) *is the variety of the ideal $\mathscr{B}_5$ generated by the first five focus quantities and is composed of the following eight irreducible components:*
1. $\mathbf{V}(J_1)$, *where* $J_1 = \langle a_{10},\ a_{-13},\ b_{10},\ 3a_{01} - b_{01}\rangle$;
2. $\mathbf{V}(J_2)$, *where* $J_2 = \langle a_{01},\ b_{3,-1},\ b_{01},\ 3b_{10} - a_{10}\rangle$;
3. $\mathbf{V}(J_3)$, *where* $J_3 = \langle a_{10},\ a_{-13},\ b_{10},\ 3a_{01} + b_{01}\rangle$;
4. $\mathbf{V}(J_4)$, *where* $J_4 = \langle a_{01},\ b_{3,-1},\ b_{01},\ 3b_{10} + a_{10}\rangle$;
5. $\mathbf{V}(J_5)$, *where* $J_5 = \langle a_{01},\ a_{-13},\ b_{10}\rangle$;
6. $\mathbf{V}(J_6)$, *where* $J_6 = \langle a_{01},\ b_{3,-1},\ b_{10}\rangle$;
7. $\mathbf{V}(J_7)$, *where* $J_7 = \langle a_{01} - 2b_{01},\ b_{10} - 2a_{10}\rangle$;
8. $\mathbf{V}(J_8)$, *where* $J_8 = \langle a_{10}a_{01} - b_{01}b_{10},\ a_{01}^4 b_{3,-1} - b_{10}^4 a_{-13},\ a_{10}^4 a_{-13} - b_{01}^4 b_{3,-1},$
$a_{10}a_{-13}b_{10}^3 - a_{01}^3 b_{01} b_{3,-1},\ a_{10}^2 a_{-13} b_{10}^2 - a_{01}^2 b_{01}^2 b_{3,-1},$
$a_{10}^3 a_{-13} b_{10} - a_{01} b_{01}^3 b_{3,-1}\rangle$.

Proof. When we compute g_{kk} for family (3.130) and (for $k \geq 2$) reduce it modulo $\mathscr{B}_{k-1}$ (Definition 1.2.15), retaining the same notation g_{kk} for the reduced quantities, we obtain

$$
\begin{aligned}
g_{11} &= -i(a_{10}a_{01} - b_{01}b_{10})\\
g_{22} &= 0\\
g_{33} &= i(2a_{10}^3 a_{-13}b_{10} - a_{10}^2 a_{-13}b_{10}^2 - 18a_{10}a_{-13}b_{10}^3 - 9a_{01}^4 b_{3,-1}\\
&\qquad + 18a_{01}^3 b_{01}b_{3,-1} + a_{01}^2 b_{01}^2 b_{3,-1} - 2a_{01}b_{01}^3 b_{3,-1} + 9a_{-13}b_{10}^4)/8\\
g_{44} &= i(14a_{10}b_{01}(2a_{10}a_{-13}b_{10}^3 + a_{01}^4 b_{3,-1} - 2a_{01}^3 b_{01}b_{3,-1} - a_{-13}b_{10}^4))/27\\
g_{55} &= -ia_{-13}b_{3,-1}(378a_{10}^4 a_{-13} + 5771a_{10}^3 a_{-13}b_{10} - 25462a_{10}^2 a_{-13}b_{10}^2\\
&\qquad + 11241a_{10}a_{-13}b_{10}^3 - 11241a_{01}^3 b_{01}b_{3,-1} + 25462a_{01}^2 b_{01}^2 b_{3,-1}\\
&\qquad - 5771a_{01}b_{01}^3 b_{3,-1} - 378b_{01}^4 b_{3,-1})/3240\\
g_{66} &= 0\\
g_{77} &= ia_{-13}^2 b_{3,-1}^2(343834a_{10}^2 a_{-13}b_{10}^2 - 1184919a_{10}a_{-13}b_{10}^3 + 506501a_{-13}b_{10}^4\\
&\qquad - 506501a_{01}^4 b_{3,-1} + 1184919a_{01}^3 b_{01}b_{3,-1} - 343834a_{01}^2 b_{01}^2 b_{3,-1})\\
g_{88} &= 0\\
g_{99} &= ia_{-13}^3 b_{3,-1}^3(2a_{10}a_{-13}b_{10}^3 - a_{-13}b_{10}^4 + a_{01}^4 b_{3,-1} - 2a_{01}^3 b_{01}b_{3,-1})
\end{aligned}
$$

The Radical Membership Test (Table 1.4) shows that both g_{77} and g_{99} lie in $\sqrt{\mathscr{B}_5}$, which suggests that $\mathbf{V}(\mathscr{B}) = \mathbf{V}(\mathscr{B}_5)$, which we will now prove.

Computing I_{Ham} using (3.85) gives J_7; setting the relevant coefficients equal to zero in the polynomials listed in Table 3.2 gives $I_{\text{sym}} = J_8$. By factoring the focus quantities through a change of parameters, as was done for the quadratic family, we find that the system $g_{11} = g_{33} = g_{44} = g_{55} = 0$ is equivalent to the eight conditions listed in the theorem. Rather than duplicating those computations, the reader can verify this by first applying the algorithm in Table 1.5 recursively to compute $I = \cap_{j=1}^{8} J_j$, then using the Radical Membership Test to verify that $\sqrt{I} = \sqrt{\mathscr{B}_5}$ (Exercise 3.36). Because each of V_1 through V_7 has a polynomial parametrization, it is irreducible; V_8 is irreducible because it is $\mathbf{V}(I_{\text{sym}})$.

We now prove that every system from V_j, $1 \le j \le 8$, has a center at the origin. First observe that if in a generator of J_1 every occurrence of a_{jk} is replaced by b_{kj} and every occurrence of b_{jk} is replaced by a_{kj}, then a generator of J_2 is obtained, and conversely. Thus any procedure that yields a first integral H for a system in V_2 also yields a first integral for the corresponding element of V_1 under the involution: if there is a formula for H in terms of (a,b), simply perform the same involution on it. Thus we need not treat V_1. Similarly, we need not treat V_3 or V_5. The varieties V_7 and V_8 need not be treated since, as already mentioned, $J_7 = I_{\text{Ham}}$ and $J_8 = I_{\text{sym}}$.

Any system from V_2 has the form

$$
\dot{x} = x - 3b_{10}x^2 - a_{-13}y^3, \qquad \dot{y} = -y + b_{10}xy,
$$

for which $\operatorname{div}\mathscr{X} = -5b_{10}x$. If either a_{-13} or b_{10} is zero, then in fact the system also comes from V_8, already known to have only systems with a center at the origin, so we need only treat the case $a_{-13}b_{10} \neq 0$. We look for a Darboux first integral or a Darboux integrating factor. A search for invariant lines by the usual method of undetermined coefficients yields only the obvious algebraic partial integral $f = y$ and its cofactor $K = -1 + b_{10}x$. Since for no choice of β can equation (3.109) be satisfied, we cannot obtain an integrating factor from just f alone, and so we look for higher-order invariant curves. We obtain $f_1 = 1 - 6b_{10}x + 9b_{10}^2x^2 + 2a_{-13}b_{10}y^3$ and its cofactor $K_1 = -6b_{10}x$. Since equation (3.109) holds with $\beta_1 = -5/6$, an analytic Darboux integrating factor for our system is $\mu = f_1^{-5/6}$. Then

$$H(x,y) = \int -\mu(x,y)P(x,y)\,dy = \int -x + \overset{(2)}{\cdots}\,dy = (-xy + \overset{(3)}{\cdots}) + c(x),$$

for some analytic $c(x)$, is a first integral in a neighborhood of the origin. Because

$$\partial H/\partial x = (-y + \overset{(2)}{\cdots}) + c'(x) = \mu(x,y)Q(x,y) = -y + \overset{(2)}{\cdots},$$

$c(x)$ begins with terms of order at least three. Thus $\Psi = -H$ is a Lyapunov first integral on a neighborhood of the origin, which is thus a center.

Any system from V_4 has the form

$$\dot{x} = x + 3b_{10}x^2 - a_{-13}y^3, \qquad \dot{y} = -y + b_{10}xy,$$

for which $\operatorname{div}\mathscr{X} = 7b_{10}x$. If either a_{-13} or b_{10} is zero, then this system also comes from V_8, so we restrict attention to the situation $a_{-13}b_{10} \neq 0$. An invariant cubic curve is the zero set of $f_1 = 1 + 3b_{10}x - a_{-13}b_{10}y^3$, whose cofactor is $K_1 = 3b_{10}x$. Since equation (3.109) holds with $\beta_1 = -7/3$, an analytic Darboux integrating factor for our system is $\mu = f_1^{-7/3}$, by means of which we easily obtain a Lyapunov first integral on a neighborhood of the origin, which is thus a center.

Any system from V_6 has the form

$$\dot{x} = x - a_{10}x^2 - a_{-13}y^3, \qquad \dot{y} = -y + b_{01}y^2.$$

This system also comes from V_8 unless $a_{10}a_{-13} \neq 0$, so we continue on the assumption that this inequality holds. A search for a Darboux first integral or integrating factor proves fruitless, so we must proceed along different lines for this case. To simplify the computations that follow, we make the change of variables $x_1 = -a_{10}x$, $y_1 = y$. Dropping the subscripts, the system under consideration is transformed into

$$\dot{x} = x + x^2 + a_{10}a_{-13}y^3, \qquad \dot{y} = -y + b_{01}y^2. \tag{3.146}$$

We look for a formal first integral expressed in the form $\Psi(x,y) = \sum_{j=1}^{\infty} v_j(x)y^j$. When this expression is inserted into equation (3.31) and terms are collected on powers of y, the functions v_j are determined recursively by the first-order linear differential equations

$$(x + x^2)v_j'(x) - jv_j(x) + a_{10}a_{-13}v_{j-3}'(x) + b_{10}(j-1)v_{j-1}(x) = 0, \tag{3.147}$$

if we define $v_j(x) \equiv 0$ for $j \in \{-2,-1,0\}$. It is easily established by mathematical induction that, making suitable choices for the constants of integration, there exist functions of the form

$$v_j(x) = \frac{P_j(x)}{(x+1)^j}$$

that satisfy (3.147), where $P_j(x)$ is a polynomial of degree j, and that we may choose $v_1(x) = x/(x+1)$. Thus system (3.146) admits a formal first integral of the form $\Psi(x,y) = xy + \cdots$, hence has a center at the origin. □

The reader may have noticed that the nonzero focus quantities and reduced focus quantities listed for systems (3.129) and (3.130) at the beginning of the proofs of Theorems 3.7.1 and 3.7.2 are homogeneous polynomials of increasing degree. While the nonzero focus quantities and reduced focus quantities need not be homogeneous in general, the fact that these are is not a coincidence. It stems from the fact that the nonlinearities in families (3.129) and (3.130) are themselves homogeneous. These ideas are developed in Exercises 3.40 and 3.41.

3.8 The Center Problem for Liénard Systems

In Section 3.5 we studied systems that are symmetric with respect to the action of a linear group followed by reversion of time. In the present section we will study a more complex symmetry, the symmetry that is the result of the action, not of a linear group of affine transformations, but of an analytic invertible transformation followed by reversion of time. This kind of symmetry is sometimes called *generalized symmetry*. We will analyze the important family of real analytic systems of the form

$$\dot{x} = y, \qquad \dot{y} = -g(x) - yf(x), \tag{3.148}$$

which are equivalent to the second-order differential equation

$$\ddot{x} + f(x)\dot{x} + g(x) = 0. \tag{3.149}$$

Any system of the form (3.148) is known as a *Liénard system*; the corresponding differential equation (3.149) is called a *Liénard equation*. Equations of this type arise frequently in the study of various mathematical models of physical, chemical, and other processes. We will not complexify system (3.148), so in this section x and y will denote real variables.

Our standing assumptions will be that the functions f and g are real analytic in a neighborhood of the origin and that

$$g(0) = 0, \qquad g'(0) > 0. \tag{3.150}$$

The condition $g(0) = 0$ is equivalent to system (3.148) having a singularity at the origin; the condition $g'(0) > 0$ is equivalent to its linear part having a positive determinant there. In such a case the origin is an antisaddle if and only if $f(0)^2 < 4g'(0)$, but we will not make use of this condition. F and G will denote the particular antiderivatives of f and g given by

$$F(x) = \int_0^x f(s)ds, \quad G(x) = \int_0^x g(s)ds. \tag{3.151}$$

We will show that a Liénard system has a center at the origin only if it possesses a generalized symmetry, and identify all *polynomial* Liénard systems having a center at the origin. We begin with two criteria for distinguishing between a center and a focus at the origin in system (3.148).

Theorem 3.8.1. *Suppose system* (3.148) *satisfies* (3.150)*. Then* (3.148) *has a center at the origin if and only if the functions* F *and* G *defined by* (3.151) *are related by* $F(x) = \Psi(G(x))$ *for some analytic function* Ψ *for which* $\Psi(0) = 0$.

Proof. By means of the so-called *Liénard transformation* $y_1 = y + F(x)$, we obtain from (3.148) the system

$$\dot{x} = y - F(x), \quad \dot{y} = -g(x), \tag{3.152}$$

where the subscript 1 in y_1 has been dropped.

Since $2G(x) = g'(0)x^2 + \cdots$, we may introduce a new variable by setting

$$u = \upsilon(x) = \operatorname{sgn} x\sqrt{2G(x)}. \tag{3.153}$$

Condition (3.150) implies that $\upsilon(x)$ is an analytic, invertible function on a neighborhood of $x = 0$ of the form $\upsilon(x) = \sqrt{g'(0)}x + O(x^2)$. Let $x = \xi(u)$ denote its inverse. The change of coordinates $u = \upsilon(x)$, $y = y$ transforms (3.152) into the form

$$\dot{u} = \frac{g(\xi(u))}{u}(y - F(\xi(u))), \qquad \dot{y} = -g(\xi(u)). \tag{3.154}$$

Because $g(\xi(u))/u = \sqrt{g'(0)} + O(u)$ is analytic and different from zero in a neighborhood of the origin, we conclude that the origin is a center for system (3.154), equivalently for systems (3.152) and (3.148), if and only if it is a center for the system

$$\dot{u} = y - F(\xi(u)), \quad \dot{y} = -u. \tag{3.155}$$

Consider the power series expansion of $F_1(u) := F(\xi(u))$, $F_1(u) = \sum_{k=1}^{\infty} a_k u^k$. We claim that the origin is a center for (3.155) if and only if

$$a_{2k-1} = 0 \quad \text{for all} \quad k \in \mathbb{N}. \tag{3.156}$$

Indeed, if (3.156) holds, then (3.155) is time-reversible, hence has a center at the origin. Conversely, suppose that not all a_s with odd index s vanish, and let the first such nonzero coefficient be a_{2m+1}. Then, as we have just seen, the origin is a center for the system

$$\dot{u} = y - \sum_{k=1}^{\infty} a_{2k}u^{2k} := y - \widehat{F}_1(u), \quad \dot{y} = -u, \tag{3.157}$$

hence, according to Theorem 3.2.9, it has a first integral $\Phi(u,y) = u^2 + y^2 + \cdots$ on a neighborhood of the origin. This Φ is a Lyapunov function for system (3.155); $\dot{\Phi}$ is

$$\frac{\partial \Phi}{\partial u}(y - \widehat{F}_1(u)) - \frac{\partial \Phi}{\partial y}u - \frac{\partial \Phi}{\partial u}(a_{2m+1}u^{2m+1} + \cdots) = -2a_{2m+1}u^{2m+2}(1 + \cdots).$$

By Theorem 2.1.4, the origin is a stable focus if $a_{2m+1} > 0$ and an unstable focus if $a_{2m+1} < 0$.

Thus we have shown that the origin is a center for (3.155), hence for (3.148), if and only if $F(\xi(u)) = h(u^2)$ for some analytic function h for which $h(0) = 0$. However, by (3.153) we have $u^2 = 2G(\xi(u))$, which means that the theorem follows with $\Psi(v) = h(2v)$. □

Theorem 3.8.2. *Suppose system* (3.148) *satisfies* (3.150). *Then it has a center at the origin if and only if there exists a function* $\zeta(x)$ *that is defined and analytic on a neighborhood of* 0 *and satisfies*

$$\zeta(0) = 0, \quad \zeta'(0) < 0 \tag{3.158}$$

and

$$F(x) = F(\zeta(x)), \quad G(x) = G(\zeta(x)), \tag{3.159}$$

where F *and* G *are the functions defined by* (3.151).

Proof. Define a real analytic function from a neighborhood of $(0,0) \in \mathbb{R}^2$ into $\mathbb{R}$ by

$$\begin{aligned}\widehat{G}(x,z) &= G(x) - G(z)\\ &= \tfrac{1}{2}g'(0)(x^2 - z^2) + \widehat{G}_3 \cdot (x^3 - z^3) + \cdots\\ &= (x-z)[\tfrac{1}{2}g'(0)(x+z) + R(x,z)],\end{aligned}$$

where $R(x,z)$ is a real analytic function that vanishes together with its first partial derivatives at $(0,0)$. Because $g'(0) \neq 0$, by the Implicit Function Theorem the equation $G(x) - G(z) = 0$ defines, in addition to $z = x$, a real analytic function $z = \zeta(x) = -x + O(x^2)$ on a neighborhood of 0 in $\mathbb{R}$. That is, the second equation in (3.159) always has a unique real analytic solution $z = \zeta(x)$ satisfying (3.158).

As in the proof of Theorem 3.8.1, define a function $u = \upsilon(x) = \operatorname{sgn}(x)\sqrt{2G(x)}$ and its inverse $x = \xi(u)$. Then $2G(\xi(-u)) = (-u)^2 = u^2 = 2G(\xi(u))$, so that $G(\xi(-u)) = G(\xi(u))$. But $\xi(-\upsilon(x)) = -x + O(x^2)$, so we conclude that in fact $z = \zeta(x) = \xi(-\upsilon(x))$.

The proof of Theorem 3.8.1 showed that (3.148) has a center if and only if $F_1(u) = F(\xi(u))$ is an even function, that is, if and only if $F(\xi(u)) = F(\xi(-u))$, equivalently, if and only if

$$F(\xi(\upsilon(x)) - F(\xi(-\upsilon(x))) = F(x) - F(\zeta(x)) = 0,$$

and the theorem follows. □

The next proposition, in conjunction with Theorem 3.8.2, shows that the generalized symmetry is the only mechanism yielding a center in Liénard systems.

Proposition 3.8.3. *Suppose system* (3.148) *satisfies* (3.150)*. Then the origin is a center for* (3.148) *only if there exists an analytic invertible transformation T of the form $T(x,y) = (\zeta(x), y)$, defined on a neighborhood of the origin, such that* (3.148) *is invariant with respect to an application of T and a reversal of time.*

Proof. Suppose the origin is a center for (3.148), let $z = \zeta(x)$ be the function given by Theorem 3.8.2, and let $\xi(z)$ be the inverse of $\zeta(x)$. Define a transformation $T(x,y)$ by $(z, y_1) = T(x,y) = (\zeta(x), y)$, so that T is an analytic invertible transformation defined on a neighborhood of $(0,0)$.

Differentiation of (3.159) yields

$$F'(x) = F'(\zeta(x)) \cdot \zeta'(x), \quad G'(x) = G'(\zeta(x)) \cdot \zeta'(x). \tag{3.160}$$

Thus under application of T we have

$$\dot{z} = \zeta'(x)\dot{x} = \zeta'(x) y_1 = \zeta'(\xi(z)) y_1$$

and, using (3.160),

$$\begin{aligned} \dot{y}_1 = \dot{y} &= -G'(x) - yF'(x) \\ &= -G'(\zeta(x))\zeta'(x) - yF'(\zeta(x))\zeta'(x) = \zeta'(\xi(z))[-g(z) - y_1 f(z)]. \end{aligned}$$

That is, under the transformation T the system becomes

$$\dot{z} = Z(z) y_1, \qquad \dot{y}_1 = Z(z)[-g(z) - y_1 f(z)]$$

for $Z : \mathbb{R} \to \mathbb{R} : z \mapsto \zeta'(\xi(z)) = -1 + \cdots$, whose orbits in a neighborhood of the origin are precisely those of (3.148) but with the sense reversed. □

We now study in more detail the situation that f and g are polynomials, in which case F and G are polynomials as well. To do so, we first recall that if k is a field and if polynomials p, $q \in k[x]$ both have full degree at least one, then their resultant, $\text{Resultant}(p,q,x)$, is a polynomial in the coefficients of p and q that takes the value zero if and only if p and q have a common factor. For p, $q \in k[x,z]$, if neither $p(x,0)$ nor $q(x,0)$ reduces to a constant, then we can regard p and q as elements of $k[x]$ with coefficients in $k[z]$ and form their resultant, $\text{Resultant}(p,q,x)$, to obtain a polynomial in $k[z]$, and it proves to be the case that p and q have a common factor in $k[x,z]$ if and only if $\text{Resultant}(p,q,x) = 0$ ([60, Chap. 3, §6]).

Proposition 3.8.4. *Suppose that in* (3.148) *f and g are polynomial functions and that* (3.150) *holds. Then the origin is a center only if the resultant with respect to x of*

$$\frac{F(x) - F(z)}{x - z} \quad \textit{and} \quad \frac{G(x) - G(z)}{x - z}$$

is equal to zero. Conversely, if the resultant with respect to x of these two polynomials is zero, and if the common factor that therefore exists vanishes at $(x,z) = (0,0)$, *then the origin is a center for system* (3.148).

Proof. Since $F(x)$ and $G(x)$ are polynomials, a solution $z = \zeta(x)$ of (3.159) satisfying (3.158) corresponds to a common factor between $F(x) - F(z)$ and $G(x) - G(z)$ other than $x - z$. Thus if such a solution $z = \zeta(x)$ exists, then

$$\text{Resultant}\left(\frac{F(x)-F(z)}{x-z}, \frac{G(x)-G(z)}{x-z}, x\right) = 0. \tag{3.161}$$

Conversely, suppose (3.161) holds and let $c(x,z)$ be the common factor of the two polynomials in (3.161). In Exercise 3.42 the reader is asked to show that if $c(0,0) = 0$, then the equation $c(x,z) = 0$ defines a function $z = \zeta(x)$ that satisfies $\zeta'(0) = -1$. Thus by Theorem 3.8.2 system (3.148) has a center at the origin. □

We close this section with a result that shows how polynomial Liénard systems with a center at the origin arise: they are precisely those polynomial systems (3.148) for which the functions F and G given by (3.151) are polynomial functions of a common polynomial function h. To establish this result, we must examine the subfield $\mathcal{K}$ of the field of rational functions generated by the polynomials $F(x)$ and $G(x)$.

In general, for an arbitrary field k we let $k(x)$ denote the field of rational functions with coefficients in k, that is,

$$k(x) = \left\{\frac{\alpha(x)}{\beta(x)} : \alpha, \beta \in k[x],\ \beta \neq 0\right\},$$

and for $h \in k[x]$, let $k(h)$ denote the smallest subfield of $k(x)$ that contains h, namely,

$$k(h) = \left\{\frac{\alpha(h(x))}{\beta(h(x))} : \alpha, \beta \in k[x],\ \beta \neq 0\right\}.$$

In particular, in reference to the Liénard system (3.148), for F and G as given by (3.151), we may first form $\mathbb{R}(F)$ and, since $\mathbb{R}(F)$ is a field, in turn form the field $\mathcal{K} := \mathbb{R}(F)(G)$. Then $\mathcal{K}$ is the smallest subfield of $\mathbb{R}(x)$ that contains both F and G or, in other words, is the subfield of $\mathbb{R}(x)$ generated by F and G.

Lemma 3.8.5. *If an analytic function* $\zeta(x)$ *defined on a neighborhood of* 0 *satisfies* (3.159), *then* $h(x) = h(\zeta(x))$ *for every* $h \in \mathcal{K}$.

Proof. The result follows immediately from the fact that any element of $\mathcal{K}$ has the form

$$\frac{A_N(x)G^N(x) + \cdots + A_0(x)}{B_M(x)G^M(x) + \cdots + B_0(x)},$$

where each A_j and B_j has the form

$$\frac{a_n F^n(x) + \cdots + a_0}{b_m F^m(x) + \cdots + b_0}$$

for $M,N,m,n \in \mathbb{N}_0$ and $a_j, b_j \in \mathbb{R}$. □

Lemma 3.8.6. *There exists a polynomial $h \in \mathbb{R}[x]$ such that $\mathcal{K} = \mathbb{R}(h)$.*

Proof. Because $\mathbb{R} \subset \mathcal{K} \subset \mathbb{R}(x)$ and $\mathcal{K}$ contains a nonconstant polynomial, the result follows immediately from the following result of abstract algebra. □

Proposition 3.8.7. *Let k and K be fields such that $k \subset K \subset k(x)$. If K contains a nonconstant polynomial, then $K = k(h)$ for some $h \in k[x]$.*

Proof. Lüroth's Theorem (see [195]) states that any subfield of $k(x)$ that strictly contains k is isomorphic to $k(x)$. Thus there is a field isomorphism $\xi : k(x) \to K$. Let $r \in K$ denote the image of the identity function $\iota(x) = x$ under ξ. Then for any element of $k(x)$,

$$\xi\left(\frac{a_n x^n + \cdots + a_0}{b_m x^m + \cdots + b_0}\right) = \frac{a_n \xi^n(x) + \cdots + a_0}{b_m \xi^m(x) + \cdots + b_0} = \frac{a_n r^n + \cdots + a_0}{b_m r^m + \cdots + b_0} \in k(r).$$

Since ξ is one-to-one and onto, we conclude that $K = k(r)$.

We know that $r = A/B$ for some polynomials $A, B \in k[x]$; we must show that B is a constant. Since the theorem is trivially true if A is a constant multiple of B, without loss of generality we may assume that A and B are relatively prime. It is well known ([195, §63]) that for any $a,b,c,d \in k$ for which $ad - bc \neq 0$,

$$k(r) = k\left(\frac{ar+b}{cr+d}\right) = k\left(\frac{aA+bB}{cA+dB}\right),$$

hence without loss of generality we may also assume that $\deg(A) > \deg(B)$ (Exercise 3.43).

Let $H \in k[x]$ be any nonconstant polynomial in K. Then because $K = k(A/B)$, we have

$$H = \frac{a_n(\frac{A}{B})^n + \cdots + a_1\frac{A}{B} + a_0}{b_m(\frac{A}{B})^m + \cdots + b_1\frac{A}{B} + b_0}$$

for some $a_j, b_j \in k$. It must be the case that $n > m$, for if $n \leq m$, then clearing fractions yields

$$H \cdot (b_m A^m + \cdots + b_1 A B^{m-1} + b_0 B^m) = a_n A^n B^{m-n} + \cdots + a_1 A B^{m-1} + a_0 B^m.$$

However, the degree of the polynomial on the left is $\deg(H) + m\deg(A)$, while the degree of the polynomial on the right is at most $m\deg(A)$. Clearing fractions thus actually yields

$$H \cdot (b_m A^m B^{n-m} + \cdots + b_1 A B^{n-1} + b_0 B^n) = a_n A^n + \cdots + a_1 A B^{n-1} + a_0 B^n.$$

If, contrary to what we wish to show, $\deg(B) > 0$, then B is a divisor of the polynomial on the left but is relatively prime to the polynomial on the right, a contradiction. Thus B is constant, as we wished to show. □

Theorem 3.8.8. *Suppose that the functions f and g in* (3.148) *are polynomials for which* (3.150) *holds, and let F and G be the functions defined by* (3.151)*. The origin is a center for system* (3.148) *if and only if there exist a polynomial $h \in \mathbb{R}[x]$ that satisfies*

$$h'(0) = 0 \quad \text{and} \quad h''(0) \neq 0 \tag{3.162}$$

and polynomials $\alpha, \beta \in \mathbb{R}[x]$ such that $F(x) = \alpha(h(x))$ and $G(x) = \beta(h(x))$.

Proof. Suppose there exist polynomials $\alpha, \beta, h \in \mathbb{R}[x]$ such that $F(x) = \alpha(h(x))$, $G(x) = \beta(h(x))$, and h satisfies (3.162), say $h(x) = a_n x^n + \cdots + a_2 x^2 + a_0$, $a_2 \neq 0$. Then for $x \neq z$, $U(x,z) = (h(x) - h(z))/(x - z)$ is the same as the polynomial function $V(x,z)$ that results by factoring $x - z$ out of the numerator of U and cancelling it with the denominator, and $V(x,z) = a_2(x+z) + \cdots$. But V satisfies $V(0,0) = 0$ and $V_x(0,0) = V_z(0,0) = a_2 \neq 0$, hence by the Implicit Function Theorem there is an analytic function ζ defined on a neighborhood of 0 such that $\zeta(0) = 0$ and in a neighborhood of $(0,0)$ in $\mathbb{R}^2$, $V(x,z) = 0$ if and only if $z = \zeta(x)$. Moreover, $\zeta'(0) = -1$. There is thus a neighborhood of 0 in which $\zeta(x) \neq x$ for $x \neq 0$, hence in which $V = 0$ is equivalent to $U = 0$. Thus $h(x) = h(\zeta(x))$ holds on a neighborhood of 0. Since $F(x) = \alpha(h(x)) = \alpha(h(\zeta(x))) = F(\zeta(x))$ and similarly $G(x) = G(\zeta(x))$, by Theorem 3.8.2 the origin is a center for system (3.148).

Conversely, assume that the origin is a center for the polynomial Liénard system (3.148), and let ζ be the analytic function defined on a neighborhood of the origin, satisfying (3.159) and (3.158), as provided by Theorem 3.8.2. If h is the polynomial provided by Lemma 3.8.6, then $F(x) = \alpha(h(x))$ and $G(x) = \beta(h(x))$ for some polynomials $\alpha, \beta \in \mathbb{R}[x]$, since both F and G lie in $\mathscr{K}$. But then by Lemma 3.8.5, $h(x) = h(\zeta(x))$, hence $h'(x) = h'(\zeta(x))\zeta'(x)$. Evaluating this latter equation at $x = 0$ and applying (3.158) yields $h'(0) = 0$. Similarly, differentiating the identity $G(x) = \beta(h(x))$ twice and evaluating at $x = 0$ yields $G''(0) = \beta'(h(0))h''(0)$; since $G''(0) = g'(0) > 0$ (by (3.150)) we have $h''(0) \neq 0$, so that h satisfies (3.162). □

Theorem 3.8.8 gives us a means for constructing polynomial Liénard systems having a center at the origin.

3.9 Notes and Complements

The center problem for polynomial systems dates back about 100 years, beginning with Dulac's 1908 study [64] of the quadratic system (3.131). He showed that for this system a first integral of the form (3.52) exists if and only if by means of a linear transformation the equation of the trajectories can be brought into one of 11 forms, each of which contains at most two parameters. He also found first integrals for all 11 forms. The center problem for quadratic systems was also solved by Kapteyn ([103, 104]) in 1911 and 1912.

Neither Dulac nor Kapteyn, however, gave explicit conditions on the coefficients of the system for the existence of a center. Thus they did not give so-called "necessary and sufficient center conditions." This problem was considered for the first time by Frommer ([74]), who appears to have been unaware of the work of Dulac and Kapteyn. He investigated not the complex system (3.129), but the corresponding real quadratic system of the form (3.111), for which calculations are much more difficult. For this reason, perhaps, his center conditions were incomplete and partially incorrect. Somewhat later correct coefficient conditions for a center for real quadratic system were obtained by Saharnikov ([168]), Sibirsky ([174, 175]), Malkin ([134]), and others ([76, 206]). For practical use we note in particular the discriminant quantities of Li Chengzhi computed from the normal form (2.75), whose successive vanishing (and sign) determines that the origin is a first-, second-, or third-order fine focus (and of what asymptotic stability) or a center (Theorem 12.2 of [202], or Theorem II.5.2 of [205]). (Recall Remark 3.1.6 for the definition of a fine focus.)

The conditions obtained by different authors for the same system often look markedly different. But these "center conditions" are simply the zero set of a system of polynomials, the focus quantities, hence the set of systems (3.131) with a center is a complex variety (a real variety in the case of real systems of the form (3.2)). This makes it natural to formulate the center problem as we did above: find an irreducible decomposition of the center variety defined by prime ideals. Then the answer will be unique in the sense that we can easily check if the conditions obtained by different authors are the same simply by computing the reduced Gröbner bases. It is thus surprising that, given the substantial amount of work devoted to this topic, the concept of the center variety apparently first appeared explicitly in the literature only relatively recently, in the paper [210] of Żołądek. The irreducible decomposition of the center variety of quadratic systems was obtained for the first time in [154].

In Section 3.4 we described the Focus Quantity Algorithm (page 128) for computing the focus quantities for family (3.3), which is based on the work in [149, 150, 153]. Of course, there are many other algorithms for computing these and similar quantities. The reader can consult, for example, [71, 77, 80, 121, 125, 129, 130, 131, 196]. In particular, an algorithm for computing focus quantities is available in the Epsilon library of Maple (see [199]).

The $p:-q$ resonant center problem has become the subject of study relatively recently. The discussion in Section 3.4 related to the focus quantities for family (3.3), leading to the Focus Quantity Algorithm, has a direct generalization to the family (3.63) containing the $p:-q$ resonant centers, leading to an analogous algorithm for the quantities $g_{kq,kp}$ of (3.68). The only change in the algorithm is that the expression for the quantity M becomes $\lfloor K(p+q)/w \rfloor$, the condition in the IF statement changes to $L_1(\nu)p = L_2(\nu)q$, and the formulas for V and $g_{kk}^{(\nu)}$ are replaced by those of Exercises 3.18 and 3.19. See [150] for a complete treatment. Other results on $p:-q$ resonant centers are obtained in [57]. The center problem for quadratic systems with a $1:-2$ resonant singular point has been solved by Fronville, Sadovskii, and Żołądek ([75, 210]).

In the section devoted to time-reversible systems we considered the simplest case of reversibility, namely, reversibility with respect to the action of the rotation group. This concept can be generalized, and it is an interesting and important problem to study the reversibility of differential equations with respect to different classes of rational and analytic transformations. See [124] and [208]. We note that this problem is also important from the point of view of physical applications; see the survey of Lamb and Roberts ([108]) and the references therein.

In Sections 3.6 and 3.7 we discussed the Darboux method of integrability as a tool for investigating the center problem. However, it also provides a way to construct elementary and Liouvillian first integrals of ordinary differential equations and is therefore also an efficient tool in the study of the important problem of the integrability of differential equations in closed form. In recent years the ideas of Darboux have been significantly developed; see [54, 55, 135, 170] and the references they contain. In particular, Jouanalou ([101]) extended the Darboux theory to polynomial systems in $\mathbb{R}^n$ and $\mathbb{C}^n$. In [144] Prelle and Singer showed that if a polynomial vector field has an elementary first integral, then that first integral can be computed using the Darboux method. Later Singer ([182]) proved that if a polynomial vector field has Liouvillian first integrals, then it has integrating factors given by Darbouxian functions. A good survey of all these results is [119].

An important problem from the point of view of practical usage of the Darboux method is the study of possible degrees of invariant algebraic curves and their properties. This problem is addressed in the papers of Campillo–Carnicer ([29]), Cerveau–Lins Neto ([30]), Tsygvintsev ([188]), and Żołądek ([211, 212]). This is an interesting subject with the interplay of methods of differential equations and algebraic geometry and is likely to be a fruitful field for future research.

A generalization of the Darboux method to higher-dimensional systems is presented in [120].

In Section 3.7 we presented the solution of the center problem for the general quadratic system. If we go further and consider the center problem for the general cubic system (3.100), we face tremendous computational difficulties. Although it appears that by using the modern computational facilities it is possible to compute enough focus quantities to determine the center variety, the polynomials obtained are so cumbersome that it is impossible to carry out a decomposition of the center variety. Thus current research is devoted to the study of various subfamilies of (3.100). One important subfamily is the cubic Liénard system $\dot{u} = -v$, $\dot{v} = u + \sum_{k+p=2}^{3} u^k v^p$. The center problem for this system has been solved by Sadovskii ([166]) and Lloyd and Pearson ([49, 122]). Another variant is to eliminate all quadratic terms, that is, to study the center problem for a planar linear center perturbed by third-degree homogeneous polynomials. This problem was studied initially by Al'muhamedov ([3]) and Saharnikov ([169]), who obtained sufficient conditions for a center. The complete solution of the problem was obtained by Malkin ([133]) and Sadovskii ([163]) for the real and complex cases, respectively; see Theorem 6.4.3. Later work on this problem appears in [58, 76, 128, 118, 207]. Similarly, one could also consider the situation in which the perturbation terms are fourth- or fifth-degree homogeneous polynomials; see [34, 35]. For results on the center problem for some particular subfamilies of the cubic system (not restricting to homogeneous nonlinearities) see [22, 70, 122, 123, 117, 167, 186, 196] and the references they contain.

The results on the center problem for the Liénard system presented in Section 3.8 are due mainly to Cherkas ([43]), Christopher ([50]), and Kukles ([105, 106]).

We have not considered the center problem in the interesting but much more difficult case that the singular point is not simple, that is, when the determinant of the linear part of the right-hand sides vanishes at the singular point. For treatment of this situation the reader can consult [9, 10, 11, 33, 99, 111, 138, 164, 165, 179]. In sharp contrast to the theory that we have presented here, when the linear part vanishes identically the center problem is not even algebraically solvable ([98]).

Exercises

3.1 Without appealing to the Hartman–Grobman Theorem, show that if $\mathbf{df}(\mathbf{0})$ has eigenvalues $\alpha \pm i\beta$ with $\alpha\beta \neq 0$, then the phase portraits of $\dot{\mathbf{u}} = \mathbf{f}(\mathbf{u})$ and $\dot{\mathbf{u}} = \mathbf{df}(\mathbf{0}) \cdot \mathbf{u}$ in a neighborhood of the origin, the latter of which is a focus, are topologically equivalent (that is, that there is a homeomorphism of a neighborhood of the origin onto its image mapping orbits of one system (as point sets) onto those of the other).
Hint. Show that $\mathbf{0}$ is the only singularity of the nonlinear system near $\mathbf{0}$. Then show that there exists a simple closed curve surrounding $\mathbf{0}$ that every nonstationary trajectory of the nonlinear system near $\mathbf{0}$ intersects exactly once.

3.2 Find a nonsingular linear transformation that changes

$$\begin{aligned} \dot{u} &= au + bv + \sum_{j+k=2}^{\infty} U_{jk} u^j v^k \\ \dot{v} &= cu + dv + \sum_{j+k=2}^{\infty} V_{jk} u^j v^k \end{aligned} \tag{3.163}$$

into (3.4) when the eigenvalues of the linear part $\begin{pmatrix} a & b \\ c & d \end{pmatrix}$ of system (3.163) are $\alpha \pm i\beta$, $\beta \neq 0$.

3.3 Prove Proposition 3.1.4.
Hint. The geometry of the return map.

3.4 A real quadratic system of differential equations that has a weak focus (one for which the real part of the eigenvalues of the linear part is zero) or a center at the origin can be placed in the form

$$\dot{u} = -v + a_{20}u^2 + a_{11}uv + a_{02}v^2, \qquad \dot{v} = u + b_{20}u^2 + b_{11}uv + b_{02}v^2.$$

a. Show that the complex form (3.19) (or (3.51)) of this family of systems is $\dot{x} = ix + c_{20}x^2 + c_{11}x\bar{x} + c_{02}\bar{x}^2$. Find the coefficients $c_{jk} \in \mathbb{C}$ in terms of the real coefficients a_{jk} and b_{jk}.
b. Confirm that the transformation $\xi : \mathbb{R}^6 \to \mathbb{R}^6$ that carries $(a_{20}, a_{11}, \dots, b_{02})$ to $(\operatorname{Re} c_{20}, \operatorname{Im} c_{20}, \dots, \operatorname{Im} c_{02})$ is an invertible linear transformation.

3.5 Show that if in the real system (3.2) U and V are polynomial functions that satisfy $\max\{\deg P, \deg V\} = n$, then the complexification (3.19) (or (3.51)) has the form $\dot{x} = ix + R(x,\bar{x})$, where R is a polynomial without constant or linear terms that satisfies $\deg R = n$, and that the transformation from $\mathbb{R}^{(n+1)(n+2)-6}$ to $\mathbb{R}^{(n+1)(n+2)-6}$ that carries the coefficients of U and V to the real and imaginary parts of those of R is an invertible linear transformation.

3.6 Show that the line $x_2 = \bar{x}_1$ is invariant under the flow of system (3.21) on $\mathbb{C}^2$. Show that if the invariant line is viewed as a 2-plane in real 4-space, then the flow on the plane is the flow of the original real system (3.4). (Part of the problem is to make these statements precise.)

3.7 [Referenced in Proposition 3.2.1 and Theorem 3.2.7, proofs.] Writing the higher-order terms in (3.22) as $X_j(x_1,x_2) = \sum_{\alpha+\beta\geq 2} X_j^{(\alpha,\beta)} x_1^\alpha x_2^\beta$, $j = 1,2$, show that a member of family (3.22) arises from a real system if and only if $\lambda_2 = \bar{\lambda}_1$ and $X_2^{(\alpha,\beta)} = \overline{X_1^{(\beta,\alpha)}}$.

3.8 Suppose that in system (3.22), $\lambda_1/\lambda_2 = -p/q$ for $p,q \in \mathbb{N}$, $\mathrm{GCD}(p,q) = 1$, and that Ψ is a formal first integral of (3.22) that does not have a constant term and begins with terms of order no higher than $p+q$. Show that Ψ must have the form $\Psi(x_1,x_2) = x_1^q x_2^p + \cdots$.
Hint. One approach is to (i) show that all partial derivatives of Ψ through order $p+q-1$ vanish at $(0,0)$; and then (ii) show that if not all partial derivatives of Ψ through order $p+q$ vanish at $(0,0)$, then $\Psi(x_1,x_2) = x_1^q x_2^p + \cdots$.

3.9 [Referenced in Theorems 3.2.7 and 3.2.9, proofs.] Let $P(u,v)$, $Q(u,v)$, and $\Psi(u,v)$ be real functions of real variables u and v and suppose that for complex variables x and y the functions

$$P\left(\frac{x+y}{2}, \frac{x-y}{2i}\right), \quad Q\left(\frac{x+y}{2}, \frac{x-y}{2i}\right), \quad \text{and} \quad \Psi\left(\frac{x+y}{2}, \frac{x-y}{2i}\right)$$

are well-defined. Show that $\Psi(u,v)$ is a formal first integral for the system

$$\begin{aligned} \dot{u} &= P(u,v) \\ \dot{v} &= Q(u,v) \end{aligned}$$

if and only if

$$\Phi(x,y) \stackrel{\text{def}}{=} \Psi\left(\frac{x+y}{2}, \frac{x-y}{2i}\right)$$

is a formal first integral for the system

$$\begin{aligned} \dot{x} &= P\left(\frac{x+y}{2}, \frac{x-y}{2i}\right) + iQ\left(\frac{x+y}{2}, \frac{x-y}{2i}\right) \\ \dot{y} &= \overline{P}\left(\frac{x+y}{2}, \frac{x-y}{2i}\right) - i\overline{Q}\left(\frac{x+y}{2}, \frac{x-y}{2i}\right) \\ &= P\left(\frac{x+y}{2}, \frac{x-y}{2i}\right) - iQ\left(\frac{x+y}{2}, \frac{x-y}{2i}\right). \end{aligned}$$

3.10 [Referenced in Theorem 3.2.7, proof.] Show that if system (3.4) has a first integral of the form $\Psi(u,v) = u^2 + v^2 + \cdots$, then the origin is a center.
Hint. Restricting Ψ to a sufficiently small neighborhood of the origin, for (u_0, v_0) sufficiently near the origin $\Psi(u_0, v_0)$ is a regular value of Ψ, hence, by an appropriate result of differential geometry and the form of Ψ, the set $\{(u,v) : \Psi(u,v) = \Psi(u_0, v_0)\}$ is an oval.

3.11 Consider the following system (analogous to an example of R. Moussu) of polynomial differential equations on $\mathbb{R}^2$:

$$\dot{u} = -v^3, \qquad \dot{v} = u^3 + u^2 v^2. \tag{3.164}$$

Note that in the absence of the quartic terms the system is Hamiltonian, hence has an analytic first integral.
a. Show that $(0,0)$ is an antisaddle of focus or center type, say by passing to polar coordinates as at the beginning of Section 3.1.
b. Show that $(0,0)$ is in fact a center by finding a line of symmetry for the orbits as point sets.
c. Show that if $H(u,v) = \sum h_{jk} u^j v^k$ is constant on orbits of (3.164), then $h_{jk} = 0$ for all $(j,k) \in \mathbb{N}_0 \times \mathbb{N}_0 \setminus \{(0,0)\}$.

3.12 Refer to Definition 3.3.3. Suppose $s \in \mathbb{N}$ is such that $\mathbf{V}(\mathscr{B}_s) = \mathbf{V}(\mathscr{B}) = V_{\mathscr{C}}$. Prove the following implication: if for some $r \geq s$ $\mathscr{B}_r$ is a radical ideal, then for every $j \geq r$, $\mathscr{B}_j = \mathscr{B}_r = \mathscr{B}$ (hence is a radical ideal).

3.13 Explain why $V_{\mathscr{C}}^R$ is a real affine variety.

3.14 Does (3.82) hold for system (3.48)? What is the structure of the focus quantities for this system?

3.15 Let $S = \{(p_1, q_1), \ldots, (p_\ell, q_\ell)\}$ be the indexing set for family (3.3) and fix a number $K \in \mathbb{N}_0^\ell$. Let $w = \min\{p_j + q_j : (p_j, q_j) \in S\}$. Show that if $v \in \mathbb{N}_0^{2\ell}$ is such that $|v| > \lfloor 2K/w \rfloor$, where $\lfloor r \rfloor$ denotes the greatest integer less than or equal to r, then $v \notin \operatorname{Supp}(g_{kk})$ for $1 \leq k \leq K$.

3.16 Prove Corollary 3.4.6.

3.17 Using the algorithm of Section 3.4, write a code in an available computer algebra system and compute the first three focus quantities of system (3.129). Check that the quantities thus obtained generate the same ideal as the quantities (3.133)–(3.135).

3.18 Imitate the proof of Theorem 3.4.2 to prove the following analogous theorem for family (3.63). (Definition 3.4.1 is identical except that it is stated for all $(j,k) \in \mathbb{N}_{-q} \times \mathbb{N}_{-p}$.)

Theorem. *Let family* (3.63) *be given, where* $p, q \in \mathbb{N}$, $\mathrm{GCD}(p,q) = 1$. *There exist a formal series* $\Psi(x,y)$ *of the form* (3.64) *and polynomials* $g_{q,p}$, $g_{2q,2p}$, $g_{3q,3p}$, $\ldots$ *in* $\mathbb{C}[a,b]$ *such that*
1. *equation* (3.68) *holds;*
2. *for every pair* $(j,k) \in \mathbb{N}_{-q} \times \mathbb{N}_{-p}$ *such that* $j + k \geq 0$, $v_{jk} \in \mathbb{Q}[a,b]$, *and* v_{jk} *is a* (j,k)*-polynomial;*
3. *for every* $k \geq 1$, $v_{kq,kp} = 0$*; and*
4. *for every* $k \geq 1$, $g_{kq,kp} \in \mathbb{Q}[a,b]$, *and* $g_{kq,kp}$ *is a* (kq,kp)*-polynomial.*

3.19 For $p,q \in \mathbb{N}$ with $\mathrm{GCD}(p,q) = 1$, modify the definition (3.76) of $V(\nu)$ so that (3.76b) becomes $V(\nu) = 0$ if $L_1(\nu)p = L_2(\nu)q$ and (3.76c) becomes

$$V(\nu) = \frac{1}{L_1(\nu)p - L_2(\nu)q} \times \left[\sum_{j=1}^{\ell} \widetilde{V}(\nu_1,\ldots,\nu_j - 1,\ldots,\nu_{2\ell})(L_1(\nu_1,\ldots,\nu_j - 1,\ldots,\nu_{2\ell}) + q) - \sum_{j=\ell+1}^{2\ell} \widetilde{V}(\nu_1,\ldots,\nu_j - 1,\ldots,\nu_{2\ell})(L_2(\nu_1,\ldots,\nu_j - 1,\ldots,\nu_{2\ell}) + p) \right]$$

(but maintain $V(0,\ldots,0) = 1$ as before). On the assumption that $V(\nu) = 0$ if $L_1(\nu) < -q$ or if $L_2(\nu) < -p$ (the analogue of Lemma 3.4.4, which is true), prove the following analogue of the first two parts of Theorem 3.4.5.

Theorem. *Let family* (3.63) *be given, where* $p,q \in \mathbb{N}$, $\mathrm{GCD}(p,q) = 1$. *Let* Ψ *be the formal series of the form* (3.64) *and let* $\{g_{kq,kp} : k \in \mathbb{N}\}$ *be the polynomials in* $\mathbb{C}[a,b]$ *given by the theorem in Exercise 3.18. Then*

1. *for* $\nu \in \mathrm{Supp}(v_{k_1,k_2})$, *the coefficient* $v^{(\nu)}_{k_1,k_2}$ *of* $[\nu]$ *in* v_{k_1,k_2} *is* $V(\nu)$;
2. *for* $\nu \in \mathrm{Supp}(g_{kq,kp})$, *the coefficient* $g^{(\nu)}_{kq,kp}$ *of* $[\nu]$ *in* $g_{kq,kp}$ *is*

$$g^{(\nu)}_{kq,kp} = -\left[\sum_{j=1}^{\ell} \widetilde{V}(\nu_1,\ldots,\nu_j - 1,\ldots,\nu_{2\ell})(L_1(\nu_1,\ldots,\nu_j - 1,\ldots,\nu_{2\ell}) + q) - \sum_{j=\ell+1}^{2\ell} \widetilde{V}(\nu_1,\ldots,\nu_j - 1,\ldots,\nu_{2\ell})(L_2(\nu_1,\ldots,\nu_j - 1,\ldots,\nu_{2\ell}) + p) \right].$$

3.20 Show that equations (3.86) are the conditions for the two types of symmetry of a real system $\dot{u} = \widetilde{U}(u,v)$, $\dot{v} = \widetilde{V}(u,v)$ with respect to the u-axis.

3.21 Prove that for the family of Example 3.5.7, $I_{\mathrm{sym}} = \langle a_{10}a_{01} - b_{10}b_{01} \rangle$, that is, that $a_{10}a_{01} - b_{10}b_{01}$ divides $[\nu] - \widehat{[\nu]}$ for every $\nu \in \mathscr{M}$.

3.22 Suppose $H = \prod_{j=1}^{s} f_j^{\alpha_j}$ is a first integral for system (3.101), where for each j, $\alpha_j \in \mathbb{C}$ and f_j is an irreducible element of $\mathbb{C}[x,y]$, and for $k \neq j$, f_k and f_j are relatively prime. Show that for each j, $f_j | \mathscr{X} f_j$.
Hint. Factor f_j from the equation $\mathscr{X} H \equiv 0$.

3.23 Prove that for the polynomial K in (3.103), $\deg K \leq m - 1$. Construct an example for which the inequality is strict.

3.24 Let $\mathscr{X}$ be the vector field corresponding to (3.101), let $f_1,\ldots,f_s$ be elements of $\mathbb{C}[x,y]$, and let $I = \langle f_1,\ldots,f_s \rangle$. This exercise is a generalization of the discussion that surrounds (3.102) and leads up to Proposition 3.6.2.

a. Prove that if $\mathscr{X} f_j \in I$ for $1 \leq j \leq s$, then $\mathbf{V}(f_1,\ldots,f_s)$ is invariant under the flow of $\mathscr{X}$. (Invariance means that if $\eta(t) = (x(t),y(t))$ is a solution of (3.101) for which $\eta(0) \in \mathbf{V}(f_1,\ldots,f_s)$, then $f_j(\eta(t)) \equiv 0$ for $1 \leq j \leq s$.)

b. Prove conversely that if $\mathbf{V}(f_1,\ldots,f_s)$ is invariant under the flow of $\mathscr{X}$, then $\mathscr{X}f_j \in \sqrt{I}$ for $1 \leq j \leq s$ (hence $\mathscr{X}f_j \in I$ for $1 \leq j \leq s$ if I is a radical ideal).

3.25 Suppose (3.101) has a Darboux first integral. Formulate conditions that imply that every trajectory of (3.101) lies in an algebraic curve. (One condition was given immediately following Definition 3.6.3.)

3.26 [Referenced in Theorem 3.6.4, proof.] Prove that the function H constructed in the proof of Darboux's Theorem (Theorem 3.6.4) is not constant.

3.27 Show that $f(u,v)$ is a complex algebraic partial integral of the real system (3.101) with cofactor $K(u,v)$ if and only if $\bar{f}(u,v)$ is a complex algebraic partial integral of the real system (3.101) with cofactor $\bar{K}(u,v)$, where the conjugation is of the coefficients of the polynomials only.

3.28 Consider the system

$$\dot{x} = x - 3x^2y + 2xy^2 + 5y^3, \qquad \dot{y} = -3y - x^3 + 4x^2y + 3xy^2 - 2y^3. \tag{3.165}$$

a. Verify by finding their cofactors that both $f_1(x,y) = 1 - x^2 - 2xy - y^2$ and $f_2(x,y) = 6y + x^3 - 3x^2y - 9xy^2 - 5y^3$ are algebraic partial integrals.

b. Show that $\mu = f_1 f_2^{-2/3}$ is an integrating factor of (3.165) on some open set $\Omega \subset \mathbb{C}^2$. What is the largest that Ω can be with respect to set inclusion?

c. Using part (b) and a computer algebra system, derive the first integral $\Phi(x,y) = (2x - x^3 - x^2y + xy^2 + y^3)f_2(x,y)^{1/3}$ of (3.165) on Ω. Explain why Φ is *not* a first integral of (3.165) on any neighborhood of $(0,0)$ in $\mathbb{C}^2$ even though it is defined on $\mathbb{C}^2$.

d. Explain why $\Psi = \frac{1}{48}\Phi^3$ *is* a first integral of (3.165) on any neighborhood of $(0,0)$. Conclude that (3.165) has a $1:-3$-resonant center at $(0,0)$.

3.29 Consider the family

$$\dot{x} = x + ax^3 + bx^2y - xy^2, \qquad \dot{y} = -3y + x^2y + \tilde{b}xy^2 + \tilde{c}y^3, \tag{3.166}$$

where the coefficients are restricted to lying in the variety $\mathbf{V}(I) \subset \mathbb{C}^4$ of the ideal $I = \langle 5b\tilde{c} + 3b + \tilde{b}\tilde{c} + 3\tilde{b}, 2a\tilde{c} + 3a + \tilde{c}, 3ab - a\tilde{b} - b - \tilde{b}\rangle$. Obvious algebraic partial integrals are x and y with their complementary factors in the polynomials in $\dot{x}$ and $\dot{y}$ as their cofactors.

a. Show that when $a \neq 1/3$, an integrating factor of (3.166) on the open set $\Omega = \mathbb{C}^2 \setminus \{(x,y) : xy = 0\}$ is $\mu(x,y) = x^\alpha y^\beta$, where $\alpha = -(9a+1)/(3a+1)$ and $\beta = -(5a+1)/(3a+1)$.

b. Using a computer algebra system, derive the first integral

$$\begin{aligned}\Phi(x,y) &= (x^3y)^{-2a/(3a+1)}\\ &\times [(2a+1)(a+1) + a(2a+1)(a+1)x^2 + 2a(2a+1)xy + a(a+1)y^2]\end{aligned}$$

of (3.166) on Ω, valid for $a \neq \{-1/3, 0\}$.

c. Show that $\Psi = [\Phi/((2a+1)(a+1))]^{-(3a+1)/2a}$ is a first integral of (3.166) on a neighborhood of $(0,0)$ in $\mathbb{C}^2$, of the form $x^3y + \cdots$, for all systems (3.166) corresponding to $S = \mathbf{V}(I) \setminus \mathbf{V}(J)$, $J = \langle a(3a+1)(2a+1)(a+1)\rangle$.

d. Show that the Zariski closure $\bar{S}$ of S is $\mathbf{V}(I)$ by computing the ideal quotient $I:J$ and confirming that $\sqrt{I} = \sqrt{I:J}$ using the Radical Membership Test.
e. Use (c) and (d) to conclude that every element of $\mathbf{V}(I)$ has a $1:-3$-resonant center at the origin.

3.30 Consider the family

$$\dot{x} = x - x^3, \qquad \dot{y} = -3y + ax^3 + x^2y + bxy^2, \qquad (a,b) \in \mathbb{C}^2. \tag{3.167}$$

a. By finding their cofactors, verify that the following five polynomials are algebraic partial integrals of (3.167):

$$f_1(x,y) = x, \qquad f_2(x,y) = 1 + x, \qquad f_3(x,y) = 1 - x,$$
$$f_4(x,y) = -2 + \left(\sqrt{1-ab} + 1\right)x^2 + bxy,$$
$$f_5(x,y) = 4 - 4x^2 - 4bxy.$$

b. Show that for $b(ab-1) \neq 0$,

$$\Phi = f_2^{\sqrt{1-ab}} f_3^{\sqrt{1-ab}} f_4^{-2} f_5 = 1 + \frac{b}{2}\sqrt{1-ab}\,x^3y + \overset{(5)}{\cdots}$$

is an analytic first integral of (3.167) on a neighborhood of $(0,0)$ in $\mathbb{C}^2$, hence that $\Psi = \frac{2}{b\sqrt{1-ab}}[\Phi - 1]$ is an analytic first integral of (3.167) on a neighborhood of $(0,0)$ in $\mathbb{C}^2$ of the form $x^3y + \cdots$.
c. Use Exercise 1.45(b) to conclude that (3.167) has a $1:-3$-resonant center at the origin for all $(a,b) \in \mathbb{C}^2$.

3.31 Show that system (3.115) in Example 3.6.11 has a stable (resp., unstable) focus at the origin when $A < 0$ (resp., $A > 0$) by changing to polar coordinates and integrating the differential equation (3.6) for the trajectories.

3.32 Verify the truth of (3.136).

3.33 In the context of Theorem 3.7.1, suppose $(a,b) \in V_3 \cap \mathbf{V}(b_{2,-1}b_{10})$.
a. Show that if $(b_{2,-1}, b_{10}) = (0,0)$, then $(a,b) \in \mathbf{V}(I_{\text{sym}})$, hence there is a center at $(0,0)$.
b. Show that if $(b_{2,-1}, b_{10}) \neq (0,0)$, then there exist irreducible algebraic partial integrals h_1, h_2, and h_3 for which there exist suitable constants α_1, α_2, and α_3 such that by Theorem 3.6.8, $H = h_1^{\alpha_1} h_2^{\alpha_2} h_3^{\alpha_3}$ is a first integral.
Hint. $\deg(h_1) = 1$, $\deg(h_2) = \deg(h_3) = 2$.

3.34 Each of the systems $\dot{x} = x - y^2$, $\dot{y} = -y + y^2$ and $\dot{x} = x + x^2 - y^2$, $\dot{y} = -y + x^2 - y^2$ corresponds to an element of the set V_2 of Theorem 3.7.1. Show that the first system has exactly two invariant lines, real or complex, and that the second has exactly one.

3.35 System (3.2) with quadratic polynomial nonlinearities can be written in the form

$$\begin{aligned} \dot{u} &= -v - bu^2 - (2c+\beta)uv - dv^2 \\ \dot{v} &= u + au^2 + (2b+\alpha)uv + cv^2. \end{aligned} \tag{3.168}$$

Necessary and sufficient conditions that there be a center at the origin as formulated by Kapteyn are that at least one of the following sets of equalities holds:

I. $a+c=b+d=0$;

II. $\alpha(a+c)=\beta(b+d)$ and $\alpha^3 a-(\alpha+3b)\alpha^2\beta+(\beta+3c)\alpha\beta^2-\beta^3 d=0$;

III. $\alpha+5(b+d)=\beta+5(a+c)=2(a^2+d^2)+ac+bd=0$.

Derive these conditions by complexifying (3.168), relating the parameters there to a_{10}, a_{01}, and so on, and using Theorem 3.7.1.

3.36 Check that in the context of Theorem 3.7.2, $\mathbf{V}(\mathscr{B}_5)=\cup_{j=1}^{8}V_j$ using the procedure described on page 156.

3.37 Prove that the origin is a center for system (3.2) if the system admits the integrating factor $(u^2+v^2)^\alpha$ with $|\alpha|\leq 2$.

3.38 Consider the family of systems

$$\begin{aligned}\dot{x} &= x-a_{10}x^2-a_{20}x^3-a_{11}x^2y-a_{02}xy^2,\\ \dot{y} &= -y+b_{01}y^2+b_{02}y^3+b_{11}xy^2+b_{20}x^2y.\end{aligned} \tag{3.169}$$

a. Show that when $a_{11}=b_{11}$ and $a_{20}+b_{20}=0$ the corresponding system (3.169) has a center at the origin.
Hint. $H(x,y)=(1-a_{10}x+b_{01}y+x^2+y^2)/(xy)+a_{11}\log xy$.

b. Find the center variety of (3.169).

3.39 Find the center variety and its irreducible decomposition for the system

$$\begin{aligned}\dot{x} &= x-a_{20}x^3-a_{11}x^2y-a_{02}xy^2-a_{-13}y^3,\\ \dot{y} &= -y+b_{02}y^3+b_{11}xy^2+b_{20}x^2y+b_{3,-1}x^3.\end{aligned}$$

Hint. See Theorem 6.4.3.

3.40 [Referenced in Proposition 4.2.12, proof.] Suppose the nonlinearities in family (3.3) are homogeneous of degree K, which means that each nonlinear term is of the same total degree K, and that a polynomial $f\in\mathbb{C}[a,b]$ is an (r,s)-polynomial with respect to the corresponding function L of Definition 3.71. (You can think of the focus quantities, but other polynomials connected with family (3.3) that we will encounter subsequently have the same property.) Show that f is homogeneous of degree $(r+s)/(K-1)$.
Hint. $L_1(\nu)+L_2(\nu)$.

3.41 Let k be a field and suppose each polynomial $f_j\in k[x_1,\dots,x_n]$ in the collection $F=\{f_1,\dots,f_s\}$ is nonzero and homogeneous (but possibly of all different degrees), and that $f\in k[x_1,\dots,x_n]$ is nonzero and homogeneous. Show that the remainder when f is reduced modulo F, under any ordering of the elements of F, if nonzero, is homogeneous. Combined with the previous exercise, this shows that when family (3.3) has homogeneous nonlinearities, the focus quantity g_{kk} and its remainder modulo $\mathscr{B}_{k-1}$ are homogeneous polynomials, if nonzero.

3.42 [Referenced in Proposition 3.8.4, proof.] Show that if the common factor $c(x,z)$ of the polynomials displayed in Proposition 3.8.4 vanishes at $(0,0)$, then condition (3.150) ensures that the equation $c(x,z)=0$ defines a function $z=\zeta(x)$ and that $\zeta'(0)=-1$.

3.43 [Referenced in Proposition 3.8.7, proof.] Let k be a field. For $A, B \in k[x]$ such that A/B is not constant, show that there exist constants $a, b, c, d \in k$ satisfying $ad - bc \neq 0$ and such that $\deg(aA + bB) > \deg(cA + dB)$.
Hint. If $\deg(A) = \deg(B)$, consider $b_n A - a_n B$, where $A = a_n x^n + \cdots + a_0$ and $B = b_n x^n + \cdots + b_0$.

3.44 Use the family of vector fields originated by C. Christopher ([50]),

$$\dot{x} = y, \qquad \dot{y} = -\left[\tfrac{1}{8} + \tfrac{1}{4}h(x)\right]h'(x) - y\left[1 + (2\lambda - 4)h^2(x)\right]h'(x),$$

where $h(x) = \frac{1}{2}(x^2 - 2x^3)$, to show that there exist polynomial Liénard systems with coexisting centers and foci, coexisting centers and fine foci, and coexisting limit cycles and centers.
Hint. Use Hopf bifurcation theory ([44, 140]).

Chapter 4
The Isochronicity and Linearizability Problems

In the previous chapter we presented methods for determining whether the antisaddle at the origin of the real polynomial system (3.2) is a center or a focus, and more generally if the singularity at the origin of the complex polynomial system (3.4) is a center. In this chapter we assume that the singularity in question is known to be a center and present methods for determining whether or not it is *isochronous*, that is, whether or not every periodic orbit in a neighborhood of the origin has the same period. A seemingly unrelated problem is that of whether the system is linearizable (see Definition 2.3.4) in a neighborhood of the origin. In fact, the two problems are intimately connected, as we will see in Section 4.2, and are remarkably parallel to the center problem.

4.1 The Period Function

A planar real analytic system with a center at the origin and nonzero linear part there can, by a suitable analytic change of coordinates and time rescaling, be written in the form (3.2):

$$\dot{u} = -v + U(u,v), \qquad \dot{v} = u + V(u,v), \tag{4.1}$$

where U and V are convergent real series that start with quadratic terms. Recall that the period annulus of the center is the largest neighborhood Ω of the origin with the property that the orbit of every point in $\Omega \setminus \{(0,0)\}$ is a simple closed curve that encloses the origin. Thus the trajectory of every point in $\Omega \setminus \{(0,0)\}$ is a periodic function, and the following definition makes sense.

Definition 4.1.1. Suppose the origin is a center for system (4.1) and that the number $r^* > 0$ is so small that the segment $\Sigma = \{(u,v) : 0 < u < r^*,\ v = 0\}$ of the u-axis lies wholly within the period annulus. For r satisfying $0 < r < r^*$, let $T(r)$ denote the least period of the trajectory through $(u,v) = (r,0) \in \Sigma$. The function $T(r)$ is the *period function* of the center (which by the Implicit Function Theorem is real analytic). If the function $T(r)$ is constant, then the center is said to be *isochronous*.

V.G. Romanovski, D.S. Shafer, *The Center and Cyclicity Problems*,
DOI 10.1007/978-0-8176-4727-8_4,

Going over to polar coordinates $x = r\cos\varphi$, $y = r\sin\varphi$, system (4.1) becomes

$$\frac{dr}{dt} = \sum_{k=1}^{\infty} \xi_k(\varphi) r^{k+1}, \qquad \frac{d\varphi}{dt} = 1 + \sum_{k=1}^{\infty} \zeta_k(\varphi) r^k, \tag{4.2}$$

where $\xi_k(\varphi)$ and $\zeta_k(\varphi)$ are homogeneous polynomials in $\sin\varphi$ and $\cos\varphi$ of degree $k+2$. Elimination of time t from (4.2) yields the equation

$$\frac{dr}{d\varphi} = \sum_{k=2}^{\infty} R_k(\varphi) r^k, \tag{4.3}$$

where $R_k(\varphi)$ are 2π-periodic functions of φ and the series is convergent for all φ and for all sufficiently small r. The initial value problem for (4.3) with the initial condition $(r,\varphi) = (r_0, 0)$ has a unique solution

$$r = r_0 + \sum_{k=2}^{\infty} u_k(\varphi) r_0^k, \tag{4.4}$$

which is convergent for all $0 \le \varphi \le 2\pi$ and all $r_0 < r^*$, for some sufficiently small $r^* > 0$; the coefficients $u_k(\varphi)$ are determined by simple quadratures using formulas (3.10) with $\alpha = 0$ and $\beta = 1$. Substituting (4.4) into the second equation of (4.2) and dropping the subscript on r yields an equation of the form

$$\frac{d\varphi}{dt} = 1 + \sum_{k=1}^{\infty} F_k(\varphi) r^k.$$

Rewriting this equation as

$$dt = \frac{d\varphi}{1 + \sum_{k=1}^{\infty} F_k(\varphi) r^k} = \left(1 + \sum_{k=1}^{\infty} \psi_k(\varphi) r^k\right) d\varphi \tag{4.5}$$

and integrating yields

$$t - \varphi = \sum_{k=1}^{\infty} \theta_k(\varphi) r^k, \tag{4.6}$$

where $\theta_k(\varphi) = \int_0^{\varphi} \psi_k(\varphi)\, d\varphi$ and the series in (4.6) converges for $0 \le \varphi \le 2\pi$ and sufficiently small $r \ge 0$. From (4.6) it follows that the least period of the trajectory of (4.1) passing through $(u,v) = (r,0)$ for $r \ne 0$ is

$$T(r) = 2\pi\left(1 + \sum_{k=1}^{\infty} T_k r^k\right), \tag{4.7}$$

where the coefficients T_k are given by the expression

$$T_k = \frac{1}{2\pi}\theta_k(2\pi) = \frac{1}{2\pi}\int_0^{2\pi} \psi_k(\varphi)\, d\varphi. \tag{4.8}$$

We now see that if the origin is an isochronous center of (4.1), then the functions $\theta_k(\varphi)$ satisfy the condition

$$\theta_k(2\pi) = 0 \tag{4.9}$$

for all $k \in \mathbb{N}$. Obviously the converse holds as well, so we have established the following result.

Theorem 4.1.2. *Suppose that in system* (4.1) *U and V are convergent real series starting with quadratic terms and that there is a center at the origin.*
1. *The period function $T(r)$ of system* (4.1) *is given by formula* (4.7).
2. *System* (4.1) *has an isochronous center at the origin if and only if* (4.9) *holds for all $k \in \mathbb{N}$.*

4.2 Isochronicity Through Normal Forms and Linearizability

In the previous section we expressed the isochronicity problem for real analytic planar systems in terms of conditions on the coefficients in an expansion of the period function. For a completely different point of view, we begin with the observation that the canonical linear center

$$\dot{u} = -v, \quad \dot{v} = u \tag{4.10}$$

is itself isochronous. Since isochronicity does not depend on the coordinates in use, certainly any system obtainable from (4.10) by an analytic change of coordinates will still be isochronous. Viewed from the other direction, any real analytic system with a center that can be reduced to (4.10) by an analytic change of coordinates (that is, any *linearizable* system according to Definition 2.3.4) must be isochronous. This discussion shows that linearizability and isochronicity are intimately connected, since it reveals that only isochronous systems are linearizable. In fact, Theorem 3.2.10 implies that the correspondence between the two families is perfect, as we now demonstrate.

Theorem 4.2.1. *The origin is an isochronous center for system* (4.1) *if and only if there is an analytic change of coordinates* (3.42) *that reduces* (4.1) *to the canonical linear center* (4.10).

Proof. By the preceding discussion we need only show that every isochronous center is linearizable. Hence suppose the origin is an isochronous center of (4.1). Then by Theorem 3.2.10 there is an analytic change of coordinates (3.42) that transforms (4.1) into the real normal form (3.43),

$$\dot{\xi} = -\eta F(\xi^2 + \eta^2), \quad \dot{\eta} = \xi F(\xi^2 + \eta^2), \tag{4.11}$$

where $F(z) = 1 + \sum_{k=1}^{\infty} \gamma_k z^k$ is analytic in a neighborhood of $z = 0$. In polar coordinates, system (4.11) is $\dot{r} = 0$, $\dot{\varphi} = F(r^2)$, for which the period function is clearly

$T(r) = 2\pi/F(r^2)$. Since the transformed system is still isochronous, $F(z) \equiv 1$, as required. □

Theorem 4.2.1 tells us that the isochronicity of a planar analytic system is equivalent to its linearizability, so instead of studying the isochronicity of planar analytic systems, we will investigate their linearizability. The advantage is that, because linearizability has a natural generalization to the complex setting, as was done in the previous chapter we can transform our real system (4.1) into a complex system on $\mathbb{C}^2$ and study the complex affine varieties corresponding to the linearizable systems. The complexification of (4.1) has the form $\dot{\mathbf{z}} = J\mathbf{z} + \mathbf{Z}(\mathbf{z})$, where J is a diagonal matrix. As a preliminary we prove the important fact that if there exists *any* normalizing transformation that linearizes such a system, then *every* normalizing transformation does so. Thus it will be no loss of generality in our investigation of linearization of such systems to impose conditions on the normalizing transformation whenever it is convenient to do so. Since the theorem is true not just in $\mathbb{C}^2$ but in $\mathbb{C}^n$, we state and prove it in this setting.

Theorem 4.2.2. *Consider the system*

$$\dot{\mathbf{z}} = J\mathbf{z} + \mathbf{Z}(\mathbf{z}), \tag{4.12}$$

where $\mathbf{z} \in \mathbb{C}^n$, J *is a diagonal matrix, and each component* $Z_j(\mathbf{z})$ *of* $\mathbf{Z}$, $1 \le j \le n$, *is a formal or convergent power series, possibly with complex coefficients, that contains no constant or linear terms. Suppose*

$$\mathbf{z} = \mathbf{y} + \widetilde{\mathbf{h}}(\mathbf{y}) \tag{4.13}$$

and

$$\mathbf{z} = \mathbf{x} + \widehat{\mathbf{h}}(\mathbf{x}) \tag{4.14}$$

are normalizing transformations (possibly merely formal) that transform (4.12) *into the respective normal forms*

$$\dot{\mathbf{y}} = J\mathbf{y} + \mathbf{Y}(\mathbf{y}) \tag{4.15}$$

and

$$\dot{\mathbf{x}} = J\mathbf{x} + \mathbf{X}(\mathbf{x}). \tag{4.16}$$

If the normal form (4.15) *is linear, that is, if* $\mathbf{Y}(\mathbf{y}) \equiv 0$, *then the normal form* (4.16) *is linear as well.*

Proof. Because (4.14) is invertible there exists a (formal) transformation

$$\mathbf{x} = \mathbf{y} + \mathbf{h}(\mathbf{y}) \tag{4.17}$$

that transforms (4.16) into (4.15). Certainly (4.15) is a normal form for (4.16) and (4.17) is a normalizing transformation that produces it. Therefore (employing the notation introduced in (2.36) and in the paragraph that follows Remark 2.3.7) for $s \in \mathbb{N}$, $s \ge 2$, any coefficient $h_m^{(\alpha)}$ of $\mathbf{h}^{(s)}$ is determined by equation (2.40):

$$[(\alpha,\kappa)-\kappa_m]h_m^{(\alpha)} = g_m^{(\alpha)} - Y_m^{(\alpha)}, \tag{4.18}$$

where $g_m^{(\alpha)}$ is a known expression depending on the coefficients of $\mathbf{h}^{(j)}$ for $j < s$. Specifically, $g_m^{(\alpha)}$ is defined by equation (2.41):

$$g_m^{(\alpha)} = \{X_m(\mathbf{y}+\mathbf{h}(\mathbf{y}))\}^{(\alpha)} - \sum_{j=1}^{n} \sum_{\substack{2\le|\beta|\le|\alpha|-1 \\ \alpha-\beta+e_j\in\mathbb{N}_0^n}} \beta_j h_m^{(\beta)} Y_j^{(\alpha-\beta+e_j)}, \tag{4.19}$$

where $\{X_m(\mathbf{y}+\mathbf{h}(\mathbf{y}))\}^{(\alpha)}$ denotes the coefficient of $\mathbf{y}^\alpha$ obtained after expanding $X_m(\mathbf{y}+\mathbf{h}(\mathbf{y}))$ in powers of $\mathbf{y}$, $e_j = (0,\dots,0,\overset{j}{1},0,\dots,0) \in \mathbb{N}_0^n$, and the summation is empty when $|\alpha| = 2$.

Assume, contrary to what we wish to show, that the normal form (4.16) is not linear. If $k \ge 2$ is such that the series expansions of the coordinate functions of $\mathbf{X}$ start with terms of order k, then there exist a coordinate index $m \in \{1,\dots,n\}$ and a multi-index γ for which $|\gamma| = k$ such that $X_m^{(\gamma)} \neq 0$. By (4.19) and the fact that $\mathbf{Y}(\mathbf{y}) \equiv 0$,

$$g_m^{(\gamma)} = \{X_m(\mathbf{y}+\mathbf{h}(\mathbf{y}))\}^{(\gamma)} - \sum_{j=1}^{n} \sum_{\substack{2\le|\beta|\le|\gamma|-1 \\ \gamma-\beta+e_j\in\mathbb{N}_0^n}} \beta_j h_m^{(\beta)} Y_j^{(\gamma-\beta+e_j)} = X_m^{(\gamma)}.$$

On the other hand, from (4.18) and the fact that m and γ must form a resonant pair we obtain

$$0 \cdot h_m^{(\gamma)} = g_m^{(\gamma)} - Y_m^{(\gamma)} = X_m^{(\gamma)} \neq 0,$$

a contradiction. □

Corollary 4.2.3. *Suppose A is a diagonal matrix. System $\dot{\mathbf{x}} = A\mathbf{x} + \mathbf{X}(\mathbf{x})$ on $\mathbb{C}^2$ is linearizable according to Definition 2.3.4 if there exists a merely formal normalizing transformation $\mathbf{x} = \mathbf{y} + \mathbf{h}(\mathbf{y})$ that places it in the normal form $\dot{\mathbf{y}} = A\mathbf{y}$.*

Proof. Suppose there is a formal normalizing transformation that reduces the system to $\dot{\mathbf{y}} = A\mathbf{y}$. Then by the theorem the distinguished normalizing transformation does, too. But $\Psi(y_1,y_2) = y_1y_2$ is a first integral for the normalized system, hence by Theorem 3.2.5(1) the distinguished normalizing transformation is convergent. □

If we apply the complexification procedure described at the beginning of Section 3.2 to (4.1), we obtain system (3.21) with $\alpha = 0$ and $\beta = 1$, where x_1 and x_2 are complex variables:

$$\begin{aligned} \dot{x}_1 &= ix_1 + X_1(x_1,x_2) \\ \dot{x}_2 &= -ix_2 + X_2(x_1,x_2) \end{aligned}, \qquad X_2(x_1,\bar{x}_1) = \overline{X_1(x_1,\bar{x}_1)}. \tag{4.20}$$

We have already seen (see Proposition 3.2.2) that the resonant pairs for (4.20) are $(1,(k+1,k))$ and $(2,(k,k+1))$, $k \in \mathbb{N}$, so that when we apply a normalizing transformation (2.33), namely

$$x_1 = y_1 + \sum_{j+k\geq 2} h_1^{(j,k)} y_1^j y_2^k, \quad x_2 = y_2 + \sum_{j+k\geq 2} h_2^{(j,k)} y_1^j y_2^k, \tag{4.21}$$

to reduce system (4.20) to the normal form (3.27), we obtain

$$\dot{y}_1 = y_1(i + Y_1(y_1 y_2)), \quad \dot{y}_2 = y_2(-i + Y_2(y_1 y_2)), \tag{4.22}$$

where

$$Y_1(y_1 y_2) = \sum_{j=1}^{\infty} Y_1^{(j+1,j)} (y_1 y_2)^j \quad \text{and} \quad Y_2(y_1 y_2) = \sum_{j=1}^{\infty} Y_2^{(j,j+1)} (y_1 y_2)^j. \tag{4.23}$$

System (4.20) is linearized by the transformation (4.21) precisely when (4.22) reduces to $\dot{y}_1 = iy_1$, $\dot{y}_2 = -iy_2$. Be that as it may, in any case the first equation in the normal form (4.22) is given by

$$\dot{y}_1 = y_1(i + \tfrac{1}{2}[G(y_1 y_2) + H(y_1 y_2)]), \tag{4.24}$$

where G and H are the functions of (3.28). If the normalizing transformation was chosen subject to the condition in Proposition 3.2.1(2), then by Theorem 3.2.7 the origin is a center for (4.1) if and only if in the normal form (4.24) $G \equiv 0$, in which case H has purely imaginary coefficients (Remark 3.2.8). It is convenient to define $\widetilde{H}$ by

$$\widetilde{H}(w) = -\tfrac{1}{2} i H(w).$$

By Proposition 3.2.1 system (4.22) is the complexification of a real system; we obtain two descriptions of it in complex form by replacing every occurrence of y_2 by $\bar{y}_1$ in each equation of (4.22). Setting $y_1 = re^{i\varphi}$, we obtain from them

$$\dot{r} = \tfrac{1}{2r}(\dot{y}_1 \bar{y}_1 + y_1 \dot{\bar{y}}_1) = 0, \qquad \dot{\varphi} = \tfrac{i}{2r^2}(y_1 \dot{\bar{y}}_1 - \dot{y}_1 \bar{y}_1) = 1 + \widetilde{H}(r^2). \tag{4.25}$$

Integrating the expression for $\dot{\varphi}$ in (4.25) yields (since r is constant in t)

$$T(r) = \frac{2\pi}{1 + \widetilde{H}(r^2)} = 2\pi \left(1 + \sum_{k=1}^{\infty} p_{2k} r^{2k}\right) \tag{4.26}$$

for some coefficients p_{2k} that will be examined later in this section and will play a prominent role in Chapter 6. Recall from (3.28a) that we write the function H as $H(w) = \sum_{k=1}^{\infty} H_{2k+1} w^k$. Similarly, write $\widetilde{H}(w) = \sum_{k=1}^{\infty} \widetilde{H}_{2k+1} w^k$. The center is isochronous if and only if $p_{2k} = 0$ for $k \geq 1$ or, equivalently, $H_{2k+1} = 0$ for $k \geq 1$. We call p_{2k} the *kth isochronicity quantity*.

The expression for T in (4.26) pertains to the polar coordinate expression (4.25) for the real system whose complexification is the normalization (4.22) of the complexification (4.20) of the original system (4.1), which has polar coordinate expression (4.2). If the polar distance in (4.25) is denoted R, then R is an analytic function of r of the form $R = r + \cdots$, so a comparison of (4.7) and (4.26), which now reads $T(R) = 2\pi\left(1 + \sum_{k=1}^{\infty} p_{2k} R^{2k}\right)$, immediately yields the following property.

Proposition 4.2.4. *The first nonzero coefficient of the expansion* (4.7) *is the coefficient of an even power of r.*

The discussion thus far in this chapter applies to any planar analytic system of the form (4.1). We now turn our attention to families of *polynomial* systems of the form (4.1), which, as in Section 3.2, we complexify to obtain a family of polynomial systems of equations on $\mathbb{C}^2$ that have the form

$$\begin{aligned} \dot{x}_1 = \widetilde{P}(x_1,x_2) = \quad & i\left(x_1 - \sum_{(p,q)\in S} a_{pq} x_1^{p+1} x_2^q\right) \\ \dot{x}_2 = \widetilde{Q}(x_1,x_2) = -& i\left(x_2 - \sum_{(p,q)\in S} b_{qp} x_1^q x_2^{p+1}\right), \end{aligned} \tag{4.27}$$

where the coefficients of $\widetilde{P}$ and $\widetilde{Q}$ are complex, where $S \subset \mathbb{N}_{-1} \times \mathbb{N}_0$ is a finite set, every element (p,q) of which satisfies $p+q \geq 1$, and where $b_{qp} = \bar{a}_{pq}$ for all $(p,q) \in S$. This, of course, is the same expression as (3.3) and (3.50). When family (4.27) arises as the complexification of a real family, the equality $b_{qp} = \bar{a}_{pq}$ holds for all $(p,q) \in S$, but we will not restrict ourselves to this condition in what follows. In such a case the function $\widetilde{H}$ and the isochronicity quantities p_{2k} as implicitly defined by the second equality in (4.26) exist, although there is no period function T. In any event, the coefficients p_{2k} defined by (4.26) and $\widetilde{H}_{2k+1}$ of $\widetilde{H} = -\frac{1}{2}iH$ are now polynomials in the parameters $(a,b) \in E(a,b) = \mathbb{C}^{2\ell}$.

The coefficients $Y_1^{(k+1,k)}$ and $Y_2^{(k,k+1)}$ of the series (4.23) in the normal form (4.22) are now elements of the ring $\mathbb{C}[a,b]$ of polynomials. They generate an ideal

$$\mathscr{Y} := \left\langle Y_1^{(j+1,j)},\ Y_2^{(j,j+1)} : j \in \mathbb{N}\right\rangle \subset \mathbb{C}[a,b]. \tag{4.28}$$

For any $k \in \mathbb{N}$ we set $\mathscr{Y}_k = \langle Y_1^{(j+1,j)},\ Y_2^{(j,j+1)} : j = 1,\dots,k\rangle$. The normal form of a particular system (a^*,b^*) is linear when all the coefficients

$$\left\{Y_1^{(j+1,j)}(a^*,b^*),\ Y_2^{(j,j+1)}(a^*,b^*) : j \in \mathbb{N}\right\}$$

are equal to zero. Thus we have the following definition.

Definition 4.2.5. Suppose a normal form (4.22) arises from system (4.27) by means of a normalizing transformation (2.33) or (4.21), and let $\mathscr{Y}$ be the ideal (4.28). The variety $V_{\mathscr{L}} := \mathbf{V}(\mathscr{Y})$ is called the *linearizability variety* of system (4.27).

As was the case with the center variety, we must address the question of whether the linearizability variety is actually well-defined, that is, that the variety $V_{\mathscr{L}}$ does not depend on the particular choice of the resonant coefficients of the normalizing transformation. In the current situation, however, no elaborate argument is needed:

the correctness of Definition 4.2.5 is an immediate consequence of Theorem 4.2.2, that if any normalizing transformation linearizes (4.27), then all normalizing transformations do.

Remark 4.2.6. If the system

$$\begin{aligned} \dot{x}_1 = \widetilde{P}(x_1,x_2) &= \quad x_1 - \sum_{(p,q)\in S} a_{pq} x_1^{p+1} x_2^q \\ \dot{x}_2 = \widetilde{Q}(x_1,x_2) &= -x_2 + \sum_{(p,q)\in S} b_{qp} x_1^q x_2^{p+1} \end{aligned} \tag{4.29}$$

is transformed into the linear system $\dot{y}_1 = y_1$, $\dot{y}_2 = -y_2$ by transformation (4.21), then (4.27) is reduced to $\dot{y}_1 = iy_1$, $\dot{y}_2 = -iy_2$ by the same transformation. Conversely, if (4.21) linearizes (4.27), then it also linearizes (4.29). Therefore systems (4.27) and (4.29) are equivalent with regard to the problem of linearizability. Thus in what follows we typically state results for both systems but provide proofs for just one or the other.

The following proposition gives another characterization of $V_{\mathscr{L}}$, based on the idea that linearizability and isochronicity must be equivalent. Form

$$\mathscr{H} = \langle H_{2j+1} : j \in \mathbb{N} \rangle = \langle \widetilde{H}_{2j+1} : j \in \mathbb{N} \rangle . \tag{4.30}$$

For each $k \in \mathbb{N}$ we also set $\mathscr{H}_k = \langle H_{2j+1} : j = 1,\ldots,k \rangle = \langle \widetilde{H}_{2j+1} : j = 1,\ldots,k \rangle$. The corresponding variety $\mathbf{V}(\mathscr{H})$, when intersected with the center variety, picks out the generalization of isochronous centers to the complex setting. Thus define

$$V_{\mathscr{I}} = \mathbf{V}(\mathscr{H}) \cap V_{\mathscr{C}} .$$

This variety, the *isochronicity variety*, should be the same as $V_{\mathscr{L}}$, and we now show that it is. This result thus shows that the variety $V_{\mathscr{I}}$ is well-defined, independently of the normalizing transformation (4.21).

Proposition 4.2.7. *With reference to family* (4.27)*, the sets* $V_{\mathscr{I}} = \mathbf{V}(\mathscr{H}) \cap V_{\mathscr{C}}$ *and* $V_{\mathscr{L}}$ *are the same.*

Proof. Let a set of coefficients $Y_1^{(k+1,k)}$, $Y_2^{(k,k+1)}$, $k \in \mathbb{N}$, arising from a normalizing transformation that satisfies the condition of Proposition 3.2.1(2) be given. By (3.28b), $H_{2k+1} = Y_1^{(k+1,k)} - Y_2^{(k,k+1)}$, which immediately implies that $V_{\mathscr{L}} \subset V_{\mathscr{I}}$. On the other hand, if $(a,b) \in V_{\mathscr{C}}$, then by Theorem 3.2.5(1) the function G computed from the $Y_1^{(k+1,k)}$ and $Y_2^{(k,k+1)}$ vanishes identically. Since, again by (3.28b),

$$Y_1^{(k+1,k)} = \frac{G_{2k+1} + H_{2k+1}}{2} \quad \text{and} \quad Y_2^{(k,k+1)} = \frac{G_{2k+1} - H_{2k+1}}{2}, \tag{4.31}$$

this yields $V_{\mathscr{I}} \subset V_{\mathscr{L}}$. $\square$

The following theorem describes some of the properties of the coefficients $h_m^{(\alpha)} \in \mathbb{C}[a,b]$ of the normalizing transformation (4.21) and of the resonant terms $Y_1^{(k+1,k)}$, $Y_2^{(k,k+1)} \in \mathbb{C}[a,b]$ of the resulting normal form (4.22) that are similar to the properties of the coefficients $v_{j-1,k-1}$ of the function (3.52) given by Theorem 3.4.2. The theorem is true in general, without reference to linearization, hence could have been presented in Chapter 3. We have saved it for now because of its particular connection to the idea of linearization in view of Definition 4.2.5 and Proposition 4.2.7. We will use the notation that was introduced in the paragraph preceding Definition 3.4.1 of (j,k)-polynomials. Note in particular that the theorem implies that, for $m \in \{1,2\}$ and $\alpha \in \mathbb{N}_0^2$, as polynomials in the indeterminates a_{pq}, b_{qp}, like $X_m^{(\alpha)}$, $Y_m^{(\alpha)}$ has pure imaginary coefficients, while the coefficients of $h_m^{(\alpha)}$ are real.

Theorem 4.2.8. *Let a set S, hence a family* (4.27) *(family* (3.3)*), be given, and let* (4.21) *be any normalizing transformation whose resonant coefficients* $h_1^{(k+1,k)}$, $h_2^{(k,k+1)} \in \mathbb{C}(a,b)$, $k \in \mathbb{N}$, *are chosen so that they are* (k,k)*-polynomials with coefficients in* $\mathbb{Q}$.

1. *For every* $(j,k) \in \mathbb{N}_0^2$ *with* $j+k \geq 2$, $h_1^{(j,k)}$ *is a* $(j-1,k)$*-polynomial and* $h_2^{(j,k)}$ *is a* $(j,k-1)$*-polynomial, each with coefficients in* $\mathbb{Q}$.
2. *For all* $k \in \mathbb{N}$, *the polynomials* $iY_1^{(k+1,k)}$ *and* $iY_2^{(k,k+1)}$ *are* (k,k)*-polynomials with coefficients in* $\mathbb{Q}$.
3. *If the resonant coefficients* $h_1^{(k+1,k)}$ *and* $h_2^{(k,k+1)}$ *are chosen so as to satisfy* $h_2^{(k,k+1)} = \widehat{h}_1^{(k+1,k)}$, *then*

$$h_2^{(k,j)} = \widehat{h}_1^{(j,k)} \text{ for all } (j,k) \in \mathbb{N}_0^2 \text{ with } j+k \geq 2$$

and

$$Y_2^{(k,k+1)} = \widehat{Y}_1^{(k+1,k)} \text{ for all } k \in \mathbb{N}_0,$$

where for $f \in \mathbb{C}[a,b]$ *the conjugate* $\widehat{f}$ *is given by Definition 3.4.3.*

Proof. In the proof it will be convenient to treat point (2) and the part of point (3) that pertains to $Y_m^{(\alpha)}$ as if they were stated for all the coefficients in the normal form, including the nonresonant ones, which of course are zero. By $i\mathbb{Q}$ we denote the set of all purely imaginary elements of the field of Gaussian rational numbers, that is, $i\mathbb{Q} = \{iq : q \in \mathbb{Q}\}$.

The vector κ of eigenvalues from Definition 2.3.5 is $\kappa = (\kappa_1, \kappa_2) = (i, -i)$, so that for $\alpha \in \mathbb{N}_0^2$, $|\alpha| \geq 2$, we have that $(\alpha, \kappa) - \kappa_1 = i(\alpha_1 - \alpha_2 - 1) \in i\mathbb{Q}$ and that $(\alpha, \kappa) - \kappa_2 = i(\alpha_1 - \alpha_2 + 1) \in i\mathbb{Q}$. The proof of points (1) and (2) will be done simultaneously for $h_m^{(\alpha)}$ and $Y_m^{(\alpha)}$ by induction on $|\alpha|$.

Basis step. Suppose $\alpha \in \mathbb{N}_0^2$ has $|\alpha| = 2$. Since there are no resonant pairs (m, α), $Y_m^{(\alpha)} = 0$, so (2) holds. The coefficients $h_m^{(\alpha)}$ are determined by (2.40) (which is (4.18)), where $g_m^{(\alpha)}$ is given by (2.41) (which is (4.19)), in this case (that is, the basis step) without the sum. Since $\mathbf{X}$ and $\mathbf{h}$ begin with quadratic terms,

$\{\mathbf{X}(\mathbf{y}+\mathbf{h}(\mathbf{y}))\}^{(\alpha)} = X_m^{(\alpha)}$, hence $h_m^{(\alpha)} = [(\alpha,\kappa)-\kappa_m]^{-1}X_m^{(\alpha)}$. Recalling the notation of Section 3.4, and writing $\alpha=(j,k)$, $X_1^{(j,k)} = -ia_{j-1,k} = -i[\nu]$ for the string $\nu=(0,\ldots,1,\ldots,0)$ with the 1 in the position c such that $(j-1,k)=(p_c,q_c)$. But $L(\nu)=(p_c,q_c)$, so $h_1^{(j,k)}$ is a $(j-1,k)$-polynomial with coefficients in $\mathbb{Q}$. An analogous argument shows that $h_2^{(j,k)}$ is a $(j,k-1)$-polynomial with coefficients in $\mathbb{Q}$, so (1) holds for $|\alpha|=2$.

Inductive step. Suppose points (1) and (2) hold for all $h_m^{(\alpha)}$ and $Y_m^{(\alpha)}$, $m\in\{1,2\}$, $2\le|\alpha|\le s$, and fix $\alpha=(j,k)\in\mathbb{N}_0^2$ with $|\alpha|=s+1$. If m and α form a nonresonant pair, then $Y_m^{(\alpha)}=0$, so (2) holds, and $h_m^{(\alpha)} = C\cdot g_m^{(\alpha)}$ for $C\in i\mathbb{Q}$. If m and α form a resonant pair, then $h_m^{(\alpha)}$ and its companion resonant coefficient in $\mathbf{h}$ are selected arbitrarily, up to the condition that they satisfy the conclusion in point (1) and that their coefficients lie in $\mathbb{Q}$, and $Y_m^{(\alpha)} = g_m^{(\alpha)}$. Thus point (2) will be established in either case if we can show that $g_1^{(j,k)}$ is a $(j-1,k)$-polynomial and $g_2^{(j,k)}$ is a $(j,k-1)$-polynomial, each with coefficients in $i\mathbb{Q}$. For the proof it is convenient to introduce the six constant polynomials in $\mathbb{C}[a,b]$ defined by $h_1^{(1,0)} = h_2^{(0,1)} = 1$ and $h_1^{(0,0)} = h_1^{(0,1)} = h_2^{(0,0)} = h_1^{(1,0)} = 0$. Since the support of each of the last four is empty and $\mathrm{Supp}(h_1^{(1,0)}) = \mathrm{Supp}(h_2^{(0,1)}) = (0,\ldots,0)\in\mathbb{N}_0^{2\ell}$, they satisfy the conclusions of the theorem. They allow us to write

$$X_m(\mathbf{y}+\mathbf{h}(\mathbf{y})) = \sum_{\substack{c+d\ge 2\\ (c,d)\in\mathbb{N}_0^2}} X_m^{(c,d)} \Big(\sum_{\substack{r+t\ge 0\\ (r,t)\in\mathbb{N}_0^2}} h_1^{(r,t)} y_1^r y_2^t\Big)^c \Big(\sum_{\substack{r+t\ge 0\\ (r,t)\in\mathbb{N}_0^2}} h_2^{(r,t)} y_1^r y_2^t\Big)^d, \tag{4.32}$$

so that (before collecting on powers of y_1 and y_2) $X_m(\mathbf{y}+\mathbf{h}(\mathbf{y}))$ is a sum of terms of the form

$$X_m^{(c,d)} h_1^{(r_1,t_1)}\cdots h_1^{(r_c,t_c)} h_2^{(u_1,v_1)}\cdots h_2^{(u_d,v_d)} y_1^{r_1+\cdots+r_c+u_1+\cdots u_d} y_2^{t_1+\cdots+t_c+v_1+\cdots v_d}. \tag{4.33}$$

Consider any such term in $X_m(\mathbf{y}+\mathbf{h}(\mathbf{y}))$ for which

$$(r_1+\cdots+r_c+u_1+\cdots+u_d, t_1+\cdots+t_c+v_1+\cdots+v_d) = (j,k).$$

Then $X_1^{(c,d)} = -ia_{c-1,d} = -i[\nu]$, where $L(\nu)=(c-1,d)$, hence by the induction hypothesis for any monomial $[\mu]$ appearing in the coefficient in (4.33) when $m=1$,

$$\begin{aligned} L(\mu) &= (c-1,d)+(r_1-1,t_1)+\cdots+(r_c-1,t_c)+(u_1,v_1-1)+\cdots+(u_d,v_d-1)\\ &= (j-1,k). \end{aligned}$$

Again, $X_2^{(c,d)} = ib_{c,d-1} = i[\nu']$, where $L(\nu')=(c,d-1)$, and in a similar fashion implies that for any monomial $[\mu]$ appearing in the coefficient in (4.33) when $m=2$, $L(\mu)=(j,k-1)$. Therefore $\{X_1(\mathbf{y}+\mathbf{h}(\mathbf{y}))\}^{(j,k)}$ is a $(j-1,k)$-polynomial and $\{X_2(\mathbf{y}+\mathbf{h}(\mathbf{y}))\}^{(j,k)}$ is a $(j,k-1)$-polynomial, whose coefficients are clearly elements of $i\mathbb{Q}$.

Any monomial $[\mu]$ that appears in the sum in (4.19) is a product of a monomial from $h_m^{(\beta)}$ and (changing the name of the index on the first sum to w) a monomial from $Y_w^{(\alpha-\beta+e_w)}$, hence by the induction hypothesis, for $m=1$ and $w=1$,

$$L(\mu) = (\beta_1 - 1, \beta_2) + ((j - \beta_1 + 1) - 1, k - \beta_2) = (j-1, k)$$

and similarly for $m=1$ and $w=2$, and for $m=2$ and $w=1$,

$$L(\mu) = (\beta_1, \beta_2 - 1) + (j - \beta_1, (k - \beta_2 + 1) - 1) = (j, k-1)$$

and similarly for $m=2$ and $w=2$. Thus $g_1^{(j,k)}$ is a $(j-1,k)$-polynomial and $g_2^{(j,k)}$ is a $(j-1,k)$-polynomial, each with coefficients in $i\mathbb{Q}$, and the inductive step is complete, proving points (1) and (2).

The proof of point (3) will be done simultaneously for $h_m^{(\alpha)}$ and $\widehat{Y_m^{(\alpha)}}$ by induction on $|\alpha|$. For notational convenience we will write $\widehat{\ }[f]$ in place of $\widehat{f}$ when the expression for f is long.

Basis step. Suppose $\alpha \in \mathbb{N}_0^2$ has $|\alpha| = 2$. There are no resonant pairs (m,α), hence for $\alpha = (j,k)$,

$$\begin{aligned}
h_2^{(k,j)} &= (i(k-j+1))^{-1} X_2^{(k,j)} = (i(k-j+1))^{-1} i b_{k,j-1} \\
&= \overline{(i(j-k-1))^{-1}(-i)} b_{k,j-1} = \widehat{\ }[(i(j-k-1))^{-1}(-i) a_{j-1,k}] \\
&= \widehat{\ }[(i(j-k-1))^{-1} X_1^{j,k}] = \widehat{h}_1^{(j,k)},
\end{aligned}$$

and since $Y_m^{(\alpha)} = 0$, $Y_2^{(k,k+1)} = \widehat{Y}_1^{(k+1,k)}$ is automatic.

Inductive step. Suppose the resonant terms $h_m^{(\alpha)}$ in $\mathbf{h}$ with $|\alpha| \le s$ have been chosen so that $h_2^{(k,k+1)} = \widehat{h}_1^{(k+1,k)}$ and that the conclusion in (3) holds for all $h_m^{(\alpha)}$ and $Y_m^{(\alpha)}$ for which $2 \le |\alpha| \le s$. Fix $\alpha = (j,k) \in \mathbb{N}_0^2$ with $|\alpha| = s+1$. If $(1,(j,k))$ is a nonresonant pair, then $(2,(k,j))$ is and $Y_2^{(k,j)} = 0 = \widehat{0} = \widehat{Y}_1^{(j,k)}$, so (3) holds for $Y_1^{(j,k)}$ and $Y_2^{(k,j)}$, while $h_1^{(j,k)} = C \cdot g_1^{(j,k)}$ and $h_2^{(k,j)} = \bar{C} \cdot g_2^{(k,j)}$, $C = i(j-k-1)$. If $(1,(j,k))$ is a resonant pair, then we choose $h_1^{(j,k)}$ and the resonant $h_2^{(k,j)}$ with coefficients in $\mathbb{Q}$ and so as to meet the condition in (3), and $Y_1^{(j,k)} = g_1^{(j,k)}$ and $Y_2^{(k,j)} = g_2^{(k,j)}$. Thus point (3) will be established if we can prove that $g_2^{(k,j)} = \widehat{g}_1^{(j,k)}$.

When expression (4.32) for $X_m(\mathbf{y} + \mathbf{h}(\mathbf{y}))$ is fully expanded without collecting on powers of y_1 and y_2, it is the sum of all terms of the form given in (4.33). Thus in general $X_m(\mathbf{y} + \mathbf{h}(\mathbf{y}))^{(\beta)}$ is the sum of all products of the form

$$X_m^{(c,d)} h_1^{(r_1,t_1)} \cdots h_1^{(r_c,t_c)} h_2^{(u_1,v_1)} \cdots h_2^{(u_d,v_d)} \tag{4.34}$$

with admissible (c,d), (r_w,t_w), $(u_w,v_w) \in \mathbb{N}_0^2$ for which

$$(r_1 + \cdots + r_c + u_1 + \cdots + u_d, t_1 + \cdots + t_c + v_1 + \cdots + v_d) = (\beta_1, \beta_2). \tag{4.35}$$

By the obvious properties of conjugation, $\widehat{\ }[X_1(\mathbf{y}+\mathbf{h}(\mathbf{y}))^{(j,k)}]$ is thus the sum of the conjugates of all such products, hence of all products

$$\widehat{X}_1^{(c,d)}\widehat{h}_1^{(r_1,t_1)}\cdots\widehat{h}_1^{(r_c,t_c)}\widehat{h}_2^{(u_1,v_1)}\cdots\widehat{h}_2^{(u_d,v_d)} = X_2^{(d,c)}h_2^{(t_1,r_1)}\cdots h_2^{(t_c,r_c)}h_1^{(v_1,u_1)}\cdots h_1^{(v_d,u_d)}$$

with admissible (c,d), (r_w,t_w), $(u_w,v_w)\in\mathbb{N}_0^2$ for which (4.35) holds with (β_1,β_2) replaced by (j,k), where we have applied the induction hypothesis. But this latter sum (with precisely this condition) is exactly $X_2(\mathbf{y}+\mathbf{h}(\mathbf{y}))^{(k,j)}$, so that

$$\widehat{\ }[X_2(\mathbf{y}+\mathbf{h}(\mathbf{y}))^{(k,j)}] = X_1(\mathbf{y}+\mathbf{h}(\mathbf{y}))^{(j,k)}. \tag{4.36}$$

Again by the induction hypothesis (recall that $\alpha=(j,k)$)

$$\begin{aligned}
&\widehat{\ }\Big[\sum_{w=1}^{2}\sum_{\substack{2\le|\beta|\le|\alpha|-1\\ \alpha-\beta+e_w\in\mathbb{N}_0^2}}\beta_w h_1^{(\beta)}Y_w^{(\alpha-\beta+e_w)}\Big]\\
&= \sum_{\substack{2\le|\beta|\le|\alpha|-1\\ (j-\beta_1+1,k-\beta_2)\in\mathbb{N}_0^2}}\widehat{\ }\big[\beta_1 h_1^{(\beta_1,\beta_2)}Y_1^{(j-\beta_1+1,k-\beta_2)}\big]\\
&\qquad+\sum_{\substack{2\le|\beta|\le|\alpha|-1\\ (j-\beta_1,k-\beta_2+1)\in\mathbb{N}_0^2}}\widehat{\ }\big[\beta_2 h_1^{(\beta_1,\beta_2)}Y_2^{(j-\beta_1,k-\beta_2+1)}\big]\\
&= \sum_{\substack{2\le|\beta|\le|\alpha|-1\\ (k-\beta_2,j-\beta_1+1)\in\mathbb{N}_0^2}}\beta_1 h_2^{(\beta_2,\beta_1)}Y_2^{(k-\beta_2,j-\beta_1+1)}\\
&\qquad+\sum_{\substack{2\le|\beta|\le|\alpha|-1\\ (k-\beta_2+1,j-\beta_1)\in\mathbb{N}_0^2}}\beta_2 h_2^{(\beta_2,\beta_1)}Y_1^{(k-\beta_2+1,j-\beta_1)}
\end{aligned}$$

and, writing $\gamma=(k,j)$ and $\delta=(\beta_2,\beta_1)$,

$$=\sum_{w=1}^{2}\sum_{\substack{2\le|\delta|\le|\gamma|-1\\ \gamma-\delta+e_w\in\mathbb{N}_0^2}}\delta_w h_2^{(\delta)}Y_w^{(\gamma-\delta+e_w)}. \tag{4.37}$$

Equations (4.36) and (4.37) together imply that $g_2^{(k,j)}=\widehat{g}_1^{(j,k)}$ for $|\alpha|=j+k=s+1$, so the inductive step is complete, and the truth of (3) follows. □

Corollary 4.2.9. *Let a set S, hence a family* (4.27)*, be given and let* (4.21) *be any normalizing transformation whose resonant coefficients $h_1^{(k+1,k)}$, $h_2^{(k,k+1)}\in\mathbb{C}[a,b]$, $k\in\mathbb{N}$, are chosen so that they are (k,k)-polynomials with coefficients in $\mathbb{Q}$ such that $h_2^{(k,k+1)}=\widehat{h}_1^{(k+1,k)}$. Then the pair of polynomials $G_{2k+1}=Y_1^{(k+1,k)}+Y_2^{(k,k+1)}$ and $H_{2k+1}=Y_1^{(k+1,k)}-Y_2^{(k,k+1)}$ defined by* (3.28) *have the form*

$$G_{2k+1} = \sum_{\{\nu:L(\nu)=(k,k)\}} iq^{(\nu)}([\nu]-[\widehat{\nu}]), \qquad H_{2k+1} = \sum_{\{\nu:L(\nu)=(k,k)\}} iq^{(\nu)}([\nu]+[\widehat{\nu}]),$$

for $q^{(\nu)} \in \mathbb{Q}$. (The symbol $q^{(\nu)}$ stands for the same number in each expression.)

Proof. By point (2) of Theorem 4.2.8, the polynomial $Y_1^{(k+1,k)}$ has the form

$$Y_1^{(k+1,k)} = \sum_{\{\nu:L(\nu)=(k,k)\}} iq^{(\nu)}[\nu] \tag{4.38}$$

for $q^{(\nu)} \in \mathbb{Q}$, hence by point (3) of Theorem 4.2.8,

$$Y_2^{(k,k+1)} = \sum_{\{\nu:L(\nu)=(k,k)\}} \overline{iq^{(\nu)}}[\widehat{\nu}] = \sum_{\{\nu:L(\nu)=(k,k)\}} -iq^{(\nu)}[\widehat{\nu}]. \tag{4.39}$$

Adding (4.39) to and subtracting it from (4.38) immediately gives the result. □

Let us now return to the isochronicity quantities p_{2k}, $k \in \mathbb{N}$, defined implicitly by (4.26), where our discussion is still restricted to families of system (4.27). We do not, however, restrict to the situation $b_{qp} = \bar{a}_{pq}$, so that the complex systems under consideration are not necessarily complexifications of real systems. To find the isochronicity quantities we must first find the polynomials $\widetilde{H}_{2k+1}$, which themselves are given in terms of the polynomials $Y_1^{(k+1,k)}$ and $Y_2^{(k,k+1)}$ of the normal form (4.22). These latter polynomials can be computed by means of the Normal Form Algorithm in Table 2.1 on page 75, at least when the normalizing transformation is distinguished, which will be adequate for our purposes. We already know that $Y_1^{(k+1,k)}$ and $Y_2^{(k,k+1)}$ are (k,k)-polynomials, hence so are H_{2k+1} and $\widetilde{H}_{2k+1}$. The same is true for the isochronicity quantities.

Proposition 4.2.10. *Let a set S, hence a family* (4.27)*, be given. The isochronicity quantities p_{2k} for that family are (k,k)-polynomials.*

Proof. To get our hands on p_{2k} we must invert the series on the left-hand side of (4.26). In general, if $(1+\sum_{k=1}^{\infty} a_k x^k)^{-1} = 1+\sum_{k=1}^{\infty} b_k x^k$, then clearing the denominator $1 = 1+\sum_{k=1}^{\infty}(a_k + a_{k-1}b_1 + \cdots + b_k)x^k$, hence $b_1 = -a_1$ and for $k \geq 2$ $b_k = -a_1 b_{k-1} - a_2 b_{k-2} - \cdots - a_{k-1}b_1 - a_k$, so that the coefficients b_k can be recursively computed in terms of the a_k: $b_1 = -a_1$, $b_2 = -a_1 b_1 - a_2 = a_1^2 - a_2$, and so on. It follows easily by mathematical induction that for every term $c a_1^{\nu_1} a_2^{\nu_2} \cdots a_k^{\nu_k}$ occurring in the expression for b_k,

$$\nu_1 + 2\nu_2 + 3\nu_3 + \cdots + k\nu_k = k. \tag{4.40}$$

In our specific case, by (4.26) $a_k = \widetilde{H}_{2k+1} = -\frac{1}{2}iH_{2k+1} = -\frac{1}{2}i(Y_1^{(k+1,k)} - Y_2^{(k,k+1)})$ and $b_k = p_{2k}$. Thus p_{2k} is a sum of polynomials of the form

$$c\,(Y_1^{(2,1)} - Y_2^{(1,2)})^{\nu_1}\,(Y_1^{(3,2)} - Y_2^{(2,3)})^{\nu_2} \cdots (Y_1^{(k+1,k)} - Y_2^{(k,k+1)})^{\nu_k}.$$

By Exercise 4.2, this expression defines a (j, j)-polynomial, where j is given by $j = \nu_1 + 2\nu_2 + \cdots + k\nu_k = k$, hence p_{2k} is a (k,k)-polynomial. □

From (4.26) and the recursion formula for the inversion of series given in the proof of Proposition 4.2.10 it is immediately apparent that the polynomials p_{2k} and $\widetilde{H}_{2k+1}$ satisfy (see Definition 1.1.8)

$$p_2 = -\widetilde{H}_3 \quad \text{and} \quad p_{2k} \equiv -\widetilde{H}_{2k+1} \bmod \langle \widetilde{H}_3, \ldots, \widetilde{H}_{2k-1} \rangle \text{ for } k \geq 2, \tag{4.41}$$

which is enough to show that $\langle \widetilde{H}_3, \ldots, \widetilde{H}_{2k+1} \rangle = \langle p_2, \ldots, p_{2k} \rangle$ for all $k \in \mathbb{N}$, hence that $\langle \widetilde{H}_{2k+1} : k \in \mathbb{N} \rangle = \langle p_{2k} : k \in \mathbb{N} \rangle$. With more work we can obtain explicit expressions for the polynomials p_{2k}. The first few are (see Exercise 4.4)

$$\begin{aligned}
p_2 &= -\widetilde{H}_3 \\
&= \tfrac{i}{2}\left(Y_1^{(2,1)} - Y_2^{(1,2)}\right) \\
p_4 &= -\widetilde{H}_5 + (\widetilde{H}_3)^2 \\
&= \tfrac{i}{2}\left(Y_1^{(3,2)} - Y_2^{(2,3)}\right) - \tfrac{1}{4}\left(Y_1^{(2,1)} - Y_2^{(1,2)}\right)^2 \\
p_6 &= -\widetilde{H}_7 + 2\widetilde{H}_3\widetilde{H}_5 - (\widetilde{H}_3)^3 \\
&= \tfrac{i}{2}\left(Y_1^{(4,3)} - Y_2^{(3,4)}\right) - \tfrac{1}{2}\left(Y_1^{(2,1)} - Y_2^{(1,2)}\right)\left(Y_1^{(3,2)} - Y_2^{(2,3)}\right) \\
&\quad - \tfrac{i}{8}\left(Y_1^{(2,1)} - Y_2^{(1,2)}\right)^3 \\
p_8 &= -\widetilde{H}_9 + 2\widetilde{H}_7\widetilde{H}_3 + (\widetilde{H}_5)^2 - 3\widetilde{H}_5(\widetilde{H}_3)^2 + (\widetilde{H}_3)^4 \\
&= \tfrac{i}{2}\left(Y_1^{(5,4)} - Y_2^{(4,5)}\right) - \tfrac{1}{2}\left(Y_1^{(4,3)} - Y_2^{(3,4)}\right)\left(Y_1^{(2,1)} - Y_2^{(1,2)}\right) \\
&\quad - \tfrac{1}{4}\left(Y_1^{(3,2)} - Y_2^{(2,3)}\right)^2 - \tfrac{3i}{8}\left(Y_1^{(3,2)} - Y_2^{(2,3)}\right)\left(Y_1^{(2,1)} - Y_2^{(1,2)}\right)^2 \\
&\quad + \tfrac{1}{16}\left(Y_1^{(2,1)} - Y_2^{(1,2)}\right)^4.
\end{aligned} \tag{4.42}$$

In general, the isochronicity quantities exhibit the following structure.

Proposition 4.2.11. *Let a set S, hence a family* (4.27)*, be given. The isochronicity quantities of that system have the form*

$$p_{2k} = \tfrac{1}{2} \sum_{\{\nu : L(\nu) = (k,k)\}} p_{2k}^{(\nu)}([\nu] + [\widehat{\nu}]).$$

Proof. From (4.26) and the recursion formula for the inversion of series given in the proof of Proposition 4.2.10 it is clear that the polynomial p_{2k} is a polynomial function of $\widetilde{H}_3, \widetilde{H}_5, \ldots, \widetilde{H}_{2k+1}$. The result now follows from Corollary 4.2.9 and Exercise 4.5. □

The following proposition states a fact about the isochronicity quantities that will be needed in Chapter 6.

Proposition 4.2.12. *Suppose the nonlinear terms in family* (4.27) *are homogeneous polynomials of degree K. Then the isochronicity quantities, p_{2k} for $k \in \mathbb{N}$, are homogeneous polynomials in $\mathbb{C}[a,b]$ of degree $2k/(K-1)$.*

Proof. This fact follows immediately from Proposition 4.2.10 and Exercise 3.40. □

Although the isochronicity quantities p_{2k} are defined for all choices of the coefficients $(a,b) \in E(a,b)$, they are relevant only for $(a,b) \in V_{\mathscr{C}}$, the center variety, where their vanishing identifies isochronicity for real centers and linearizability for all centers. Thus if two of them agree at every point of the center variety, then they are equivalent with respect to the information that they provide about the linearizability of centers in the family under consideration. We are naturally led therefore to the concept of the equivalence of polynomials with respect to a variety. In general terms, let k be a field and let V be a variety in k^n. We define an equivalence relation on $k[x_1,\ldots,x_n]$ by saying that two polynomials f and g are equivalent if, for every $\mathbf{x} \in V$, $f(\mathbf{x}) = g(\mathbf{x})$ as constants in k. The set of equivalence classes is denoted $k[V]$ and is called the *coordinate ring* of the variety V. If $[[f]]$ denotes the equivalence class of the polynomial f, then it is easy to see that the operations $[[f]] + [[g]] = [[f+g]]$ and $[[f]][[g]] = [[fg]]$ are well-defined and give $k[V]$ the structure of a commutative ring. The reader is asked in Exercise 4.7 to show that f and g are equivalent under this new equivalence relation if and only if $f \equiv g \bmod \mathbf{I}(V)$. That is, f and g are in the same equivalence class in $k[V]$ if and only if they are in the same equivalence class in $k[x_1,\ldots,x_n]/\mathbf{I}(V)$. Hence there is a mapping $\varphi : k[V] \to k[x_1,\ldots,x_n]/\mathbf{I}(V)$ naturally defined by $\varphi([[f]]) = [f]$. It is an isomorphism of rings (Exercise 4.8, but see also Exercise 4.9).

Returning to the specific case of the isochronicity quantities, by way of notation let

$$P = \langle p_{2k} : k \in \mathbb{N} \rangle \subset \mathbb{C}[a,b] \qquad \text{and} \qquad \widetilde{P} = \langle [[p_{2k}]] : k \in \mathbb{N} \rangle \subset \mathbb{C}[V_{\mathscr{C}}]$$

and for $k \in \mathbb{N}$ let

$$P_k = \langle p_2,\ldots,p_{2k} \rangle \qquad \text{and} \qquad \widetilde{P}_k = \langle [[p_2]],\ldots,[[p_{2k}]] \rangle.$$

Since the equivalences given in (4.41) clearly imply that $\mathscr{H} = P$ (and that $\mathscr{H}_k = P_k$ for all $k \in \mathbb{N}$), a system that has a linearizable center at the origin corresponds to a point in the center variety at which every polynomial p_{2k} vanishes; symbolically: $V_{\mathscr{I}} = \mathbf{V}(\mathscr{H}) \cap V_{\mathscr{C}} = \mathbf{V}(P) \cap V_{\mathscr{C}}$. We will also be interested in $\mathbf{V}(P_k) \cap V_{\mathscr{C}}$. Since all points of interest lie in the center variety, on which every representative of $[[p_{2k}]]$ agrees, we are tempted to write $\mathbf{V}(\widetilde{P}) \cap V_{\mathscr{C}}$ and $\mathbf{V}(\widetilde{P}_k) \cap V_{\mathscr{C}}$ instead of $\mathbf{V}(P) \cap V_{\mathscr{C}}$ and $\mathbf{V}(P_k) \cap V_{\mathscr{C}}$. The objects $\mathbf{V}(\widetilde{P})$ and $\mathbf{V}(\widetilde{P}_k)$ are not well-defined, however, since elements of $\widetilde{P}$ are equivalence classes and, for $(a,b) \in \mathbb{C}^{2\ell} \setminus V_{\mathscr{C}}$, the value of $f(a,b)$ depends on the respresentative $f \in [[p_{2k}]]$ that is chosen. The value is independent of the representative when $(a,b) \in V_{\mathscr{C}}$, however, so we may validly define

$$V_{V_{\mathscr{C}}}(\widetilde{P}) := \{(a,b) \in V_{\mathscr{C}} : \text{if } f \in [[p_{2k}]] \text{ for some } k \in \mathbb{N} \text{ then } f(a,b) = 0\}. \qquad (4.43)$$

The definition of $V_{V_{\mathscr{C}}}(\widetilde{P}_k)$ is similar. The following proposition shows that these sets are subvarieties of the center variety and are precisely what we expect them to be.

Proposition 4.2.13. *Let $I = P_k$ for some $k \in \mathbb{N}$ or $I = P$. In the former case let $\widetilde{I} = \widetilde{P_k} = \langle [[p_2]], \ldots, [[p_{2k}]] \rangle$ and in the latter case let $\widetilde{I} = \widetilde{P} = \langle [[p_{2k}]] : k \in \mathbb{N} \rangle$. Then $V_{V_{\mathscr{C}}}(\widetilde{I}) = \mathbf{V}(I) \cap V_{\mathscr{C}}$.*

Proof. We prove the proposition when $I = P$; the ideas are the same when $I = P_k$.

Suppose $(a^*, b^*) \in V_{V_{\mathscr{C}}}(\widetilde{P})$, so that $(a^*, b^*) \in V_{\mathscr{C}}$ and $f(a^*, b^*) = 0$ for every f that lies in $[[p_{2k}]]$ for some $k \in \mathbb{N}$. Since $p_{2k} \in [[p_{2k}]]$, we have that $p_{2k}(a^*, b^*) = 0$ for all $k \in \mathbb{N}$, so $(a^*, b^*) \in \mathbf{V}(P) \cap V_{\mathscr{C}}$.

Conversely, suppose $(a^*, b^*) \in \mathbf{V}(P) \cap V_{\mathscr{C}}$ and that the polynomial $f \in \mathbb{C}[a,b]$ satisfies $[[f]] \in \widetilde{P}$. We must show that $f(a^*, b^*) = 0$ even though f need not be in P. There exist $f_{j_1}, \ldots, f_{j_s} \in \mathbb{C}[a,b]$ such that

$$[[f]] = [[f_{j_1}]][[p_{2j_1}]] + \cdots + [[f_{j_s}]][[p_{2j_s}]] = [[f_{j_1} p_{2j_1} + \cdots + f_{j_s} p_{2j_s}]].$$

By the definition of $\mathbb{C}[V_{\mathscr{C}}]$, since $(a^*, b^*) \in V_{\mathscr{C}}$ f agrees with $f_{j_1} p_{2j_1} + \cdots + f_{j_s} p_{2j_s}$ at (a^*, b^*), we have

$$\begin{aligned} f(a^*, b^*) &= f_{j_1}(a^*, b^*) p_{2j_1}(a^*, b^*) + \cdots + f_{j_s}(a^*, b^*) p_{2j_s}(a^*, b^*) \\ &= f_{j_1}(a^*, b^*) \cdot 0 + \cdots + f_{j_s}(a^*, b^*) \cdot 0 = 0 \end{aligned}$$

since $(a^*, b^*) \in V(P)$, as required. □

As previously noted, the equivalences given in (4.41) imply that $\mathscr{H}_k = P_k$ and $\mathscr{H} = P$, so that $\mathbf{V}(\mathscr{H}_k) = \mathbf{V}(P_k)$ and $\mathbf{V}(\mathscr{H}) = \mathbf{V}(P)$. The polynomials p_{2k} are ultimately defined in terms of the polynomials $Y_1^{(k+1,k)}$ and $Y_2^{(k,k+1)}$, so we now connect $\mathbf{V}(P_k)$ and $\mathbf{V}(P)$ to $\mathbf{V}(\mathscr{Y}_k)$ and $\mathbf{V}(\mathscr{Y})$ directly. (See also Exercise 4.10.)

Proposition 4.2.14. *With reference to family* (4.27),

$$\mathbf{V}(P) \cap V_{\mathscr{C}} = \mathbf{V}(\mathscr{Y}) \cap V_{\mathscr{C}} \quad \textit{and} \quad \mathbf{V}(P_k) \cap V_{\mathscr{C}} = \mathbf{V}(\mathscr{Y}_k) \cap V_{\mathscr{C}} \textit{ for all } k \in \mathbb{N}.$$

Proof. Since $H_{2k+1} = Y_1^{(k+1,k)} - Y_2^{(k,k+1)}$, if $f \in \mathscr{H}_k$, then $f \in \mathscr{Y}_k$, but not conversely. But by (4.31) and the identity $[[G_{2k+1}]] = [[0]]$ in $\mathbb{C}[V_{\mathscr{C}}]$ (see Theorem 3.2.7), we have that $f \in \mathscr{Y}_k$ implies $[[f]] \in \langle [[H_3]], \ldots, [[H_{2k+1}]] \rangle = \langle [[p_2]], \ldots, [[p_{2k}]] \rangle = \widetilde{P_k}$. Therefore if we now let $\widetilde{\mathscr{Y}_k} = \langle [[Y_1^{(2,1)}]], [[Y_2^{(1,2)}]], \ldots, [[Y_1^{(k+1,k)}]], [[Y_2^{(k,k+1)}]] \rangle$ and similarly let $\widetilde{\mathscr{Y}} = \langle [[Y_1^{(j+1,j)}]], [[Y_2^{(j,j+1)}]] : j \in \mathbb{N} \rangle$, then $\widetilde{\mathscr{Y}_k} = \widetilde{P_k}$ and $\widetilde{\mathscr{Y}} = \widetilde{P}$ in $\mathbb{C}[V_{\mathscr{C}}]$. Then for all $k \in \mathbb{N}$, using Proposition 4.2.13 for the first and last equalities,

$$\mathbf{V}(P_k) \cap V_{\mathscr{C}} = V_{V_{\mathscr{C}}}(\widetilde{P_k}) = V_{V_{\mathscr{C}}}(\widetilde{\mathscr{Y}_k}) = \mathbf{V}(\mathscr{Y}_k) \cap V_{\mathscr{C}}$$

and $\mathbf{V}$(P) $\cap V_{\mathscr{C}} = \mathbf{V}(\mathscr{Y}) \cap V_{\mathscr{C}}$ similarly. □

4.3 The Linearizability Quantities

In order to find all systems (4.27) that are linearizable by a convergent transformation of the type (4.21), one approach is to try to directly construct a linearizing transformation (4.21) and the corresponding normal form (3.27), imposing the condition that $Y_1^{(k+1,k)} = Y_2^{(k,k+1)} = 0$ for all $k \in \mathbb{N}$. Instead of doing so, we will look for the inverse of such a transformation. The motivation for this idea is that it will lead to a recursive computational formula that is closely related to formula (3.76) in Section 3.4. First we write what we might call the inverse linearizing transformation, which changes the linear system

$$\dot{y}_1 = i y_1, \quad \dot{y}_2 = -i y_2 \tag{4.44}$$

into system (4.27) as

$$y_1 = x_1 + \sum_{j+k=2}^{\infty} u_1^{(j-1,k)} x_1{}^j x_2{}^k, \quad y_2 = x_2 + \sum_{j+k=2}^{\infty} u_2^{(j,k-1)} x_1{}^j x_2{}^k, \tag{4.45}$$

where we have made the asymmetrical index shift so that the "linearizability quantities" that we ultimately obtain are indexed I_{kk} and J_{kk} rather than $I_{k,k+1}$ and $J_{k+1,k}$. In agreement with (4.45), let $u_1^{(0,0)} = u_2^{(0,0)} = 1$ and $u_1^{(-1,1)} = u_2^{(1,-1)} = 0$, so that the indexing sets are $\mathbb{N}_{-1} \times \mathbb{N}_0$ for $u_1^{(k_1,k_2)}$ and $\mathbb{N}_0 \times \mathbb{N}_{-1}$ for $u_2^{(k_1,k_2)}$. Then making the convention that $a_{pq} = b_{qp} = 0$ if $(p,q) \notin S$, when we differentiate each part of (4.45) with respect to t, apply (4.27) and (4.44), and equate the coefficients of like powers $x_1^{\alpha_1} x_2^{\alpha_2}$, the resulting equations yield the recurrence formulas

$$(k_1 - k_2) u_1^{(k_1,k_2)} = \sum_{\substack{s_1+s_2=0 \\ s_1 \geq -1, s_2 \geq 0}}^{k_1+k_2-1} [(s_1+1) a_{k_1-s_1,k_2-s_2} - s_2 b_{k_1-s_1,k_2-s_2}] u_1^{(s_1,s_2)} \tag{4.46a}$$

$$(k_1 - k_2) u_2^{(k_1,k_2)} = \sum_{\substack{s_1+s_2=0 \\ s_1 \geq 0, s_2 \geq -1}}^{k_1+k_2-1} [s_1 a_{k_1-s_1,k_2-s_2} - (s_2+1) b_{k_1-s_1,k_2-s_2}] u_2^{(s_1,s_2)}, \tag{4.46b}$$

$(k_1,k_2) \in \mathbb{N}_{-1} \times \mathbb{N}_0$ for $u_1^{(k_1,k_2)}$ and $(k_1,k_2) \in \mathbb{N}_0 \times \mathbb{N}_{-1}$ for $u_2^{(k_1,k_2)}$. (An extra detail on this computation is given in the paragraph of the proof of Theorem 4.3.2 between displays (4.50) and (4.51).) With the appropriate initialization $u_1^{(0,0)} = u_2^{(0,0)} = 1$ and $u_1^{(-1,1)} = u_2^{(1,-1)} = 0$, and the convention that $a_{pq} = b_{qp} = 0$ if $(p,q) \notin S$, mentioned above, the coefficients $u_1^{(k_1,k_2)}$ and $u_2^{(k_1,k_2)}$ of (4.45) can be computed recursively using formulas (4.46), where the recursion is on $k_1 + k_2$, beginning at $k_1 + k_2 = 0$. But at every even value of $k_1 + k_2 \geq 2$ there occurs the one pair (k_1,k_2) such that $k_1 = k_2 = k > 0$, for which (4.46) becomes $I_{kk} = J_{kk} = 0$, where

$$I_{kk} = \sum_{\substack{s_1+s_2=0 \\ s_1 \geq -1, s_2 \geq 0}}^{2k-1} [(s_1+1)a_{k-s_1,k-s_2} - s_2 b_{k-s_1,k-s_2}] u_1^{(s_1,s_2)} \tag{4.47a}$$

$$J_{kk} = \sum_{\substack{s_1+s_2=0 \\ s_1 \geq 0, s_2 \geq -1}}^{2k-1} [s_1 a_{k-s_1,k-s_2} - (s_2+1) b_{k-s_1,k-s_2}] u_2^{(s_1,s_2)}. \tag{4.47b}$$

Of course, when $k_1 = k_2 = k$, $u_1^{(k,k)}$ and $u_2^{(k,k)}$ can be chosen arbitrarily, but we typically make the choice $u_1^{(k,k)} = u_2^{(k,k)} = 0$. The process of creating the inverse of a normalizing transformation that linearizes (4.27) succeeds only if the expressions on the right-hand sides of (4.46) are equal to zero for all $k \in \mathbb{N}$. That is, for a particular member of family (4.27) corresponding to (a^*, b^*), the vanishing of the polynomials I_{kk} and J_{kk} at (a^*, b^*) for all $k \in \mathbb{N}$ is the condition that the system be linearizable.

Remark 4.3.1. If a similar procedure is applied to (4.29), then precisely the same polynomials are obtained: I_{kk} and J_{kk} the same for systems (4.27) and (4.29).

Thus, in a procedure reminiscent of the attempt in Section 3.3 to construct a first integral (3.52) for system (3.3) using (3.55) to recursively determine the coefficients $v_{k_1 k_2}$, for fixed $(a^*, b^*) \in E(a,b)$ we may attempt to construct the inverse (4.45) of a linearization for system (4.27) or (4.29) in a step-by-step procedure, beginning by choosing $u_1^{(0,0)} = u_2^{(0,0)} = 1$ and $u_1^{(-1,1)} = u_2^{(1,-1)} = 0$ and using (4.46) to recursively construct all the remaining coefficients in (4.45). The first few coefficients are determined uniquely, but for $k \in \mathbb{N}$, $u_1^{(k,k)}$ and $u_2^{(k,k)}$ exist only if I_{kk} and J_{kk} given by (4.47) are zero, in which case $u_1^{(k,k)}$ and $u_2^{(k,k)}$ may be selected arbitrarily. For $k_0 \geq 2$, the values of $I_{k_0 k_0}$ and $J_{k_0 k_0}$ seemingly depend on the choices made earlier for $u_1^{(k,k)}$ and $u_2^{(k,k)}$ for $k < k_0$. Hence, although $I_{kk} = J_{kk} = 0$ for all $k \in \mathbb{N}$ means that the procedure succeeds and system (a^*, b^*) is linearizable, it does not automatically follow that if $(I_{k_0 k_0}, J_{k_0 k_0}) \neq (0,0)$ for some k_0, then the system (a^*, b^*) is not linearizable, since it is conceivable that for different choices of $u_1^{(k,k)}$ and $u_2^{(k,k)}$ for $k < k_0$ we would have gotten $(I_{k_0 k_0}, J_{k_0 k_0}) = (0,0)$. We will now show that this is not the case but that, on the contrary, whether or not $I_{kk} = J_{kk} = 0$ for all $k \in \mathbb{N}$ is independent of the choices made for $u_1^{(k,k)}$ and $u_2^{(k,k)}$.

In order to do so, we shift our focus from a single system corresponding to a choice (a^*, b^*) of the coefficients (a,b) in (4.27) or (4.29) to the full family parametrized by $(a,b) \in E(a,b) = \mathbb{C}^{2\ell}$ that corresponds to the choice of a fixed ℓ-element index set S. Now the initialization is $u_1^{(0,0)} = u_2^{(0,0)} \equiv 1$ and $u_1^{(-1,1)} = u_2^{(1,-1)} \equiv 0$ in $\mathbb{C}[a,b]$ and for arbitrary choice of elements $u_1^{(k,k)}$ and $u_2^{(k,k)}$ in $\mathbb{C}[a,b]$, $k \in \mathbb{N}$, formulas (4.46) and (4.47) recursively define a collection of polynomials in $\mathbb{C}[a,b]$: $u_1^{(k_1,k_2)}$ for $(k_1,k_2) \in \mathbb{N}_{-1} \times \mathbb{N}_0$ but $k_1 \neq k_2$, $u_2^{(k_1,k_2)}$ for $(k_1,k_2) \in \mathbb{N}_0 \times \mathbb{N}_{-1}$ but $k_1 \neq k_2$, and I_{kk} and J_{kk} for $k \in \mathbb{N}$. As is the case with the focus quantities g_{kk} for system (3.3) and the coefficients $Y_1^{(k+1,k)}$, $Y_2^{(k,k+1)}$ of the normal form (4.22), the polynomials I_{kk}, J_{kk}

are not determined uniquely, the indefiniteness in this case arising from the freedom we have in choosing the coefficients $u_1^{(k,k)}$ and $u_2^{(k,k)}$ (in $\mathbb{C}[a,b]$) of an inverse normalizing transformation that is to linearize (4.27). But the varieties in the space of parameters $E(a,b) = \mathbb{C}^{2\ell}$ defined by these polynomials should be the same regardless of those choices. Theorem 4.2.2 (or alternatively Proposition 4.2.7) established this fact in the case of $Y_1^{(k+1,k)}$ and $Y_2^{(k,k+1)}$. The following theorem states that it is true for I_{kk} and J_{kk}.

Theorem 4.3.2. *Let a set S, hence families* (4.27) *and* (4.29)*, be given, and let I_{kk} and J_{kk}, $k \in \mathbb{N}$, be any collection of polynomials generated recursively by* (4.46) *and* (4.47) *from the initialization $u_1^{(0,0)} = u_2^{(0,0)} \equiv 1$ and $u_1^{(-1,1)} = u_2^{(1,-1)} \equiv 0$ and for some specific but arbitrary choice of $u_1^{(k,k)}$ and $u_2^{(k,k)}$, $k \in \mathbb{N}$ (see Remark 4.3.1). Then the linearizability variety of the systems* (4.27) *and* (4.29) *(see Remark 4.2.6) as given by Definition 4.2.5 (and Proposition 4.2.7) coincides with the variety* $\mathbf{V}(\langle I_{kk}, J_{kk} : k \in \mathbb{N}\rangle)$.

Proof. The reasoning is identical whether family (4.27) or (4.29) is under consideration, so we work with just (4.29). For convenience we let $\tilde{u}_1^{(k,k)}$, $\tilde{u}_2^{(k,k)}$, $\tilde{I}_{kk}$, and $\tilde{J}_{kk}$ denote the specific elements of $\mathbb{C}[a,b]$ that were specified in the statement of the theorem.

Suppose system $(a^*,b^*) \in \mathbf{V}(\langle \tilde{I}_{kk}, \tilde{J}_{kk} : k \in \mathbb{N}\rangle)$. Then the recursive process using (4.46) to create a transformation that will transform

$$\dot{y}_1 = y_1, \qquad \dot{y}_2 = -y_2 \tag{4.48}$$

into (4.29) for $(a,b) = (a^*,b^*)$ succeeds when the constants $u_1^{(k,k)}$ and $u_2^{(k,k)}$ are chosen to be $\tilde{u}_1^{(k,k)}(a^*,b^*)$ and $\tilde{u}_2^{(k,k)}(a^*,b^*)$, respectively, $k \in \mathbb{N}$; call the resulting formal transformation $\mathbf{U}$. Then the formal inverse $\mathbf{U}^{-1}$ is a normalizing transformation that linearizes (4.29) for $(a,b) = (a^*,b^*)$. By Corollary 4.2.3, system (4.29) for $(a,b) = (a^*,b^*)$ is therefore linearizable, and so $(a^*,b^*) \in \mathbf{V}(\mathscr{Y})$.

Conversely, suppose $(a^*,b^*) \in \mathbf{V}(\mathscr{Y})$. This means there exists a normalizing transformation $\mathbf{H}$ that transforms (4.29) for $(a,b) = (a^*,b^*)$ into (4.48). Then $\mathbf{H}^{-1}$ is a change of coordinates of the form (4.45) that transforms (4.48) back into (4.29) for $(a,b) = (a^*,b^*)$. This latter transformation corresponds to a choice of constants $u_1^{(k,k)}$ and $u_2^{(k,k)}$ for which the corresponding constants I_{kk} and J_{kk} are all zero. Now begin the recursive procedure of constructing a transformation of the form (4.45) for system (a^*,b^*), but at every step using for our choice of the constants $u_1^{(k,k)}$ and $u_2^{(k,k)}$ the values $\tilde{u}_1^{(k,k)}(a^*,b^*)$ and $\tilde{u}_2^{(k,k)}(a^*,b^*)$. Suppose, contrary to what we wish to show, that for some (least) $K \in \mathbb{N}$, $\tilde{I}_{KK}(a^*,b^*) \neq 0$ or $\tilde{J}_{KK}(a^*,b^*) \neq 0$, so the recursive process halts, and consider the polynomial transformation $\mathbf{T}$ constructed so far:

$$y_1 = x_1 + \sum_{\substack{j+k=2 \\ (j,k)\neq(K+1,K)}}^{2K+1} u_1^{(j-1,k)} x_1^j x_2^k, \qquad y_2 = x_2 + \sum_{\substack{j+k=2 \\ (j,k)\neq(K,K+1)}}^{2K+1} u_2^{(j,k-1)} x_1^j x_2^k. \tag{4.49}$$

We will check how close $\mathbf{T}^{-1}$ comes to linearizing (a^*,b^*) by examining $\dot{y}_1 - y_1$ and $\dot{y}_2 + y_2$. The key computational insight is to maintain the old coordinates until the last step, avoiding having to actually work with $\mathbf{T}^{-1}$. We start with $\dot{y}_2 + y_2$, for which we will do the computation in detail. Differentiation of the second equation in (4.49) and application of (4.29) yields

$$\begin{aligned}\dot{y}_2 + y_2 = & \sum_{\substack{j+k=2\\(j,k)\neq(K,K+1)}}^{2K+1} (1+j-k)u_2^{(j,k-1)} x_1^j x_2^k + \sum_{(p,q)\in S} b_{qp} x_1^q x_2^{p+1} \\ & + \sum_{\substack{j+k=2\\(j,k)\neq(K,K+1)}}^{2K+1} u_2^{(j,k-1)} \Big[\sum_{(p,q)\in S} k\, b_{qp} x_1^{q+j} x_2^{p+k} \Big] \\ & - \sum_{\substack{j+k=2\\(j,k)\neq(K,K+1)}}^{2K+1} u_2^{(j,k-1)} \Big[\sum_{(p,q)\in S} j\, a_{pq} x_1^{p+j} x_2^{q+k} \Big] .\end{aligned} \tag{4.50}$$

Setting $D =: \{(j,k) \in \mathbb{N}_0 \times \mathbb{N}_{-1} : 2 \le j+k \le r+s-2\} \cup \{(r,s-1)\}$, for $(r,s) \in \mathbb{N}^2$, $r+s \ge 2$, a contribution to the coefficient of $x_1^r x_2^s$ in (4.50) is made by $u_2^{(k_1,k_2)}$ if and only if $(k_1,k_2) \in D$. There is complete cancellation in the coefficient of $x_1^r x_2^s$ provided $(K,K) \notin D$. That is certainly true if $r+s-1 < 2K$. For $r+s = 2K+1$, if $(r,s) \neq (K,K+1)$, then $(K,K) \notin D$, but if $(r,s) = (K,K+1)$, then the last pair listed in D is (K,K). Thus the only term present in (4.50) is that corresponding to $x_1^K x_2^{K+1}$. To find its coefficient, we repeat the reasoning that is used to derive (4.46b) from the analogue of (4.50): identify, for each pair $(r,s) \in \mathbb{N}^2$, $r+s \ge 2$, the contribution of each sum in (4.50) to the coefficient of $x_1^r x_2^s$ (only in deriving (4.46b) we set the resulting sum to zero). Here the first sum contributes nothing to the coefficient of $x_1^r x_2^s$ and the second sum contributes precisely b_{KK}. In the third sum, for any (j,k), $(p+j,q+k) = (r,s)$ forces $(p,q) = (r-j,s-k)$; $j+k$ is maximized when $p+q$ is minimized, so $\max(j+k) + \min(p+q) = \max(j+k) + 1 = r+s$, and the contribution is $\sum_{j+k=2}^{2K} k u_2^{(j,k-1)} b_{K-j,K-k+1}$. Similarly, the fourth sum contributes $-\sum_{j+k=2}^{2K} j u_2^{(j,k-1)} a_{K-j,K-k+1}$, so that ultimately we obtain, using nothing more about $\mathbf{T}^{-1}$ than that it must have the form $(x_1,x_2) = \mathbf{T}(y_1,y_2) = (y_1 + \cdots, y_2 + \cdots)$, that $\dot{y}_2 + y_2 = -J_{KK} y_1^K y_2^{K+1} + \cdots$. Similarly, $\dot{y}_1 - y_1 = I_{KK} y_1^{K+1} y_2^K + \cdots$. We conclude that the transformation $\mathbf{x} = \mathbf{T}^{-1}(\mathbf{y}) = \mathbf{y} + \cdots$ transforms system (a^*,b^*) into

$$\begin{aligned}\dot{y}_1 &= \ \ y_1 + \tilde{I}_{KK}(a^*,b^*) y_1^{K+1} y_2^K + r_1(y_1,y_2)\\ \dot{y}_2 &= -y_2 - \tilde{J}_{KK}(a^*,b^*) y_1^K y_2^{K+1} + r_2(y_1,y_2),\end{aligned} \tag{4.51}$$

where $r_1(y_1,y_2)$ and $r_2(y_1,y_2)$ are analytic functions beginning with terms of order at least $2K+2$. By Proposition 3.2.2, (4.51) is a normal form through order $2K+1$ but is not necessarily a normal form since we do not know that higher-order nonresonant terms have been eliminated. However, we can freely change any term of order $2K+2$ or greater in $\mathbf{T}^{-1}$ without changing terms in (4.51) of order less than or equal to $2K+1$, hence there exists a normalizing transformation $\mathbf{x} = \mathbf{S}(\mathbf{y}) = \mathbf{y} + \mathbf{s}(\mathbf{y})$ that agrees with $\mathbf{T}^{-1}$ through order $2K+1$, hence whose inverse agrees with $\mathbf{T}$ through order $2K+1$, and which produces a normal form that agrees with (4.51) through order $2K+1$. But then the assumption that $(\tilde{I}_{KK}(a^*,b^*), \tilde{J}_{KK}(a^*,b^*)) \neq (0,0)$ leads to a contradiction: for we know that $\mathbf{H}$ linearizes (4.29) for $(a,b) = (a^*,b^*)$, hence by Theorem 4.2.2 $\mathbf{S}$ does. Thus $(\tilde{I}_{KK}(a^*,b^*), \tilde{J}_{KK}(a^*,b^*)) = (0,0)$, and $(a^*,b^*) \in \mathbf{V}(\langle I_{kk}, J_{kk} : k \in \mathbb{N}\rangle)$. □

Theorem 4.3.2 justifies the following definition.

Definition 4.3.3. Let a set S, hence families (4.27) and (4.29), be given. For any choice of the polynomials $u_1^{(k,k)}$ and $u_2^{(k,k)}$, $k \in \mathbb{N}$, in $\mathbb{C}[a,b]$, the polynomials I_{kk} and J_{kk} defined recursively by (4.46) and by (4.47) (starting with the initialization $u_1^{(0,0)} = u_2^{(0,0)} \equiv 1$ and $u_1^{(-1,1)} = u_2^{(1,-1)} \equiv 0$ and the convention that $a_{pq} = b_{qp} = 0$ if $(p,q) \notin S$) are the *kth linearizability quantities*. The ideal $\langle I_{kk}, J_{kk} : k \in \mathbb{N}\rangle$ that they generate is the *linearizability ideal* and is denoted $\mathscr{L}$.

To recapitulate, with regard to isochronicity and linearizability we have one variety with several characterizations and names:

$$\mathbf{V}(\mathscr{H}) \cap V_{\mathscr{C}} = V_{\mathscr{I}} = V_{\mathscr{L}} = \mathbf{V}(\mathscr{Y}) = \mathbf{V}(\mathscr{L}).$$

The first and third equalities are definitions and the second and fourth are by Proposition 4.2.7 and Theorem 4.3.2, respectively.

With a proper choice of the coefficients $u_1^{(k,k)}$ and $u_2^{(k,k)}$, the quantities J_{kk} can be computed immediately from the quantities I_{kk}, as described by the following proposition. As always, $\widehat{f}$ denotes the involution of Definition 3.4.3.

Proposition 4.3.4. *Let a set S, hence families* (4.27) *and* (4.29), *be given.*
1. If, for all $k \in \mathbb{N}$, $u_1^{(k,k)}$ and $u_2^{(k,k)}$ are chosen so as to satisfy $u_2^{(k,k)} = \widehat{u}_1^{(k,k)}$, then $u_2^{(k,j)} = \widehat{u}_1^{(j,k)}$ for all $(j,k) \in \mathbb{N}_{-1} \times \mathbb{N}_0$.
2. If, for all $k \in \mathbb{N}$, $u_1^{(k,k)}$ and $u_2^{(k,k)}$ are chosen as in (1), then $J_{kk} = -\widehat{I}_{kk}$ for all $k \in \mathbb{N}$.

Proof. The first part follows directly from formula (4.46) by induction on $j+k$. (Exercise 4.11.) The second part follows by a direct application of part (1) of the proposition to (4.47). □

Note that the usual choices $u_1^{(k,k)} = 0$ and $u_2^{(k,k)} = 0$ satisfy the condition in part (1) of the proposition. Moreover, since clearly $u_1^{(k,k)}$ and $u_2^{(k,k)}$ have coefficients that lie in $\mathbb{Q}$, the same is true of I_{kk}, hence, because its coefficients are real, $\widehat{I}_{kk}$ can be obtained from I_{kk} merely by replacing every monomial $[\nu]$ that appears in I_{kk} by the monomial $[\hat{\nu}]$.

The properties of the focus quantities and the coefficients of the function Ψ as described in Theorem 3.4.2 have analogues for the linearizability quantities I_{kk} and J_{kk} and for the coefficients of the transformation (4.45). The precise statements are given in the following theorem.

Theorem 4.3.5. *Let a set S, hence families (4.27) and (4.29), be given.*
1. *For every $(j,k) \in \mathbb{N}_{-1} \times \mathbb{N}_0$ (respectively, for every $(j,k) \in \mathbb{N}_0 \times \mathbb{N}_{-1}$) the coefficient $u_1^{(j,k)}$ (respectively, $u_2^{(j,k)}$) of the transformation (4.45) is a (j,k)-polynomial with coefficients in $\mathbb{Q}$.*
2. *For every $k \in \mathbb{N}$, the linearizability quantities I_{kk} and J_{kk} are (k,k)-polynomials with coefficients in $\mathbb{Q}$.*

Proof. As just noted it is immediate from (4.46) that $u_1^{(j,k)}$, $u_2^{(j,k)}$, I_{kk}, and J_{kk} lie in $\mathbb{Q}[a,b]$. The proof that they are (j,k)- and (k,k)-polynomials is precisely the same inductive argument for the proof of points (2) and (4) of Theorem 3.4.2. □

Theorem 3.4.5 also has an analogue for linearizability. We adopt the notation introduced in the paragraph preceding Definition 3.4.1, including letting ℓ denote the cardinality of the index set S. To state the result we must first generalize the function V on $\mathbb{N}^{2\ell}$ given by (3.76).

Let a set S, hence a family (4.27), be given, and fix $(m,n) \in \{(0,1),(1,0),(1,1)\}$. For any $\nu \in \mathbb{N}_0^{2\ell}$ define $V_{(m,n)}(\nu) \in \mathbb{Q}$ recursively, with respect to $|\nu| = \nu_1 + \cdots + \nu_{2\ell}$, as follows:

$$V_{(m,n)}((0,\ldots,0)) = 1; \tag{4.52a}$$

for $\nu \neq (0,\ldots,0)$,

$$V_{(m,n)}(\nu) = 0 \qquad \text{if} \qquad L_1(\nu) = L_2(\nu); \tag{4.52b}$$

and when $L_1(\nu) \neq L_2(\nu)$,

$$\begin{aligned} V_{(m,n)}(\nu) &= \frac{1}{L_1(\nu) - L_2(\nu)} \\ &\times \left[\sum_{j=1}^{\ell} \widetilde{V}_{(m,n)}(\nu_1,\ldots,\nu_j - 1,\ldots,\nu_{2\ell})(L_1(\nu_1,\ldots,\nu_j - 1,\ldots,\nu_{2\ell}) + m) \right. \\ &\left. - \sum_{j=\ell+1}^{2\ell} \widetilde{V}_{(m,n)}(\nu_1,\ldots,\nu_j - 1,\ldots,\nu_{2\ell})(L_2(\nu_1,\ldots,\nu_j - 1,\ldots,\nu_{2\ell}) + n) \right], \end{aligned} \tag{4.52c}$$

where

$$\widetilde{V}_{(m,n)}(\eta) = \begin{cases} V_{(m,n)}(\eta) & \text{if } \eta \in \mathbb{N}_0^{2\ell} \\ 0 & \text{if } \eta \in \mathbb{N}_{-1}^{2\ell} \setminus \mathbb{N}_0^{2\ell}. \end{cases}$$

The function $V_{(1,1)}$ is the function defined by (3.76). $V_{(1,0)}$ pertains to I_{kk} and $V_{(0,1)}$ pertains to J_{kk}, but because of the relationship described in Proposition 4.3.4(2), we will need to consider only $V_{(1,0)}$, for which we now prove that the analogue of Lemma 3.4.4 is true.

Lemma 4.3.6. *Suppose $\nu \in \mathbb{N}_0^{2\ell}$ is such that either $L_1(\nu) < -1$ or $L_2(\nu) < -1$. Then $V_{(1,0)}(\nu) = 0$.*

Proof. The proof is by induction on $|\nu|$. The basis step is practically the same as that in the proof of Lemma 3.4.4, as are the parts of the inductive step in the situations $L_1(\nu) < -2$, $L_2(\nu) < -2$, and $L_1(\nu) = -2$, so we will not repeat them. We are left with showing that if the lemma is true for $|\nu| \le m$, and if ν is such that $|\nu| = m+1$, $L_2(\nu) = -2$, and $L_1(\nu) \ge -1$ (else we are in a previous case), then $V_{(1,0)}(\nu) = 0$. For this part of the inductive step the old proof does not suffice. It does show that $L_2(\mu) = L_2(\nu) - q_j = -2 - q_j \le -2$ when μ is the argument of any term in the first sum in (4.52c), so that by the induction hypothesis, $V_{(1,0)}(\mu) = 0$, which implies $\widetilde{V}_{(1,0)}(\mu) = 0$, while $L_2(\mu) = L_2(\nu) - p_j = -2 - p_j$ if μ is the argument of any term in the second sum in (4.52c), so that if $p_j \ge 0$, then the induction hypothesis applies to give $\widetilde{V}_{(1,0)}(\mu) = 0$. Thus we know that if $L_2(\nu) = -2$ (and $L_1(\nu) \ge -1$, to avoid a known case), then expression (4.52c) for $V_{(1,0)}(\nu)$ reduces to

$$\begin{aligned} V_{(1,0)}(\nu) = & \frac{-1}{L_1(\nu) - L_2(\nu)} \\ & \times \sum_{\substack{j=\ell+1 \\ p_j=-1}}^{2\ell} \widetilde{V}_{(1,0)}(\nu_1,\ldots,\nu_j - 1,\ldots,\nu_{2\ell}) L_2(\nu_1,\ldots,\nu_j - 1,\ldots,\nu_{2\ell}). \end{aligned} \tag{4.53}$$

To show that $V_{(1,0)}(\nu) = 0$, we consider the iterative process by which $V_{(1,0)}(\nu)$ is evaluated. It consists of m additional steps: each summand in (4.53), call it $\widetilde{V}_{(1,0)}(\mu_0)L_2(\mu_0)$, is replaced by a sum of up to 2ℓ terms, each of which has the form $\widetilde{V}_{(1,0)}(\mu_1)M_1(\mu_1)L_2(\mu_0)$, where μ_1 is derived from μ_0 by decreasing exactly one entry by 1 and where $M_1(\mu_1)$ is either $L_1(\mu_1) + 1$ or $L_1(\mu_1)$, then each of these terms is replaced by a sum of up to 2ℓ terms similarly, and so on until ultimately (4.53) is reduced to a sum of terms each of which is of the form

$$C \cdot \widetilde{V}_{(1,0)}(0,\ldots,0) M_m(\mu_m) M_{m-1}(\mu_{m-1}) \cdots M_2(\mu_2) M_1(\mu_1) L_2(\mu_0), \tag{4.54}$$

where $C \in \mathbb{Q}$, where $M_k(\mu_k) = L_1(\mu_k) + 1$ or $M_k(\mu_k) = L_2(\mu_k)$, and where μ_{k+1} is derived from μ_k by decreasing exactly one entry by 1.

Consider the sequence of values $L_2(\mu_0), L_2(\mu_1), \ldots, L_2(\mu_m)$ that is created in forming any such term from a summand in (4.53). For some $j \in \{\ell+1,\ldots,2\ell\}$, $L_2(\mu_0) = L_2(\nu) - p_j = -2 + 1 = -1$; $L_2(\mu_m) = L_2(0,\ldots,0) = 0$. On the rth step, supposing $\mu_r = \mu_{r-1} + (0,\ldots,0,\overset{r}{-1},0,\ldots,0)$, the value of L_2 decreases by $q_r \ge 0$ if $1 \le r \le \ell$, decreases by $p_{2\ell-r+1}$ if $r > \ell$ and $p_{2\ell-r+1} \ge 0$, and increases by 1 if $r > \ell$ and $p_{2\ell-r+1} = -1$. Let w denote the number of the step of the reduction process on which the value of L_2 changes to 0 for the last time; that is, $L_2(\mu_{w-1}) \ne 0$, $L_2(\mu_k) = 0$ for $k \ge w$. If $L_2(\mu_{w-1}) < 0$, then $L_2(\mu_w)$ is obtained from $L_2(\mu_{w-1})$ by an increase of 1, hence the value of the index j of the entry that decreased in order to form μ_w exceeds ℓ, hence $M_w(\mu_w) = L_2(\mu_w) = 0$. If $L_2(\mu_{w-1}) > 0$, then the value of L_2 must have increased across 0 on some earlier sequence of steps. Thus for some $v < w$, $L_2(\mu_{v-1})$ increased to $L_2(\mu_v) = 0$ by an increase of 1 unit, so that on that step the index j of the entry in μ_{v-1} that decreased to form μ_v exceeded

ℓ, so $M_\nu(\mu_\nu) = L_2(\mu_\nu) = 0$. Thus for at least one index k in the product (4.54), $M_k(\mu_k) = L_2(\mu_k) = 0$. Hence $V_{(1,0)}(\nu)$ evaluates to a sum of zeros, and the inductive step is complete, proving the lemma. □

Here then is the analogue of Theorem 3.4.5 for linearizability.

Theorem 4.3.7. *Let a set S, hence families* (4.27) *and* (4.29)*, be given. Define four constant polynomials by* $u_1^{(-1,1)} = u_2^{(1,-1)} = 0$, $u_1^{(0,0)} = u_2^{(0,0)} = 1$, *and by recursion on* $j+k$ *generate polynomials* $u_1^{(j,k)}$ *and* $u_2^{(j,k)}$ *in* $\mathbb{Q}[a,b]$ *using equation* (4.46)*, where the choices* $u_1^{(k,k)} = 0$ *and* $u_2^{(k,k)} = 0$ *are always made. For* $k \in \mathbb{N}$*, let* I_{kk} *and* J_{kk} *be the polynomials defined by equation* (4.47)*. Let* $\widehat{f}$ *denote the conjugate of f as given by Definition 3.4.3.*

1. The coefficient of $[\nu]$ *in the polynomial* $u_1^{(j,k)}$ *and the coefficient of* $[\widehat{\nu}]$ *in the polynomial* $u_2^{(k,j)}$ *are equal to* $V_{(1,0)}(\nu_1, \nu_2, \ldots, \nu_{2\ell})$.

2. The linearizability quantities I_{kk} *and* J_{kk} *of family* (4.27) *are given by*

$$I_{kk} = \sum_{\{\nu : L(\nu)=(k,k)\}} I_{kk}^{(\nu)}[\nu] \qquad \text{and} \qquad J_{kk} = \sum_{\{\nu : L(\nu)=(k,k)\}} J_{kk}^{(\nu)}[\nu],$$

where

$$\begin{aligned} I_{kk}^{(\nu)} = &\sum_{k=1}^{\ell} \widetilde{V}_{(1,0)}(\nu_1,\ldots,\nu_j-1,\ldots,\nu_{2\ell})(L_1(\nu_1,\ldots,\nu_j-1,\ldots,\nu_{2\ell})+1) \\ &- \sum_{j=\ell+1}^{2\ell} \widetilde{V}_{(1,0)}(\nu_1,\ldots,\nu_j-1,\ldots,\nu_{2\ell})L_2(\nu_1,\ldots,\nu_j-1,\ldots,\nu_{2\ell}) \end{aligned} \tag{4.55}$$

and $J_{kk}^{(\nu)} = -\overline{I}_{kk}^{(\widehat{\nu})}$.

Proof. The proofs of the assertion made about $u_1^{(j,k)}$ in point (1) and of the assertion made about I_{kk} in point (2) are identical to the proofs of the corresponding points in the proof of Theorem 3.4.5, where Lemma 4.3.6 is used in place of Lemma 3.4.4 at one point. The assertions concerning $u_2^{(j,k)}$ and J_{kk} are just Proposition 4.3.4, based on the choices $u_1^{(k,k)} = 0$ and $u_2^{(k,k)} = 0$ and properties of the conjugation. □

Thus we see that the linearizability quantities can be computed by practically the same formulas that we obtained for the focus quantities in Section 3.4. An algorithm for their computation analogous to that displayed in Table 3.1 on page 128 for the focus quantities is given in Table 4.1 on page 199. An implementation in Mathematica is in the Appendix. As always, if we are interested in linearizability conditions for a real polynomial family of the form (4.1), we can obtain them by computing linearizability conditions for the family of systems (4.27) on $\mathbb{C}^2$ that arises by complexification of the original real family and then replacing every occurrence of b_{kj} by $\bar{a}_{jk}$.

Linearizability Quantities Algorithm

Input:

$K \in \mathbb{N}$
Ordered set $S = \{(p_1,q_1),\ldots,(p_\ell,q_\ell)\} \subset (\{-1\} \times \mathbb{N}_0)^2$
satisfying $p_j + q_j \geq 1$, $1 \leq j \leq \ell$

Output:

Linearizability quantities I_{kk}, J_{kk}, $1 \leq k \leq K$, for family (4.27)

Procedure:

$w := \min\{p_1+q_1,\ldots,p_\ell+q_\ell\}$
$M := \lfloor \frac{2K}{w} \rfloor$
$I_{11} := 0; \ldots, I_{KK} := 0;$
$J_{11} := 0; \ldots, J_{KK} := 0;$
$V_{(1,0)}(0,\ldots,0) := 1;$
FOR $m = 1$ TO M DO
FOR $\nu \in \mathbb{N}_0^{2\ell}$ such that $|\nu| = m$
DO
Compute $L(\nu)$ using (3.71)
Compute $V_{(1,0)}(\nu)$ using (4.52)
IF
$L_1(\nu) = L_2(\nu)$
THEN
Compute $I^{(\nu)}_{L(\nu)}$ using (4.55)
$I_{L(\nu)} := I_{L(\nu)} + I^{(\nu)}_{L(\nu)}[\nu]$
FOR $\nu \in \mathbb{N}_0^{2\ell}$ such that $|\nu| = m$
DO
Compute $L(\nu)$ using (3.71)
IF
$L_1(\nu) = L_2(\nu)$
THEN
$J_{L(\nu)} := -I_{L(\hat{\nu})} + J^{(\nu)}_{L(\nu)}[\nu]$

Table 4.1 The Linearizability Quantities Algorithm

4.4 Darboux Linearization

In Chapter 3 we saw that the problem of finding the center variety of system (3.3) splits into two parts: in the first part we compute an initial string of some number r of the focus quantities until it appears that $\mathbf{V}(\mathscr{B}) = \mathbf{V}(\mathscr{B}_K)$ for some number K, and find the minimal decomposition into irreducible components of the variety $\mathbf{V}(\mathscr{B}_K)$ so obtained (thus deriving a collection of necessary conditions for a center); in the second part we prove that any system from any component of $\mathbf{V}(\mathscr{B}_K)$ actually has a center at the origin, typically by constructing a first integral, and thus proving that the necessary conditions for a center derived in the first part are also sufficient.

The method for finding the linearizability variety of (3.3) (which is the same as (4.27)) is analogous: we begin by computing the first few linearizability quantities and find the irreducible decomposition of the variety obtained, then we check that all systems from the variety are linearizable. In this section we present one of the most efficient tools for performing such a check, namely, the method of Darboux linearization. In the definition that follows, the coordinate transformation in question is actually the inverse of what we have heretofore called the linearization of the original system (see Section 2.3 and in particular the discussion surrounding Definition 2.3.4). We will continue to use this mild abuse of language for the remainder of this section.

Definition 4.4.1. For $\mathbf{x} = (x_1, x_2) \in \mathbb{C}^2$, a *Darboux linearization* of a polynomial system of the form (2.27), $\dot{\mathbf{x}} = A\mathbf{x} + \mathbf{X}(\mathbf{x})$, is an analytic change of variables

$$y_1 = Z(x_1, x_2), \quad y_2 = W(x_1, x_2) \tag{4.56}$$

whose inverse linearizes (2.27) and is such that $Z(x_1, x_2)$ and $W(x_1, x_2)$ are of the form

$$Z(x_1, x_2) = \prod_{j=0}^{m} f_j^{\alpha_j}(x_1, x_2) = x_1 + Z'(x_1, x_2)$$

$$W(x_1, x_2) = \prod_{j=0}^{n} g_j^{\beta_j}(x_1, x_2) = x_2 + W'(x_1, x_2),$$

where f_j, $g_j \in \mathbb{C}[x_1, x_2]$, α_j, $\beta_j \in \mathbb{C}$, and Z' and W' begin with terms of order at least two. A *generalized Darboux linearization* is a transformation (4.56) for which the functions $Z(x_1, x_2)$ and $W(x_1, x_2)$ are Darboux functions (see Definition 3.6.7). A system is *Darboux linearizable* (respectively, *generalized Darboux linearizable*) if it admits a Darboux (respectively, generalized Darboux) linearization.

Theorem 4.4.2. *Fix a polynomial system of the form* (4.29).

1. The system is Darboux linearizable if and only if there exist $s+1 \geq 1$ algebraic partial integrals $f_0, \ldots, f_s$ with corresponding cofactors $K_0, \ldots, K_s$ and $t+1 \geq 1$ algebraic partial integrals $g_0, \ldots, g_t$ with corresponding cofactors $L_0, \ldots, L_t$ with the following properties:

a. $f_0(x_1, x_2) = x_1 + \cdots$ but $f_j(0,0) = 1$ for $j \geq 1$;

b. $g_0(x_1, x_2) = x_2 + \cdots$ but $g_j(0,0) = 1$ for $j \geq 1$; and

c. there are $s+t$ constants $\alpha_1, \ldots, \alpha_s, \beta_1, \ldots, \beta_t \in \mathbb{C}$ such that

$$K_0 + \alpha_1 K_1 + \cdots + \alpha_s K_s = 1 \quad \text{and} \quad L_0 + \beta_1 L_1 + \cdots + \beta_t L_t = -1\,. \tag{4.57}$$

The Darboux linearization is then given by

$$y_1 = Z(x_1, x_2) = f_0 f_1^{\alpha_1} \cdots f_s^{\alpha_s}, \qquad y_2 = W(x_1, x_2) = g_0 g_1^{\beta_1} \cdots g_t^{\beta_t}\,.$$

2. *The system is generalized Darboux linearizable if and only if the same conditions as in part (1) hold, with the following modification: either (i) $s \geq 1$ and f_1 is an exponential factor rather than an algebraic partial integral or (ii) $t \geq 1$ and g_1 is an exponential factor rather than an algebraic partial integral (or both (i) and (ii) hold), and* (4.57) *holds with $\alpha_1 = 1$ (if $s \geq 1$) and $\beta_1 = 1$ (if $t \geq 1$).*

Proof. The vector field $\mathscr{X}$ is a derivation: for smooth functions f and g,

$$\mathscr{X}(fg) = f\mathscr{X}g + g\mathscr{X}f, \quad \mathscr{X}\frac{f}{g} = \frac{g\mathscr{X}f - f\mathscr{X}g}{g^2}, \quad \mathscr{X}e^f = e^f\mathscr{X}f, \tag{4.58}$$

which may be verified using the definition of $\mathscr{X}f$ and straightforward computations.

Suppose that for a polynomial system (4.29) there exist $s+1 \geq 1$ algebraic partial integrals $f_0, \dots, f_s$ and $t+1 \geq 1$ algebraic partial integrals $g_0, \dots, g_t$ that satisfy the conditions (a), (b), and (c). Form the mapping

$$y_1 = Z(x_1,x_2) = f_0 f_1{}^{\alpha_1} \cdots f_s{}^{\alpha_s}, \qquad y_2 = W(x_1,x_2) = g_0 g_1{}^{\beta_1} \cdots g_t{}^{\beta_t}, \tag{4.59}$$

which by the conditions imposed is analytic and by the Inverse Function Theorem has an analytic inverse $x_1 = U(y_1,y_2)$, $x_2 = V(y_1,y_2)$ on a neighborhood of the origin in $\mathbb{C}^2$. For ease of exposition we introduce $\alpha_0 = 1$. Then differentiation of the first equation in (4.59) with respect to t yields

$$\begin{aligned} \dot{y}_1 = \mathscr{X}(f_0{}^{\alpha_0} \cdots f_s{}^{\alpha_s}) &= \sum_{j=0}^{s} f_0{}^{\alpha_0} \cdots \alpha_j f_j{}^{\alpha_j - 1} \mathscr{X}(f_j) \cdots f_s{}^{\alpha_s} \\ &= \sum_{j=0}^{s} \alpha_j K_j f_0{}^{\alpha_0} \cdots f_s{}^{\alpha_s} \\ &= f_0{}^{\alpha_0} \cdots f_s{}^{\alpha_s} \sum_{j=0}^{s} \alpha_j K_j = Z = y_1 . \end{aligned} \tag{4.60}$$

Similarly, $\dot{y}_2 = -y_2$, so the system has been linearized by the transformation (4.59).

If f_1 or g_1 exists and is an exponential factor meeting the conditions of the theorem, the proof of sufficiency is identical. In either case, the pair of identities $Z(x_1,x_2) = x_1 + \cdots$ and $W(x_1,x_2) = x_2 + \cdots$ is forced by the lowest-order terms of the original system and its linearization. Note that this forces the presence of at least one algebraic partial integral, to play the role of f_0. Moreover, there is no loss of generality in assuming $\alpha_1 = 1$ and $\beta_1 = 1$ since a power can be absorbed into the exponential factor.

Conversely, suppose there exists a generalized Darboux linearization of systems (3.3) and (3.69), and let the first component be

$$y_1 = Z(x_1,x_2) = e^{\frac{f}{g}} h_1{}^{\gamma_1} \cdots h_s{}^{\gamma_s} .$$

Without loss of generality we may assume that the polynomials h_j are irreducible and relatively prime and that f and g are relatively prime. Then using (4.58),

$$\dot{y}_1 = \mathscr{X}Z = e^{\frac{f}{g}}\left(\mathscr{X}\frac{f}{g}\right)\prod_{j=1}^{s} h_j{}^{\gamma_j} + e^{\frac{f}{g}}\sum_{k=1}^{s}\Big(\prod_{\substack{j=1\\ j\neq k}}^{s} h_j{}^{\gamma_j}\Big)\gamma_k h_k{}^{\gamma_k-1}\mathscr{X}h_k = e^{\frac{f}{g}}\prod_{j=1}^{s} h_j{}^{\gamma_j}.$$

For any index value w, each term on each side of the last equation contains $h_w{}^{\gamma_w-1}$ as a factor, hence we may divide through by $e^{f/g}h_1{}^{\gamma_1-1}\cdots h_s{}^{\gamma_s-1}$ to obtain

$$\left(\mathscr{X}\frac{f}{g}\right)h_1\cdots h_s + \sum_{k=1}^{s}\Big(\prod_{\substack{j=1\\ j\neq k}}^{s} h_j\Big)\gamma_k\mathscr{X}h_k = h_1\cdots h_s.$$

For any index value w, every term except the summand corresponding to $k = w$ contains h_w as a factor, hence is divisible by h_w. Thus the summand corresponding to $k = w$ is divisible by h_w as well, and since the polynomials h_j are relatively prime, h_w divides $\mathscr{X}h_w$. This means that h_w is an algebraic partial integral, say with cofactor K_w, and the equation reduces to

$$\mathscr{X}\frac{f}{g} + \sum_{k=1}^{s}\gamma_k K_k = 1.$$

Using the expression for $\mathscr{X}\frac{f}{g}$ given in (4.58) and multiplying by g^2, we obtain

$$(g\mathscr{X}f - f\mathscr{X}g) + g^2\sum_{k=1}^{s}\gamma_k K_k = g^2,$$

which implies that the term $-f\mathscr{X}g$ is divisible by g, say $f\mathscr{X}g = fgL$, and upon dividing by g again, we have

$$(\mathscr{X}f - fL) + g\sum_{k=1}^{s}\gamma_k K_k = g.$$

But now the term in parentheses must be divisible by g, say $\mathscr{X}f - fL = gK$, and upon one more division by g we have

$$K + \sum_{k=1}^{s}\gamma_k K_k = 1. \tag{4.61}$$

Moreover,

$$\mathscr{X}\frac{f}{g} = \frac{g\mathscr{X}f - f\mathscr{X}g}{g^2} = \frac{g\mathscr{X}f - fgL}{g^2} = \frac{\mathscr{X}f - fL}{g} = \frac{gK}{g} = K$$

and, by (4.61), $\deg(K) \le \max_{1\le j\le s}\{\deg(K_j)\} \le m-1$, where m is the degree of the original system, so $e^{f/g}$ is an exponential factor with cofactor K.

Since $Z(x_1,x_2) = x_1 + \cdots$ and $W(x_1,x_2) = x_2 + \cdots$, $e^{f/g} = 1 + \cdots$, and all the h_j are polynomials, for exactly one index value w the polynomial h_w must have the form $h_w(x_1,x_2) = x_1 + \cdots$, γ_w must be 1, and for $j \neq w$, h_j must have the form $h_j(x_1,x_2) = 1 + \cdots$. Thus, by (4.61), condition (a) and the first condition in (4.57) hold, with h_w playing the role of f_0.

If $f = 0$, then every line of the argument remains true without modification.

The discussion for the second component of the linearizing transformation is identical, since the minus sign has no bearing on the argument, except to reverse the sign on the right-hand side of equation (4.61), thereby yielding the second condition in display (4.57). □

Remark. Note that a single algebraic partial integral can serve both as one of the f_j and as one of the g_j. This fact is illustrated by system (4.69) below.

The following two theorems show that even if we are unable to find sufficiently many algebraic partial integrals to construct a linearizing transformation for system (4.27) by means of Theorem 4.4.2, it is sometimes possible to construct a linearizing transformation if, in addition, we use a first integral of the system, which must exist since we are operating on the assumption that system (4.27) has a center at the origin. In the first theorem we suppose that we can find algebraic partial integrals (and possibly an exponential factor) satisfying just one or the other of the conditions (4.57). For simplicity we state the theorem only for the case that we have just $f_0,\ldots,f_s$ meeting the first equation in (4.57).

Theorem 4.4.3. *Suppose system* (4.29) *has a center at the origin, hence possesses a formal first integral* $\Psi(x_1,x_2)$ *of the form* $\Psi = x_1x_2 + \cdots$, *and that there exist algebraic partial integrals and possibly an exponential factor,* $f_0,\ldots,f_s$, *that meet condition (1.a) in Theorem 4.4.2 and satisfy the first of equations* (4.57), *possibly as modified in (2) of Theorem 4.4.2. Then system* (4.29) *is linearized by the transformation*

$$\begin{aligned} y_1 &= Z(x_1,x_2) = f_0 \prod_{j=1}^{s} f_j^{\alpha_j} = x_1 + \cdots, \\ y_2 &= W(x_1,x_2) = \frac{\Psi}{Z(x_1,x_2)} = x_2 + \cdots. \end{aligned} \tag{4.62}$$

Proof. Recall from Corollary 3.2.6 that if a formal first integral Ψ of the form (3.52) exists, then there exists an analytic first integral of the same form, which we still denote by Ψ. Condition 1(a) in Theorem 4.4.2 ensures that the transformation (4.62) is analytic and has an analytic inverse on a neighborhood of the origin in $\mathbb{C}^2$. The computation (4.60) is valid and gives $\dot{y}_1 = \mathscr{X}Z = Z = y_1$. As in the proof of Theorem 4.4.2, set $\alpha_0 = 1$. Then by (4.58) and the fact that Ψ is a first integral,

$$\dot{y}_2 = \frac{\prod_{j=0}^{s} f_j^{\alpha_j}\mathscr{X}\Psi - \Psi\mathscr{X}\prod_{j=0}^{s} f_j^{\alpha_j}}{\left(\prod_{j=0}^{s} f_j^{\alpha_j}\right)^2} = \frac{-\Psi\prod_{j=0}^{s} f_j^{\alpha_j}}{\left(\prod_{j=0}^{s} f_j^{\alpha_j}\right)^2} = -y_2\,. \ \square$$

In the second theorem one of the conditions on system (4.29) is that the coordinate axes be invariant curves through the center. In the context of systems of differential equations on $\mathbb{C}^2$, such a curve is termed a *separatrix* of the center, which underscores the contrast with real systems.

Theorem 4.4.4. *Suppose that system* (4.29) *has a center at the origin, hence possesses a formal first integral* $\Psi(x_1,x_2)$ *of the form* $\Psi = x_1x_2 + \cdots$, *that* $\widetilde{P}$ *contains* x_1 *as a factor and* $\widetilde{Q}$ *contains* x_2 *as a factor, and that there exist* s *algebraic partial integrals and exponential factors* $f_1,\ldots,f_s$ *with corresponding cofactors* $K_1,\ldots,K_s$ *and* t *algebraic partial integrals and exponential factors* $g_1,\ldots,g_t$ *with corresponding cofactors* L_j *with the following properties:*
a. $f_j(0,0) = 1$ *for* $1 \le j \le s$;
b. $g_j(0,0) = 1$ *for* $1 \le j \le t$;
c. there exist $s+t+2$ *constants* a, b, $\alpha_1,\ldots,\alpha_s,\beta_1,\ldots,\beta_t \in \mathbb{C}$ *such that*

$$(1-a)x_1^{-1}\widetilde{P} - ax_2^{-1}\widetilde{Q} + \sum_{j=1}^{s} \alpha_j K_j = 1 \tag{4.63a}$$

and

$$-bx_1^{-1}\widetilde{P} + (1-b)x_2^{-1}\widetilde{Q} + \sum_{j=1}^{t} \alpha_j L_j = -1\,. \tag{4.63b}$$

Then (4.29) *is linearized by the transformation*

$$\begin{aligned} y_1 &= Z(x_1,x_2) = x_1^{1-a}x_2^{-a}\Psi^a f_1^{\alpha_1}\cdots f_s^{\alpha_s} \\ y_2 &= W(x_1,x_2) = x_1^{-b}x_2^{1-b}\Psi^b g_1^{\beta_1}\cdots g_t^{\beta_t}\,. \end{aligned} \tag{4.64}$$

Proof. Conditions (4.63) imply that $\widetilde{P}$ does not contain any term of the form x_2^k and that $\widetilde{Q}$ does not contain any term of the form x_1^k. By an inductive argument using (3.55), it follows that any first integral (3.52) of the system must have the form

$$\Psi(x_1,x_2) = x_1x_2\Big(1 + \sum_{\substack{k+j=1\\ k,j\ge 0}}^{\infty} v_{k,j}x_1^k x_2^j\Big).$$

This fact together with conditions (a) and (b) implies that the mapping (4.64) is of the form $y_1 = Z(x_1,x_2) = x_1 + \cdots$, $y_2 = W(x_1,x_2) = x_2 + \cdots$, hence is analytic and has an analytic inverse on a neighborhood of the origin in $\mathbb{C}^2$. Computations just like (4.60) and using (4.63) show that (4.64) linearizes (4.29). □

As was the case with Theorem 4.4.2, a single function can play the role of one of the functions f_j and one of the functions g_j, as is the case in the following example.

Example. Consider the system

$$\begin{aligned} \dot{x} &= x(1 - 6b_{10}x + 8b_{10}^2x^2 - 2b_{02}y^2) \\ \dot{y} &= -y(1 - b_{10}x - b_{02}y^2) \end{aligned} \tag{4.65}$$

on $\mathbb{C}^2$ with $b_{10}b_{02} \neq 0$. This system has algebraic partial integrals

$$h_1 = 1 - 4b_{10}x + 8b_{02}b_{10}xy^2, \quad h_2 = 1 - 12b_{10}x + 48b_{10}^2x^2 - 64b_{10}^3x^3 + 24b_{02}b_{10}xy^2$$

with respective cofactors

$$M_1 = -4b_{10}x + 8b_{10}^2x^2, \quad M_2 = -12b_{10}x + 24b_{10}^2x^2.$$

Since $3M_1 + M_2 \equiv 0$, by Theorem 3.6.8 a Darboux first integral is given by the analytic function $\Phi(x,y) = h_1^{-3}h_2 = 1 + 192b_{02}b_{10}^2x^2y^2 + \cdots$. Thus

$$\Psi(x,y) = \frac{1}{\sqrt{192\, b_{02}\, b_{10}^2}}\sqrt{f_2 f_1^{-3} - 1}$$

is a first integral of the form (3.52). Taking $f_1 = g_1 = h_1$ and $f_2 = g_2 = h_2$, equations (4.63) are in this instance a pair of systems of linear equations with infinitely many solutions, among them $a = 2$, $\alpha_1 = 1$, $\alpha_2 = 0$ and $b = -1$, $\beta_1 = -1$, $\beta_2 = 0$. Thus, by Theorem 4.4.4, system (4.65) is linearized by the inverse of the transformation

$$z = \frac{f_1(x,y)\Psi^2(x,y)}{xy^2} = x + \cdots, \qquad w = \frac{xy^2}{f_1(x,y)\Psi(x,y)} = y + \cdots.$$

Two more examples of linearizable polynomial systems are given in Exercises 4.12 and 4.13. In the next section we apply the theory developed in this section to the question of linearizability of quadratic systems.

4.5 Linearizable Quadratic Centers

This section is devoted to finding the linearizability variety for the full family of quadratic systems of the form (4.29) (= (3.69)):

$$\begin{aligned} \dot{x} &= \quad x - a_{10}x^2 - a_{01}xy - a_{-12}y^2 \\ \dot{y} &= -y + b_{2,-1}x^2 + b_{10}xy + b_{01}y^2 . \end{aligned} \tag{4.66}$$

This is the same family for which the center variety was obtained in Section 3.7, as described in Theorem 3.7.1.

Theorem 4.5.1. *The linearizability variety of family* (4.66) *consists of the following nine irreducible components:*

1. $V_1 = \mathbf{V}(J_1)$ *where* $J_1 = \langle a_{01}, b_{01}, b_{2,-1}, a_{10} + 2b_{10}\rangle$;
2. $V_2 = \mathbf{V}(J_2)$ *where* $J_2 = \langle a_{10}, a_{-12}, b_{10}, 2a_{01} + b_{01}\rangle$;
3. $V_3 = \mathbf{V}(J_3)$ *where* $J_3 = \langle a_{01}, a_{-12}, b_{01}\rangle$;
4. $V_4 = \mathbf{V}(J_4)$ *where* $J_4 = \langle b_{10}, b_{2,-1}, a_{10}\rangle$;
5. $V_5 = \mathbf{V}(J_5)$ *where* $J_5 = \langle -7b_{10}^2 + 12b_{2,-1}b_{01}, 49a_{-12}b_{10} + 18b_{01}^2, 14a_{-12}b_{2,-1} + 3b_{10}b_{01}, 7a_{01} + 6b_{01}, 6a_{10} + 7b_{10}\rangle$;
6. $V_6 = \mathbf{V}(J_6)$ *where* $J_6 = \langle 15b_{10}^2 + 4b_{2,-1}b_{01}, 25a_{-12}b_{10} - 6b_{01}^2, 10a_{-12}b_{2,-1} + 9b_{10}b_{01}, 5a_{01} + 2b_{01}, 2a_{10} + 5b_{10}\rangle$;
7. $V_7 = \mathbf{V}(J_7)$ *where* $J_7 = \langle b_{2,-1}, a_{-12}, a_{01} + b_{01}, a_{10} + b_{10}\rangle$;

8. $V_8 = \mathbf{V}(J_8)$ *where* $J_8 = \langle a_{01}, b_{10}, b_{2,-1} \rangle$;
9. $V_9 = \mathbf{V}(J_9)$ *where* $J_9 = \langle b_{10}, a_{01}, a_{-12} \rangle$.

Proof. In analogy to what was done in Section 3.7 when we found the center variety for this family, we compute the first few pairs of linearizability quantities until a pair or two are found to lie in the radical of the ideal generated by the earlier pairs. The actual computation, which was done by means of the algorithm based on Theorem 4.3.7, led us to suspect that the first three pairs of linearizability quantities, which we will not list here, form a basis of $\mathscr{L}$, indicating that $\mathbf{V}(\mathscr{L}) = \mathbf{V}(\mathscr{L}_3)$. Using the Singular routine `minAssGTZ` (which computes the minimal associated primes of polynomial ideals by means of the algorithm of [81]), we found that the minimal associate primes of $\mathscr{L}_3$ are the nine ideals written out in the right-hand sides of the equations for the V_j above. Thus $\mathbf{V}(\mathscr{L}) \subset \cup_{j=1}^{9} V_j$.

To prove the reverse inclusion, we have to show that for every system from V_j for $j \in \{1,\ldots,9\}$, there is a transformation $z = x + \cdots$, $w = y + \cdots$ that reduces (4.66) to the linear system $\dot{z} = z$, $\dot{w} = -w$. We will find that in every case either Theorem 4.4.2 or Theorem 4.4.3 applies. To begin, we observe that each variety V_1 through V_9 lies in some irreducible component of the center variety $V_{\mathscr{C}}$ as identified in Theorem 3.7.1 (Exercise 4.14), so that there must exist an analytic first integral Ψ of the form $\Psi(x,y) = xy + \cdots$ for each of the corresponding systems. We also observe that the polynomials defining V_2, V_4, and V_9 are conjugate to the polynomials defining V_1, V_3, and V_8, respectively, so by reasoning analogous to that presented in the proof of Theorem 3.7.2, it is sufficient to consider systems from just the components V_1, V_3, V_5, V_6, V_7, and V_8.

The component V_1. Systems from the component V_1 have the form

$$\dot{x} = x + 2b_{10}x^2 - a_{-12}y^2, \quad \dot{y} = -y + b_{10}xy. \tag{4.67}$$

In an attempt to apply Theorem 4.4.2, we search for algebraic partial integrals, beginning with those of degree one, whose cofactors have degree at most one. Applying the technique of undetermined coefficients (first described in Example 3.6.10), we find that any algebraic partial integral of (4.67) of degree one that is valid for the whole family (and not just for a special case, such as $a_{-12} = 0$) has the form $h_0 = cy$, $c \in \mathbb{R} \setminus \{0\}$, with corresponding cofactor $M_0 = -1 + b_{10}x$. A search for algebraic partial integrals of the second degree, for which the cofactor M will be a polynomial of degree at most one, yields (up to multiplication by a nonzero constant) the two polynomials $h_1 = 1 + 2b_{10}x - a_{-12}b_{10}y^2$ and $h_2 = x - (a_{-12}y^2)/3$ for system (4.67), with the respective cofactors $M_1 = 2b_{10}x$ and $M_2 = 1 + 2b_{10}x$. The form of h_2 as $h_2(x,y) = x + \cdots$ suggests that we let it play the role of f_0 in Theorem 4.4.2 and attempt to find α_j solving the first equation in (4.57). Since $M_2 + (-1)M_1 = 1$, we choose h_1 for f_1 and $\alpha_1 = -1$. The form of h_0 as $h_0(x,y) = y + \cdots$ suggests that we attempt to find β_j solving the second equation in (4.57). Since $M_o + (-1/2)M_1 = -1$, we choose h_1 for g_1 and $\beta_1 = -1/2$. Thus, by Theorem 4.4.2, any system in V_1 is linearized by the inverse of the transformation

$$z = h_2(x,y)[h_1(x,y)]^{-1} = \frac{x - (a_{-12}y^2)/3}{1 + 2b_{10}x - a_{-12}b_{10}y^2} = x + \cdots$$
$$w = h_0(x,y)[h_1(x,y)]^{-1/2} = \frac{y}{\sqrt{1 + 2b_{10}x - a_{-12}b_{10}y^2}} = y + \cdots.$$

To finish the proof for component V_1, see Exercise 4.16.

The component V_3. Systems from the component V_3 have the form

$$\dot{x} = x - a_{10}x^2, \quad \dot{y} = -y + b_{2,-1}x^2 + b_{10}xy. \tag{4.68}$$

The first equation is independent of y, so we may apply the one-dimensional analogue of Theorem 4.4.2: first-degree algebraic partial integrals are $f_0(x) = x$ and $f_1(x) = 1 - a_{10}x$ with corresponding cofactors $K_0(x) = 1 - a_{10}x$ and $K_1(x) = -a_{10}x$ that satisfy $K_0 + \alpha_1 K_1 = 1$ when $\alpha_1 = -1$ so that $\dot{x} = x - a_{10}x^2$ is transformed into $\dot{z} = z$ by $z = Z(x) = f_0(x)[f_1(x)]^{-1} = x/(1 - a_{10}x)$. Since a system (a,b) from V_3 need not be in $\mathbf{V}(J_2) \cup \mathbf{V}(J_3)$ for the irreducible components $\mathbf{V}(J_2)$ and $\mathbf{V}(J_3)$ of the center variety, as identified in Theorem 3.7.1, we do not expect to find algebraic partial integrals from which we could build an explicit first integral. Thus simply noting that any system (4.68) lies in $\mathbf{V}(I_{\text{sym}})$ (component V_4 in Theorem 3.7.1), although it is not time-reversible except in the trivial case that it is already linear (see Exercises 4.14 and 4.15), we know that it must possess an analytic first integral $\Psi = xy + \cdots$, hence, by Theorem 4.4.3, is linearized by

$$z = Z(x,y) = \frac{x}{1 - a_{10}x}$$
$$w = W(x,y) = \Psi(x,y)\frac{f_1(x,y)}{f_0(x,y)} = \Psi(x,y)\frac{1 - a_{10}x}{x}.$$

The component V_5. Systems from V_5 for which $a_{01}b_{10} \neq 0$ have the form

$$\dot{x} = x + \tfrac{7}{6}b_{10}x^2 - a_{01}xy + \frac{a_{01}^2}{2b_{10}}y^2, \quad \dot{y} = -y - \frac{b_{10}^2}{2a_{01}}x^2 + b_{10}xy - \tfrac{7}{6}a_{01}y^2. \tag{4.69}$$

By the usual method of undetermined coefficients, we ascertain that any such system has the irreducible algebraic partial integrals and cofactors

$$\begin{aligned}
h_1 &= 1 + \tfrac{2}{3}b_{10}x + \tfrac{2}{3}a_{01}y & M_1 &= \tfrac{2}{3}(b_{10}x - a_{01}y)\\
h_2 &= 6b_{10}x + b_{10}^2x^2 + 2a_{01}b_{10}xy + a_{01}^2y^2 & M_2 &= 1 + \tfrac{4}{3}b_{10}x - \tfrac{4}{3}a_{01}y\\
h_3 &= 6a_{01}y + b_{10}^2x^2 + 2a_{01}b_{10}xy + a_{01}^2y^2 & M_3 &= -1 + \tfrac{4}{3}b_{10}x - \tfrac{4}{3}a_{01}y.
\end{aligned}$$

Since $1 \cdot M_2 + (-2)M_1 = 1$ and $1 \cdot M_3 + (-2)M_1 = -1$ and $\tilde{h}_2 = h_2/(6b_{10})$ and $\tilde{h}_3 = h_3/(6a_{01})$ are algebraic partial integrals with respective cofactors M_2 and M_3, Theorem 4.4.2 applies with $f_0 = \tilde{h}_2$, $f_1 = h_0$, $g_0 = \tilde{h}_3$, and $g_1 = h_1$ to yield the Darboux linearization

$$z = \frac{1}{6b_{10}} h_2 h_1^{-2} = \frac{6b_{10}x + b_{10}^2 x^2 + 2a_{01}b_{10}xy + a_{01}^2 y^2}{6b_{10}(1 + \frac{2}{3}b_{10}x + \frac{2}{3}a_{01}y)^2}$$

$$w = \frac{1}{6a_{01}} h_3 h_1^{-2} = \frac{b_{10}^2 6a_{01}y + x^2 + 2a_{01}b_{10}xy + a_{01}^2 y^2}{6a_{01}(1 + \frac{2}{3}b_{10}x + \frac{2}{3}a_{01}y)^2}$$

for systems from V_5 for which $a_{01}b_{10} \neq 0$. Since $\overline{V_5 \setminus \mathbf{V}(a_{01}b_{10})} = V_5$, as can be shown either by computing the quotient of the corresponding ideals or by following the strategy outlined in Exercise 4.18 for the analogous equality that arises for component V_1, we conclude that a linearization exists for every system from V_5.

The arguments for the components V_6, V_7, and V_8 are similar and are left as Exercise 4.21. □

4.6 Notes and Complements

The history of isochronicity is as old as the history of clocks based on some sort of periodic motion, such as the swinging of a pendulum. Based on his knowledge that the cycloid is a tautochrone (a frictionless particle sliding down a wire in the shape of a cycloid reaches the lowest point in the same amount of time, regardless of its starting position) and that the evolute of a cycloid is a cycloid, in the 17th century Huygens designed and built a pendulum clock with cycloidal "cheeks." This is probably the earliest example of a nonlinear isochronous system.

Interest in isochronicity in planar systems of ordinary differential equations was renewed in the second half of the 20th century. In the early 1960s a criterion for isochronicity was obtained by Urabe ([189, 190]) for the simple but physically important "kinetic + potential" Hamiltonian system $\dot{x} = H_x = y$, $\dot{y} = -H_y = -g(x)$, where $f(x)$ and $g(x)$ are smooth functions in a neighborhood of the origin and $H(x,y) = y^2/2 + \int_0^x g(s)\,ds$. Urabe's method can also be applied to study isochronicity in the system

$$\dot{x} = y, \qquad \dot{y} = -g(x) - f(x)y^2$$

and some families of polynomial systems ([47, 48, 162]). Efficient criteria for isochronicity in Liénard systems

$$\dot{x} = y, \qquad \dot{y} = -g(x) - f(x)y \tag{4.70}$$

have been obtained by Sabatini ([160]) and Christopher and Devlin ([52]). In [52] the authors classified all isochronous polynomial Liénard systems (4.70) of degree 34 or less.

The method of Darboux linearization presented in this chapter indeed is based on an idea of Darboux, but was applied to the linearization of differential equations only in 1995, in the work of Mardešić, Rousseau, and Toni ([136]). See such works as [57, 135] for further developments. Theorem 4.5.1 on the linearization of the quadratic system (4.66) is due to Christopher and Rousseau ([58]). The problem of isochronicity for real quadratic systems was solved earlier by Loud ([127]). Systems with only homogeneous cubic nonlinearities were treated by Pleshkan ([141]). Necessary and sufficient conditions for linearizability of the system $\dot{x} = x + P(x,y)$, $\dot{y} = -y + Q(x,y)$, where P and Q are homogeneous polynomials of degree five, were obtained in [148] .

The problems of isochronicity and linearizability for certain families of time-reversible polynomial systems have been considered in [21, 32, 36, 40, 145]. There are also many works concerning the problems of linearizability and isochronicity for various particular families of polynomial systems.

In this chapter we gave the proof that isochronicity of a planar system with a center is equivalent to its linearizability. Isochronicity has also been characterized in terms of commuting systems. Vector fields $\mathscr{X}$ and $\mathscr{Z}$ are said to commute if their Lie bracket $[\mathscr{X}, \mathscr{Z}]$ is identically zero. It was proved by Ladis ([107]; see also [2, 161, 193]) that system (4.1) has an isochronous center at the origin if and only if there exists a holomorphic system $\dot{u} = u + M(u,v)$, $\dot{v} = v + N(u,v)$ such that the associated vector fields commute. This result was generalized by Giné and Grau ([82]), who showed that the smooth (respectively, analytic) system $\dot{\mathbf{x}} = \mathbf{f}(\mathbf{x})$ with a non-degenerate singular point at the origin and with the associated smooth (respectively, analytic) vector field $\mathscr{X} = \sum_{j=1}^n f_j(\mathbf{x})\partial/\partial x_j$ is linearizable if and only if there exists a smooth (respectively, analytic) vector field $\mathscr{Y} = \sum_{j=1}^n (x_j + o(|\mathbf{x}|^2))\partial/\partial x_j$ such that $[\mathscr{X}, \mathscr{Y}] \equiv 0$. Along with Darboux linearization, the construction of a commuting system is a powerful method for proving the isochronicity or linearizability of particular families of polynomial systems.

The concept of isochronicity can also be extended to foci. Briefly put, a focus of an analytic system is isochronous if there is a local analytic change of variables η for which $d\eta(\mathbf{0})$ is the identity and is such that in a subsequent change to polar coordinates the $\dot{\varphi}$ equation has no r dependence. The reader is referred to [6, 7, 83, 84] and the references they contain.

Finally, there is also the concept of strong isochronicity: a system with an antisaddle at the origin is strongly isochronous of order n if in polar coordinates (r, φ) there exist n rays $L_j = \{(r, \varphi) : r \geq 0,\ \varphi = \varphi_0 + 2j\pi/n\}$, $j = 0, \ldots, n-1$, such that the time required for any trajectory sufficiently near the origin to pass from L_j to L_{j+1} is $2\pi/n$. If the system is strongly isochronous of order $n = 2$ with respect to the initial polar ray $\varphi_0 = \pi/2$, then it is called strongly isochronous. Strong isochronicity has been investigated by Amel'kin and his coworkers (see [4, 7] and references therein).

For a survey on the problem of isochronicity, consult [38].

Exercises

4.1 Find the error in the following argument: Every quadratic system is linearizable. For suppose family (4.27) has only quadratic nonlinearities. There are no resonant terms, hence the normal form is $\dot{y}_1 = iy_1$, $\dot{y}_2 = -iy_2$.

4.2 [Referenced in Proposition 4.2.10, proof.] Suppose f is a (k,k)-polynomial, g is a (j,j)-polynomial, and $r \in \mathbb{N}$. Show that f^r is an (rk,rk)-polynomial and that fg is a $(k+j,k+j)$-polynomial.

4.3 a. Show that any (k,k)-polynomial $f = \sum_{\{\nu:L(\nu)=(k,k)\}} f^{(\nu)}[\nu]$ can be written in the form $f = \sum_{\{\nu:L(\nu)=(k,k)\}} (f^{(\nu)}[\nu] + f^{(\widehat{\nu})}[\widehat{\nu}])$.
Hint. Let $\nu_1,\ldots,\nu_m$ be those elements of $\{\nu : L(\nu) = (k,k)\}$ for which $\widehat{\nu} \neq \nu$ and let $\mu_1,\ldots,\mu_n$ be those that are self-conjugate. Write out $f+f$ and pair each term with the term with the conjugate monomial.
b. Derive Corollary 3.4.6 from (a), Theorem 3.4.2(4), and (3.78a).

4.4 Follow the procedure outlined in the proof of Proposition 4.2.10 to derive the expressions for the first four isochronicity quantities given by (4.42).

4.5 [Referenced in Proposition 4.2.11, proof.] Prove that if polynomials f and g have the form $f = \sum_{\nu \in F} f^{(\nu)}([\nu]+[\widehat{\nu}])$ and $g = \sum_{\nu \in F} g^{(\nu)}([\nu]+[\widehat{\nu}])$ for some finite indexing set F, then for all $r,s \in \mathbb{N}_0$, the polynomial $f^r g^s$ has the same form.
Hint. It is enough to demonstrate the result just for the two cases $(r,s) = (r,0)$ and $(r,s) = (1,1)$. In the former case recall that $[\nu]^k = [k\nu]$ and $[\nu][\mu] = [\nu\mu]$ and use the Binomial Theorem.

4.6 a. Use the Normal Form Algorithm in Table 2.1 on page 75 to compute the first few coefficients $Y_1^{(k+1,k)}$ for the family of systems (4.27) with a full set of homogeneous cubic nonlinearities; this is the complexification of the system whose complex form is given by (6.50). Compare your answer to the list given in Section 6.4 on page 292.
b. Insert the results of the computations in part (a) into (4.42) to obtain the first few polynomials p_{2k} explicitly in terms of the parameters (a,b).

4.7 Let k be a field and V a variety in k^n. Prove that two polynomials f and g in $k[x_1,\ldots,x_n]$ are in the same equivalence class in $k[V]$ if and only if they are in the same equivalence class in $k[x_1,\ldots,x_n]/\mathbf{I}(V)$.

4.8 In the context of the previous problem, prove that the mapping φ from $k[V]$ to $k[x_1,\ldots,x_n]/\mathbf{I}(V)$ defined by $\varphi([[f]]) = [f]$ is an isomorphism of rings.

4.9 In contrast with the previous problem, show that in $k[x]$ if $I = \langle x^2 \rangle$, then $k[\mathbf{V}(I)]$ and $k[x]/I$ are not isomorphic.
Hint. One is an integral domain.

4.10 Derive the equality $\mathbf{V}(P) \cap V_{\mathscr{C}} = \mathbf{V}(\mathscr{Y})$, a slightly improved version of the first equality in Proposition 4.2.14, directly from Proposition 4.2.7 and the identity $\mathscr{H} = P$.

4.11 Supply the inductive argument for the proof of part (1) of Proposition 4.3.4.

4.12 Prove that the system

$$\dot{x} = x - a_{13}x^2y^3 - a_{04}xy^4 - y^5, \quad \dot{y} = -y + b_{13}xy^4 + b_{04}y^5$$

is linearizable.
Hint. Show that there exists a Lyapunov first integral for this system of the form $\Psi(x,y) = \sum_{k=1}^{\infty} g_k(x)y^k$, where $g_1(x) = x$, $g_2(x) = x^2$, and $g_k(x)$ are polynomials of degree k, and that a linearization for the second equation of the system

can be constructed in the form $w = \sum_{k=1}^{\infty} f_k(x)y^k$, where $f_k(x)$, $k = 2,3,\ldots$, are polynomials of degree $k-1$ and $f_1(x) \equiv 1$.

4.13 Find a linearization of the system

$$\dot{x} = x - x^5 - \frac{4}{b}x^4y - xy^4, \quad \dot{y} = -y + x^4y + bxy^4 + y^5.$$

Hint. Use Theorem 4.4.4.

4.14 With reference to the family (4.66) of quadratic systems, show that the components V_8 and V_9 of the linearizability variety $\mathbf{V}(\mathscr{L})$, as identified in Theorem 4.5.1, lie in the irreducible component $\mathbf{V}(J_2)$ of the center variety $V_{\mathscr{C}}$, as identified in Theorem 3.7.1. Show that V_1 and V_2 lie in the component $\mathbf{V}(J_3)$ of $V_{\mathscr{C}}$ and that the remaining components of $\mathbf{V}(\mathscr{L})$ lie in $\mathbf{V}(J_4) = \mathbf{V}(I_{\text{sym}})$ of $V_{\mathscr{C}}$.

4.15 Using the previous exercise, apply Theorem 3.5.8(2) to systems in family (4.66) corresponding to component V_3 in Theorem 4.5.1 to obtain further examples of systems in $\mathbf{V}(I_{\text{sym}})$ that are not time-reversible.

4.16 a. In the proof of Theorem 4.5.1, explain why the argument that every system in V_1 is linearizable is not valid when $b_{10} = 0$.
b. Show that the result stated in that part of the proof, that the inverse of the mapping $z = h_2/h_1$, $w = h_0/h_1^{1/2}$ is a linearization, nevertheless is valid when $b_{10} = 0$.

4.17 a. Use Theorem 3.6.8 and the algebraic partial integrals found in the proof of Theorem 4.5.1, that every element of V_1 is linearizable, to construct an explicit first integral Ψ for systems in V_1, and by means of it use Theorem 4.4.3 to rederive the linearization of elements of V_1.
b. When $b_{10} = 0$, the function h_1 is no longer an algebraic partial integral. Is the mapping Ψ still a first integral?

4.18 In the proof of Theorem 4.5.1, that every system in V_1 is linearizable, both the proof in the text using Theorem 4.4.2 and the proof in Exercise 4.17 using Theorem 4.4.3, the case $b_{10} = 0$ was anomalous, but we were able to verify the linearizability of systems in V_1 directly because the expressions simplified significantly when $b_{10} = 0$. For a different and somewhat more general approach to the situation, show that a linearization must exist for systems in V_1 with $b_{10} = 0$ (not finding it explicitly) in the following steps. Let $E(a,b) = \mathbb{C}^6$ be the parameter space for system (4.66), with the usual topology.
a. Show that in $E(a,b)$, $C\ell(V_1 \setminus \mathbf{V}(b_{10})) = V_1$.
b. Use the result of part (a) and Exercise 1.42 to show that $V_1 \subset \overline{V_1 \setminus \mathbf{V}(\langle b_{10}\rangle)}$.
c. Prove that $\overline{V_1 \setminus \mathbf{V}(\langle b_{10}\rangle)} = V_1$.
d. Use part (c) and Proposition 1.3.20 to conclude that a linearization exists for every system from V_1, including those for which $b_{10} = 0$.

4.19 In the context of the previous exercises, give an alternate proof that systems in V_1 for which $b_{10} = 0$ are linearizable based on Theorem 4.4.3 and the fact that the $\dot{y}$ equation is already linear.

4.20 Consider the family of systems of the form (4.27) that have only homogeneous cubic nonlinearities

$$\begin{aligned} \dot{x} &= \quad i(x - a_{20}x^3 - a_{11}x^2y - a_{02}xy^2 - a_{-13}y^3) \\ \dot{y} &= -i(y - b_{3,-1}x^3 - b_{20}x^2y - b_{11}xy^2 - b_{02}y^3)\,. \end{aligned} \tag{4.71}$$

a. Find the first four pairs of linearizability quantities for family (4.71).
b. Using the Radical Membership Test, verify that the third pair found in part (a) does not lie in the radical of the ideal generated by the previous pairs, but that the fourth pair does. (If you are so inclined, you could compute the fifth pair and verify that it lies in the ideal generated by the previous pairs, too.)
c. From part (a) you have the ideal $\mathscr{L}_4$ and in part (b) a computation that suggests that $\mathbf{V}(\mathscr{L}) = \mathbf{V}(\mathscr{L}_4)$. Using the Singular routine `primdecGTZ` or the like, compute the primary decomposition of $\mathscr{L}_4$, $\mathscr{L}_4 = \cap_{j=1}^{w} Q_j$, for some $w \in \mathbb{N}$ and the associated prime ideals $P_j = \sqrt{Q_j}$.
d. From the previous two parts you have, as in the proof of Theorem 4.5.1,

$$\mathbf{V}(\mathscr{L}_4) = \mathbf{V}\big(\sqrt{\mathscr{L}_4}\big) = \cup_{j=1}^{w} \mathbf{V}(P_j),$$

a minimal decomposition of $\mathbf{V}(\mathscr{L}_4)$, and you know that $\mathbf{V}(\mathscr{L}) \subset \mathbf{V}(\mathscr{L}_4)$. The reverse inclusion is true. Confirm that if $(a,b) \in \mathbf{V}(P_j)$, then the corresponding system (4.71) is linearizable for as many of the P_j as you can.

Hint. See Theorem 6.4.4 for the decomposition in part (c).

4.21 [Referenced in the proof of Theorem 4.5.1.] Find linearizing transformations for quadratic systems (4.66) from the components V_6, V_7, and V_8 of Theorem 4.5.1.

4.22 Although, by Exercise 4.14 above, $V_1 \subset \mathbf{V}(J_3) \subset V_{\mathscr{C}}$, show that none of its elements possesses an irreducible algebraic partial integral of degree three, but rather every algebraic partial integral of degree at most three is a product of the algebraic partial integrals listed in the proof of Theorem 4.5.1 for component V_1.

Chapter 5
Invariants of the Rotation Group

In Section 3.5 we stated the conjecture that the center variety of family (3.3), or equivalently of family (3.69), always contains the variety $\mathbf{V}(I_{\text{sym}})$ as a component. This variety $\mathbf{V}(I_{\text{sym}})$ always contains the set $\mathscr{R}$ that corresponds to the time-reversible systems within family (3.3) or (3.69), which, when they arise through the complexification of a real family (3.2), generalize systems that have a line of symmetry passing through the origin. In Section 3.5 we had left incomplete a full characterization of $\mathscr{R}$. To derive it we are led to a development of some aspects of the theory of invariants of complex systems of differential equations. Using this theory, we will complete the characterization of $\mathscr{R}$ and show that $\mathbf{V}(I_{\text{sym}})$ is actually its Zariski closure, the smallest variety that contains it. In the final section we will also apply the theory of invariants to derive a sharp bound on the number of axes of symmetry of a real planar system of differential equations.

We will consider polynomial systems on $\mathbb{C}^2$ in a form that is a bit more general than (3.69), namely, systems of the form

$$\begin{aligned} \dot{x} &= -\sum_{(p,q)\in\widetilde{S}} a_{pq}x^{p+1}y^q = P(x,y), \\ \dot{y} &= \sum_{(p,q)\in\widetilde{S}} b_{qp}x^q y^{p+1} = Q(x,y), \end{aligned} \tag{5.1}$$

where the index set $\widetilde{S} \subset \mathbb{N}_{-1} \times \mathbb{N}_0$ is a finite set and each of its elements (p,q) satisfies $p+q \geq 0$. As before, if ℓ is the cardinality of the set $\widetilde{S}$, we use the abbreviated notation $(a,b) = (a_{p_1,q_1}, a_{p_2,q_2}, \ldots, a_{p_\ell,q_\ell}, b_{q_\ell,p_\ell}, \ldots, b_{q_2,p_2}, b_{q_1,p_1})$ for the ordered vector of coefficients of system (5.1), let $E(a,b)$ (which is just $\mathbb{C}^{2\ell}$) denote the parameter space of (5.1), and let $\mathbb{C}[a,b]$ denote the polynomial ring in the variables a_{pq} and b_{qp}. The only difference between (3.3) and (3.69) on the one hand and (5.1) on the other is that in the definition of the index set $\widetilde{S}$ for family (5.1) it is required that $p+q \geq 0$, whereas in the definition of S for families (3.3) and (3.69) the inequality $p+q \geq 1$ must hold for all pairs (p,q). Thus the linear part in (5.1) is

$$\dot{x} = -a_{00}x - a_{-1,1}y, \quad \dot{y} = b_{1,-1}x + b_{00}y,$$

V.G. Romanovski, D.S. Shafer, *The Center and Cyclicity Problems*,
DOI 10.1007/978-0-8176-4727-8_5,

whereas heretofore we have restricted attention to systems with diagonal linear part, which includes the complexification of any real system $\dot{\mathbf{u}} = \mathbf{f}(\mathbf{u})$ with $\mathbf{f}(\mathbf{0}) = \mathbf{0}$ and for which the eigenvalues of $d\mathbf{f}(\mathbf{0}) = \mathbf{0}$ are purely imaginary.

5.1 Properties of Invariants

We begin this section by sketching out how an examination of the condition for reversibility naturally leads to an investigation of rotations of phase space, and identify a condition for reversibility that is expressed in the language of invariants. This will serve to orient the reader to the development of the ideas in this section.

Condition (3.96) for reversibility of a system of the form (5.1), when written out in detail and with γ expressed in polar form $\gamma = \rho e^{i\theta}$, is

$$\begin{aligned} b_{q_1p_1} &= \rho^{p_1-q_1} e^{i(p_1-q_1)\theta} a_{p_1q_1} \\ &\vdots \\ b_{q_\ell p_\ell} &= \rho^{p_\ell-q_\ell} e^{i(p_\ell-q_\ell)\theta} a_{p_\ell q_\ell} \\ a_{p_\ell q_\ell} &= \rho^{q_\ell-p_\ell} e^{i(q_\ell-p_\ell)\theta} b_{q_\ell p_\ell} \\ &\vdots \\ a_{p_1q_1} &= \rho^{q_1-p_1} e^{i(q_1-p_1)\theta} b_{q_1p_1} . \end{aligned} \tag{5.2}$$

Actually, condition (3.96) gives just the first ℓ equations of (5.2); we have solved each equation for the coefficient a_{pq} and adjoined these ℓ new equations to the original ℓ equations for reasons that will soon be apparent.

When ρ is 1, equation (5.2) is suggestive of a rotation of coordinate axes, and in fact when the dual rotation of coordinates (5.3) below is done in $\mathbb{C}^2$, then, as we will see, the effect on the coefficient vector $(a,b) \in \mathbb{C}^2$ in system (5.1) is multiplication of (a,b) on the left by a diagonal matrix U_θ whose entries are the exponentials in (5.2). The opposite directions of rotation in (5.3) is natural because when a real system is complexified the second component is obtained by conjugation of the first. Thus when $\rho = 1$ the right-hand side of (5.2) is $U_\theta \cdot (a,b)$. If we associate the vector $\zeta = (\zeta_1, \ldots, \zeta_{2\ell}) = (p_1 - q_1, \ldots, p_{2\ell} - q_{2\ell}, q_{2\ell} - p_{2\ell}, \ldots, q_1 - p_1)$ to family (5.1) and for $\rho \in \mathbb{R}^+$ and any vector $\mathbf{c} = (c_1, \ldots, c_{2\ell})$ let $\rho^\zeta \mathbf{c}$ denote the vector $(\rho^{\zeta_1} c_1, \ldots, \rho^{\zeta_{2\ell}} c_{2\ell})$, then the right-hand side of (5.2) is $\rho^\zeta U_\theta \cdot (a,b)$. Consequently, letting $(\hat{b}, \hat{a})$ denote the involution of (a,b) given by reversing the order of its entries, the condition for reversibility of system (5.1) is the existence of $\rho \in \mathbb{R}^+$ and $\theta \in [0, 2\pi)$ such that $(\hat{b}, \hat{a}) = \rho^\zeta U_\theta \cdot (a,b)$. Generalizing somewhat, we are thus led to investigate conditions under which, for fixed vectors $\mathbf{c}, \mathbf{d} \in E(a,b) = \mathbb{C}^{2\ell}$, the equation $\sigma^{-\zeta}\mathbf{c} = U_\varphi \cdot \mathbf{d}$ has a solution $(\sigma, \varphi) \in \mathbb{R}^+ \times [0, 2\pi)$. The answer (Theorem 5.1.15) is most easily expressed in terms of invariants of the rotation group (Definition 5.1.4): $\mathbf{c}$ and $\mathbf{d}$ must have any nonzero entries in the same positions (which we know immediately from (3.96) that $\mathbf{c}$ and $\hat{\mathbf{c}}$ do for $\mathbf{c} \in \mathscr{R}$), and must yield the same value on any unary and binary invariant (Definition 5.1.9), hence on any invariant (Theorem 5.1.19).

Definition 5.1.1. Let k be a field, let G be a group of $n \times n$ matrices with elements in k, and for $A \in G$ and $\mathbf{x} \in k^n$ let $A \cdot \mathbf{x}$ denote the usual action of G on k^n. A polynomial $f \in k[x_1, \ldots, x_n]$ is *invariant under* G if $f(\mathbf{x}) = f(A \cdot \mathbf{x})$ for every $A \in G$. The polynomial f is also called an *invariant* of G. An invariant is *irreducible* if it does not factor as a product of polynomials that are themselves invariants (although it could very well factor).

Example 5.1.2. Let $B = \left(\begin{smallmatrix} 0 & -1 \\ 1 & 0 \end{smallmatrix}\right)$ and let I_2 denote the 2×2 identity matrix. Then the set $C_4 = \{I_2, B, B^2, B^3\}$ is a group under multiplication (Exercise 5.1), and for the polynomial $f(\mathbf{x}) = f(x_1, x_2) = \frac{1}{2}(x_1^2 + x_2^2)$ we have $f(\mathbf{x}) = f(B \cdot \mathbf{x})$, $f(\mathbf{x}) = f(B^2 \cdot \mathbf{x})$, and $f(\mathbf{x}) = f(B^3 \cdot \mathbf{x})$. Thus f is an invariant of the group C_4. Of course, when $k = \mathbb{R}$, B is simply the group of rotations by multiples of $\pi/2$ radians (mod 2π) about the origin in $\mathbb{R}^2$, and f is an invariant because its level sets are circles centered at the origin, which are unchanged by such rotations.

Generalizing the example, consider the group of rotations

$$x' = e^{-i\varphi}x, \quad y' = e^{i\varphi}y \tag{5.3}$$

of the phase space $\mathbb{C}^2$ of (5.1). Viewing the action of an element of the group as a coordinate transformation, in (x', y')-coordinates system (5.1) has the form

$$\dot{x}' = -\sum_{(p,q)\in\widetilde{S}} a(\varphi)_{pq} x'^{p+1} y'^{q}, \quad \dot{y}' = \sum_{(p,q)\in\widetilde{S}} b(\varphi)_{qp} x'^{q} y'^{p+1},$$

where the coefficients of the transformed system are

$$a(\varphi)_{p_j q_j} = a_{p_j q_j} e^{i(p_j - q_j)\varphi}, \quad b(\varphi)_{q_j p_j} = b_{q_j p_j} e^{i(q_j - p_j)\varphi}, \tag{5.4}$$

for $j = 1, \ldots, \ell$. Once the index set $\widetilde{S}$ has been ordered in some manner, for any fixed angle φ the equations in (5.4) determine an invertible linear mapping U_φ of the space $E(a,b)$ of parameters of (5.1) onto itself, which we will represent as the block diagonal $2\ell \times 2\ell$ matrix

$$U_\varphi = \begin{pmatrix} U_\varphi^{(a)} & 0 \\ 0 & U_\varphi^{(b)} \end{pmatrix},$$

where $U_\varphi^{(a)}$ and $U_\varphi^{(b)}$ are diagonal matrices that act on the coordinates a and b, respectively.

Example 5.1.3. For the family of systems

$$\dot{x} = -a_{00}x - a_{-11}y - a_{20}x^3, \quad \dot{y} = b_{1,-1}x + b_{00}y + b_{02}y^3 \tag{5.5}$$

$\widetilde{S}$ is the ordered set $\{(0,0),(-1,1),(2,0)\}$, and equation (5.4) gives the collection of $2\ell = 6$ equations

$$\begin{array}{lll} a(\varphi)_{00} = a_{00}e^{i(0-0)\varphi} & a(\varphi)_{-11} = a_{-11}e^{i(-1-1)\varphi} & a(\varphi)_{20} = a_{20}e^{i(2-0)\varphi} \\ b(\varphi)_{00} = b_{00}e^{i(0-0)\varphi} & b(\varphi)_{1,-1} = b_{1,-1}e^{i(1-(-1))\varphi} & b(\varphi)_{02} = b_{02}e^{i(0-2)\varphi} \end{array}$$

so that

$$U_\varphi \cdot (a,b) = \begin{pmatrix} U_\varphi^{(a)} & 0 \\ 0 & U_\varphi^{(b)} \end{pmatrix} \cdot (a,b)^T$$

$$= \begin{pmatrix} 1 & 0 & 0 & 0 & 0 & 0 \\ 0 & e^{-i2\varphi} & 0 & 0 & 0 & 0 \\ 0 & 0 & e^{i2\varphi} & 0 & 0 & 0 \\ 0 & 0 & 0 & e^{-i2\varphi} & 0 & 0 \\ 0 & 0 & 0 & 0 & e^{i2\varphi} & 0 \\ 0 & 0 & 0 & 0 & 0 & 1 \end{pmatrix} \cdot \begin{pmatrix} a_{00} \\ a_{-11} \\ a_{20} \\ b_{02} \\ b_{1,-1} \\ b_{00} \end{pmatrix} = \begin{pmatrix} a_{00} \\ a_{-11}e^{-i2\varphi} \\ a_{20}e^{i2\varphi} \\ b_{02}e^{-i2\varphi} \\ b_{1,-1}e^{i2\varphi} \\ b_{00} \end{pmatrix}.$$

Thus here

$$U_\varphi^{(a)} = \begin{pmatrix} 1 & 0 & 0 \\ 0 & e^{-i2\varphi} & 0 \\ 0 & 0 & e^{i2\varphi} \end{pmatrix} \quad \text{and} \quad U_\varphi^{(b)} = \begin{pmatrix} e^{-i2\varphi} & 0 & 0 \\ 0 & e^{i2\varphi} & 0 \\ 0 & 0 & 1 \end{pmatrix}.$$

Note that $U_\varphi^{(a)}$ and $U_\varphi^{(b)}$ do not really depend on a and b; rather, the notation simply indicates that $U_\varphi^{(a)}$ acts on the vector composed of the coefficients of the first equation of (5.1) and $U_\varphi^{(b)}$ acts on the vector composed of coefficients of the second equation of (5.1).

We will usually write (5.4) in the short form

$$(a(\varphi), b(\varphi)) = U_\varphi \cdot (a,b) = (U_\varphi^{(a)} \cdot a, U_\varphi^{(b)} \cdot b).$$

The set $U = \{U_\varphi : \varphi \in \mathbb{R}\}$ is a group, a subgroup of the group of invertible $2\ell \times 2\ell$ matrices with entries in k, under multiplication. In the context of U the group operation corresponds to following one rotation with another.

Definition 5.1.4. The group $U = \{U_\varphi : \varphi \in \mathbb{R}\}$ is the *rotation group* of family (5.1). A polynomial invariant of the group U is termed an *invariant of the rotation group* or, more simply, an *invariant.*

Since the terminology is so similar, care must be taken not to confuse the rotation group U of family (5.1) with the group of rotations (5.3) of the phase plane, which is not associated with any particular family of systems of differential equations.

The rotation group U of family (5.1) acts on $E(a,b) = \mathbb{C}^{2\ell}$ by multiplication on the left. We wish to identify all polynomial invariants of this group action. The polynomials in question are elements of $\mathbb{C}[a,b]$. They identify polynomial expressions in the coefficients of elements of family (5.1) that are unchanged under a rotation of coordinates. Since U_φ changes only the coefficients of polynomials, a polynomial $f \in \mathbb{C}[a,b]$ is an invariant of the group U if and only if each of its terms is an invariant, so it suffices to find the invariant monomials. By (5.4), for $\nu \in \mathbb{N}_0^{2\ell}$, the image of the corresponding monomial $[\nu] = a_{p_1q_1}^{\nu_1} \cdots a_{p_\ell q_\ell}^{\nu_\ell} b_{q_\ell p_\ell}^{\nu_{\ell+1}} \cdots b_{q_1p_1}^{\nu_{2\ell}} \in \mathbb{C}[a,b]$ under U_φ is the monomial

$$\begin{aligned}
a(\varphi)_{p_1q_1}^{\nu_1} &\cdots a(\varphi)_{p_\ell q_\ell}^{\nu_\ell} b(\varphi)_{q_\ell p_\ell}^{\nu_{\ell+1}} \cdots b(\varphi)_{q_1p_1}^{\nu_{2\ell}} \\
&= a_{p_1q_1}^{\nu_1} e^{i\varphi\nu_1(p_1-q_1)} \cdots a_{p_\ell q_\ell}^{\nu_\ell} e^{i\varphi\nu_\ell(p_\ell-q_\ell)} \\
&\qquad \times b_{q_\ell p_\ell}^{\nu_{\ell+1}} e^{i\varphi\nu_{\ell+1}(q_\ell-p_\ell)} \cdots b_{q_1p_1}^{\nu_{2\ell}} e^{i\varphi\nu_{2\ell}(q_1-p_1)} \\
&= e^{i\varphi[\nu_1(p_1-q_1)+\cdots+\nu_\ell(p_\ell-q_\ell)+\nu_{\ell+1}(q_\ell-p_\ell)+\cdots+\nu_{2\ell}(q_1-p_1)]} \\
&\qquad \times a_{p_1q_1}^{\nu_1} \cdots a_{p_\ell q_\ell}^{\nu_\ell} b_{q_\ell p_\ell}^{\nu_{\ell+1}} \cdots b_{q_1p_1}^{\nu_{2\ell}}.
\end{aligned} \tag{5.6}$$

The quantity in square brackets in the exponent in the first term is $L_1(\nu) - L_2(\nu)$, where $L(\nu) = (L_1(\nu), L_2(\nu))$ is the linear operator on $\mathbb{N}_0^{2\ell}$ defined with respect to the ordered set $\widetilde{S}$ by (3.71):

$$\begin{aligned}
L(\nu) &= (L_1(\nu), L_2(\nu)) \\
&= \nu_1(p_1,q_1) + \cdots + \nu_\ell(p_\ell,q_\ell) + \nu_{\ell+1}(q_\ell,p_\ell) + \cdots + \nu_{2\ell}(q_1,p_1) \\
&= (p_1\nu_1 + \cdots + p_\ell\nu_\ell + q_\ell\nu_{\ell+1} + \cdots + q_1\nu_{2\ell}, \\
&\qquad q_1\nu_1 + \cdots + q_\ell\nu_\ell + p_\ell\nu_{\ell+1} + \cdots + p_1\nu_{2\ell}).
\end{aligned}$$

Thus the monomial $[\nu]$ is an invariant if and only if $L_1(\nu) = L_2(\nu)$. As in display (3.98) of Section 3.5, we define the set $\mathcal{M}$ by

$$\mathcal{M} = \{\nu \in \mathbb{N}_0^{2\ell} : L(\nu) = (j,j) \text{ for some } j \in \mathbb{N}_0\}, \tag{5.7}$$

which has the structure of a monoid under addition. (Recall that a *monoid* is a set M together with a binary operation $*$ that is associative and for which there is an identity element ι: for all a, b, $c \in M$, $a*(b*c) = (a*b)*c$ and $a*\iota = \iota*a = a$.) We have established the following proposition.

Proposition 5.1.5. *The monomial $[\nu]$ is invariant under the rotation group U of (5.1) if and only if $L_1(\nu) = L_2(\nu)$, that is, if and only if $\nu \in \mathcal{M}$.*

Since, for any $\nu \in \mathbb{N}_0^{2\ell}$, $L_1(\nu) - L_2(\nu) = -(L_1(\hat{\nu}) - L_2(\hat{\nu}))$, the monomial $[\nu]$ is invariant under U if and only if its conjugate $[\hat{\nu}]$ is.

Before considering an example, we wish to make the following point. In order to find all $\nu \in \mathbb{N}_0^{2\ell}$ such that $\nu \in \mathcal{M}$, we would naturally express the condition that $L_1(\nu) = L_2(\nu) = j \in \mathbb{N}_0$ as $L_1(\nu) - L_2(\nu) = 0$, which, written out completely, is

$$\begin{aligned}
L_1(\nu) - L_2(\nu) = (p_1 - q_1)\nu_1 + (p_2 - q_2)\nu_2 + \cdots + (p_\ell - q_\ell)\nu_\ell& \\
+ (q_\ell - p_\ell)\nu_{\ell+1} + \cdots + (q_1 - p_1)\nu_{2\ell} = 0&,
\end{aligned} \tag{5.8}$$

and look for solutions of this latter equation. Certainly, if $\nu \in \mathscr{M}$, then ν solves (5.8). On the other hand, if $\nu \in \mathbb{N}_0^{2\ell}$ solves (5.8), that does not a priori guarantee that $\nu \in \mathscr{M}$, since membership in $\mathscr{M}$ requires that the common value of $L_1(\nu)$ and $L_2(\nu)$ be nonnegative. The following proposition asserts that it always is. The proof is outlined in Exercise 5.3.

Proposition 5.1.6. *The set of all solutions in* $\mathbb{N}_0^{2\ell}$ *of equation* (5.8) *coincides with the monoid* $\mathscr{M}$ *defined by equation* (5.7).

Example 5.1.7. We will find all the monomials of degree at most three that are invariant under the rotation group U for the family of systems (5.5) of Example 5.1.3. Since $\widetilde{S} = \{(0,0),(-1,1),(2,0)\}$, for $\nu \in \mathbb{N}_0^6$,

$$\begin{aligned} L(\nu) &= \nu_1\,(0,0) + \nu_2\,(-1,1) + \nu_3\,(2,0) + \nu_4\,(0,2) + \nu_5\,(1,-1) + \nu_6\,(0,0) \\ &= (-\nu_2 + 2\nu_3 + \nu_5, \nu_2 + 2\nu_4 - \nu_5), \end{aligned}$$

so that equation (5.8) reads

$$-2\nu_2 + 2\nu_3 - 2\nu_4 + 2\nu_5 = 0. \tag{5.9}$$

$\deg([\nu]) = 0$. The monomial 1, corresponding to $\nu = 0 \in \mathbb{N}_0^6$, is of course always an invariant.

$\deg([\nu]) = 1$. In this case $\nu = (0,\ldots,0,\overset{j}{1},0,\ldots,0) \in \mathbb{N}_0^6$ for some j. Clearly (5.9) holds if and only if $\nu = e_1$ or $\nu = e_6$, yielding $a_{00}^1 a_{-11}^0 a_{20}^0 b_{02}^0 b_{1,-1}^0 b_{00}^0 = a_{00}$ and $a_{00}^0 a_{-11}^0 a_{20}^0 b_{02}^0 b_{1,-1}^0 b_{00}^1 = b_{00}$, respectively.

$\deg([\nu]) = 2$. If $\nu = 2e_j$ and satisfies (5.9), then $j = 1$ or $j = 6$, yielding a_{00}^2 and b_{00}^2, respectively. If $\nu = e_j + e_k$ for $j < k$, then (5.9) holds if and only if either $(j,k) = (1,6)$ or one of j and k corresponds to a term in (5.9) with a plus sign and the other to a term with a minus sign, hence $(j,k) \in P := \{(2,3),(2,5),(3,4),(4,5)\}$. The former case gives $a_{00}b_{00}$; the latter case gives

$$\begin{aligned}
\nu &= (0,1,1,0,0,0) \quad \text{yielding} \quad a_{00}^0 a_{-11}^1 a_{20}^1 b_{02}^0 b_{1,-1}^0 b_{00}^0 = a_{-11}\,a_{20} \\
\nu &= (0,1,0,0,1,0) \quad \text{yielding} \quad a_{00}^0 a_{-11}^1 a_{20}^0 b_{02}^0 b_{1,-1}^1 b_{00}^0 = a_{-11}\,b_{1,-1} \\
\nu &= (0,0,1,1,0,0) \quad \text{yielding} \quad a_{00}^0 a_{-11}^0 a_{20}^1 b_{02}^1 b_{1,-1}^0 b_{00}^0 = a_{20}\,b_{02} \\
\nu &= (0,0,0,1,1,0) \quad \text{yielding} \quad a_{00}^0 a_{-11}^0 a_{20}^0 b_{02}^1 b_{1,-1}^1 b_{00}^0 = b_{02}\,b_{1,-1}.
\end{aligned}$$

$\deg([\nu]) = 3$. If all but one entry in ν is a zero and the nonzero entry is a 3, then clearly (5.9) holds precisely for $\nu = 3e_1$ and $\nu = 3e_6$, yielding a_{00}^3 and b_{00}^3, respectively. If $\nu = 2e_j + e_k$, then it is clear from considerations of parity that (5.9) holds if and only if $(j,k) \in \{(1,6),(6,1)\}$, yielding $a_{00}^2 b_{00}$ and $a_{00}b_{00}^2$. If $\nu = e_j + e_k + e_m$ for distinct j, k, and m, then again by parity considerations we get either $j = 1$ and (k,m) is in the set P of pairs specified in the previous case, or $j = 6$ and $(k,m) \in P$. Thus we get finally the four pairs from the previous case concatenated with a_{00} and the same four pairs concatenated with b_{00}.

To summarize, we have found that the full set of monomial invariants of degree at most three for family (5.5) is

degree 0: 1

degree 1: $a_{00},\ b_{00}$

degree 2: $a_{00}^2,\ b_{00}^2, a_{00}b_{00},\ a_{-11}a_{20},\ a_{-11}b_{1,-1}, a_{20}b_{02},\ b_{02}b_{1,-1}$

degree 3: $a_{00}^3,\ b_{00}^3, a_{00}^2b_{00},\ a_{00}b_{00}^2, a_{00}a_{-11}a_{20},\ a_{00}a_{-11}b_{1,-1}, a_{00}a_{20}b_{02},$
$a_{00}b_{02}b_{1,-1},\ b_{00}a_{-11}a_{20},\ b_{00}a_{-11}b_{1,-1}, b_{00}a_{20}b_{02},\ b_{00}b_{02}b_{1,-1}.$

The following definition arises from the characterization of $\mathscr{M}$ that was given by Proposition 5.1.6.

Definition 5.1.8. Fix an ordered index set $\widetilde{S}$. Then for the corresponding family (5.1) we define:
the *characteristic vector*:

$$\zeta = (p_1 - q_1, \dots, p_\ell - q_\ell, q_\ell - p_\ell, \dots, q_1 - p_1) \in \mathbb{Z}^{2\ell}, \tag{5.10}$$

the *characteristic number*:

$$\mathrm{GCD}(\zeta) = \mathrm{GCD}(p_1 - q_1, \dots, p_\ell - q_\ell) \in \mathbb{N}_0$$

when $\zeta \neq (0, \dots, 0)$, and 1 otherwise, and
the *reduced characteristic vector*:

$$\kappa = \frac{1}{\mathrm{GCD}(\zeta)}\,\zeta.$$

It will be convenient to introduce the notation $\mathbf{z} = (z_1, z_2, \dots, z_{2\ell})$ to denote a generic ordered vector $(a,b) = (a_{p_1,q_1}, a_{p_2,q_2}, \dots, a_{p_\ell,q_\ell}, b_{q_\ell,p_\ell}, \dots, b_{q_2,p_2}, b_{q_1,p_1})$ of coefficients of system (5.1), regarded as variables in a polynomial, so that

$$z_j = \begin{cases} a_{p_j,q_j} & \text{if } 1 \le j \le \ell \\ b_{q_{2\ell-j+1},p_{2\ell-j+1}} & \text{if } \ell+1 \le j \le 2\ell. \end{cases} \tag{5.11}$$

A pair of variables z_r and z_s in $\mathbf{z}$, $1 \le r < s \le 2\ell$, are *conjugate variables* provided there exists $u \in \{1, \dots, \ell\}$ such that $z_r = z_u$ and $z_s = z_{2\ell-u+1}$, so that in terms of the original variable names, $(z_r, z_s) = (a_{p_uq_u}, b_{q_up_u})$.

Definition 5.1.9. A *unary* invariant monomial is an invariant monomial that depends on only one variable, or on only one variable and its conjugate variable. A *binary* invariant monomial is an invariant monomial that depends on two nonconjugate variables, and possibly on one or both of their conjugate variables.

To illustrate, from Example 5.1.7 four invariant monomials for family (5.5) are $f_1 = a_{00}^2$, $f_2 = a_{-11}b_{1,-1}$, $f_3 = a_{-11}a_{20}$, and $f_4 = a_{00}a_{-11}b_{1,-1}$. Obviously, f_1 is a unary invariant, and so is f_2, since a_{-11} and $b_{1,-1}$ are conjugate variables; f_3 and f_4 are binary invariants because the two variables that appear in f_3 are not conjugate, while of the three variables appearing in f_4, two are conjugate.

Of course, if, for a general family (5.1), the characteristic vector $\zeta = (0,\ldots,0)$ then every monomial is invariant. In such a case the *irreducible* unary invariant monomials are all a_{pq} and all b_{qp}, and there are no irreducible binary invariant monomials. The following proposition identifies the irreducible unary and binary invariant monomials in the nontrivial case.

Proposition 5.1.10. *Fix the ordered index set $\widetilde{S}$ and let ζ be the characteristic vector (Definition 5.1.8) of the corresponding family* (5.1)*. Suppose $\zeta \neq (0,\ldots,0)$.*

1. *The unary irreducible invariant monomials of family* (5.1) *are all the monomials of the form a_{pp}, b_{pp}, and $a_{pq}b_{qp}$ for $p \neq q$.*
2. *The binary irreducible invariant monomials of family* (5.1) *are all the monomials of the form*

$$z_r^{|\zeta_s|/\mathrm{GCD}(\zeta_r,\zeta_s)} z_s^{|\zeta_r|/\mathrm{GCD}(\zeta_r,\zeta_s)}, \tag{5.12}$$

where z_r and z_s are defined by (5.11)*, and r and s are any pair from $\{1,2,\ldots,2\ell\}$ such that $\zeta_r\,\zeta_s < 0$ and z_r and z_s are not conjugate variables ($r+s \neq 2\ell+1$).*

Proof. (1) Propositions 5.1.5 and 5.1.6 imply that for $\nu = (0,\ldots,0,\overset{s}{1},0,\ldots,0)$, $[\nu] = z_s$ is a unary invariant if and only if the corresponding coordinate of the characteristic vector ζ is equal to zero, that is, if and only if the corresponding coefficient of system (5.1), $a_{p_sq_s}$ or $b_{q_sp_s}$, satisfies $p_s = q_s$. Similarly, for $\nu = (0,\ldots,\overset{r}{\mu},\ldots,\overset{s}{\eta},\ldots,0)$ with $s = 2\ell - r + 1$, by Proposition 5.1.5 and (5.8) the corresponding monomial

$$z_r^{\mu} z_s^{\eta} = a_{p_rq_r}^{\mu} b_{q_rp_r}^{\eta} \tag{5.13}$$

is a unary invariant if and only if (μ,η) is a solution in $\mathbb{N}\times\mathbb{N}$ of the equation $\mu(p_r - q_r) + \eta(q_r - p_r) = 0$. If $p_r = q_r$, then $a_{p_rq_r}$ and $b_{q_rp_r}$ are unary invariants, so the monomial is not irreducible. Therefore $\mu = \eta$, and the only irreducible invariant of the form (5.13) is $a_{pq}b_{qp}$, $p \neq q$.

Clearly, no invariant monomial containing three or more distinct variables can be a unary invariant, since at most two of the variables can be conjugate.

(2) By Propositions 5.1.5 and 5.1.6, for nonconjugate coefficients z_r and z_s, $z_r^{\mu} z_s^{\eta}$ is a binary invariant monomial if and only if $\mu\,\zeta_r + \eta\,\zeta_s = 0$ and $\mu\,\eta > 0$ (else the invariant is actually unary). If $\zeta_r = 0$, then $\zeta_s = 0$ is forced, in which case it follows directly from the definition that each of z_r^{μ} and z_s^{η} is a unary invariant monomial, so that $z_r^{\mu} z_s^{\eta}$ is not an irreducible invariant.

Thus for an irreducible binary invariant of the form $z_r^{\mu} z_s^{\eta}$ we must have $\zeta_r\,\zeta_s \neq 0$. In fact, it must be the case that $\zeta_r\,\zeta_s < 0$, since $\mu,\,\eta \in \mathbb{N}$. One solution to the equation $\mu\,\zeta_r + \eta\,\zeta_s = 0$ is obviously $(\mu,\eta) = (|\zeta_s|/G, |\zeta_s|/G)$, where $G = \mathrm{GCD}(\zeta_r,\zeta_s) > 0$, so that (5.12) gives a binary invariant. In Exercise 5.4 the reader is asked to show that if (μ,η) is any solution to $\mu\,\zeta_r + \eta\,\zeta_s = 0$, then there exists $a \in \mathbb{N}$ such that

$$z_r^{\mu} z_s^{\eta} = \left(z_r^{|\zeta_s|/G} z_s^{|\zeta_r|/G}\right)^a,$$

so that (5.12) is irreducible and is the only irreducible binary invariant monomial of the form $z_r^{\mu} z_s^{\eta}$ involving nonconjugate coefficients z_r and z_s.

Treatment of the invariant monomials that contain two conjugate variables and a distinct third variable, or contain two distinct pairs of conjugate variables, is left as Exercise 5.5. □

Remark 5.1.11. In terms of the original variables $a_{p_j q_j}$ and $b_{q_j p_j}$, the second statement of the proposition is that the binary irreducible invariants are (Exercise 5.6): for any $(r,s) \in \mathbb{N}^2$ for which $1 \le r, s \le \ell$ and $(p_r - q_r)(p_s - q_s) < 0$:

$$a_{p_r q_r}^{|p_s - q_s|/\mathrm{GCD}(p_r - q_r, p_s - q_s)} a_{p_s q_s}^{|p_r - q_r|/\mathrm{GCD}(p_r - q_r, p_s - q_s)}$$

and

$$b_{q_r p_r}^{|p_s - q_s|/\mathrm{GCD}(p_r - q_r, p_s - q_s)} b_{q_s p_s}^{|p_r - q_r|/\mathrm{GCD}(p_r - q_r, p_s - q_s)}$$

and for any $(r,s) \in \mathbb{N}^2$ for which $1 \le r, s \le \ell$, $r \ne s$, and $(p_r - q_r)(p_s - q_s) > 0$:

$$a_{p_r q_r}^{|p_s - q_s|/\mathrm{GCD}(p_r - q_r, p_s - q_s)} b_{q_s p_s}^{|p_r - q_r|/\mathrm{GCD}(p_r - q_r, p_s - q_s)}.$$

Example 5.1.12. By part (1) of Proposition 5.1.10, all the unary irreducible invariant monomials for family (5.5) are a_{00}, b_{00}, $a_{-11} b_{1,-1}$, and $a_{20} b_{02}$. We find all the binary irreducible invariant monomials for this family in two ways.
(i) Using Proposition 5.1.10(2) directly, we first identify all pairs (r,s), $1 \le r \le 6$, $1 \le s \le 6$, with $r \ne s$ (else $\zeta_r \zeta_s \not< 0$) and $r + s \ne 2 \cdot 3 + 1 = 7$ and compute $\zeta_r \zeta_s$ for each:

$$\begin{aligned}
(1,2), (1,3), (1,4), (1,5):\ & \zeta_1 \zeta_s = (0)\zeta_s = 0 \\
(2,3):\ & \zeta_2 \zeta_3 = (-2)(2) = -4 \\
(2,4):\ & \zeta_2 \zeta_4 = (-2)(-2) = 4 \\
(2,6):\ & \zeta_2 \zeta_6 = (-2)(0) = 0 \\
(3,5):\ & \zeta_3 \zeta_5 = (2)(2) = 4 \\
(3,6):\ & \zeta_3 \zeta_6 = (2)(0) = 0 \\
(4,5):\ & \zeta_4 \zeta_5 = (-2)(2) = -4 \\
(4,6), (5,6):\ & \zeta_r \zeta_6 = \zeta_r(0) = 0.
\end{aligned}$$

Working with the two pairs $(2,3)$ and $(4,5)$ for which $\zeta_r \zeta_s < 0$, we obtain

$$z_2^{|-2|/2} z_3^{|2|/2} = z_2 z_3 = a_{-11} a_{20} \qquad \text{and} \qquad z_4^{|-2|/2} z_5^{|2|/2} = z_4 z_5 = b_{02} b_{1,-1}.$$

(ii) Using Remark 5.1.11:
(a) we identify all pairs (r,s), $1 \le r, s \le 3$, with $r \ne s$ (else $(p_r - q_r)(p_s - q_s) \not< 0$) and compute $(p_r - q_r)(p_s - q_s)$ for each:

$$\begin{aligned}
(1,2):\ & (p_1 - q_1)(p_2 - q_2) = (0)(-2) = 0 \\
(1,3):\ & (p_1 - q_1)(p_3 - q_3) = (0)(2) = 0 \\
(2,3):\ & (p_2 - q_2)(p_3 - q_3) = (-2)(2) = -4,
\end{aligned}$$

hence

$$a_{-11}^{|2|/2} a_{20}^{|-2|/2} = a_{-11} a_{20} \qquad \text{and} \qquad b_{1,-1}^{|2|/2} b_{02}^{|-2|/2} = b_{1.-1} b_{02};$$

(b) we identify all pairs (r,s), $1 \le r, s \le 3$, with $r \ne s$ and $(p_r - q_r)(p_s - q_s) > 0$: the computation already done shows that there are none.

Unary and binary invariants are of particular importance because, up to a natural technical condition on the positions of zero entries, if all unary and binary invariants agree on a pair of coefficient vectors in the parameter space $E(a,b)$ of family (5.1), then all invariants agree on the two vectors of coefficients, which we will prove later.

In order to obtain the main result on the solution of the equation $\sigma^{-\zeta}\mathbf{c} = U_\varphi \cdot \mathbf{d}$ mentioned in the introductory paragraphs of this section, we will need a pair of technical lemmas. The first pertains to the equation

$$\begin{aligned} L_1(\nu) - L_2(\nu) = (p_1 - q_1)\nu_1 + \cdots + (p_\ell - q_\ell)\nu_\ell \\ + (q_\ell - p_\ell)\nu_{\ell+1} + \cdots + (q_1 - p_1)\nu_{2\ell} = \mathrm{GCD}(\zeta), \end{aligned} \tag{5.14a}$$

which can be written in the equivalent form

$$\zeta_1 \nu_1 + \cdots + \zeta_\ell \nu_\ell + \zeta_{\ell+1}\nu_{\ell+1} + \cdots + \zeta_{2\ell}\nu_{2\ell} = \mathrm{GCD}(\zeta). \tag{5.14b}$$

Lemma 5.1.13. *If $\zeta \in \mathbb{Z}^{2\ell}$ and $\zeta \ne (0,\ldots,0)$, then equation* (5.14) *has a solution $\nu \in \mathbb{N}_0^{2\ell}$ for which $\nu_j \ge 0$ for all j and $\nu_j > 0$ for at least one j.*

Proof. By Exercise 5.7, the equation $(p_1 - q_1)t_1 + \cdots + (p_\ell - q_\ell)t_\ell = \mathrm{GCD}(\zeta)$ has a solution $t \in \mathbb{Z}^\ell$. Let $t^+ \in \mathbb{N}_0^\ell$ be the positive part of t, defined by $(t^+)_j = \max(t_j, 0)$, and let $t^- \in \mathbb{N}_0^\ell$ be the negative part of t, defined by $(t^-)_j = \max(-t_j, 0)$, so that $t = t^+ - t^-$. Let $\hat{t}^-$ be the involution of t^- defined by $(\hat{t}^-)_j = (t^-)_{\ell-j}$, and form the doubly long string $\nu \in \mathbb{N}_0^{2\ell}$ by concatenating t^+ and $\hat{t}^-$: $\nu = (t^+, \hat{t}^-)$. Then $|\nu| > 0$ and writing out (5.14) in detail shows that it is satisfied by this choice of ν. □

To state and prove the second lemma and the remaining results of this section most simply, we introduce the following notation. We will denote a specific element of the coefficient set $E(a,b) = \mathbb{C}^{2\ell}$ of family (5.1) by $\mathbf{c}$ or $\mathbf{d}$. (The notation $(\mathbf{c}^{(a)}, \mathbf{c}^{(b)})$ analogous to the notation $(U_\varphi^{(a)}, U_\varphi^{(b)})$ employed for an element U_φ of the rotation group might be more appropriate, but is overly cumbersome.) For $\mathbf{c} \in \mathbb{C}^{2\ell}$ we let $R(\mathbf{c})$ denote the set of indices of the nonzero coordinates of the vector $\mathbf{c}$.

Lemma 5.1.14. *Fix the ordered index set $\widetilde{S}$. For $\mathbf{c}$ and $\mathbf{d}$ in $E(a,b)$, suppose that $R(\mathbf{c}) = R(\mathbf{d})$ and that for every unary or binary invariant monomial $J(a,b)$ of family* (5.1) *the condition*

$$J(\mathbf{c}) = J(\mathbf{d}) \tag{5.15}$$

holds. Then for any $r, s \in R(\mathbf{c}) = R(\mathbf{d})$,

$$\left(\frac{d_r}{c_r}\right)^{\kappa_s} = \left(\frac{d_s}{c_s}\right)^{\kappa_r}, \tag{5.16}$$

where κ_j denotes the jth entry in the reduced characteristic vector κ (Definition 5.1.8).

Proof. Suppose elements $\mathbf{c}$ and $\mathbf{d}$ of $E(a,b)$ satisfy the hypotheses of the lemma and that r and s lie in $R(\mathbf{c})$, so that none of c_r, c_s, d_r, and d_s is zero. If $r = s$, then certainly (5.16) holds, so we suppose that $r \neq s$.

If $\kappa_s = 0$ (that is, $p_s - q_s = 0$), then z_s (which is equal to $a_{p_s p_s}$ or to $b_{p_{2\ell-s+1} p_{2\ell-s+1}}$) is a unary invariant. By (5.15), $c_s = d_s$ and so (5.16) holds. The case $\kappa_r = 0$ is similar.

Now suppose $\kappa_r \kappa_s < 0$. Consider the monomial defined by (5.12). If z_r and z_s are not conjugate variables, then it is a binary invariant. If they *are* conjugate variables, then (relabelling if necessary so that $r < s$) since now $|\zeta_r| = |\zeta_s|$, it is a power of the unary irreducible invariant $a_{p_r q_r} b_{q_r p_r}$, hence is a unary invariant. Either way, by hypothesis (5.15) it agrees on $\mathbf{c}$ and $\mathbf{d}$, so that

$$c_r^{|\kappa_s|} c_s^{|\kappa_r|} = d_r^{|\kappa_s|} d_s^{|\kappa_r|} \tag{5.17}$$

(where we have raised each side of (5.15) to the power $\mathrm{GCD}(\zeta_s, \zeta_r)/\mathrm{GCD}(\zeta)$). Considering each of the two possibilities $\kappa_r < 0 < \kappa_s$ and $\kappa_r > 0 > \kappa_s$, we see that (5.17) yields (5.16).

Finally, consider the case $\kappa_r \kappa_s > 0$. Suppose κ_r and κ_s are both positive. Since $\kappa_j = -\kappa_{2\ell-j+1}$ for all $j \in \{1, \dots, 2\ell\}$ and $r \neq s$, so that z_r and z_s cannot be conjugate variables, by Lemma 5.1.10

$$z_r^{|\zeta_{2\ell-s+1}|/\mathrm{GCD}(\zeta_r, \zeta_{2\ell-s+1})} z_{2\ell-s+1}^{|\zeta_r|/\mathrm{GCD}(\zeta_r, \zeta_{2\ell-s+1})} = z_r^{\zeta_s/\mathrm{GCD}(\zeta_r, \zeta_{2\ell-s+1})} z_{2\ell-s+1}^{\zeta_r/\mathrm{GCD}(\zeta_r, \zeta_{2\ell-s+1})}$$

is a binary invariant. Thus if we raise it to the power $\mathrm{GCD}(\zeta_r, \zeta_{2\ell-s+1})/\mathrm{GCD}(\zeta)$, we find that $z_r^{\kappa_s} z_{2\ell-s+1}^{\kappa_r}$ is a binary invariant, hence, by (5.15),

$$c_r^{\kappa_s} c_{2\ell-s+1}^{\kappa_r} = d_r^{\kappa_s} d_{2\ell-s+1}^{\kappa_r}. \tag{5.18}$$

Since by Lemma 5.1.10 $z_s z_{2\ell-s+1}$ is a unary invariant, by (5.15) we have that $c_s^{-\kappa_r} c_{2\ell-s+1}^{-\kappa_r} = d_s^{-\kappa_r} d_{2\ell-s+1}^{-\kappa_r}$, which, when multiplied times (5.18), yields (5.16). The case that κ_r and κ_s are both negative is similar. □

Suppose $\sigma \in \mathbb{R} \setminus \{0\}$, $\mathbf{c} \in E(a,b)$, $\nu \in \mathbb{N}_0^{2\ell}$, and $\zeta \in \mathbb{Z}^{2\ell}$ is the characteristic vector of family (5.1). We make the following notational conventions:

$$\begin{aligned} \sigma^{\zeta} &= (\sigma^{\zeta_1}, \dots, \sigma^{\zeta_{2\ell}}) \\ \sigma^{-\zeta}\mathbf{c} &= (\sigma^{-\zeta_1} c_1, \dots, \sigma^{-\zeta_{2\ell}} c_{2\ell}) \\ \mathbf{c}^{\nu} &= [\nu]|_{\mathbf{c}} = c_1^{\nu_1} \cdots c_{2\ell}^{\nu_{2\ell}}. \end{aligned}$$

Theorem 5.1.15. *Fix the ordered index set $\widetilde{S}$, hence the family* (5.1), *the characteristic vector ζ, and rotation group $U = \{U_\varphi : \varphi \in \mathbb{R}\}$ for the family* (5.1). *For $\mathbf{c}, \mathbf{d} \in E(a,b)$, there exist $\sigma \in \mathbb{R}^+$ and $\varphi \in [0, 2\pi)$ such that*

$$\sigma^{-\zeta}\mathbf{c} = U_\varphi \cdot \mathbf{d} \tag{5.19}$$

if and only if $R(\mathbf{c}) = R(\mathbf{d})$ *and, for all unary and binary invariants* $J(a,b)$ *of family* (5.1), *condition* (5.15) *holds.*

Proof. Suppose there exists a solution $\sigma_0 > 0$ and $\varphi_0 \in [0, 2\pi)$ to (5.19). By definition of $\sigma_0^{-\zeta}\mathbf{c}$ and $U_{\varphi_0} \cdot \mathbf{d}$ we have $c_j = \sigma_0^{\zeta_j} e^{i\zeta_j\varphi_0} d_j$ for every j, $1 \leq j \leq 2\ell$, hence $R(\mathbf{c}) = R(\mathbf{d})$. Let $[\nu]$ be any invariant monomial. Then by Propositions 5.1.5 and 5.1.6, (5.8) holds for ν, which also reads $\zeta_1\nu_1 + \cdots + \zeta_{2\ell}\nu_{2\ell} = 0$ and immediately yields

$$\begin{aligned}
[\nu]|_{\mathbf{c}} = \mathbf{c}^\nu = c_1^{\nu_1} \cdots c_{2\ell}^{\nu_{2\ell}} &= (\sigma_0^{\nu_1\zeta_1} e^{i\nu_1\zeta_1\varphi_0} d_1) \cdots (\sigma_0^{\nu_{2\ell}\zeta_{2\ell}} e^{i\nu_{2\ell}\zeta_{2\ell}\varphi_0} d_{2\ell}) \\
&= d_1^{\nu_1} \cdots d_{2\ell}^{\nu_{2\ell}} \sigma_0^{(\zeta_1\nu_1+\cdots+\zeta_{2\ell}\nu_{2\ell})} e^{i(\zeta_1\nu_1+\cdots+\zeta_{2\ell}\nu_{2\ell})\varphi_0} = \mathbf{d}^\nu = [\nu]|_{\mathbf{d}}.
\end{aligned}$$

Thus all invariant monomials agree on $\mathbf{c}$ and $\mathbf{d}$, hence all invariants do.

Conversely, suppose that $\mathbf{c}, \mathbf{d} \in E(a,b)$ satisfy $R(\mathbf{c}) = R(\mathbf{d})$ and that every unary and every binary invariant monomial for family (5.1) agree on $\mathbf{c}$ and $\mathbf{d}$. We must show that there exist $\sigma_0 \in \mathbb{R}^+$ and $\varphi_0 \in [0, 2\pi)$ solving the system of equations

$$c_r = d_r \sigma^{\zeta_r} e^{i\zeta_r\varphi} \qquad \text{for} \qquad r = 1, \ldots, 2\ell, \tag{5.20}$$

that is, such that

$$c_r = d_r \sigma_0^{\zeta_r} e^{i\zeta_r\varphi_0} \qquad \text{for} \qquad r = 1, \ldots, 2\ell. \tag{5.21}$$

If $\zeta = (0, \ldots, 0)$, then our assumptions on $\mathbf{c}$ and $\mathbf{d}$ force $\mathbf{c} = \mathbf{d}$ (Exercise 5.8) and (5.21) holds with $\sigma_0 = 1$ and $\varphi_0 = 0$. Hence suppose $\zeta \neq (0, \ldots, 0)$. First we treat the case that every coordinate in each of the vectors $\mathbf{c}$ and $\mathbf{d}$ is nonzero. Let $\mu \in \mathbb{N}_0^{2\ell}$ be the solution to (5.14) guaranteed by Lemma 5.1.13 to exist, and consider the equation, for unknown σ and φ,

$$\mathbf{c}^\mu = \mathbf{d}^\mu \sigma^{\mathrm{GCD}(\zeta)} e^{i\mathrm{GCD}(\zeta)\varphi}.$$

By assumption, $\mathbf{c}^\mu$ and $\mathbf{d}^\mu$ are specific nonzero numbers in $\mathbb{C}$, so we may write this as

$$\left(\frac{\mathbf{c}^\mu}{\mathbf{d}^\mu}\right) = \sigma^{\mathrm{GCD}(\zeta)} e^{i\mathrm{GCD}(\zeta)\varphi}.$$

But the left-hand side is a nonzero complex number $r_0 e^{i\theta_0}$, $r_0 \in \mathbb{R}^+$, $\theta_0 \in [0, 2\pi)$, hence $\sigma_0 = r_0^{1/\mathrm{GCD}(\zeta)}$ and $\varphi_0 = \theta_0/\mathrm{GCD}(\zeta)$ satisfy

$$\mathbf{c}^\mu = \mathbf{d}^\mu \sigma_0^{\mathrm{GCD}(\zeta)} e^{i\mathrm{GCD}(\zeta)\varphi_0}. \tag{5.22}$$

We will now show that σ_0 and φ_0 are as required.

Fix $r \in \{1, \ldots, 2\ell\}$, raise each side of (5.22) to the power κ_r, and use the identity $\mathrm{GCD}(\zeta)\kappa_r = \zeta_r$ to obtain

$$(\mathbf{c}^\mu)^{\kappa_r} = (\mathbf{d}^\mu)^{\kappa_r} \sigma_0^{\zeta_r} e^{i\zeta_r\varphi_0}. \tag{5.23}$$

For each $s \in \{1,\dots,2\ell\}$, form (5.16), raise each side to the power μ_s, and multiply the resulting expressions together to obtain

$$\left(\frac{d_r}{c_r}\right)^{\kappa_1\mu_1}\cdots\left(\frac{d_r}{c_r}\right)^{\kappa_{2\ell}\mu_{2\ell}}=\left(\frac{d_1}{c_1}\right)^{\kappa_r\mu_1}\cdots\left(\frac{d_{2\ell}}{c_{2\ell}}\right)^{\kappa_r\mu_{2\ell}},$$

whence

$$\frac{d_r^{\kappa_1\mu_1}\cdots d_r^{\kappa_{2\ell}\mu_{2\ell}}}{d_1^{\kappa_r\mu_1}\cdots d_{2\ell}^{\kappa_r\mu_{2\ell}}}=\frac{c_r^{\kappa_1\mu_1}\cdots c_r^{\kappa_{2\ell}\mu_{2\ell}}}{c_1^{\kappa_r\mu_1}\cdots c_{2\ell}^{\kappa_r\mu_{2\ell}}}.$$

But by (5.14b), which μ solves, $\kappa_1\mu_1+\cdots+\kappa_{2\ell}\mu_{2\ell}=\frac{\zeta_1\mu_1}{\mathrm{GCD}(\zeta)}+\cdots+\frac{\zeta_{2\ell}\mu_{2\ell}}{\mathrm{GCD}(\zeta)}=1$, so the last expression simplifies to

$$\frac{d_r}{(\mathbf{d}^\mu)^{\kappa_r}}=\frac{c_r}{(\mathbf{c}^\mu)^{\kappa_r}},$$

which, when solved for $(\mathbf{c}^\mu)^{\kappa_r}$ and inserted into (5.23), gives the equation with index r in system (5.21), as required.

Now suppose that there are zero elements in the vector $\mathbf{c}$; by hypothesis, the corresponding elements in the vector $\mathbf{d}$ are zero. If all ζ_r corresponding to the nonzero elements c_r and d_r are equal to zero, then system (5.20) is satisfied for any φ and $\sigma \neq 0$ (in this case the corresponding polynomials z_r are unary invariants). Otherwise, in order to see that there is a solution of the system composed of the remaining equations of (5.20),

$$c_r = d_r\sigma^{\zeta_r}e^{i\zeta_r\varphi} \qquad \text{for} \qquad r\in R(\mathbf{c}),$$

we proceed as above, but with the characteristic vector ζ replaced by the vector $\zeta(\mathbf{c})$ that is obtained from ζ by striking all entries ζ_r for which $r \notin R(\mathbf{c})$ (so that we also have $\mathrm{GCD}(\zeta(\mathbf{c}))$ in place of $GCE(\zeta)$ and $\kappa(\mathbf{c})$ in place of κ). Since Lemmas 5.1.13 and 5.1.14 are still valid, the proof goes through as before. □

Corollary 5.1.16. *Fix the ordered index set* $\widetilde{S}$, *hence the family* (5.1), *the characteristic vector* ζ, *and rotation group* U_φ *for the family* (5.1). *For* $\mathbf{c},\mathbf{d}\in E(a,b)$, *there exists* $\varphi\in[0,2\pi)$ *such that*

$$\mathbf{c} = U_\varphi \cdot \mathbf{d} \tag{5.24}$$

if and only if $R(\mathbf{c}) = R(\mathbf{d})$, $|c_r| = |d_r|$ *for* $r = 1,\dots,2\ell$, *and, for all unary and binary invariants* $J(a,b)$, *condition* (5.15) *holds.*

Proof. We repeat the proof of the theorem almost verbatim. The difference is that in this case, because of the condition on the moduli of the components of $\mathbf{c}$ and $\mathbf{d}$, we have $|\mathbf{c}^\mu| = |\mathbf{d}^\mu|$, so that in the discussion leading up to (5.22), $r_0 = 1$ and in place of (5.22) we obtain the equation

$$\mathbf{c}^\mu = \mathbf{d}^\mu e^{i\mathrm{GCD}(\zeta)\varphi_0}.$$

The remainder of the proof applies without modification to show that φ_0 is as required. □

Proposition 5.1.17. *Fix the ordered index set $\widetilde{S}$, hence the family* (5.1), *the characteristic vector ζ, and rotation group U_φ for the family* (5.1).

1. *Suppose $\zeta \neq (0,\ldots,0)$ and $\mathbf{c},\mathbf{d} \in E(a,b)$ satisfy $R(\mathbf{c}) = R(\mathbf{d})$ and $J(\mathbf{c}) = J(\mathbf{d})$ for all unary and binary invariants $J(a,b)$. Then $(\sigma_0,\varphi_0) \in \mathbb{R}^+ \times [0,2\pi)$ satisfies*

$$\sigma_0^{-\zeta}\mathbf{c} = U_{\varphi_0}\mathbf{d} \tag{5.25}$$

if and only if for every solution $\nu = \mu$ of (5.14), *equation* (5.22) *holds:*

$$\mathbf{c}^\mu = \mathbf{d}^\mu \sigma_0^{\mathrm{GCD}(\zeta)} e^{i\mathrm{GCD}(\zeta)\varphi_0}. \tag{5.26}$$

2. *If in part (1) the condition $R(\mathbf{c}) = R(\mathbf{d})$ is replaced by the more restrictive condition $|c_r| = |d_r|$ for $r = 1,\ldots,2\ell$, then $\varphi_0 \in [0,2\pi)$ satisfies*

$$\mathbf{c} = U_{\varphi_0} \cdot \mathbf{d} \tag{5.27}$$

if and only if for every solution $\nu = \mu$ of (5.14),

$$\mathbf{c}^\mu = \mathbf{d}^\mu e^{i\mathrm{GCD}(\zeta)\varphi_0}. \tag{5.28}$$

Proof. As to part (1), the paragraph in the proof of Theorem 5.1.15 that follows (5.22) showed that if (σ_0,φ_0) satisfies (5.26) then it satisfies (5.25). Conversely, if (5.25) holds, and if $\nu = \mu$ is any solution of (5.14), then for each $r \in \{1,\ldots,2\ell\}$ raise (5.21), the rth component of (5.25), to the power μ_r and multiply all these terms together. Because $\nu = \mu$ solves (5.14), the resulting product reduces to (5.26).

The proof of statement (2) is practically the same. □

Remark 5.1.18. A *basis* or *generating set* of a monoid $(M,+)$ written additively is a set B such that every element of M is a finite sum of elements of B, where summands need not be distinct. If for $\beta \in B$ we define 0β to be the identity in M, and for $c \in \mathbb{N}$ and $\beta \in B$ define $c\beta$ to be the c-fold sum $b + \cdots + b$ (c summands), then B is a basis of M if and only if every element of M is an $\mathbb{N}_0$-linear combination of elements of B. It is clear that if (5.15) holds for all unary and binary invariants from a basis of the monoid $\mathscr{M}$, then it holds for all unary and binary invariants (since every unary and binary invariant is a sum of unary and binary invariants from a basis of $\mathscr{M}$). Hence in the statements of Lemma 5.1.14, Theorem 5.1.15, Corollary 5.1.16, and Theorem 5.1.17 it is sufficient to require that condition (5.15) hold not for *all* unary and binary invariants but only for unary and binary invariants from a basis of $\mathscr{M}$.

The precise statement concerning how agreement of unary and binary invariants on a pair of coefficient vectors forces agreement of all invariants on the pair is the following.

Theorem 5.1.19. *Fix the ordered index set* $\widetilde{S}$. *For* $\mathbf{c}$ *and* $\mathbf{d}$ *in* $E(a,b)$, *suppose that* $R(\mathbf{c}) = R(\mathbf{d})$ *and that for every unary or binary invariant monomial* J *of family* (5.1), $J(\mathbf{c}) = J(\mathbf{d})$. *Then* $J(\mathbf{c}) = J(\mathbf{d})$ *holds for all invariants* $J(z)$.

Proof. It is sufficient to consider only invariant monomials. Thus let $[\nu]$ be any invariant monomial and suppose $\mathbf{c}$ and $\mathbf{d}$ satisfy the hypotheses of the theorem. Then by Theorem 5.1.15 there exist $\sigma \in \mathbb{R}^+$ and $\varphi \in [0, 2\pi)$ such that $\sigma^{-\zeta}\mathbf{c} = U_\varphi \cdot \mathbf{d}$, which reads $c_j = \sigma^{\zeta_j} e^{i\zeta_j\varphi} d_j$ for all $j = 1, 2, \ldots, 2\ell$. The last two sentences of the first paragraph of the proof of Theorem 5.1.15, copied verbatim, complete the proof. □

We can now complete the proof of Theorem 3.5.8 of Section 3.5. For a coefficient vector $(a,b) = (a_{p_1q_1}, \ldots, a_{p_\ell q_\ell}, b_{q_\ell p_\ell}, \ldots, b_{q_1p_1}) \in E(a,b)$, recall that $(\hat{b}, \hat{a})$ denotes the involution of (a,b) defined by

$$(\hat{b}, \hat{a}) = (b_{q_1p_1}, \ldots, b_{q_\ell p_\ell}, a_{p_\ell q_\ell}, \ldots, a_{p_1q_1}). \tag{5.29}$$

Theorem 5.1.20 (Theorem 3.5.8). *Fix an ordered index set* S, *hence family* (3.3). *The set* $\mathscr{R} \subset E(a,b)$ *of all time-reversible systems in family* (3.3) *satisfies:*
1. $\mathscr{R} \subset \mathbf{V}(I_{\text{sym}})$;
2. $\mathbf{V}(I_{\text{sym}}) \setminus \mathscr{R}$
$= \{(a,b) \in \mathbf{V}(I_{\text{sym}}) :$ *there exists* $(p,q) \in S$ *with* $a_{pq}b_{qp} = 0$ *but* $a_{pq} + b_{qp} \neq 0\}$.

Proof. We already know (page 134) that point (1) holds. As to point (2), the statement that there exists a index pair $(p,q) \in S$ such that $a_{pq}b_{qp} = 0$ but $a_{pq} + b_{qp} \neq 0$ is precisely the statement that for some $(p,q) \in S$ exactly one of a_{pq} and b_{qp} is zero, which in our current language is precisely the statement that $R(a,b) \neq R(\hat{b}, \hat{a})$. Thus if we denote by $\mathscr{D}$ the set $\mathscr{D} := \{(a,b) \in \mathbf{V}(I_{\text{sym}}) : R(a,b) \neq R(\hat{b}, \hat{a})\}$, we must show that $\mathbf{V}(I_{\text{sym}}) \setminus \mathscr{R} = \mathscr{D}$. The inclusion $\mathscr{D} \subset \mathbf{V}(I_{\text{sym}}) \setminus \mathscr{R}$ follows directly from the characterization of $\mathscr{R}$ given by (3.96) (Exercise 5.9). To establish the inclusion $\mathbf{V}(I_{\text{sym}}) \setminus \mathscr{R} \subset \mathscr{D}$, we demonstrate the truth of the equivalent inclusion $\mathbf{V}(I_{\text{sym}}) \setminus \mathscr{D} \subset \mathscr{R}$. Hence suppose system $(a,b) \in \mathbf{V}(I_{\text{sym}})$ satisfies $R(a,b) = R(\hat{b}, \hat{a})$. By definition of I_{sym}

$$[\nu]|_{(a,b)} = [\hat{\nu}]|_{(a,b)} \tag{5.30}$$

for all $\nu \in \mathscr{M}$. But by Proposition 5.1.5 the set $\{[\nu] : \nu \in \mathscr{M}\}$ contains all unary and binary invariant monomials of system (3.3). Thus, since $[\hat{\nu}]|_{(a,b)} = [\nu]|_{(\hat{b},\hat{a})}$, (5.30) means that $J(a,b) = J(\hat{b}, \hat{a})$ for all such invariants. Thus by Theorem 5.1.15 the system of equations $\sigma^{-\zeta} \cdot (\hat{b}, \hat{a}) = U_\varphi(a,b)$ in unknowns σ and φ has a solution $(\sigma, \varphi) = (\rho, \theta) \in \mathbb{R}^+ \times [0, 2\pi)$. When written out in detail, componentwise, this is precisely (5.2), that is, is (3.96) with $\alpha = \rho e^{i\theta}$, so $(a,b) \in \mathscr{R}$. □

We have developed the ideas in this section using the rotation group of family (5.1). Generalizing to a slightly more general group of transformations yields a pleasing result that connects time-reversibility to orbits of a group action. We end this section with an exposition of this idea.

Consider the group of transformations of the phase space $\mathbb{C}^2$ of (5.1) given by

$$x' = \eta x, \quad y' = \eta^{-1} y \tag{5.31}$$

for $\eta \in \mathbb{C} \setminus \{0\}$; it is isomorphic to a subgroup of $SL(2,\mathbb{C})$. In (x',y')-coordinates system (5.1) has the form

$$\dot{x}' = \sum_{(p,q)\in\widetilde{S}} a(\eta)_{pq} x'^{p+1} y'^{q}, \quad \dot{y}' = \sum_{(p,q)\in\widetilde{S}} b(\eta)_{qp} x'^{q} y'^{p+1},$$

where the coefficients of the transformed system are

$$a(\eta)_{p_j q_j} = a_{p_j q_j} \eta^{q_j - p_j}, \quad b(\eta)_{q_j p_j} = b_{q_j p_j} \eta^{p_j - q_j} \tag{5.32}$$

for $j = 1, \ldots, \ell$. Let U_η denote both an individual transformation (5.32), which we write in the short form $(a(\eta), b(\eta)) = U_\eta \cdot (a,b)$, and the full group of transformations for all $\eta \in \mathbb{C} \setminus \{0\}$. As before, a polynomial $f \in \mathbb{C}[a,b]$ is an invariant of the group U_η if and only if each of its terms is an invariant.

In analogy with (5.6), by (5.32), for $\nu \in \mathbb{N}_0^{2\ell}$, the image of the corresponding monomial $[\nu] = a_{p_1q_1}^{\nu_1} \cdots a_{p_\ell q_\ell}^{\nu_\ell} b_{q_\ell p_\ell}^{\nu_{\ell+1}} \cdots b_{q_1p_1}^{\nu_{2\ell}} \in \mathbb{C}[a,b]$ under U_η is the monomial

$$\begin{aligned} U_\eta \cdot [\nu] &= a(\eta)_{p_1q_1}^{\nu_1} \cdots a(\eta)_{p_\ell q_\ell}^{\nu_\ell} b(\eta)_{q_\ell p_\ell}^{\nu_{\ell+1}} \cdots b(\eta)_{q_1p_1}^{\nu_{2\ell}} \\ &= \eta^{\zeta\cdot\nu} a_{p_1q_1}^{\nu_1} \cdots a_{p_\ell q_\ell}^{\nu_\ell} b_{q_\ell p_\ell}^{\nu_{\ell+1}} \cdots b_{q_1p_1}^{\nu_{2\ell}} \\ &= \eta^{\zeta\cdot\nu}[\nu]. \end{aligned} \tag{5.33}$$

Thus exactly as with the rotation group U_φ, a monomial $[\nu]$ is invariant under the action of U_η if and only if $\zeta \cdot \nu = 0$, that is, if and only if $\nu \in \mathcal{M}$.

Written in the language of the action of U_η, condition (5.2) for time-reversibility is that there exist $\eta = \rho e^{i\theta}$, $\rho \neq 0$, such that $(\hat{b}, \hat{a}) = U_\eta \cdot (a,b)$, which is the same as (5.19) with $\mathbf{c} = (\hat{b}, \hat{a})$, $\mathbf{d} = (a,b)$, and $\eta = \sigma e^{i\varphi}$. Letting $\mathcal{O}$ be an orbit of the group action, $\mathcal{O} = \{(a(\eta), b(\eta)) : \eta \in \mathbb{C} \setminus \{0\}\} \subset E(a,b) = \mathbb{C}^{2\ell}$, this means that for $(a,b) \in \mathcal{O}$ the system (a,b) is time-reversible if and only if $(\hat{b}, \hat{a}) \in \mathcal{O}$ as well. We claim that if an orbit $\mathcal{O}$ contains one time-reversible system, then every system that it contains is time-reversible. For suppose $(a_0, b_0) \in \mathcal{O}$ is time-reversible. Then for some $\eta_0 = \sigma e^{i\theta}$, equation (5.19) holds with $\mathbf{c} = (a_0, b_0)$ and $\mathbf{d} = (\hat{b}_0, \hat{a}_0)$, so that by Theorem 5.1.15,

$$R(a_0, b_0) = R(\hat{b}_0, \hat{a}_0). \tag{5.34}$$

Now let any element (a_1, b_1) of $\mathcal{O}$ be given, and let $\eta_1 = \sigma_1 e^{i\varphi_1} \neq 0$ be such that $(a_1, b_1) = U_{\eta_1} \cdot (a_0, b_0)$. It is clear that (5.34) implies

$$R(a_1, b_1) = R(\hat{b}_1, \hat{a}_1). \tag{5.35}$$

If $[\nu]$ is any invariant of the U_η action, then because $[\nu]$ is constant on $\mathcal{O}$,

$$[\nu]|_{(a_0,b_0)} = [\nu]|_{(\hat{b}_0,\hat{a}_0)} = [\hat{\nu}]|_{(a_0,b_0)}$$

and because $[\nu]$ is unchanged under the action of U_{η_1} this implies that

$$[v]|_{U_{\eta_1}\cdot(a_0,b_0)} = [\hat{v}]|_{U_{\eta_1}\cdot(a_0,b_0)} = [\hat{v}]|_{(a_1,b_1)} = [v]|_{(\hat{b}_1,\hat{a}_1)},$$

which is $[v]|_{(a_1,b_1)} = [v]|_{(\hat{b}_1,\hat{a}_1)}$. This shows that every U_η invariant agrees on (a_1,b_1) and $(\hat{b}_1,\hat{a}_1)$, hence because the U_η and U_φ invariants are the same, every U_φ invariant does. This fact together with (5.35) implies by Theorem 5.1.15 that there exists $\eta_2 = \sigma_2 e^{i\varphi_2}$ such that (5.19) holds, so that $(\hat{b}_1,\hat{a}_1) \in \mathscr{O}$, as required.

We say that the orbit $\mathscr{O}$ of the U_η group action is invariant under the involution (5.29) if $(a,b) \in \mathscr{O}$ implies $(\hat{b},\hat{a}) \in \mathscr{O}$. We have proven the first point in the following theorem. The second point is a consequence of this discussion and Theorem 5.2.4 in the next section.

Theorem 5.1.21. *Let an ordered index set $\widetilde{S}$, hence a family* (5.1)*, be fixed, and let U_η be the group of transformations* (5.31)*, $\eta \in \mathbb{C}\setminus\{0\}$.*
1. *The set of orbits of U_η is divided into two disjoint subsets: one subset lies in and fills up the set $\mathscr{R}$ of time-reversible systems; the other subset lies in and fills up $E(a,b)\setminus\mathscr{R}$.*
2. *The symmetry variety $\mathbf{V}(I_{\text{sym}})$ is the Zariski closure of the set of orbits of the group U_η that are invariant under the involution* (5.29)*.*

5.2 The Symmetry Ideal and the Set of Time-Reversible Systems

The necessary and sufficient condition that a system of the form (5.1) be time-reversible, that there exist $\gamma \in \mathbb{C}\setminus\{0\}$ such that

$$b_{qp} = \gamma^{p-q} a_{pq} \qquad \text{for all} \qquad (p,q) \in \widetilde{S} \tag{5.36}$$

(which is just condition (3.96) applied to (5.1)), is neither a fixed polynomial description of the set $\mathscr{R}$ of reversible systems (since the number γ varies from system to system in $\mathscr{R}$) nor a polynomial parametrization of $\mathscr{R}$. When we encountered this situation in Section 3.5, we introduced the monoid

$$\mathscr{M} = \{v \in \mathbb{N}_0^{2\ell} : L(v) = (j,j) \text{ for some } j \in \mathbb{N}_0\}$$

(where $L(v) = (L_1(v), L_2(v))$ is the linear operator on $\mathbb{N}_0^{2\ell}$ defined with respect to the ordered set $\widetilde{S}$ by (3.71)) and the symmetry ideal

$$I_{\text{sym}} \stackrel{\text{def}}{=} \langle [v] - [\hat{v}] : v \in \mathscr{M} \rangle \subset \mathbb{C}[a,b]$$

and showed that $\mathscr{R} \subset \mathbf{V}(I_{\text{sym}})$. In this section we will show that $\mathbf{V}(I_{\text{sym}})$ is the Zariski closure of $\mathscr{R}$ by exploiting the implicitization theorems from the end of Section 1.4. We begin by introducing a parametrization of the set $\mathscr{R}$: for $(t_1,\ldots,t_\ell) \in \mathbb{C}^\ell$ and $\gamma \in \mathbb{C}\setminus\{0\}$, condition (5.36) is equivalent to

$$a_{p_j q_j} = t_j, \quad b_{q_j p_j} = \gamma^{p_j - q_j} t_j \quad \text{for} \quad j = 1,\ldots,\ell. \tag{5.37}$$

Refer to equations (1.56) and (1.57) and to Theorem 1.4.15. We define for $1 \le r \le \ell$:

$$g_r(t_1,\ldots,t_\ell,\gamma) \equiv 1,$$

for $\ell+1 \le r \le 2\ell$:

$$g_r(t_1,\ldots,t_\ell,\gamma) = \begin{cases} \gamma^{q_{2\ell-r+1}-p_{2\ell-r+1}} & \text{if } p_{2\ell-r+1}-q_{2\ell-r+1} \le 0 \\ 1 & \text{if } p_{2\ell-r+1}-q_{2\ell-r+1} > 0, \end{cases}$$

for $1 \le r \le \ell$:

$$f_r(t_1,\ldots,t_\ell,\gamma) = t_r,$$

for $\ell+1 \le r \le 2\ell$:

$$f_r(t_1,\ldots,t_\ell,\gamma) = \begin{cases} t_{2\ell-r+1} & \text{if } p_{2\ell-r+1}-q_{2\ell-r+1} \le 0 \\ \gamma^{p_{2\ell-r+1}-q_{2\ell-r+1}} t_{2\ell-r+1} & \text{if } p_{2\ell-r+1}-q_{2\ell-r+1} > 0; \end{cases}$$

set

$$g = g_1 \cdots g_{2\ell}$$

$$x_r = \begin{cases} a_{p_r q_r} & \text{if } 1 \le r \le \ell \\ b_{q_{2\ell-r+1} p_{2\ell-r+1}} & \text{if } \ell+1 \le r \le 2\ell; \end{cases}$$

and let

$$H = \langle 1 - tg, g_r x_r - f_r : 1 \le r \le 2\ell \rangle \subset \mathbb{C}[t,t_1,\ldots,t_\ell,\gamma,a,b]. \tag{5.38}$$

Then by Theorem 1.4.15

$$\overline{\mathscr{R}} = \mathbf{V}(\mathscr{I}), \quad \text{where} \quad \mathscr{I} = H \cap \mathbb{C}[a,b]. \tag{5.39}$$

We will now show that $\mathscr{I}$ and I_{sym} are the same ideal. To do so we will need the following preliminary result.

Lemma 5.2.1. *The ideal $\mathscr{I}$ is a prime ideal that contains no monomials and has a reduced Gröbner basis consisting solely of binomials.*

Proof. By Theorem 1.4.17 the ideal $\mathscr{I}$ is prime, because, as shown in the proof of that theorem, H is the kernel of the ring homomorphism

$$\psi : \mathbb{C}[x_1,\ldots,x_{2\ell},t_1,\ldots,t_\ell,\gamma,t] \to \mathbb{C}(t_1,\ldots,t_\ell,\gamma)$$

defined by

$$t_k \to t_k, \quad x_j \to f_j(t_1,\ldots,t_\ell,\gamma)/g_j(t_1,\ldots,t_\ell,\gamma), \quad t \to 1/g(t_1,\ldots,t_\ell,\gamma),$$

$k = 1,\ldots,\ell$, $j = 1,\ldots,\ell$, which, in the notation of the coefficients $a_{p_jq_j}$, $b_{q_jp_j}$, is $\psi : \mathbb{C}[a,b,t_1,\ldots,t_\ell,\gamma,t] \to \mathbb{C}(t_1,\ldots,t_\ell,\gamma)$ defined by

$$t_k \to t_k, \quad a_{p_jq_j} \to t_j, \quad b_{q_jp_j} \to \gamma^{p_j-q_j}t_j, \quad t \to 1/g(t_1,\ldots,t_\ell,\gamma),$$

$k = 1,\ldots,\ell$, $j = 1,\ldots,\ell$. Simply evaluating ψ on any monomial shows that it is impossible to get zero, so there is no monomial in H, hence none in $\mathscr{I}$, which is a subset of H.

As defined by (5.38), the ideal H has a basis consisting of binomials. Order the variables $t > t_1 > \cdots > t_\ell > \gamma > a_{p_1q_1} > \cdots > b_{q_1p_1}$ and fix the lexicographic term order with respect to this ordering of variables on the polynomial ring $\mathbb{C}[t,t_1,\ldots,t_\ell,\gamma,a,b]$. Suppose a Gröbner basis G is constructed with respect to this order using Buchberger's Algorithm (Table 1.3 on page 21). Since the S-polynomial of any two binomials is either zero, a monomial, or a binomial, as is the remainder upon division of a binomial by a binomial, and since H contains no monomials, G consists solely of binomials. Then by the Elimination Theorem, Theorem 1.3.2, $G_1 := G \cap \mathbb{C}[a,b]$ is a Gröbner basis of $\mathscr{I}$ consisting solely of binomials. A minimal Gröbner basis G_2 is obtained from G_1 by discarding certain elements of G_1 and rescaling the rest (see the proof of Theorem 1.2.24), and a reduced Gröbner basis G_3 is obtained from G_2 by replacing each element of G_2 by its remainder upon division by the remaining elements of G_2 (see the proof of Theorem 1.2.27). Since these remainders are binomials, G_3 is a reduced Gröbner basis of $\mathscr{I}$ that consists solely of binomials. □

Theorem 5.2.2. *Let an index set* $\widetilde{S}$, *hence a family* (5.1), *be given. The ideal* I_{sym} *coincides with the ideal* $\mathscr{I}$ *defined by* (5.38) *and* (5.39).

Proof. Let $f \in I_{\text{sym}} \subset \mathbb{C}[a,b]$ be given, so that f is a finite linear combination, with coefficients in $\mathbb{C}[a,b]$, of binomials of the form $[\nu] - [\hat{\nu}]$, where $\nu \in \mathscr{M}$. To show that $f \in \mathscr{I}$ it is clearly sufficient to show that any such binomial is in $\mathscr{I}$. By definition of the mapping ψ,

$$\begin{aligned}\psi([\nu] - [\hat{\nu}]) &= t_1^{\nu_1}\cdots t_\ell^{\nu_\ell}(\gamma^{p_\ell-q_\ell}t_\ell)^{\nu_{\ell+1}}\cdots(\gamma^{p_1-q_1}t_1)^{\nu_{2\ell}}\\ &\quad - t_1^{\nu_{2\ell}}\cdots t_\ell^{\nu_{\ell+1}}(\gamma^{p_\ell-q_\ell}t_\ell)^{\nu_\ell}\cdots(\gamma^{p_1-q_1}t_1)^{\nu_1}\\ &= t_1^{\nu_1}\cdots t_\ell^{\nu_\ell}t_1^{\nu_{2\ell}}\cdots t_\ell^{\nu_{\ell+1}}(\gamma^{\nu_1\zeta_1+\cdots+\nu_\ell\zeta_\ell} - \gamma^{\nu_{2\ell}\zeta_1+\cdots+\nu_{\ell+1}\zeta_\ell}).\end{aligned} \tag{5.40}$$

Since $\nu \in \mathscr{M}$, $\zeta_1\nu_1 + \cdots + \zeta_{2\ell}\nu_{2\ell} = 0$. But $\zeta_j = -\zeta_{2\ell-j+1}$ for $1 \le j \le 2\ell$, so

$$\zeta_1\nu_1 + \cdots + \zeta_\ell\nu_\ell = -\zeta_{\ell+1}\nu_{\ell+1} - \cdots - \zeta_{2\ell}\nu_{2\ell} = \zeta_\ell\nu_{\ell+1} + \cdots + \zeta_1\nu_{2\ell}$$

and the exponents on γ in (5.40) are the same. Thus $[\nu] - [\hat{\nu}] \in \ker(\psi) = H$, hence $[\nu] - [\hat{\nu}] \in H \cap \mathbb{C}[a,b] = \mathscr{I}$, as required.

Now suppose $f \in \mathscr{I} = H \cap \mathbb{C}[a,b] \subset \mathbb{C}[a,b]$. Because by Lemma 5.2.1 $\mathscr{I}$ has a basis consisting wholly of binomials, it is enough to restrict to the case that f is binomial, $f = a_\alpha[\alpha] + a_\beta[\beta]$. Using the definition of ψ and collecting terms,

$$\psi(a_\alpha[\alpha] + a_\beta[\beta])$$
$$= a_\alpha t_1^{\alpha_1+\alpha_{2\ell}} \cdots t_\ell^{\alpha_\ell+\alpha_{\ell+1}} \gamma^{\zeta_\ell \alpha_{\ell+1}+\cdots+\zeta_1\alpha_{2\ell}} + a_\beta t_1^{\beta_1+\beta_{2\ell}} \cdots t_\ell^{\beta_\ell+\beta_{\ell+1}} \gamma^{\zeta_\ell \beta_{\ell+1}+\cdots+\zeta_1\beta_{2\ell}}.$$

Since $H = \ker(\psi)$ this is the zero polynomial, so

$$a_\beta = -a_\alpha \tag{5.41a}$$
$$\alpha_j + \alpha_{2\ell-j+1} = \beta_j + \beta_{2\ell-j+1} \quad \text{for} \quad j = 1,\ldots,\ell \tag{5.41b}$$
$$\zeta_\ell \alpha_{\ell+1} + \cdots + \zeta_1\alpha_{2\ell} = \zeta_\ell \beta_{\ell+1} + \cdots + \zeta_1\beta_{2\ell}. \tag{5.41c}$$

Imitating the notation of Section 5.1, for $\nu \in \mathbb{N}_0^{2\ell}$ let $R(\nu)$ denote the set of indices j for which $\nu_j \neq 0$. First suppose that $R(\alpha) \cap R(\beta) = \varnothing$. It is easy to check that condition (5.41b) forces $\beta_j = \alpha_{2\ell-j+1}$ for $j = 1,\ldots,2\ell$, so that $\beta = \hat{\alpha}$. But then because $\zeta_j = -\zeta_{2\ell-j+1}$ for $1 \le j \le 2\ell$, condition (5.41c) reads

$$-\zeta_{\ell+1}\alpha_{\ell+1} - \cdots - \zeta_{2\ell}\alpha_{2\ell} = \zeta_\ell\alpha_\ell + \cdots + \zeta_1\alpha_1$$

or $\zeta_1\alpha_1 + \cdots + \zeta_{2\ell}\alpha_{2\ell} = 0$, so $\alpha \in \mathcal{M}$. Thus $f = a_\alpha([\alpha] - [\hat{\alpha}])$ and $\alpha \in \mathcal{M}$, so $f \in I_{\text{sym}}$.

If $R(\alpha) \cap R(\beta) \neq \varnothing$, then $[\alpha]$ and $[\beta]$ contain common factors, corresponding to the common indices of some of their nonzero coefficients. Factoring out the common terms, which form a monomial $[\mu]$, we obtain $f = [\mu](a_\alpha[\alpha'] + a_\beta[\beta'])$, where $R(\alpha') \cap R(\beta') = \varnothing$. Since by Lemma 5.2.1 the ideal $\mathcal{I}$ is prime and contains no monomial, we conclude that $a_\alpha[\alpha'] + a_\beta[\beta'] \in \mathcal{I}$, hence, by the first case, that $a_\alpha[\alpha'] + a_\beta[\beta'] \in I_{\text{sym}}$, hence that $f \in I_{\text{sym}}$. □

Lemma 5.2.1 and Theorem 5.2.2 together immediately yield the following theorem, which includes Theorem 3.5.9.

Theorem 5.2.3. *Let an ordered index set $\widetilde{S}$, hence a family* (5.1)*, be given. The ideal I_{sym} is a prime ideal that contains no monomials and has a reduced Gröbner basis consisting solely of binomials.*

Together with (5.39), Theorem 5.2.2 provides the following description of time-reversible systems.

Theorem 5.2.4. *Let an ordered index set $\widetilde{S}$, hence a family* (5.1)*, be given. The variety of the ideal I_{sym} is the Zariski closure of the set $\mathcal{R}$ of time-reversible systems in family (5.1).*

A generating set or basis $\mathcal{N}$ of $\mathcal{M}$ (see Remark 5.1.18) is *minimal* if, for each $\nu \in \mathcal{N}$, $\mathcal{N} \setminus \{\nu\}$ is not a generating set. A minimal generating set, which need not be unique (Exercise 5.10), is sometimes referred to as a *Hilbert basis* of $\mathcal{M}$. The fact that by Proposition 5.1.6 the monoid $\mathcal{M}$ is $V \cap \mathbb{N}_0^{2\ell}$, where V is the vector subspace of $\mathbb{R}^{2\ell}$ determined by the same equation (5.8) that determines $\mathcal{M}$, can lead to some confusion. As elements of $\mathbb{R}^{2\ell}$, the elements of the Hilbert Basis of $\mathcal{M}$ span V but are not necessarily a vector space basis of V. See Exercises 5.11 and 5.12 concerning this comment and other facts about Hilbert bases of $\mathcal{M}$.

The next result describes the reduced Gröbner basis of the ideal I_{sym} in more detail and shows how to use it to construct a Hilbert basis of the monoid $\mathscr{M}$.

Theorem 5.2.5. *Let G be the reduced Gröbner basis of I_{sym} with respect to any term order.*
1. Every element of G has the form $[\nu]-[\hat{\nu}]$, where $\nu \in \mathscr{M}$ and $[\nu]$ and $[\hat{\nu}]$ have no common factors.
2. The set

$$\begin{aligned}\mathscr{H} = &\{\mu,\hat{\mu} : [\mu]-[\hat{\mu}] \in G\}\\ &\cup \{\mathbf{e}_j+\mathbf{e}_{2\ell-j+1} : j=1,\dots,\ell \text{ and } \pm([\mathbf{e}_j]-[\mathbf{e}_{2\ell-j+1}]) \notin G\},\end{aligned}$$

where $\mathbf{e}_j = (0,\dots,0,\overset{j}{1},0,\dots,0)$, is a Hilbert basis of $\mathscr{M}$.

Proof. Let $g = a_\alpha[\alpha]+a_\beta[\beta]$ be an element of the reduced Gröbner basis of I_{sym}. The reasoning in the part of the proof of Theorem 5.2.2 that showed that $\mathscr{I} \subset I_{\text{sym}}$ shows that any binomial in I_{sym} has the form $[\eta]([\nu]-[\hat{\nu}])$, where $\nu \in \mathscr{M}$ and $R(\nu)\cap R(\hat{\nu}) = \varnothing$. Thus if $R(\alpha)\cap R(\beta)=\varnothing$, then $g = a([\nu]-[\hat{\nu}])$, and $a=1$ since G is reduced. If $R(\alpha)\cap R(\beta)\neq\varnothing$, then $g=[\eta]([\nu]-[\hat{\nu}]) = [\eta]h$ and $h \in I_{\text{sym}}$ since $\nu \in \mathscr{M}$. But then there exists $g_1 \in G$ such that $\mathrm{LT}(g_1)$ divides $\mathrm{LT}(h)$, which implies that $\mathrm{LT}(g_1)$ divides $\mathrm{LT}(g)$, which is impossible since $g \in G$ and G is reduced. This proves point (1).

It is an immediate consequence of the definition of $\mathscr{M}$ that $\mathbf{e}_j+\mathbf{e}_{2\ell-j+1} \in \mathscr{M}$ for $j=1,\dots,\ell$. If, for some j, $1\le j\le \ell$, $[\mathbf{e}_j]-[\mathbf{e}_{2\ell-j+1}] \in G$, then $\mathbf{e}_j$ and $\mathbf{e}_{2\ell-j+1}$ are both in $\mathscr{H}$, hence $\mathbf{e}_j+\mathbf{e}_{2\ell-j+1} \in \operatorname{Span}\mathscr{H}$, so that $\mathscr{H}$ is a basis of $\mathscr{M}$ if

$$\mathscr{H}^+ = \{\mu,\hat{\mu} : [\mu]-[\hat{\mu}] \in G\} \cup \{\mathbf{e}_j+\mathbf{e}_{2\ell-j+1} : j=1,\dots,\ell\}$$

is, and it is more convenient to work with $\mathscr{H}^+$ in this regard.

Hence suppose ν is in $\mathscr{M}$. If $\hat{\nu}=\nu$, then $\nu_j = \nu_{2\ell-j+1}$ for $j=1,\dots,\ell$, so that $\nu = \sum_{j=1}^{\ell}\nu_j(\mathbf{e}_j+\mathbf{e}_{2\ell-j+1})$. Thus ν could fail to be a finite sum of elements of $\mathscr{H}^+$ only if $\nu\neq\hat{\nu}$, hence only if $[\nu]-[\hat{\nu}]\neq 0$. Suppose, contrary to what we wish to show, that the set F of elements of $\mathscr{M}$ that cannot be expressed as a finite $\mathbb{N}_0$-linear combination of elements of $\mathscr{H}^+$ is nonempty. Then F has a least element μ with respect to the total order on $\mathbb{N}_0^{2\ell}$ that corresponds to (and in fact is) the term order on $\mathbb{C}[a,b]$. The binomial $[\mu]-[\hat{\mu}]$ is in I_{sym} and is nonzero, hence its leading term $[\hat{\mu}]$ is divisible by the leading term $[\alpha]$ of some element $[\alpha]-[\hat{\alpha}]$ of G. Performing the division yields

$$[\hat{\mu}]-[\mu] = [\beta]([\alpha]-[\hat{\alpha}])+[\hat{\alpha}]([\beta]-[\hat{\beta}]),$$

which, adopting the notation $\mathbf{x} = (x_1,\dots,x_{2\ell}) = (a_{p_1q_1},\dots,b_{q_1p_1})$ of the definition of the ideal H defining $\mathscr{I}$, is more intuitively expressed as

$$\mathbf{x}^{\hat{\mu}}-\mathbf{x}^{\mu} = \mathbf{x}^{\beta}(\mathbf{x}^{\alpha}-\mathbf{x}^{\hat{\alpha}})+\mathbf{x}^{\hat{\alpha}}(\mathbf{x}^{\beta}-\mathbf{x}^{\hat{\beta}}),$$

for some $\beta \in \mathbb{N}_0^{2\ell}$ that satisfies $\alpha + \beta = \hat{\mu}$. But then $\hat{\alpha} + \hat{\beta} = \mu$, so that $\hat{\beta} < \mu$ (see Exercise 1.15) and $\hat{\beta} = \mu - \hat{\alpha} \in \mathscr{M}$. Thus $\hat{\beta}$ cannot be in F, whose least element is μ, so $\hat{\beta}$ is a finite $\mathbb{N}_0$-linear combination of elements of $\mathscr{H}^+$. But then so is $\mu = \hat{\alpha} + \hat{\beta}$, a contradiction. Thus F is empty, as required.

It remains to show that $\mathscr{H}$ is minimal. We do this by showing that no element of $\mathscr{H}$ is an $\mathbb{N}_0$-linear combination of the remaining elements of $\mathscr{H}$. To fix notation we suppose that the reduced Gröbner basis of I_{sym} is $G = \{g_1, \ldots, g_r\}$, where in standard form (leading term first) g_j is $g_j = [\mu_j] - [\hat{\mu}_j]$, $1 \le j \le r$.

Fix $\mu_k \in \mathscr{H}$ for which either $[\mu_k] - [\hat{\mu}_k] \in G$ or $[\hat{\mu}_k] - [\mu_k] \in G$. It is impossible that $\mu_k = \sum_{j=1}^{\ell} c_j(\mathbf{e}_j + \mathbf{e}_{2\ell-j+1})$ (with $c_j = 0$ if $\mathbf{e}_j + \mathbf{e}_{2\ell-j+1} \notin \mathscr{H}$), since this equation implies that $\hat{\mu}_k = \mu_k$, hence that $g_k = ([\mu_k] - [\hat{\mu}_k]) = 0$, which is not true. It is also impossible that

$$\mu_k = \sum_{\substack{j=1\\ j\neq k}}^{r} a_j\mu_j + \sum_{j=1}^{r} b_j\hat{\mu}_j + \sum_{j=1}^{\ell} c_j(\mathbf{e}_j + \mathbf{e}_{2\ell-j+1}),$$

which is equivalent to

$$\hat{\mu}_k = \sum_{\substack{j=1\\ j\neq k}}^{r} a_j\hat{\mu}_j + \sum_{j=1}^{r} b_j\mu_j + \sum_{j=1}^{\ell} c_j(\mathbf{e}_{2\ell-j+1} + \mathbf{e}_j).$$

For if some $a_j \neq 0$, then $[\mu_k]$ is divisible by the leading term $[\mu_j]$ of g_j (and $j \neq k$), while if some $b_j \neq 0$ then $[\hat{\mu}_k]$ is divisible by the leading term $[\mu_j]$ of g_j (and $j \neq k$ since, by point (1) of the theorem, $[\mu_k]$ and $[\hat{\mu}_k]$ have no common factors), contrary to the fact that G is reduced. Clearly this argument also shows that no $\hat{\mu}_k$ is an $\mathbb{N}_0$-linear combination of the remaining elements of $\mathscr{H}$.

Finally, suppose $k \in \{1, \ldots, \ell\}$ is such that

$$\mathbf{e}_k + \mathbf{e}_{2\ell-k+1} = \sum_{\theta_j \in \mathscr{H}} a_j\theta_j \tag{5.42}$$

for $a_j \in \mathbb{N}_0$ (we are not assuming that k is such that $\mathbf{e}_k + \mathbf{e}_{2\ell-k+1} \in \mathscr{H}$). Because all a_j and all entries in all θ_j are nonnegative no cancellation of entries is possible. Thus either the sum has just one summand, $\mathbf{e}_k + \mathbf{e}_{2\ell-k+1}$, or it has exactly two summands, $\mathbf{e}_k$ and $\mathbf{e}_{2\ell-k+1}$. In the former case $\mathbf{e}_k + \mathbf{e}_{2\ell-k+1} \in \mathscr{H}$; in the latter case both $\mathbf{e}_k$ and $\mathbf{e}_{2\ell-k+1} = \hat{\mathbf{e}}_k$ are in $\mathscr{H}$, so that by definition of $\mathscr{H}$ either $[\mathbf{e}_k] - [\mathbf{e}_{2\ell-k+1}] \in G$ or $[\mathbf{e}_{2\ell-k+1}] - [\mathbf{e}_k] \in G$, hence $\mathbf{e}_k + \mathbf{e}_{2\ell-k+1} \notin \mathscr{H}$. Thus if $\mathbf{e}_k + \mathbf{e}_{2\ell-k+1} \in \mathscr{H}$, then (5.42) holds only if the right-hand side reduces to $1 \cdot (\mathbf{e}_k + \mathbf{e}_{2\ell-k+1})$. □

Theorem 1.3.2 provides an algorithm for computing a generating set for the ideal $\mathscr{I}$ and, therefore, for the ideal I_{sym}. Using Theorem 5.2.5, we obtain also a Hilbert basis of the monoid $\mathscr{M}$. The complete algorithm is given in Table 5.1 on page 235.

Example 5.2.6. Consider the ordered set S of indices $S = \{(0,1), (-1,3)\}$ corresponding to the ordered parameter set $\{a_{01}, a_{-13}, b_{3,-1}, b_{10}\}$ of family (5.1). In this

Algorithm for computing I_{sym} and a Hilbert basis of $\mathscr{M}$

Input:

An ordered index set $\widetilde{S} = \{(p_1,q_1),\ldots,(p_\ell,q_\ell)\}$
specifying a family of systems (5.1)

Output:

A reduced Gröbner basis G for the ideal I_{sym}
and a Hilbert basis $\mathscr{H}$ for the monoid $\mathscr{M}$ for family (5.1)

Procedure:

1. Compute the reduced Gröbner basis G_H for H defined by (5.38) with respect to lexicographic order with $t > t_1 > \cdots > t_\ell > \gamma > a_{p_1q_1} > \cdots > b_{q_1p_1}$.
2. $G := G_H \cap \mathbb{C}[a,b]$.
3. $\mathscr{H}$ is the set defined in point (2) of Theorem 5.2.5.

Table 5.1 Algorithm for Computing I_{sym} and a Hilbert Basis of $\mathscr{M}$

case $\ell = 2$, $L : \mathbb{N}_0^4 \to \mathbb{Z}^2$ is the map

$$\begin{aligned} L(\nu_1,\nu_2,\nu_3,\nu_4) &= \nu_1(0,1) + \nu_2(-1,3) + \nu_3(3,-1) + \nu_4(1,0) \\ &= (-\nu_2 + 3\nu_3 + \nu_4, \nu_1 + 3\nu_2 - \nu_3), \end{aligned}$$

and by Proposition 5.1.6, $\mathscr{M}$ is the set of all $\nu = (\nu_1,\nu_2,\nu_3,\nu_4) \in \mathbb{N}_0^4$ such that

$$\nu_1 + 4\nu_2 - 4\nu_3 - \nu_4 = 0\,. \tag{5.43}$$

When we compute the quantities defined in the display between equations (5.37) and (5.38), we obtain the polynomials

$$\begin{aligned} g_1(t_1,t_2,\gamma) &= 1 & \qquad f_1(t_1,t_2,\gamma) &= t_1 \\ g_2(t_1,t_2,\gamma) &= 1 & f_2(t_1,t_2,\gamma) &= t_2 \\ g_3(t_1,t_2,\gamma) &= \gamma^4 & f_3(t_1,t_2,\gamma) &= t_2 \\ g_4(t_1,t_2,\gamma) &= \gamma & f_4(t_1,t_2,\gamma) &= t_1 \end{aligned}$$

and the polynomial and variables

$$g(t_1,t_2,\gamma) = \gamma^5 \quad x_1 = a_{01} \quad x_2 = a_{-13} \quad x_3 = b_{3,-1} \quad x_4 = b_{10}.$$

Thus the ideal defined by (5.38) lies in $\mathbb{C}[t,t_1,t_2,\gamma,a_{01},a_{-13},b_{3,-1},b_{10}]$ and is

$$H = \langle 1 - t\gamma^5, a_{01} - t_1, a_{-13} - t_2, \gamma^4 b_{3,-1} - t_2, \gamma b_{10} - t_1 \rangle.$$

Using practically any readily available computer algebra system, we can compute the reduced Gröbner basis for H with respect to lexicographic order with the variables ordered by $t > t_1 > t_2 > \gamma > a_{01} > a_{-13} > b_{3,-1} > b_{10}$ (step 1 of the algorithm) as the set of polynomials

$$
\begin{array}{lll}
b_{3,-1}a_{01}^4 - b_{10}^4 a_{-1,3} & \gamma b_{3,-1}a_{01}^3 - a_{-13}b_{10}^3 & \gamma b_{10} - a_{01} \\
\gamma^2 b_{3,-1}a_{01}^2 - b_{10}^2 a_{-13} & \gamma^3 b_{3,-1}a_{01} - b_{10}a_{-13} & \gamma^4 b_{3,-1} - a_{-13} \\
t a_{-13}a_{01} - b_{10}b_{3,-1} & \gamma t a_{-13} - b_{3,-1} & t_2 - a_{-13} \\
t a_{-13}^2 - \gamma^3 b_{3,-1}^2 & \gamma^2 t a_{01}^3 - b_{10}^3 & t_1 - a_{01} \\
\gamma t a_{01}^4 - b_{10}^4 & t a_{01}^5 - b_{10}^5 & \gamma^5 t - 1 \\
\gamma^3 t a_{01}^2 - b_{10}^2 & \gamma^4 t a_{01} - b_{10}. &
\end{array}
$$

The reduced Gröbner basis of the fourth elimination ideal (step 2 of the algorithm) is composed of just those basis elements that do not contain any of t, t_1, t_2, or γ, hence is simply $\{b_{3,-1}a_{01}^4 - b_{10}^4 a_{-13}\}$. This is the reduced Gröbner basis of I_{sym}. Finally (step 3), by Theorem 5.2.5, since $[\mathbf{e}_1] = a_{01}^1 a_{-13}^0 b_{3,-1}^0 b_{10}^0 = a_{01}$ and so on, a Hilbert basis of H is

$$\{(4,0,1,0),(0,1,0,4),(1,0,0,1),(0,1,1,0)\}.$$

Any string (v_1, v_2, v_3, v_4) in $\mathbb{N}_0^4$ that satisfies (5.43) is an $\mathbb{N}_0$-linear combination of these four 4-tuples.

We now use the algorithm to prove Theorem 3.5.10, that is, to compute the generators of the ideal I_{sym} for system (3.100):

$$
\begin{aligned}
\dot{x} &= x - a_{10}x^2 - a_{01}xy - a_{-12}y^2 - a_{20}x^3 - a_{11}x^2y - a_{02}xy^2 - a_{-13}y^3 \\
\dot{y} &= -y + b_{2,-1}x^2 + b_{10}xy + b_{01}y^2 + b_{3,-1}x^3 + b_{20}x^2y + b_{11}xy^2 + b_{02}y^3.
\end{aligned}
$$

Proof of Theorem 3.5.10. We first compute a reduced Gröbner basis of the ideal of (5.38) for our family, namely the ideal

$$
\begin{aligned}
H = \langle & 1-\gamma^{10}t, a_{10}-t_1, a_{01}-t_2, a_{-12}-t_3, a_{20}-t_4, a_{11}-t_5, a_{02}-t_6, a_{-13}-t_7, \\
& \gamma^4 b_{3,-1}-t_7, \gamma^2 b_{20}-t_6, \gamma b_{11}-t_5, b_{02}-\gamma^2 t_4, \gamma^3 b_{2,-1}-t_3, \gamma b_{10}-t_2, b_{01}-\gamma t_1\rangle,
\end{aligned}
$$

with respect to lexicographic order with

$$
\begin{aligned}
& t > t_1 > t_2 > t_3 > t_4 > t_5 > t_6 > t_7 > \gamma > a_{10} > a_{01} > a_{-12} \\
& \quad > a_{20} > a_{11} > a_{02} > a_{-13} > b_{3,-1} > b_{20} > b_{11} > b_{02} > b_{2,-1} > b_{10} > b_{01}.
\end{aligned}
$$

We obtain a list of polynomials that is far too long to be presented here, but which the reader should be able to compute using practically any computer algebra system. According to the second step of Algorithm 5.1, in order to obtain a basis of I_{sym} we simply select from the list those polynomials that do not depend on t, t_1, t_2, t_3, t_4,

t_5, t_6, t_7, or γ. These are precisely the polynomials presented in Table 3.2. Note how they exhibit the structure described in point (1) of Theorem 5.2.5. □

5.3 Axes of Symmetry of a Plane System

As defined in Section 3.5, a line L is an axis of symmetry of the real system

$$\dot{u} = \widetilde{U}(u,v), \quad \dot{v} = \widetilde{V}(u,v) \tag{5.44}$$

if as point sets the orbits of the system are symmetric with respect to L. In this section by an *axis of symmetry* we mean an axis *passing through the origin*, at which system (5.44) is assumed to have a singularity. We will also assume that $\widetilde{U}$ and $\widetilde{V}$ are polynomials. Note that the eigenvalues of the linear part at the origin are not assumed to be purely imaginary, however, which is the reason for the enlarged index set $\widetilde{S}$ employed in this chapter. The ideas developed in Section 5.1 lead naturally to a necessary condition on a planar system in order for it to possess an axis of symmetry and allow us to derive a bound on the number of axes of symmetry possible in the phase portrait of a polynomial system, expressed in terms of the degrees of the polynomials.

As was done in Section 3.5, we write (5.44) as a single complex differential equation by writing $x = u + iv$ and differentiating with respect to t. We obtain

$$\frac{dx}{dt} = \sum_{(p,q)\in\widetilde{S}} a_{pq} x^{p+1} \bar{x}^q = \widetilde{P}(x,\bar{x}), \tag{5.45}$$

where $\widetilde{P} = \widetilde{U} + i\widetilde{V}$, $\widetilde{U}$ and $\widetilde{V}$ evaluated at $((x_1+\bar{x}_1)/2, (x_1-\bar{x}_1)/(2i))$. When we refer below to equation (5.45), we will generally have in mind that it is equivalent to system (5.44).

Writing (5.44) in the complex form (5.45) is just the first step in the process of complexifying (5.44). Thus, letting a denote the vector whose components are the coefficients of the polynomial $\widetilde{P}$, it is natural to write the parameter space for (5.45) as $E(a) = \mathbb{C}^\ell$. We let $E_P(a) \subset E(a)$ denote the set of systems (5.45) for which the original polynomials $\widetilde{U}$ and $\widetilde{V}$ in (5.44) have no nonconstant common factors. When we refer to "system a" we will mean system (5.45) with coefficients given by the components of the vector a.

The choice of the ℓ-element index set $\widetilde{S}$ not only specifies the real family in the complex form (5.45) but simultaneously specifies the full family (5.1) on $\mathbb{C}^2$, of which (5.45) is the first component evaluated on the complex line $y = \bar{x}$. Thus we will use the notation $U_\varphi^{(a)}$ for a rotation of coordinates in $\mathbb{C}$, corresponding to the first equation in display (5.3), for family (5.45).

In Lemma 3.5.3 we proved that existence of either time-reversible or mirror symmetry with respect to the u-axis is characterized by the condition $a_0 = \pm\bar{a}_0$, which is thus a sufficient condition that the u-axis be an axis of symmetry for system a_0.

Consider now for $\varphi_0 \in [0, 2\pi)$ the line $x = re^{i\varphi_0}$. Adapting the notation from the beginning of Section 5.1 as just mentioned, after a rotation in $\mathbb{C}$ through angle $-\varphi_0$ about the origin we obtain from system a_0 the system $a_0' = U_{\varphi_0}^{(a)} \cdot a_0$ (observe that in (5.3) the first rotation is through $-\varphi$). If $a_0' = \pm \bar{a}_0'$, then the line $x = re^{i\varphi_0}$ is an axis of symmetry of (5.45). Thus, because it is always true that

$$\overline{U_{\varphi_0}^{(a)} \cdot a_0} = U_{-\varphi_0}^{(a)} \cdot \bar{a}_0, \tag{5.46}$$

Lemma 3.5.3(2) yields the following result.

Lemma 5.3.1. *Fix an ordered index set $\widetilde{S}$, hence the real family in complex form* (5.45), *the complex family* (5.1), *and the rotation group U_φ for family* (5.1). *If*

$$U_{-\varphi_0}^{(a)} \cdot \bar{a}_0 = U_{\varphi_0}^{(a)} \cdot a_0 \tag{5.47}$$

or

$$U_{-\varphi_0}^{(a)} \cdot \bar{a}_0 = -U_{\varphi_0}^{(a)} \cdot a_0, \tag{5.48}$$

then the line with complex equation $x = re^{i\varphi_0}$ is an axis of symmetry of system (5.45) *with coefficient vector a_0.*

Condition (5.47) characterizes mirror symmetry with respect to the line $x = re^{i\varphi_0}$; condition (5.48) characterizes time-reversible symmetry with respect to the line $x = re^{i\varphi_0}$. It is conceivable, however, that a line $x = re^{i\varphi_0}$ be an axis of symmetry for a system a_0 yet neither condition hold. If such were the case for a system a_0, then it is apparent that the system would still have to possess either mirror or time-reversible symmetry on a neighborhood N of any regular point in its phase portrait, hence on the saturation of N under the flow. Thus the phase portrait would have to be decomposed into a union of disjoint open regions on each of which either (5.47) or (5.48) holds, separated by curves of singularities. This suggests that more general symmetry can be excluded if $\widetilde{U}$ and $\widetilde{V}$ are relatively prime polynomials. The following lemma confirms this conjecture.

Lemma 5.3.2. *Fix an ordered index set $\widetilde{S}$, hence the real family in complex form* (5.45), *the complex family* (5.1), *and the rotation group U_φ for the family* (5.1). *If $a_0 \in E_P(a)$ and the line $x = re^{i\varphi_0}$ is an axis of symmetry of system* (5.45), *then either* (5.47) *or* (5.48) *holds.*

Proof. The u-axis is an axis of symmetry of (5.44) if and only if condition (3.86) holds. Let $a_0' = U_{\varphi_0} \cdot a_0$ and write the corresponding system (5.44) as $\dot{u} = \widetilde{U}'(u,v)$, $\dot{v} = \widetilde{V}'(u,v)$. It is easy to see that $a_0' \in E_P(a)$ as well (Exercise 5.15). For system a_0' the u-axis is an axis of symmetry, so condition (3.86) is satisfied, hence for all (u,v) for which $\widetilde{V}'(u,v) \neq 0$,

$$-\frac{\widetilde{V}'(u,-v)}{\widetilde{V}'(u,v)} \widetilde{U}'(u,v) \equiv \widetilde{U}'(u,-v). \tag{5.49}$$

Because $\widetilde{U}$ and $\widetilde{V}$ are relatively prime polynomials, (5.49) can be satisfied only if the expression

$$p(u,v) := -\frac{\widetilde{V}'(u,-v)}{\widetilde{V}'(u,v)}$$

defines a polynomial. Thus from (5.49) we have that $p(u,v)\widetilde{U}'(u,v) \equiv \widetilde{U}'(u,-v)$. But $\widetilde{U}'(u,v)$ and $\widetilde{U}'(u,-v)$ are polynomials of the same degree, so $p(u,v)$ must be identically constant. That is, there exists a real number $k \neq 0$ such that

$$\widetilde{U}'(u,-v) \equiv k\widetilde{U}'(u,v), \quad \widetilde{V}'(u,-v) \equiv -k\widetilde{V}'(u,v).$$

This implies that $\widetilde{P}(x,\bar{x}) \equiv k\overline{\widetilde{P}(x,\bar{x})}$, whence $a'_0 = k\bar{a}'_0$. But then $\bar{a}'_0 = ka'_0$, hence $a'_0 = k^2 a'_0$. Since $a'_0 \neq 0$, $k = \pm 1$. Thus $a'_0 = \pm\bar{a}'_0$, which by (5.46) implies that for system a_0 one of conditions (5.47) and (5.48) must hold. □

The condition that a_0 be in $E_P(a)$, that is, that $\widetilde{U}$ and $\widetilde{V}$ be relatively prime polynomials, is essential to the truth of Lemma 5.3.2. A counterexample otherwise is given by the system $\dot{u} = -u(v-1)^2(v+1)$, $\dot{v} = v(v-1)^2(v+1)$, whose trajectories are symmetric with respect to the u-axis as point sets (that is, without regard to the direction of flow), but which exhibits mirror symmetry in the strip $|v| < 1$ but time-reversible symmetry in the strips $|v| > 1$. Even so, this restriction on $\widetilde{U}$ and $\widetilde{V}$ is no real hindrance to finding all axes of symmetry of a general system a_0. We simply identify the common factors F of $\widetilde{U}$ and $\widetilde{V}$ and remove them, creating a new system a'_0, all of whose axes of symmetry are specified by conditions (5.47) and (5.48). If the zero set Z of F is symmetric with respect to an axis of symmetry L' of system a'_0, then L' is an axis of symmetry of the original system a_0; otherwise it is not. Conversely, if L is an axis of symmetry of system a_0, then just a few moments' reflection shows us that it is an axis of symmetry with respect to system a'_0, since trajectories of a'_0 are unions of trajectories of a_0 (Exercise 5.16).

The following example illustrates the use of Lemmas 5.3.1 and 5.3.2 and the computations involved.

Example 5.3.3. We will locate the axes of symmetry for the real system

$$\dot{u} = -2uv + u^3 - 3uv^2, \quad \dot{v} = -u^2 + v^2 + 3u^2v - v^3. \tag{5.50}$$

First we express the system in complex form by differentiating $x = u + iv$ and making the substitutions $u = (x+\bar{x})/2$ and $v = (x-\bar{x})/(2i)$ to obtain

$$\dot{x} = x^3 - i\bar{x}^2. \tag{5.51}$$

This system is a particular element of the family $\dot{x} = -a_{-12}\bar{x}^2 - a_{20}x^3$ on $\mathbb{C}$, which in turn lies in the family

$$\dot{x} = -a_{-12}y^2 - a_{20}x^3, \quad \dot{y} = b_{20}x^2y + b_{2,-1}x^2 \tag{5.52}$$

on $\mathbb{C}^2$. Thus here $\widetilde{S} = \{(-1,2),(2,0)\}$, so that $\zeta = (-3,2,-2,3)$. Hence because $U_\varphi = \operatorname{diag}(e^{i\zeta_1\varphi}, e^{i\zeta_2\varphi}, e^{i\zeta_3\varphi}, e^{i\zeta_4\varphi})$, equations (5.47) and (5.48) are

$$\begin{pmatrix} e^{3\varphi_0 i} & 0 \\ 0 & e^{-2\varphi_0 i} \end{pmatrix} \begin{pmatrix} \bar{a}_{-1,2} \\ \bar{a}_{20} \end{pmatrix} = \pm \begin{pmatrix} e^{-3\varphi_0 i} & 0 \\ 0 & e^{2\varphi_0 i} \end{pmatrix} \begin{pmatrix} a_{-1,2} \\ a_{20} \end{pmatrix}. \tag{5.53}$$

Choosing the plus sign, one collection of lines of symmetry $x = re^{i\varphi_0}$ arises in correspondence with solutions φ_0 of the system of equations (observing the sign convention in (5.52), which reflects that of (5.1))

$$\begin{aligned} e^{6\varphi_0 i} &= \frac{a_{-12}}{\bar{a}_{-12}} = \frac{i}{-i} = -1 &&: \quad \varphi_0 \in \{\tfrac{\pi}{6}, \tfrac{\pi}{2}, \tfrac{5\pi}{6}, \tfrac{7\pi}{6}, \tfrac{3\pi}{2}, \tfrac{11\pi}{6}\} \\ e^{4\varphi_0 i} &= \frac{\bar{a}_{20}}{a_{20}} = \frac{-1}{-1} = 1 &&: \quad \varphi_0 \in \{0, \tfrac{\pi}{2}, \pi, \tfrac{3\pi}{2}\}. \end{aligned}$$

The sole solution is the line corresponding to $\varphi = \pi/2$, the v-axis. This is a mirror symmetry of the real system.

Choosing the minus sign in (5.53), a second collection of lines of symmetry $x = re^{i\varphi_0}$ arises in correspondence with solutions φ_0 of the system of equations

$$\begin{aligned} e^{6\varphi_0 i} &= 1 &&: \quad \varphi_0 = \{0, \tfrac{\pi}{3}, \tfrac{2\pi}{3}, \pi, \tfrac{4\pi}{3}, \tfrac{5\pi}{3}\} \\ e^{4\varphi_0 i} &= -1 &&: \quad \varphi_0 = \{\tfrac{\pi}{4}, \tfrac{3\pi}{4}, \tfrac{5\pi}{4}, \tfrac{7\pi}{4}\}. \end{aligned}$$

There is no solution, so the real system does not possess a time-reversible symmetry.

We will now describe axes of symmetry in terms of invariants of the rotation group. If we are interested in finding lines of symmetry of a single specific system, as we just did in Example 5.3.3, there is nothing to be gained by this approach. But by applying the theory developed in Section 5.1, we will be able to derive a single equation whose solutions correspond to axes of symmetry, rather than a collection of equations whose common solutions yield the axes, as was the case with the example. This will allow us to readily count solutions, and thus obtain the bound in Corollary 5.3.6 below. As before, a circumflex accent denotes the involution of a vector; that is, if $d = (d_1, \dots, d_\ell)$, then $\hat{d} = (d_\ell, \dots, d_1)$. Clearly $\hat{\hat{d}} = d$.

Theorem 5.3.4. *Fix the ordered index set* $\widetilde{S}$, *hence the real family in complex form* (5.45), *the complex family* (5.1), *and the rotation group* U_φ *for the family* (5.1). *Fix* $a_0 \in E(a)$. *If, for all unary and binary invariants* $J(a,b)$ *of family* (5.1),

$$J(\bar{a}_0, \hat{a}_0) = J(a_0, \bar{\hat{a}}_0) \tag{5.54}$$

holds, or if, for all unary and binary invariants $J(a,b)$ *of family* (5.1),

$$J(\bar{a}_0, \hat{a}_0) = J(-(a_0, \bar{\hat{a}}_0)) \tag{5.55}$$

holds, then system (5.45) *with coefficient vector* a_0 *has an axis of symmetry. Conversely, if* $a_0 \in E_P(a)$ *and system* a_0 *has an axis of symmetry, then either* (5.54) *or* (5.55) *holds.*

Proof. Note at the outset that for $a_0 \in E(a)$, (5.47) holds if and only if

$$(\bar{a}_0, \hat{a}_0) = U_{2\varphi_0} \cdot (a_0, \bar{\hat{a}}_0) \tag{5.56}$$

holds and (5.48) holds if and only if

$$(\bar{a}_0, \hat{a}_0) = -U_{2\varphi_0} \cdot (a_0, \bar{\hat{a}}_0) \tag{5.57}$$

holds. Thus for $a_0 \in E_P(a)$ the truth of (5.56) or (5.57) is equivalent to the line $x = re^{i\varphi_0}$ being an axis of symmetry and is sufficient for the line $x = re^{i\varphi_0}$ being an axis of symmetry for any $a_0 \in E(a)$.

Fix $a_0 \in E(a)$ and suppose (5.54) holds. Note that $R(a_0, \bar{\hat{a}}_0) = R(\bar{a}_0, \hat{a}_0)$ and the corresponding coordinates of the vectors $(a_0, \bar{\hat{a}}_0)$ and $(\bar{a}, \hat{a})$ have the same moduli. Hence, by Corollary 5.1.16, (5.54) implies that there exists a φ_0 such that (5.56) holds, or equivalently $U_{-\varphi_0} \cdot (\bar{a}_0, \hat{a}_0) = U_{\varphi_0} \cdot (a_0, \bar{\hat{a}}_0)$, which in expanded form is $(U^{(a)}_{-\varphi_0} \cdot \bar{a}_0, U^{(b)}_{-\varphi_0} \cdot \hat{a}_0) = (U^{(a)}_{\varphi_0} \cdot a_0, U^{(b)}_{\varphi_0} \cdot \bar{\hat{a}}_0)$. Thus, by Lemma 5.3.1, $x = re^{i\varphi_0}$ is an axis of symmetry of system a_0. The proof that (5.55) implies (5.48) is similar.

Suppose now that for $a_0 \in E_P(a)$ the corresponding system (5.45) has an axis of symmetry. Then by Lemma 5.3.2 either (5.47) or (5.48) holds. Suppose (5.47) holds with $\varphi = \varphi_0$. Then $a_0 = U^{(a)}_{-2\varphi_0} \bar{a}_0$ and from (5.4) we see that $\bar{\hat{a}}_0 = U^{(b)}_{-2\varphi_0} \cdot \hat{a}_0$. Thus $(a_0, \bar{\hat{a}}_0) = U_{-2\varphi_0} \cdot (\bar{a}_0, \hat{a}_0)$. But then for any invariant $J(a,b)$ of the rotation group for family (5.1), since $U_{-2\varphi_0}$ is in the group, $J(\bar{a}_0, \hat{a}_0) = J(U_{-2\varphi_0} \cdot (\bar{a}_0, \hat{a}_0)) = J(a_0, \bar{\hat{a}}_0)$, so (5.54) holds for all invariants. Similarly, if (5.48) holds, then (5.55) holds for all invariants J of family (5.1). □

Theorem 5.3.5. *Let an ordered index set* $\widetilde{S}$, *hence a real family of systems* (5.44) *in complex form* (5.45) *and a complex family* (5.1) *on* $\mathbb{C}^2$, *be given. Let a specific vector of coefficients* $a_0 = (a_{p_1 q_1}, \ldots, a_{p_\ell q_\ell}) \in E_P(a)$ *be given.*

1. *Suppose* $a_{p_j q_j} \neq 0$ *for* $1 \leq j \leq \ell$. *Then the number of axes of symmetry of system* a_0, *that is, of the corresponding system* (5.45) *with vector* a_0 *of coefficients, is* $\mathrm{GCD}(\zeta)$ *if exactly one of conditions* (5.54) *and* (5.55) *holds and is* $2\mathrm{GCD}(\zeta)$ *if both of them hold, where* ζ *is the characteristic vector of family* (5.44), *which is defined by* (5.10).
   a. *When* (5.54) *holds, a collection of axes of symmetry* $x = re^{i\varphi}$ *in* $\mathbb{C}$ *corresponds to solutions* $\varphi = \varphi_0$ *of the equation*

$$[\mu]|_{(\bar{a}_0, \hat{a}_0)} = [\mu]|_{(a_0, \bar{\hat{a}}_0)} e^{2i\mathrm{GCD}(\zeta)\varphi_0}, \tag{5.58}$$

   where μ *is any solution* $\nu = \mu$ *of* (5.14); *the same collection is determined by any such* μ. *If* (5.55) *does not hold, then there are no other axes of symmetry.*
   b. *When* (5.55) *holds, a collection of axes of symmetry* $x = re^{i\varphi}$ *in* $\mathbb{C}$ *corresponds to solutions* $\varphi = \varphi_0$ *of the equation*

$$[\mu]|_{(\bar{a}_0,\hat{a}_0)} = [\mu]|_{-(a_0,\check{a}_0)} e^{2i\text{GCD}(\zeta)\varphi_0}, \tag{5.59}$$

where μ is any solution $\nu = \mu$ of (5.14)*; the same collection is determined by any such μ. If* (5.54) *does not hold, then there are no other axes of symmetry.*

2. *Define $\chi : E(a) \to \mathbb{C}^{2\ell}$ by*

$$\chi(a) := \zeta(a,\hat{a}) = (\zeta_1 a_{p_1q_1}, \ldots, \zeta_\ell a_{p_\ell q_\ell}, \zeta_{\ell+1}\bar{a}_{p_\ell q_\ell}, \ldots, \zeta_{2\ell}\bar{a}_{p_1q_1}).$$

Suppose $\chi(a_0) = (0,\ldots,0)$. If either of conditions (5.54) *and* (5.55) *holds, then for every $\varphi_0 \in [0,2\pi)$ the line $x = re^{i\varphi_0}$ is an axis of symmetry of system* (5.45).

Proof. Suppose $a_0 \in E_P(a)$. By Lemmas 5.3.1 and 5.3.2, axes of symmetry are determined by (5.47) and (5.48), which, as noted in the first paragraph of the proof of Theorem 5.3.4, are equivalent to (5.56) and (5.57), respectively. By Proposition 5.1.17 with $\mathbf{c} = (\bar{a}_0,\hat{a}_0)$ and $\mathbf{d} = (a_0,\check{a}_0) = \bar{\mathbf{c}}$, (5.56) (respectively, (5.57)) holds if and only if (5.58) (respectively, (5.59)) does, for any solution $\nu = \mu$ of (5.14). But no entry in a_0 is zero, so neither side in equations (5.58) and (5.59) can be zero, hence each has exactly 2GCD(ζ) solutions in $[0,2\pi)$, none in common and all occurring in pairs φ_0 and $\varphi_0 + \pi$. Thus each determines GCD(ζ) lines of symmetry through the origin, all distinct from any that are determined by the other condition. This proves point (1).

Now suppose $a_0 \in E_P(a)$ and $\chi(a_0) = (0,\ldots,0)$. Denote a_0 by $(a_1,\ldots,a_\ell)$, so that the latter equation is $(\zeta_1 a_1,\ldots,\zeta_\ell a_\ell,\zeta_{\ell+1}\bar{a}_\ell,\ldots,\zeta_{2\ell}\bar{a}_1) = (0,\ldots,0)$. For any index $j \le \ell$ for which $\zeta_j = 0$, the monomial $a_{p_jq_j}$ is an invariant polynomial for family (5.1), so that (5.54) (respectively, (5.55)) holds for all unary invariants only if $a_j = \bar{a}_j$ (respectively, $a_j = -\bar{a}_j$). For $\varphi_0 \in [0,2\pi)$, the matrix representing $U^{(a)}_{\varphi_0}$ is the diagonal matrix with the number $e^{i\zeta_j\varphi_0}$ in position (j,j), which is 1 when $\zeta_j = 0$. Thus because by hypothesis $a_j = 0$ if $\zeta_j \neq 0$, (5.47) (respectively, (5.48)) holds for every φ_0, and the result follows from Lemma 5.3.1. □

It is apparent from the proof of point (1) that we can allow zero entries in a_0 and still use equations (5.58) and (5.59) to identify lines of symmetry as long as we restrict to solutions μ of (5.14) that have zero entries where a_0 does and apply the convention $0^0 = 1$, although this observation matters only if we are trying to locate common axes of symmetry in an entire family in which certain variable coefficients are allowed to vanish, or perhaps show that no axes of symmetry can exist. We also note explicitly that the sets of solutions of equations (5.58) and (5.59) are disjoint.

An important consequence of Theorem 5.3.5 is that we can derive a bound on the number of lines of symmetry in terms of the degrees of the polynomials in system (5.44).

Corollary 5.3.6. *Suppose that in system* (5.44) *$\widetilde{U}$ and $\widetilde{V}$ are polynomials for which $\max(\deg(\widetilde{U}),\deg(\widetilde{V})) = n$. If the system does not have infinitely many axes of symmetry, then it has at most $2n+2$ axes of symmetry.*

Proof. Suppose system (5.44) has at least one axis of symmetry, and let a_0 denote the vector of coefficients in its complex form (5.45). If the characteristic vector ζ

of the corresponding family (5.1) is $(0,\ldots,0)$, then every line through the origin is an axis of symmetry (Exercise 5.17). Otherwise, since for $(p,q) \in \widetilde{S}$ we have $-1 \leq p \leq n-1$ and $0 \leq q \leq n$, every component ζ_j of ζ of the corresponding family (5.1) satisfies $|\zeta_j| < n+1$. Therefore $\mathrm{GCD}(\zeta) \leq n+1$. If $a_0 \in E_P(a)$, then it follows immediately that the number of axes of symmetry is less than or equal to $2(n+1)$. If $a_0 \notin E_P(a)$ then we divide out the common factors in $\widetilde{U}$ and $\widetilde{V}$ and recall the fact that has already been noted that the number of axes of symmetry of the system thereby obtained is not exceeded by the number of axes of symmetry of the original system a_0. □

The bound given in the corollary is sharp (see Exercise 5.18).

Example 5.3.7. We return to the real system (5.50) and now locate the axes of symmetry using Theorem 5.3.5. As before we must derive the complex form (5.51) on $\mathbb{C}$ and view it as the first component of a specfic member of the family (5.52) on $\mathbb{C}^2$, for which $\widetilde{S} = \{(-1,2),(2,0)\}$, so that $\zeta = (-3,2,-2,3)$ and $\mathrm{GCD}(\zeta) = 1$. The next step in this approach is to find all the unary and binary irreducible invariant monomials of family (5.52). Applying Proposition 5.1.10, the unary irreducible invariant monomials are $J_1 = J_1(a_{-12},a_{20},b_{02},b_{2,-1}) = a_{-12}b_{2,-1}$ and $J_2 = a_{20}b_{02}$; the binary irreducible invariant monomials are $J_3 = a_{-12}^2 a_{20}^3$ and $J_4 = b_{02}^3 b_{2,-1}^2$. Observing the sign convention in (5.52), which reflects that of (5.1), the vector whose components are the coefficients of system (5.51) is $a_0 = (i,-1)$, so

$$\begin{aligned}
(\bar{a}_0,\hat{a}_0) &= (-i,-1,-1,i)\\
(a_0,\hat{\bar{a}}_0) &= (i,-1,-1,-i)\\
-(a_0,\hat{\bar{a}}_0) &= (-i,1,1,i).
\end{aligned}$$

Then

$$\begin{aligned}
J_1(\bar{a}_0,\hat{a}_0) &= (-i)(i) = 1\\
J_1(a_0,\hat{\bar{a}}_0) &= (i)(-i) = 1\\
J_1(-(a_0,\hat{\bar{a}}_0)) &= (-i)(i) = 1,
\end{aligned}$$

so that both (5.54) and (5.55) hold for J_1. Similar computations show that they both hold for J_2 but that only (5.54) holds for J_3 and J_4. Thus because (5.54) holds for all invariants of family (5.52) but (5.55) does not, all axes of symmetry of the real family will be picked out as solutions of the single equation (5.58), for any solution μ of (5.14), and (5.59) need not be considered.

Equation (5.14) is $-3\nu_1+2\nu_2-2\nu_3+3\nu_4 = 1$, for which we choose the solution $\mu = (0,0,1,1)$. Then (5.58) is

$$(-i)^0(-1)^0(-1)^1(i)^1 = (i)^0(-1)^0(-1)^1(-i)^1 e^{2i\varphi},$$

or $e^{2i\varphi_0} = -1$, and this single equation yields all the axes of symmetry $x = re^{i\varphi}$ of (5.50), namely the lines corresponding to $\varphi = \pi/2$ and $\varphi = 3\pi/2$, both of which are, of course, the v-axis.

We close this section with a result on axes of symmetry of a system (5.44) for which the origin is of focus or center type. Suppose the lowest-order terms in either of $\widetilde{U}$ and $\widetilde{V}$ have degree m; note that we allow $m > 1$. Then (5.44) has the form $\dot{u} = \widetilde{U}_m(u,v) + \cdots$, $\dot{v} = \widetilde{V}_m(u,v) + \cdots$, where each of $\widetilde{U}_m$ and $\widetilde{V}_m$ is either identically zero or a homogeneous polynomial of degree m, at least one of them is not zero, and omitted terms are of order at least $m+1$. Since (5.44) is a polynomial system, the origin is of focus or center type if and only if there are no directions of approach to it, which, by Theorem 64 of §20 of [12], is true if and only if the homogeneous polynomial $u\widetilde{V}_m(u,v) - v\widetilde{U}_m(u,v)$ vanishes only at the origin. Making the replacements $\widetilde{U}(u,v) = \widetilde{U}((x+\bar{x})/2, (x-\bar{x})/(2i)) = \operatorname{Re}\widetilde{P}(x,\bar{x})$, $\widetilde{V}(u,v) = \operatorname{Im}\widetilde{P}(x,\bar{x})$, $u = (x+\bar{x})/2$, and $v = (x-\bar{x})/(2i)$, the condition that the origin be of focus or center type for (5.44), expressed in terms of the vector of coefficients of its representation in complex form (5.45), is

$$\sum_{p+q=m-1} \left(a_{pq}x^{p+1}\bar{x}^{q+1} - \bar{a}_{pq}\bar{x}^{p+1}x^{q+1}\right) = 0 \quad \text{if and only if} \quad x = 0. \tag{5.60}$$

If the rotation indicated in the left equation in display (5.3) is performed for $\varphi = \varphi_0$, then condition (5.60) becomes

$$\sum_{p+q=m-1} \left(a(\varphi_0)_{pq}x'^{p+1}\bar{x}'^{q+1} - \bar{a}(\varphi_0)_{pq}\bar{x}'^{p+1}x'^{q+1}\right) = 0 \quad \text{if and only if} \quad x' = 0. \tag{5.61}$$

Proposition 5.3.8. *Let an ordered index set* $\widetilde{S}$, *hence a real family of systems* (5.44) *in complex form* (5.45) *and a complex family* (5.1) *on* $\mathbb{C}^2$, *be given. If for system* $a_0 \in E_P(a)$ *the origin is of focus or center type, then equation* (5.47) *has no solutions and condition* (5.54) *does not hold.*

Proof. Suppose, contrary to what we wish to show, that there exists a solution φ_0 to (5.47), that is, that $U^{(a)}_{-\varphi_0} \cdot \bar{a}_0 = U^{(a)}_{\varphi_0} \cdot a_0$. Then using (5.46) to eliminate the minus sign that is attached to φ_0 on the left, we obtain $\overline{U^{(a)}_{\varphi_0} \cdot a_0} = U^{(a)}_{\varphi_0} \cdot a_0$. But then because $U^{(a)}_{\varphi_0} \cdot a_0 = (a(\varphi_0)_{p_1q_1}, \ldots, a(\varphi_0)_{p_\ell q_\ell})$, this implies that $\overline{a(\varphi_0)_{pq}} = a(\varphi_0)_{pq}$ for all $(p,q) \in \widetilde{S}$. But then the sum in (5.61) vanishes for any x' satisfying $x' = \bar{x}' \neq 0$, a contradiction. Thus (5.47) has no solutions.

Again suppose that condition (5.54) is satisfied. Then by Corollary 5.1.16 there exists some $\varphi = 2\varphi_0$ such that $(\bar{a}_0, \hat{a}_0) = U^{(a)}_{2\varphi_0} \cdot (a_0, \bar{\hat{a}}_0)$, so that $\bar{a}_0 = U^{(a)}_{2\varphi_0} \cdot a_0$ or $U^{(a)}_{-\varphi_0} \cdot \bar{a}_0 = U^{(a)}_{\varphi_0} \cdot a_0$, that is, (5.47) holds, just shown to be impossible. □

5.4 Notes and Complements

The theory of algebraic invariants of ordinary differential equations presented in this chapter was developed in large part by K. S. Sibirsky and his collaborators, although many other investigators have played a role. The general theory, which applies to analytic systems in n variables, goes far beyond what we have discussed here. The action of any subgroup of the n-dimensional general linear group, not just the rotation group, is allowed. Moreover, in addition to polynomial invariants in the coefficients of the system of differential equations, which we have studied here, there are two other broad classes of objects of interest, the analogous comitants, which are polynomials in both the coefficients and the variables, and the syzygies, which are polynomial identities in the invariants and comitants. A standard reference is [178], in which a full development, and much more than we could indicate here, may be found.

The algorithm presented in Section 5.2 is from [102, 150].

Exercises

5.1 Show that the set C_4 of Example 5.1.2 is a group under multiplication.

5.2 Repeat Example 5.1.3 for the family (3.131) of all systems of the form (3.69) with quadratic nonlinearities.

5.3 Prove Proposition 5.1.6.
Hint. If $\nu \in \mathbb{N}_0^{2\ell}$ satisfies (5.7), then clearly it satisfies (5.8). Establish the converse in two steps: (i) rearrange (5.8) to conclude that $L_1(\nu) = L_2(\nu) = n$ for some $n \in \mathbb{Z}$; (ii) show that $n \in \mathbb{N}_0$ by writing out the right-hand side of the equation $2n = L_1(\nu) + L_2(\nu)$ and collecting on $\nu_1, \dots, \nu_{2\ell}$. You must use the restrictions on the elements of $(p_j, q_j) \in \widetilde{S}$.

5.4 [Referenced in the proof of Proposition 5.1.10.] Show that if $(\mu, \eta) \in \mathbb{N}^2$ and $(\zeta_r, \zeta_s) \in \mathbb{Z}^2 \setminus \{(0,0)\}$ satisfy $\mu\,\zeta_r + \eta\,\zeta_s = 0$, then there exists $a \in \mathbb{N}$ such that

$$\mu = \left(\frac{|\zeta_s|}{\mathrm{GCD}(\zeta_r, \zeta_s)}\right) \cdot a \qquad \text{and} \qquad \eta = \left(\frac{|\zeta_r|}{\mathrm{GCD}(\zeta_r, \zeta_s)}\right) \cdot a\,.$$

5.5 [Referenced in the proof of Proposition 5.1.10.] Complete the proof of part (2) of Proposition 5.1.10 by showing that
a. any invariant monomial of the form $z_r^{\mu} z_s^{\eta} z_t^{\theta}$, where z_r and z_t are conjugate variables and z_s is distinct from z_r and z_t, factors as a product of unary and binary invariant monomials, and
b. any invariant monomial of the form $z_r^{\mu} z_s^{\eta} z_t^{\theta} z_w^{\gamma}$, where z_r and z_t are conjugate variables, z_s and z_w are conjugate variables, and z_s is distinct from z_r and z_t, factors as a product of unary and binary invariant monomials.

5.6 Derive Remark 5.1.11 from Proposition 5.1.10.

5.7 [Referenced in the proof of Lemma 5.1.13.] The greatest common divisor of a set of numbers $\{p_1, \dots, p_s\}$ in $\mathbb{Z}$ is the unique number $d \in \mathbb{N}$ such that d divides each p_j (that is, leaves remainder zero when it is divided into p_j) and is itself divisible by any number that divides all the p_j. It can be found by repeated application of the Euclidean Algorithm (the algorithm in Table 1.10 on page 52 with suitable change in terminology). In analogy with Exercise 1.10, show that if $d = \mathrm{GCD}(p_1, \ldots, p_s)$, then there exist $\alpha_1, \ldots, \alpha_s \in \mathbb{Z}$ (not unique) such that $d = \alpha_1 p_1 + \cdots \alpha_s p_s$. (Compare with Exercise 1.10.)

5.8 [Referenced in the proof of Theorem 5.1.15.] Show that if for family (5.1) the characteristic vector $\zeta \in \mathbb{Z}^{2\ell}$ is zero, $\zeta = (0, \ldots, 0)$, then every unary and binary invariant agrees on specfic elements $\mathbf{c}$ and $\mathbf{d}$ of $E(a,b)$ only if $\mathbf{c} = \mathbf{d}$.

5.9 [Referenced in the proof of Theorem 3.5.8, end of Section 5.1.] For $\mathscr{D}$ as defined in the proof of Theorem 3.5.8, show how the inclusion $\mathscr{D} \subset \mathbf{V}(I_{\mathrm{sym}}) \setminus \mathscr{R}$ follows directly from the characterization (3.96) of $\mathscr{R}$.

5.10 Find a second Hilbert Basis for the monoid $\mathscr{M}$ of Example 5.2.6.

5.11 a. Consider the set $V = \{(x_1, x_2, x_3, x_4) : x_1 + 4x_2 - 4x_3 - x_4 = 0\}$, the vector subspace of $\mathbb{R}^4$ that is determined by the same equation (5.43) that determines the monoid $\mathscr{M}$ of Example 5.2.6. Show that the Hilbert Basis $\mathscr{H}$ of $\mathscr{M}$ is not a vector space basis of V.

b. In particular, find an element of $\mathscr{M}$ that has at least two representations as an $\mathbb{N}_0$-linear combination of elements of $\mathscr{H}$.

5.12 [Referenced in the proof of Proposition 6.3.5.] Show that in general, for an ordered set $\widetilde{S}$, the corresponding monoid $\mathscr{M}$, and a Hilbert Basis $\mathscr{H}$ of $\mathscr{M}$, any element μ of $\mathscr{M}$ has at most finitely many representations as an $\mathbb{N}_0$-linear combination of elements of $\mathscr{H}$.

5.13 Consider the family

$$\begin{aligned} \dot{x} &= \quad i(x - a_{20}x^3 - a_{11}x^2y - a_{02}xy^2 - a_{-13}y^3) \\ \dot{y} &= -i(y - b_{3,-1}x^3 - b_{20}x^2y - b_{11}xy^2 - b_{02}y^3). \end{aligned} \tag{5.62}$$

a. Find the symmetry variety of family (5.62).

b. Find a Hilbert basis for the monoid $\mathscr{M}$ for family (5.62).

Hint. See Theorem 6.4.3 and the proof of Lemma 6.4.7 for the answers.

5.14 Find the symmetry variety of the system

$$\begin{aligned} \dot{x} &= \quad x - a_{30}x^4 - a_{21}x^3y - a_{12}x^2y^2 - a_{03}xy^3 - a_{-14}y^4 \\ \dot{y} &= -y + b_{4,-1}x^4 + b_{30}x^3y + b_{21}x^2y^2 + b_{12}xy^3 + b_{0,3}y^4. \end{aligned}$$

5.15 [Referenced in the proof of Lemma 5.3.2.] Let a_0 and a_0' be vectors whose components are the coefficients of two systems of the form (5.45) and that satisfy $a_0' = U_{\varphi_0} \cdot a_0$ for some φ_0. Show that either both a and a' are in $E_P(a)$ or that neither is.

5.16 Give a precise argument to show that if the common factor F of $\widetilde{U}$ and $\widetilde{V}$ of a system (5.44) is discarded, then any axis of symmetry of the original system is an axis of symmetry of the new system thereby obtained.

5.17 [Referenced in the proof of Corollary 5.3.6.] Suppose system (5.44) has at least one axis of symmetry and that the characteristic vector ζ of the corresponding family (5.1) is $(0, \ldots, 0)$. Show that every line through the origin is an axis of symmetry.

5.18 Show that the system $\dot{u} = \mathrm{Re}(u - iv)^n$, $\dot{v} = \mathrm{Im}(u - iv)^n$ has $2n + 2$ axes of symmetry.

5.19 Find the axes of symmetry of the system $\dot{x} = x - x^3 + 2x\bar{x}^2$.

5.20 Find the axes of symmetry of the system

$$\dot{u} = -6u^2v - 8uv^2 + 6v^3, \quad \dot{v} = 6u^3 + 8u^2v - 6uv^2.$$

Note the collapsing of systems (5.47) and (5.48) into single equations.

Chapter 6
Bifurcations of Limit Cycles and Critical Periods

In this chapter we consider systems of ordinary differential equations of the form

$$\dot{u} = \widetilde{U}(u,v), \quad \dot{v} = \widetilde{V}(u,v), \tag{6.1}$$

where u and v are real variables and $\widetilde{U}(u,v)$ and $\widetilde{V}(u,v)$ are polynomials for which $\max(\deg \widetilde{U}, \deg \widetilde{V}) \leq n$. The second part of the sixteenth of Hilbert's well-known list of open problems posed in the year 1900 asks for a description of the possible number and relative locations of limit cycles (isolated periodic orbits) occurring in the phase portrait of such polynomial systems. The minimal *uniform* bound $H(n)$ on the number of limit cycles for systems (6.1) (for some fixed n) is now known as the nth *Hilbert number*.

Despite the simplicity of the statement of the problem not much progress has been made even for small values of n, and even then a number of important results have later been shown either to be false or to have faulty proofs. For many years it was widely believed, for example, that $H(2) = 3$, but around 1980 Chen and Wang ([39]) and Shi ([172]) constructed examples of quadratic systems (systems (6.1) with $n = 2$) having at least four limit cycles. About the same time the correctness of Dulac's proof of a fundamental preliminary to Hilbert's 16th problem, that any *fixed* polynomial system has but a finite number of limit cycles, was called into question (the proof was later shown to be faulty). Examining the question for quadratic systems, in 1983 Chicone and Shafer ([46]) proved that a fixed quadratic system has only a finite number of limit cycles in any bounded region of the phase plane. In 1986 Bamón ([14]) and Romanovski ([152]) extended this result to the whole phase plane, thereby establishing the correctness of Dulac's theorem in the quadratic case. A few years later Dulac's theorem was proved for an arbitrary polynomial system by Ecalle ([69]) and Il'yashenko ([99]). Even so, as of this writing no uniform bound on the number of limit cycles in polynomial systems of fixed degree is known. That is, it is unknown whether or not $H(n)$ is even finite except in the trivial cases $n = 0$ and $n = 1$.

Two fundamental concepts used in addressing the problem of estimating $H(n)$ are the twin ideas of a limit periodic set and of the cyclicity of a such a set, ideas

V.G. Romanovski, D.S. Shafer, *The Center and Cyclicity Problems*,
DOI 10.1007/978-0-8176-4727-8_6,

used by Bautin in the seminal paper [17] in which he proved that $H(2) \geq 3$. To define them, we consider a family of systems (6.1) with coefficients drawn from a specified parameter space $\mathscr{E}$ equipped with a topology. A *limit periodic set* is a point set Γ in the phase portrait of the system (6.1) that corresponds to some choice e_0 of the parameters that has the property that a limit cycle can be made to bifurcate from Γ under a suitable but arbitrarily small change in the parameters. That is, for any neighborhood U of Γ in $\mathbb{R}^2$ and any neighborhood N of e_0 in $\mathscr{E}$ there exists $e_1 \in N$ such that the system corresponding to parameter choice e_1 has a limit cycle lying wholly within U. The limit periodic set Γ has *cyclicity* c with respect to $\mathscr{E}$ if and only if for any choice e of parameters in a neighborhood of e_0 in $\mathscr{E}$ the corresponding system (6.1) has at most c limit cycles wholly contained in a neighborhood of Γ, and c is the smallest number with this property. Examples of limit periodic sets are singularities of focus or center type, periodic orbits (not necessarily limit cycles), and the set formed by a saddle point and a pair of its stable and unstable separatrices that comprise the same point set (a "homoclinic loop"). More specifically, consider a system of the form (6.11) below with $\beta \neq 0$ and $|\alpha|$ small. If α is made to cross 0 from negative to positive, then the singularity at the origin changes from a stable to an unstable focus, and typically a limit cycle surrounding the origin is created or destroyed in what is called a Hopf bifurcation. Roussarie showed ([155]) that if it could be established (again for n fixed) that every limit periodic set for family (6.1) has finite cyclicity (under a natural compactification of the parameter and phase spaces; see Section 6.5), then it would follow that $H(n)$ is finite. This program seems feasible at least for quadratic systems, for which it is currently under way.

In this chapter we consider the problem of the cyclicity of a simple singularity of system (6.1) (that is, one at which the determinant of the linear part is nonzero), a problem that is known as the *local 16th Hilbert problem.* We describe a general method based on ideas of Bautin to treat this and similar bifurcation problems, and apply the method to resolve the cyclicity problem for singular points of quadratic systems and the problem of bifurcations of critical periods in the period annulus of centers for a family of cubic systems.

6.1 Bautin's Method for Bifurcation Problems

Bautin's approach to bifurcation of limit cycles from singularities of vector fields is founded on properties of zeros of analytic functions of several variables depending on parameters, which we address in this section. Let $\mathscr{E}$ be a subset of $\mathbb{R}^n$ and let $\mathscr{F} : \mathbb{R} \times \mathscr{E} \to \mathbb{R} : (z, \theta) \mapsto \mathscr{F}(z, \theta)$ be an analytic function, which we will write in a neighborhood of $z = 0$ in the form

$$\mathscr{F}(z, \theta) = \sum_{j=0}^{\infty} f_j(\theta) z^j, \tag{6.2}$$

where, for $j \in \mathbb{N}_0$, $f_j(\theta)$ is an analytic function and for any $\theta^* \in \mathscr{E}$ the series (6.2) is convergent in a neighborhood of $(z, \theta) = (0, \theta^*)$. In all situations of interest to us we will be concerned solely with the number of *positive* solutions (for any fixed parameter value θ^*) of the equation $\mathscr{F}(z, \theta^*) = 0$ in a neighborhood of $z = 0$ in $\mathbb{R}$. Thus we define the multiplicity of the parameter value θ^* with respect to $\mathscr{E}$ as follows.

Definition 6.1.1. For any $\theta^* \in \mathscr{E}$ and any sufficiently small $\varepsilon > 0$, let $z(\theta, \varepsilon)$ denote the number of isolated zeros of $\mathscr{F}(z, \theta)$ in the interval $(0, \varepsilon) \subset \mathbb{R}$. The point $\theta^* \in \mathscr{E}$ is said to have *multiplicity* c with respect to the space $\mathscr{E}$ at the origin in $\mathbb{R}$ if there exist positive constants δ_0 and ε_0 such that, for every pair of numbers δ and ε that satisfy $0 < \delta < \delta_0$ and $0 < \varepsilon < \varepsilon_0$,

$$\max\{z(\theta, \varepsilon) : |\theta - \theta^*| \leq \delta\} = c,$$

where $|\cdot|$ denotes the usual Euclidean norm on $\mathbb{R}^n$.

There are two possibilities that are of interest to us in regard to the flatness of $\mathscr{F}(z, \theta^*)$ at $z = 0$:
(i) there exists $m \in \mathbb{N}_0$ such that $f_0(\theta^*) = \cdots = f_m(\theta^*) = 0$ but $f_{m+1}(\theta^*) \neq 0$;
(ii) $f_j(\theta^*) = 0$ for all $j \in \mathbb{N}_0$.
In the first case it is not difficult to see that the multiplicity of θ^* is at most m (for example, see Corollary 6.1.3 below). Case (ii) is much more subtle, but there is a method for its treatment suggested by Bautin in [17]. We first sketch out the method in the case that the functions f_j are polynomial functions of θ. The idea is to find a basis of the ideal $\langle f_0(\theta), f_1(\theta), f_2(\theta), \ldots \rangle$ in the ring of polynomials $\mathbb{R}[\theta]$. By the Hilbert Basis Theorem (Theorem 1.1.6) there is always a finite basis. Adding polynomials f_j as necessary in order to fill out an initial string of the sequence $\{f_0, f_1, \ldots\}$, we may choose the first $m+1$ polynomials $\{f_0(\theta), \ldots, f_m(\theta)\}$ for such a basis. Using this basis we can then write the function $\mathscr{F}$ in the form

$$\mathscr{F}(z, \theta) = \sum_{j=0}^{m} f_j(\theta)(1 + \Phi_j(z, \theta))z^j, \tag{6.3}$$

where $\Phi_j(0, \theta) = 0$ for $j = 0, 1, \ldots, m$. Therefore the function $\mathscr{F}(z, \theta)$ behaves like a polynomial in z of degree m near $\theta = \theta^*$, hence can have at most m zeros for any θ in a neighborhood of θ^*, as will be shown in Proposition 6.1.2.

We will see below that the local 16th Hilbert problem is just the problem of the multiplicity of the function $\mathscr{P}(\rho) = \mathscr{R}(\rho) - \rho$, where $\mathscr{R}(\rho)$ is the Poincaré return map (3.13). In the case that $\mathscr{F}$ is the derivative of the period function $T(\rho)$ defined by (4.26) and $\mathscr{E}$ is the center variety of system (6.1), we have the so-called *problem of bifurcations of critical periods*, which we will consider in Section 6.4.

By the discussion in the preceding paragraphs, to investigate the multiplicity of $\mathscr{F}(z, \theta)$ we will need to be able to count isolated zeros of functions expressed in the form of (6.3). The following proposition is our tool for doing so.

Proposition 6.1.2. *Let $Z : \mathbb{R} \times \mathbb{R}^n \to \mathbb{R}$ be a function that can be written in the form*

$$Z(z,\theta) = f_1(\theta)z^{j_1}(1+\psi_1(z,\theta)) + \cdots + f_s(\theta)z^{j_s}(1+\psi_s(z,\theta)), \tag{6.4}$$

where $j_u \in \mathbb{N}$ for $u = 1,\ldots,s$ and $j_1 < \cdots < j_s$, and where $f_j(\theta)$ and $\psi_j(z,\theta)$ are real analytic functions on $\{(z,\theta) : |z| < \varepsilon \text{ and } |\theta - \theta^*| < \delta\}$, for some positive real numbers δ and ε, and $\psi_j(0,\theta^*) = 0$ for $j = 1,\ldots,s$. Then there exist numbers ε_1 and δ_1, $0 < \varepsilon_1 \le \varepsilon$ and $0 < \delta_1 \le \delta$, such that for each fixed θ satisfying $|\theta - \theta^*| < \delta_1$, the equation

$$Z(z,\theta) = 0, \tag{6.5}$$

regarded as an equation in z alone, has at most $s-1$ isolated solutions in the interval $0 < z < \varepsilon_1$.

Proof. Let δ_1 and ε_1 be such that $0 < \delta_1 < \delta$ and $0 < \varepsilon_1 < \varepsilon$ and $|\psi_j(z,\theta)| < 1$ if $|z| \le \varepsilon_1$ and $|\theta - \theta^*| \le \delta_1$ for $j = 1,\ldots,s$. Let $B(\theta^*,\delta_1)$ denote the closed ball in $\mathbb{R}^n$ of radius δ_1 centered at θ^*. For each $j \in \{1,\ldots,s\}$, f_j is not the zero function, else the corresponding term is not present in (6.4). We begin by defining the set V_0 by $V_0 := \{\theta \in B(\theta^*,\delta_1) : f_j(\theta) = 0 \text{ for all } j = 1,\ldots,s\}$, a closed, proper subset of $B(\theta^*,\delta_1)$. For $\theta_0 \in V_0$, as a function of z, $Z(z,\theta_0)$ vanishes identically on $(0,\varepsilon_0)$, so the proposition holds for $\theta_0 \in V_0$.

For any $\theta_0 \in B(\theta^*,\delta_1) \setminus V_0$, let $u \in \{1,\ldots,s\}$ be the least index for which $f_u(\theta_0) \neq 0$. Then $Z(z,\theta_0) = f_u(\theta_0)z^{j_u} + z^{j_u+1}g(z,\theta_0)$, where $g(z,\theta_0)$ is a real analytic function on $[-\varepsilon_1,\varepsilon_1]$. Thus the j_uth derivative of $Z(z,\theta_0)$ is nonzero at $z = 0$, so $Z(z,\theta_0)$ is not identically zero, hence has a finite number $S_0(\theta_0)$ of zeros in $(0,\varepsilon_1)$.

Let $V_1 = \{\theta \in B(\theta^*,\delta_1) : f_j(\theta) = 0 \text{ for } j = 2,\ldots,s\}$; $V_1 \supset V_0$. For $\theta_0 \in V_1$, if $f_1(\theta_0) = 0$, then $Z(z,\theta_0)$ vanishes identically on $(0,\varepsilon_1)$; if $f_1(\theta_0) \neq 0$, then as a function of z, $Z(z,\theta_0) = f_1(\theta_0)z^{j_1}(1+\psi_1(z,\theta_0))$ has no zeros in $(0,\varepsilon_1)$. Either way the proposition holds for $\theta \in V_1$.

For $\theta \in B(\theta^*,\delta_1) \setminus V_1$, we divide $Z(z,\theta)$ by $z^{j_1}(1+\psi_1(z,\theta))$ to form a real analytic function $\widetilde{Z}^{(1)}(z,\theta)$ of z on $[-\varepsilon_1,\varepsilon_1]$, then differentiate with respect to z to obtain a real analytic function $Z^{(1)}(z,\theta)$ of z on $[-\varepsilon_1,\varepsilon_1]$ that can be written in the form

$$\begin{aligned} Z^{(1)}(z,\theta) = f_2(\theta)(j_2-j_1)z^{j_2-j_1-1}(1+\psi_2^{(1)}(z,\theta)) + \cdots \\ + f_s(\theta)(j_s-j_1)z^{j_s-j_1-1}(1+\psi_s^{(1)}(z,\theta)), \end{aligned}$$

where $\psi_j^{(1)}(0,\theta^*) = 0$ for $j = 2,\ldots,s$. As a function of z, $\widetilde{Z}^{(1)}(z,\theta)$ has the same number $S_0(\theta)$ of zeros in $(0,\varepsilon_1)$ as does $Z(z,\theta)$. As a function of z, $Z^{(1)}(z,\theta)$ is not identically zero, hence has a finite number $S_1(\theta)$ of zeros in $(0,\varepsilon_1)$. By Rolle's Theorem, $\widetilde{Z}^{(1)}(z,\theta)$ has at most one more zero in $(0,\varepsilon_1)$ than does $Z^{(1)}(z,\theta)$, so $S_0(\theta) \le S_1(\theta) + 1$.

The function $Z^{(1)}(z,\theta)$ is of the same form as $Z(z,\theta)$ (incorporating the nonzero constant $j_2 - j_1$ into $f_2(\theta)$ and taking, if necessary, ε_1 smaller in order to satisfy the condition $|\psi_j^{(1)}(z,\theta)| < 1$ if $|z| \le \varepsilon_1$), so we may repeat the same procedure: define $V_2 = \{\theta \in B(\theta^*,\delta_1) : f_j(\theta) = 0 \text{ for } j = 3,\ldots,s\}$, which contains V_1 and on which

$Z^{(1)}$, as a function of z, either vanishes identically or has no zeros in $(0, \varepsilon_1)$, then for $\theta \in B(\theta^*, \delta_1) \setminus V_2$ divide $Z^{(1)}(z, \theta)$ by $z^{j_2 - j_1 - 1}(1 + \psi_2^{(1)}(z, \theta))$ to form $\widetilde{Z}^{(2)}(z, \theta)$, which has the same number $S_1(\theta)$ of zeros in $(0, \varepsilon_1)$ as $Z^{(1)}(z, \theta)$, and finally differentiate $\widetilde{Z}^{(2)}(z, \theta)$ with respect to z to obtain a real analytic function $Z^{(2)}(z, \theta)$ that, for each $\theta \in B(\theta^*, \delta_1) \setminus V_2$, has, as a function of z, a finite number $S_2(\theta)$ of zeros in $(0, \varepsilon_1)$, and $S_1(\theta) \leq S_2(\theta) + 1$, so that $S_0(\theta) \leq S_2(\theta) + 2$.

Taking $Z^{(0)}(z, \theta) = Z(z, \theta)$ and repeating the process for a total of $s - 1$ iterations, we obtain a sequence of sets $V_0 \subset V_1 \subset \cdots \subset V_{s-1}$ and of functions $Z^{(j)}(z, \theta)$, defined and analytic on $[-\varepsilon_1, \varepsilon_1]$ for $\theta \in B(\theta^*, \delta_1) \setminus V_j$, $j = 0, \ldots, s-1$, with the property that the proposition holds on V_j, $Z^{(j)}(z, \theta)$ has, as a function of z, a finite number $S_j(\theta)$ of zeros in $(0, \varepsilon_1)$, and $S_0(\theta) \leq S_j(\theta) + j$. In particular, the proposition holds for $\theta \in V_{s-1}$, and for $\theta \in B(\theta^*, \delta_1) \setminus V_{s-1}$ $S_0(\theta) \leq S_{s-1}(\theta) + (s-1)$. But we can write $Z^{(s-1)}(z, \theta) = f_s(\theta)(j_s - j_{s-1}) \cdots (j_s - j_2)(j_s - j_1) z^{j_s - j_{s-1} - 1}(1 + \psi_s^{(s-1)}(z, \theta))$ for some function $\psi_s^{(s-1)}$ satisfying $\psi_s^{(s-1)}(0, \theta^*) = 0$. As a function of z, $Z^{(s-1)}(z, \theta)$ is not identically zero, hence has no zeros in $(0, \varepsilon_1)$. Thus $S_0(\theta) \leq s - 1$, so the proposition holds for all $\theta \in B(\theta^*, \delta_1)$. □

Remark. If we are interested in the number of isolated solutions of (6.5) in the interval $(-\varepsilon_1, \varepsilon_1)$, then we obtain the upper bound $2s - 1$: by the proposition at most $s - 1$ isolated solutions in $(0, \varepsilon)$, by the same argument as in the proof of the proposition at most $s - 1$ isolated solutions in $(-\varepsilon_1, 0)$, plus a possible solution at $z = 0$.

Corollary 6.1.3. *Suppose the coefficients of the function $\mathscr{F}$ of* (6.2) *satisfy*

$$f_0(\theta^*) = \cdots = f_m(\theta^*) = 0, \; f_{m+1}(\theta^*) \neq 0.$$

Then the multiplicity of θ^* is less than or equal to m.

Proof. Because $f_{m+1}(\theta^*) \neq 0$, in a neighborhood of θ^* we can write the function $\mathscr{F}$ in (6.2) in the form

$$\mathscr{F}(z, \theta) = \sum_{j=0}^{m} f_j(\theta) z^j + f_{m+1}(\theta) z^{m+1}(1 + \psi(z, \theta)).$$

Now apply Proposition 6.1.2. □

Proposition 6.1.2 enables to us to count zeros of functions of the form (6.4), but functions of interest to us naturally arise in the form of (6.2). In order to use Proposition 6.1.2, we must rearrange the terms of our series. In doing so it is expedient to work over $\mathbb{C}$ rather than $\mathbb{R}$, so we next review some terminology and facts related to functions of several complex variables, and to series of the form $\sum_{\alpha \in \mathbb{N}_0^n} a_\alpha (\mathbf{x} - \mathbf{c})^\alpha$, where the a_α are real or complex, that will be needed to proceed (see, for example, [87, 90, 96]). Let k denote the field $\mathbb{R}$ or $\mathbb{C}$. A *polydisk* in k^n centered at $\mathbf{c} = (c_1, \ldots, c_n)$ and of *polyradius* $r = (r_1, \ldots, r_n)$ is the open set $\{(x_1, \ldots, x_n) : |x_j - c_j| < r_j \text{ for } j = 1, \ldots, n\}$. Because the set $\mathbb{N}_0^n$ of multi-indices can be totally ordered in many ways, the question arises as to what limit precisely is being taken in the definition of the convergence of a series of the form $\sum_{\alpha \in \mathbb{N}_0^n} a_\alpha (\mathbf{x}-\mathbf{c})^\alpha$. For a formal answer that avoids reference to any order on $\mathbb{N}_0^n$ the reader is referred to the first section in [87]. However, convergence in any sense at a point $\mathbf{b} = (b_1, \ldots, b_n)$ implies absolute convergence on the open polydisk centered at $\mathbf{c}$ and of polyradius $(|c_1 - b_1|, \ldots, |c_n - b_n|)$ (this is Abel's Lemma; see the first section of [96]), so that the existence and value of the sum of the series are independent of the ordering of its terms. A *germ* of an analytic function at a point $\theta^* \in k^n$ is an equivalence class of analytic functions under the relation f is equivalent to g if there is a neighborhood of θ^* on which f and g agree. We denote by $\mathscr{G}_{\theta^*}$ the ring (under the natural addition and multiplication described in Exercise 6.1) of germs of analytic functions of θ at the point $\theta^* \in k^n$, which is a Noetherian ring that is isomorphic to the ring of convergent power series in n variables over k. If f is an analytic function on some open neighborhood of θ^* in k^n, we denote by $\mathbf{f}$ the element of $\mathscr{G}_{\theta^*}$ induced by f. (The context will make it clear when the use of boldface type indicates a mapping into $\mathbb{R}^n$ or $\mathbb{C}^n$ for $n > 1$ and when it indicates the germ of a function into $\mathbb{R}$ or $\mathbb{C}$.) The following statement is a special case ($p = 1$) of Theorem II.D.2 of [90].

Theorem 6.1.4. *Let U be an open subset of $\mathbb{C}^n$, let θ^* be a point in U, let $g_1, \ldots, g_s$ be holomophic functions on U, and let $\mathbf{g}_1, \ldots, \mathbf{g}_s$ be the germs they induce at θ^*. Let $I = \langle \mathbf{g}_1, \ldots, \mathbf{g}_s \rangle$. Then there exist a polydisk $P \subset U$, centered at θ^*, and a constant $\gamma > 0$ such that for any function f that is holomorphic on P and such that $\mathbf{f} \in I$, there exist functions $h_1, \ldots, h_s$ that are holomorphic on P and are such that $f = \sum_{j=1}^s h_j g_j$ on P and $\|h_i\|_P \le \gamma \|f\|_P$, for $j = 1, \ldots, s$, where $\|\cdot\|_P$ is the supremum norm for continuous functions on P, $\|f\|_P = \sup_{z \in P} |f(z)|$.*

Another point that arises when we change (6.2) into the form (6.4) by means of a basis of the ideal generated by the coefficient functions $f_j(\theta)$ is that the order of the coefficient functions, as determined by their indices, is important. It will be of particular importance in the proof of the lemma on rearrangement of the series that the basis of the ideal $\langle f_j : j \in \mathbb{N}_0 \rangle$ involved have the property that it include the first nonzero function and any function f_j that is independent of all the functions with lower indices, in the sense that it is not a linear combination of them. For example, for the ordered collection $\{f^3, f^2, f\}$ and the corresponding ideal I, $I = \langle f^3, f \rangle$ and $I = \langle f \rangle$ but neither of the bases specified by these expressions meets the condition. We separate out this property of certain bases in the following definition.

Definition 6.1.5. Let k be a field and let $\{f_0, f_1, f_2, \ldots\}$ be an ordered set of polynomials in $k[x_1, \ldots, x_n]$. Suppose J is the least index for which f_J is not the zero polynomial. A basis B of the ideal $I = \langle f_j : j \in \mathbb{N}_0 \rangle$ *satisfies the retention condition* if

(a) $f_J \in B$, and

(b) for $j \ge J+1$, if $f_j \notin \langle f_0, \ldots, f_{j-1} \rangle$, then $f_j \in B$.

The *minimal basis of I with respect to the retention condition* is the basis constructed in the following way: beginning with $B = \{f_J\}$, sequentially check successive elements f_j, starting with $j = J+1$, and add f_j to B if and only if $f_j \notin \langle B \rangle$.

The procedure decribed in the definition produces an ascending chain of ideals, hence must terminate in finitely many steps, since the ring $k[x_1,\dots,x_n]$ is Noetherian. The basis constructed in this fashion is minimal among all those that satisfy the retention condition in that it contains as few elements as possible.

When we write only "a basis B of the ideal $I = \langle f_1, f_2, \dots\rangle$ that satisfies the retention condition" or "the minimal basis" in this regard, omitting mention of an ordering, then it is to be understood that the set of functions in question is ordered as they are listed when I is described. With these preliminaries we may now state and prove the lemma on rearrangements of series of the form (6.2).

Lemma 6.1.6. *Let $\mathscr{F}(z,\theta)$ be a series of the form* (6.2) *that converges on a set $U = \{(z,\theta) : |z| < \varepsilon$ and $|\theta - \theta^*| < \delta\} \subset \mathbb{R} \times \mathbb{R}^n$, let $\mathbf{f}_j$ denote the germ of f_j at θ^* in the ring of germs $\mathscr{G}_{\theta^*}$ of complex analytic functions at θ^* when θ^* is regarded as an element of $\mathbb{C}^n$, and suppose there exists a basis B of the ideal $I = \langle \mathbf{f}_0, \mathbf{f}_1, \mathbf{f}_2, \dots\rangle$ in $\mathscr{G}_{\theta^*}$ that consists of m germs $\mathbf{f}_{j_1}, \dots, \mathbf{f}_{j_m}$, $j_1 < \dots < j_m$ and that satisfies the retention condition. Then there exist positive numbers ε_1 and δ_1, $0 < \varepsilon_1 \le \varepsilon$ and $0 < \delta_1 \le \delta$, and m analytic functions $\psi_{j_q}(z,\theta)$ for which $\psi_{j_q}(0,\mathbf{0}) = 0$, $q \in \{1,\dots,m\}$, such that*

$$\mathscr{F}(z,\theta) = \sum_{q=1}^{m} f_{j_q}(\theta)\left(1 + \psi_{j_q}(z,\theta)\right) z^{j_q} \tag{6.6}$$

holds on the set $U_1 = \{(z,\theta) : |z| < \varepsilon_1$ and $|\theta - \theta^*| < \delta_1\}$.

Proof. Allow z and θ to be complex. The series that defines $\mathscr{F}$ still converges on U, or more precisely on $U^{\mathbb{C}} := \{(z,\theta) : |z| < \varepsilon \text{ and } |\theta - \theta^*| < \delta\} \subset \mathbb{C} \times \mathbb{C}^n$. Let $R = \min\{\varepsilon/2, \delta/2, 1\}$, and let $M = \sup\{|\mathscr{F}(z,\theta)| : |z| \le R \text{ and } |\theta - \theta^*| \le R\} < \infty$. Since, for any fixed $\theta \in \mathbb{C}^n$ such that $|\theta - \theta^*| \le R$, the series for $\mathscr{F}(z,\theta)$ converges on $\{z \in \mathbb{C} : |z| \le R\}$, the Cauchy Inequalities state that $|f_j(\theta)| \le M/R^j$ for all $j \in \mathbb{N}_0$ so that if $W = \{\theta : |\theta - \theta^*| \le R\} \subset \mathbb{C}^n$, then

$$\|f_j\|_W \le \frac{M}{R^j}\,. \tag{6.7}$$

Applying Theorem 6.1.4 to the ideal $I = \langle \mathbf{f}_1, \mathbf{f}_2, \dots\rangle = \langle \mathbf{f}_{j_1}, \dots \mathbf{f}_{j_m}\rangle$, we obtain the existence of a polydisk $P \subset W \subset \mathbb{C}^n$ centered at $\theta^* \in \mathbb{R}^n$, a positive real constant γ, and for each $j \in \mathbb{N}_0$ m analytic functions $h_{j,1}, h_{j,2}, \dots, h_{j,m}$ on P such that

$$f_j = h_{j,1} f_{j_1} + h_{j,2} f_{j_2} + \dots + h_{j,m} f_{j_m}$$

and

$$\|h_{j,u}\|_P \le \gamma \|f_j\|_P\,, \tag{6.8}$$

for $j \in \mathbb{N}_0$ and $u = 1, \dots, m$. Thus series (6.2) is

$$\mathscr{F}(z,\theta) = \sum_{j=0}^{\infty} \left(\sum_{q=1}^{m} h_{j,q}(\theta) f_{j_q}(\theta) \right) z^j\,. \tag{6.9}$$

We wish to rearrange the terms in this series. Rearrangement is permissible if the series converges absolutely, which we now show to be true. (Since the series is not a power series we cannot appeal directly to Abel's Lemma.) Let σ denote the minimum component in the polyradius of P, so $\sigma \le R$. Then for $j \ge j_m$ and θ satisfying $|\theta - \theta^*| < \sigma$,

$$\begin{aligned}
|h_{j,q}(\theta) f_{j_q}(\theta) z^j| &\le \gamma \|f_j\|_P \frac{M}{R^{j_q}} |z|^j \quad \text{(by (6.7) and (6.8), since } \theta \in P \subset W) \\
&\le \gamma \|f_j\|_W \frac{M}{R^j} |z|^j \quad \text{(since } P \subset W,\ j \ge j_m, \text{ and } R \le 1) \\
&\le \gamma \frac{M^2}{R^{2j}} |z|^j, \quad \text{(by (6.7))}
\end{aligned}$$

which means that convergence is absolute for $|z| < \varepsilon_1 := R^2$ and $|\theta - \theta^*| < \delta_1 := \sigma$. Thus we may rewrite (6.9) as

$$\mathscr{F}(z,\theta) = \sum_{j=0}^{j_m} \left(\sum_{q=1}^{m} h_{j,q}(\theta) f_{j_q}(\theta) \right) z^j + \sum_{q=1}^{m} \left(\sum_{j=j_m+1}^{\infty} h_{j,q}(\theta) z^j \right) f_{j_q}(\theta). \tag{6.10}$$

Suppose $r \in \{0, 1, \dots, j_m\}$. If $r = j_q$ for some $q \in \{1, \dots, m\}$, then

$$f_r(\theta) z^r = (0 \cdot f_{j_1}(\theta) + \cdots + 1 \cdot f_{j_q}(\theta) + \cdots + 0 \cdot f_{j_m}(\theta)) z^{j_q} = f_{j_q}(\theta) z^{j_q}.$$

If $j_q < r < j_{q+1}$ for some $q \in \{1, \dots, m-1\}$ (the only other case, since $f_r = 0$ for $r < j_1$), then since B satisfies the retention condition, there exist functions $u_{r,1}(\theta), \dots, u_{r,q}(\theta)$, each one analytic on a neighborhood in $\mathbb{C}^n$ of $\theta^* \in \mathbb{R}^n$, such that

$$f_r(\theta) = u_{r,1}(\theta) f_{j_1}(\theta) + \cdots + u_{r,q}(\theta) f_{j_q}(\theta),$$

hence

$$f_r(\theta) z^r = (u_{r,1}(\theta) z^{r-j_1}) f_{j_1}(\theta) z^{j_1} + \cdots + (u_{r,q}(\theta) z^{r-j_q}) f_{j_q}(\theta) z^{j_q}.$$

Thus

$$\sum_{j=0}^{j_m} \left(\sum_{q=1}^{m} h_{j,q}(\theta) f_{j_q}(\theta) \right) z^j = \sum_{q=1}^{m} f_{j_q}(\theta)(1 + \widetilde{\psi}_{j_q}(z,\theta)) z^{j_q}$$

for some functions $\widetilde{\psi}(z,\theta)_{j_q}$ that are analytic on a neighborhood of $(0,\theta^*)$ in $\mathbb{C} \times \mathbb{C}^n$ and satisfy $\widetilde{\psi}(0,0) = 0$. Using (6.10) to incorporate higher-order terms in z into the functions $\widetilde{\psi}_{j_q}$, for possibly smaller ε_1 and δ_1 we have that (6.6) holds on $U_1^{\mathbb{C}} = \{(z,\theta) : |z| < \varepsilon_1 \text{ and } |\theta - \theta^*| < \delta_1\} \subset \mathbb{C} \times \mathbb{C}^n$. We now make the simple observation that if real numbers $\mu, \mu_1, \dots, \mu_m$ and complex numbers $\xi_1, \dots, \xi_m$ satisfy $\mu = \mu_1 \xi_1 + \cdots + \mu_m \xi_m$, then $\mu = \mu_1 \operatorname{Re} \xi_1 + \cdots + \mu_m \operatorname{Re} \xi_m$. Thus because the functions $f_{j_1}, \dots, f_{j_m}$ have real coefficients, and $\mathscr{F}(z,\theta)$ is real when z and θ are real, if each ψ_{j_q} is replaced by its real part, (6.6) still holds on $U_1 := U_1^{\mathbb{C}} \cap \mathbb{R} \times \mathbb{R}^n$. □

Theorem 6.1.7. *Let $\mathscr{F}(z,\theta)$ be a series of the form (6.2) that converges on a set $\{(z,\theta) : |z| < \varepsilon$ and $|\theta - \theta^*| < \delta\} \subset \mathbb{R} \times \mathbb{R}^n$, and let $\mathbf{f}_j$ denote the germ of f_j at θ^* in the ring of germs $\mathscr{G}_{\theta^*}$ of complex analytic functions at θ^* when θ^* is regarded as an element of $\mathbb{C}^n$. Suppose the ideal $I = \langle \mathbf{f}_0, \mathbf{f}_1, \mathbf{f}_2, \dots \rangle$ in $\mathscr{G}_{\theta^*}$ has a basis B that consists of m germs $\mathbf{f}_{j_1}, \dots, \mathbf{f}_{j_m}$, $j_1 < \cdots < j_m$ and satisfies the retention conditon. Then there exist numbers ε_1 and δ_1, $0 < \varepsilon_1 \leq \varepsilon$ and $0 < \delta_1 \leq \delta$, such that for each fixed θ satisfying $|\theta - \theta^*| < \delta_1$, the equation $\mathscr{F}(z,\theta) = 0$, regarded as an equation in z alone, has at most $m-1$ isolated solutions in the interval $(0, \varepsilon_1)$.*

Proof. According to Lemma 6.1.6 we can represent the function $\mathscr{F}(z,\theta)$ of (6.2) in the form

$$\mathscr{F}(z,\theta) = \sum_{q=1}^{m} f_{j_q}(\theta)\left(1 + \psi_{j_q}(z,\theta)\right) z^{j_q}.$$

The conclusion now follows by Proposition 6.1.2. □

6.2 The Cyclicity Problem

In this section we use the results just derived to develop a method for counting the maximum number of limit cycles that can bifurcate from a simple focus or center of system (6.1), which we abbreviate in this paragraph as $\dot{\mathbf{u}} = \mathbf{f}_0(\mathbf{u})$. (Recall that a singularity $\mathbf{u}_0$ of (6.1) is called *simple* or *nondegenerate* if $\det d\mathbf{f}_0(\mathbf{u}_0) \neq 0$.) Backing up a bit, suppose $\mathbf{u}_0$ is an arbitrary singularity of system (6.1). The discussion in the second paragraph of Section 2.2 implies that if $\mathbf{u}_0$ is hyperbolic then it has cyclicity zero. It fails to be hyperbolic if either $\det d\mathbf{f}_0(\mathbf{u}_0) = 0$ or $\det d\mathbf{f}_0(\mathbf{u}_0) > 0$ but $\operatorname{Tr} d\mathbf{f}_0(\mathbf{u}_0) = 0$. If we remove the hyperbolicity by letting $\det d\mathbf{f}_0(\mathbf{u}_0) = 0$, then $\mathbf{u}_0$ need not be isolated from other singularities, and when it is isolated it is possible that it either splits into more than one singularity or disappears entirely under arbitrarily small perturbation of $\mathbf{f}_0$, under any sensible topology on the set of polynomial $\mathbf{f}$. (See Exercises 6.3 through 6.6.) Thus it is natural to consider the situation $\operatorname{Tr} d\mathbf{f}_0(\mathbf{u}_0) = 0$ and $\det d\mathbf{f}_0(\mathbf{u}_0) > 0$, so that $\mathbf{u}_0$ is simple, and by Remark 3.1.7 is a focus or a center. Moreover, *any* focus or center of a quadratic system is simple (Lemma 6.3.2), so the theory developed here will cover quadratic antisaddles completely.

Thus suppose $\mathbf{u}_0$ is a singularity of system (6.1) at which the trace of the linear part is zero and the determinant of the linear part is positive, a simple but nonhyperbolic focus or center. By a translation to move $\mathbf{u}_0$ to the origin and a linear transformation to place $d\mathbf{f}_0(\mathbf{u}_0)$ in Jordan normal form, system (6.1) can be written in the form (3.2). We must allow the trace of the linear part to become nonzero under perturbation, however, so we will consider the family (3.4) for which the theory of the difference map $\mathscr{P}$ and the Lyapunov numbers was developed in Section 3.1. Under the time rescaling $\tau = \beta t$ equation (3.4) takes the simpler form

$$\begin{aligned} \dot{u} &= \lambda u - v + P(u,v) \\ \dot{v} &= u + \lambda v + Q(u,v), \end{aligned} \tag{6.11a}$$

where $\lambda = \alpha/\beta$ and where P and Q are polynomials with $\max\{\deg P, \deg Q\} = n$, say $P(u,v) = \sum_{j+k=2}^{n} A_{jk}u^j v^k$ and $Q(u,v) = \sum_{j+k=2}^{n} B_{jk}u^j v^k$. Introducing the complex coordinate $x = u + iv$ in the usual way expresses (6.11a) in the complex form

$$\dot{x} = \lambda x + i\left(x - \sum_{(p,q)\in S} a_{pq}x^{p+1}\bar{x}^q\right), \tag{6.11b}$$

where $S \subset \mathbb{N}_{-1} \times \mathbb{N}_0$ is a finite set, every element (p,q) of which satisfies the condition $p+q \geq 1$. The choice of labels on the equations is meant to underscore the fact that they are two representations of the same family of systems. Moreover, $\operatorname{Re} a_{pq} \in \mathbb{Q}[A,B]$ and $\operatorname{Im} a_{pq} \in \mathbb{Q}[A,B]$. When $\lambda = 0$, these equations reduce to

$$\begin{aligned} \dot{u} &= -v + P(u,v) \\ \dot{v} &= u + Q(u,v) \end{aligned} \tag{6.12a}$$

(which is precisely (3.2)) and

$$\dot{x} = i\left(x - \sum_{(p,q)\in S} a_{pq}x^{p+1}\bar{x}^q\right). \tag{6.12b}$$

We will use just $(\lambda,(A,B))$ to stand for the full coefficient string $(\lambda, A_{20}, \ldots, B_{0n})$ in $\mathbb{R} \times \mathbb{R}^{(n+1)(n+2)-6}$ and just a to stand for the full coefficient string $(a_{p_1,q_1}, \ldots, a_{p_\ell,q_\ell})$ in $\mathbb{C}^\ell$. Thus we let $E(\lambda,(A,B))$, $E(\lambda,a)$, $E(A,B)$, and $E(a)$ denote the space of parameters of families (6.11a), (6.11b), (6.12a), and (6.12b), respectively. These are just $\mathbb{R} \times \mathbb{R}^{(n+1)(n+2)-6}$, $\mathbb{R} \times \mathbb{C}^\ell$, and so on.

The precise definition of the cyclicity of the singularity of (6.11) is the following, which could also be expressed in terms of the parameters $(\lambda,(A,B))$. We repeat that the natural context for perturbation of an element of family (6.12) is family (6.11), hence the choice of the parameter λ in the parameter space in the definition.

Definition 6.2.1. For parameters (λ,a), let $n((\lambda,a),\varepsilon))$ denote the number of limit cycles of the corresponding system (6.11) that lie wholly within an ε-neighborhood of the origin. The singularity at the origin for system (6.11) with fixed coefficients $(\lambda^*,a^*) \in E(\lambda,a)$ has *cyclicity c with respect to the space $E(\lambda,a)$* if there exist positive constants δ_0 and ε_0 such that for every pair ε and δ satisfying $0 < \varepsilon < \varepsilon_0$ and $0 < \delta < \delta_0$,

$$\max\{n((\lambda,a),\varepsilon)) : |(\lambda,a) - (\lambda^*,a^*)| < \delta\} = c\,.$$

Refer to the discussion in Section 3.1 surrounding system (3.4). For $\rho \in \mathbb{R}$, we have a first return map $\mathscr{R}(\rho)$ determined on the positive portion of the u-axis as specified in Definition 3.1.3 for systems of the form (3.4). By (3.13), $\mathscr{R}(\rho)$ has series expansion

$$\mathscr{R}(\rho) = \tilde{\eta}_1\rho + \eta_2\rho^2 + \eta_3\rho^3 + \cdots, \tag{6.13}$$

where $\tilde{\eta}_1$ and the η_k for $k \geq 2$ are real analytic functions of the parameters $(\lambda,(A,B))$ of system (6.11), and in particular $\tilde{\eta}_1 = e^{2\pi\lambda}$, as explained in the sentence following Definition 3.1.3. Since isolated zeros of the difference function

$$\mathscr{P}(\rho) = \mathscr{R}(\rho) - \rho = \eta_1\rho + \eta_2\rho^2 + \eta_3\rho^3 + \cdots \tag{6.14}$$

correspond to limit cycles of system (6.11), the cyclicity of the origin of the system corresponding to $(\lambda^*, (A^*, B^*)) \in E(\lambda, (A, B))$ is equal to the multiplicity of the function $\mathscr{P}(\rho)$ at $(\lambda^*, (A^*, B^*))$. Thus the behavior of the Lyapunov numbers η_k, $k \in \mathbb{N}$, holds the key to the cyclicity of the origin for system (6.11). For example, we see already that if the point $(\lambda^*, (A^*, B^*))$ in the space of parameters of system (6.11) is such that $\lambda^* = \alpha^*/\beta^* \neq 0$, then $\eta_1 = \tilde{\eta}_1 - 1 \neq 0$, so the expansion of $\mathscr{P}$ shows that no limit cycle can bifurcate from the origin under small perturbation. This is in precise agreement with the fact that $(0,0)$ is a hyperbolic focus in this situation.

We asserted in Section 3.1 that, like the focus quantities g_{kk} for the complexification (3.3) of (6.12a) and the coefficients G_{2k+1} of the function G defined by (3.28) in terms of the distinguished normal form (3.27) of the complexification, the Lyapunov numbers are polynomials in the parameters (A, B) of system (6.12a). We will prove this assertion now. But since the Lyapunov numbers are also defined for the larger family of systems of the form (6.11), there are actually two sets of "Lyapunov numbers" of interest: those arising in the context of system (6.12a) and those arising in the context of system (6.11a), the natural setting in which perturbations from an element of (6.12a) take place in determination of the cyclicity of the antisaddle at the origin (Definition 6.2.1). In the context of the parametrized families (6.11) or (6.12), the Lyapunov numbers are referred to as the Lyapunov "quantities." All that we will need to know about the Lyapunov quantities for the more inclusive family (6.11) is that they are real analytic functions of the parameters $(\lambda, (A, B))$. This follows immediately from the analyticity of the solution $f(\varphi, \varphi_0, r_0)$ of the initial value problem (3.7) with initial conditions $r = r_0$ and $\varphi = \varphi_0$, the fact that $r = 0$ is not a singular point but a regular solution of (3.7), and analyticity of the evaluation map, since the Poincaré first return map is nothing other than the evaluation of $f(\varphi, 0, r_0)$ at $\varphi = 2\pi$. The actual form of the Lyapunov quantities in terms of the parameters $(\lambda, (A, B))$ for family (6.11) is described in Exercise 6.7. It is not difficult to see that if different choices are made for the initial conditions in the initial value problems that determine the functions w_k, which gives rise to the nonuniqueness of the Lyapunov quantities, the same proof is valid.

Proposition 6.2.2. *The Lyapunov quantities specified by Definition 3.1.3 for family* (6.12a) *are polynomial functions of the parameters* (A, B) *with coefficients in* $\mathbb{R}$.

Proof. It is clear from an examination of equations (3.5), (3.6), and (3.7), where, for the time-rescaled systems (6.11) and (6.12), α and β are replaced by λ and 1, respectively, that for each $k \in \mathbb{N}$, R_k is a polynomial in $\cos\varphi$, $\sin\varphi$, the coefficients of P and Q (that is, (A, B)), and λ, with integer coefficients.

We claim that for all $k \in \mathbb{N}$, the function $w_k(\varphi)$ defined by (3.9) is a polynomial in $\cos\varphi$, $\sin\varphi$, φ, and (A, B), with rational coefficients. By (3.7), $R_1 \equiv 0$ so by (3.10) and (3.11), the initial value problem that w_1 solves uniquely is $w_1' = 0$, $w_1(0) = 1$, so $w_1(\varphi) \equiv 1$. If the claim holds for $w_1, \ldots, w_{j-1}$, then by (3.10) and (3.11), the initial value problem that determines w_j uniquely is of the form $w_j'(\varphi) = S_j(\varphi)$, $w_j(0) = 0$, where S_j does not involve w_j and is a polynomial in $\cos\varphi$, $\sin\varphi$, φ, and (A,B), with rational coefficients. Simply integrating both sides shows that the claim holds for w_j, hence by mathematical induction the claim holds for w_k for all $k \in \mathbb{N}$. (The first integration of an even power of $\cos\varphi$ or $\sin\varphi$ produces a rational constant times φ as one term.) Then $\eta_1 = w_1(2\pi) - 1 \equiv 0$ and for $k \geq 2$ $\eta_k = w_k(2\pi)$ is a polynomial in (A,B) with real coefficients, since the sines and cosines evaluate to 0 and 1 and powers of φ evaluate to powers of 2π. □

By Theorems 3.1.5 and 3.3.5, the Lyapunov quantities and the focus quantities pick out the centers in family (6.12); together with Theorem 3.2.7 and Remark 3.3.6 they also distinguish stable and unstable weak foci. These facts indicate that there must be an intimate connection between them and suggests that it should be possible to use the focus quantities to investigate the cyclicity of simple foci and centers. This is true, and is important because the focus quantities are so much easier to work with than the Lyapunov quantities. The precise nature of the relationship is given in the next theorem. In anticipation of that, it will be helpful to review briefly how we derived the focus quantities for the real system (6.12a). The first step was to express (6.12a) in the complex form (6.12b) and then adjoin to that equation its complex conjugate. By regarding $\bar{x}$ as an independent variable, this pair of differential equations became a system of ordinary differential equations on $\mathbb{C}^2$, the complexification (3.50) of (6.12), which we duplicate here:

$$\dot{x} = i\Big(x - \sum_{(p,q)\in S} a_{pq}x^{p+1}y^q\Big), \quad \dot{y} = -i\Big(y - \sum_{(p,q)\in S} b_{qp}x^q y^{p+1}\Big), \tag{6.15}$$

with $b_{qp} = \bar{a}_{pq}$. Letting $\mathscr{X}$ denote the vector field on $\mathbb{C}^2$ associated to any system on $\mathbb{C}^2$ of this form, not necessarily the complexification of a real system, hence not necessarily satisfying the condition $b_{qp} = \bar{a}_{pq}$, we then applied $\mathscr{X}$ to a formal series Ψ given by

$$\Psi(x,y) = xy + \sum_{j+k\geq 3} v_{j-1,k-1}x^j y^k. \tag{6.16}$$

Recursively choosing the coefficients $v_{j-1,k-1}$ in an attempt to make all coefficients of $\mathscr{X}\Psi = \sum_{j+k\geq 1}^{\infty} g_{j,k}x^{j+1}y^{k+1}$ vanish, we obtained functions $g_{kk} \in \mathbb{C}[a,b]$ such that $ig_{kk} \in \mathbb{Q}[a,b]$ ($i = \sqrt{-1}$) and

$$\mathscr{X}\Psi = g_{11}(xy)^2 + g_{22}(xy)^3 + g_{33}(xy)^4 + \cdots. \tag{6.17}$$

Thus whereas η_k is a polynomial in the original real coefficients (A,B) of (6.12a), in fact g_{kk} is a polynomial in the complex coefficients (a,b) of the complexification (6.15). To make a proper comparison, we must express g_{kk} in terms of the parameters (A,B). This is possible because the coefficients (a,b) of the complexification satisfy $b = \bar{a}$ and $g_{kk}(a,\bar{a}) \in \mathbb{R}$ for all $a \in \mathbb{C}^\ell$ (Remark 3.4.7), and because $\operatorname{Re} a_{pq}$ and $\operatorname{Im} a_{pq}$ are polynomials (with rational coefficients) in the original coefficients (A,B), so that

$$g_{kk}^{\mathbb{R}}(A,B) := g_{kk}(a(A,B),\bar{a}(A,B)) \tag{6.18}$$

is a polynomial in (A,B) with rational coefficients. The distinction between g_{kk} and $g_{kk}^{\mathbb{R}}$ might be best understood by working through Exercise 6.9. We note, however, that this distinction is never made in the literature; one must simply keep in mind the context in which the quantity g_{kk} arises.

We remark that if the time rescaling to obtain (6.11a) from (3.4) with $\alpha = 0$ is not done, so that β is present, then π is replaced by π/β everywhere in the statement of the theorem.

Theorem 6.2.3. *Let η_k be the Lyapunov quantities for system* (6.12a) *with regard to the antisaddle at the origin, let g_{kk} be the focus quantities for the complexification* (6.15) *of* (6.12a)*, and let $g_{kk}^{\mathbb{R}}$ denote the polynomial functions defined by* (6.18)*. Then $\eta_1 = \eta_2 = 0$, $\eta_3 = \pi g_{11}^{\mathbb{R}}$, and for $k \in \mathbb{N}$, $k \geq 2$, $\eta_{2k} \in \langle g_{11}^{\mathbb{R}},\ldots,g_{k-1,k-1}^{\mathbb{R}}\rangle$ and $\eta_{2k+1} - \pi g_{kk}^{\mathbb{R}} \in \langle g_{11}^{\mathbb{R}},\ldots,g_{k-1,k-1}^{\mathbb{R}}\rangle$ in $\mathbb{R}[A,B]$.*

Proof. The idea is to compare the change $\mathscr{P}(\rho)$ in position along the positive u-axis in one turn around the singularity to the change in the value of the function Ψ expressed by (6.16), computing the change in Ψ by integrating its derivative along solutions of (6.12a), which naturally generates the focus quantities according to (6.17). In reality, Ψ is defined for $(x,y) \in \mathbb{C}^2$, but we evaluate it on $(x,\bar{x})$, the invariant plane that contains the phase portrait of (6.11); see Exercise 3.7. Since the system on $\mathbb{C}^2$ to which Ψ and the focus quantities g_{kk} pertain is the complexification of a real system, it satisfies $b = \bar{a}$, so the focus quantities g_{kk} are actually the quantities $g_{kk}^{\mathbb{R}}$. Since Ψ might not converge, hence not actually define a function, we work instead with the truncation of the series defining Ψ at some sufficiently high level,

$$\Psi_N(x,\bar{x}) := x\bar{x} + \sum_{j+k=3}^{2N+1} v_{j-1,k-1} x^j \bar{x}^k .$$

Fix an initial point on the positive u-axis with polar coordinates $(r,\varphi) = (\rho,0)$ and complex coordinate $x = u + iv = \rho + i0 = \rho$. In one turn about the singularity, time increases by some amount $\tau = \tau(\rho)$ and the change in Ψ_N is

$$\begin{aligned}
\Delta\Psi_N(\rho,\rho) &= \int_0^{\tau} \tfrac{d}{dt}\,[\Psi_N(x(t),\bar{x}(t))]\,dt \\
&= \int_0^{\tau} \sum_{k=1}^{N} g_{kk}^{\mathbb{R}}(x(t)\bar{x}(t))^{k+1} + o(|x(t)|^{2N+2})\,dt \\
&= \int_0^{\tau} \sum_{k=1}^{N} g_{kk}^{\mathbb{R}}|x(t)|^{2k+2} + o(|x(t)|^{2N+2})\,dt .
\end{aligned}$$

Now change the variable of integration from t to the polar angle φ. By (3.9) and (3.12), keeping in mind that we have rescaled time to make $(\alpha,\beta) = (0,1)$, we have $|x(t)| = r(t) = \rho + w_2(\varphi)\rho^2 + w_3(\varphi)\rho^3 + \cdots$. By the second equation of (3.5), $d\varphi/dt = 1 + \sum_{k=1}^{\infty} u_k(\varphi) r^k$, so that

$$dt = \frac{1}{1+\sum_{k=1}^{\infty} u_k(\varphi)[\rho + w_2(\varphi)\rho^2 + \cdots]^k}\, d\varphi = (1 + \widetilde{u}_1(\varphi)\rho + \widetilde{u}_2(\varphi)\rho^2 + \cdots)d\varphi.$$

Thus

$$\begin{aligned}
&\triangle\Psi_N(\rho,\rho) \\
&= \int_0^{2\pi} \sum_{k=1}^{N} g_{kk}^{\mathbb{R}}(\rho + w_2(\varphi)\rho^2 + \cdots)^{2k+2}(1 + \widetilde{u}_1(\varphi)\rho + \widetilde{u}_2(\varphi)\rho^2 + \cdots)\, d\varphi \\
&\qquad\qquad + o(\rho^{2N+2}) \\
&= \sum_{k=1}^{N} \left[2\pi g_{kk}^{\mathbb{R}} \rho^{2k+2} + g_{kk}^{\mathbb{R}}(f_{k,1}\rho^{2k+3} + f_{k,2}\rho^{2k+4} + \cdots) \right] + o(\rho^{2N+2}).
\end{aligned}$$

Turning our attention to $\triangle\rho$, for any value of $\rho > 0$ we have a positive real number ξ defined by the function $\xi = f(\rho) = \Psi(\rho,\rho) = \rho^2 + V_3\rho^3 + V_4\rho^4 + \cdots$, which has an inverse $\rho = g(\xi)$. By Taylor's Theorem with remainder, there exists $\tilde{\xi}$ between ξ and $\xi + \varepsilon$ such that $g(\xi+\varepsilon) = g(\xi) + g'(\xi)\varepsilon + \frac{1}{2!}g''(\tilde{\xi})\varepsilon^2$. Let $\tilde{\rho} = g(\tilde{\xi})$. Using the formulas for the first and second derivatives of g as the inverse of f and inverting the series obtained, we have that for $\varepsilon = \triangle\Psi_N$,

$$\begin{aligned}
\triangle\rho &= \frac{1}{2\rho + 3V_3\rho^2 + \cdots}\triangle\Psi_N + \frac{1}{2!}\left(-\frac{2 + 6V_3\tilde{\rho} + \cdots}{(2\tilde{\rho} + 3V_3\tilde{\rho}^2 + \cdots)^3} \right)(\triangle\Psi_N)^2 \\
&= \left(\tfrac{1}{2\rho} + c_0 + c_1\rho + c_2\rho^2 + \cdots \right) \\
&\qquad \times \left(\sum_{k=1}^{N} \left[2\pi g_{kk}^{\mathbb{R}}\rho^{2k+2} + g_{kk}^{\mathbb{R}}(f_{k,1}\rho^{2k+3} + \cdots) \right] + o(\rho^{2N+2}) \right) \\
&\quad + \left(-\tfrac{1}{8\tilde{\rho}^3} + d_{-2}\tfrac{1}{\tilde{\rho}^2} + d_{-1}\tfrac{1}{\tilde{\rho}} + d_0\tilde{\rho} + \cdots \right) \\
&\qquad \times \left(\sum_{k=1}^{N} (g_{kk}^{\mathbb{R}})^2 (d_{k,1}\rho^{4k+4} + \cdots) + o(\rho^{2N+4}) \right).
\end{aligned}$$

Since $\triangle\Psi_N$ is of order four or higher in ρ, it is apparent that $\tilde{\rho}$ is of order ρ. Thus

$$\triangle\rho = \sum_{k=1}^{N} \left[\pi g_{kk}^{\mathbb{R}}\rho^{2k+1} + g_{kk}^{\mathbb{R}}(\tilde{f}_{k,1}\rho^{2k+2} + \tilde{f}_{k,2}\rho^{2k+3} + \cdots) \right] + o(\rho^{2N+1}). \tag{6.19}$$

As indicated just after Definition 3.1.3, $\eta_1 = e^{2\pi\alpha/\beta} - 1$, so the hypothesis $\lambda = 0$ implies that $\eta_1 = 0$. Then by Proposition 3.1.4 $\eta_2 = 0$. This gives the first two conclusions of the proposition. Since $\triangle\rho = \mathscr{P}(\rho)$, (6.19) then reads

$$\begin{aligned}\eta_3\rho^3+\eta_4\rho^4+\eta_5\rho^5+\cdots=&\ \pi g_{11}^{\mathbb{R}}\rho^3+g_{11}^{\mathbb{R}}(\tilde{f}_{1,1}\rho^4+\tilde{f}_{1,2}\rho^5+\cdots)\\&+\pi g_{22}^{\mathbb{R}}\rho^5+g_{22}^{\mathbb{R}}(\tilde{f}_{2,1}\rho^6+\tilde{f}_{2,2}\rho^7+\cdots)\\&+\pi g_{33}^{\mathbb{R}}\rho^7+g_{33}^{\mathbb{R}}(\tilde{f}_{3,1}\rho^8+\tilde{f}_{3,2}\rho^9+\cdots)\\&+\cdots\\&+\pi g_{NN}^{\mathbb{R}}\rho^{2N+1}+g_{NN}^{\mathbb{R}}(\tilde{f}_{N,1}\rho^{2N+2}+\tilde{f}_{N,2}\rho^{2N+3}+\cdots)\\&+o(\rho^{2N+1}).\end{aligned}$$

Thus $\eta_3=\pi g_{11}^{\mathbb{R}}$ and given $k\in\mathbb{N}$, the choice $N=k$ shows that the last pair of assertions of the proposition holds for η_4 through η_{2k+1}. □

Corollary 6.2.4. *Let η_k be the Lyapunov quantities for system* (6.12a) *with regard to the antisaddle at the origin, let g_{kk} be the focus quantities for the complexification* (6.15) *of* (6.12a)*, and let $g_{kk}^{\mathbb{R}}$ denote the polynomial function defined by* (6.18)*. Then*

$$\langle g_{11}^{\mathbb{R}},g_{22}^{\mathbb{R}},g_{33}^{\mathbb{R}},\ldots\rangle=\langle\eta_1,\eta_2,\eta_3,\ldots\rangle=\langle\eta_3,\eta_5,\eta_7,\ldots\rangle$$

in $\mathbb{R}[A,B]$. For any $(A^*,B^*)\in E(A,B)$, the corresponding equalities hold true for the corresponding germs and their ideals in $\mathscr{G}_{(A^*,B^*)}$.

Proof. The proof is left to the reader as Exercise 6.10. □

The second corollary to the theorem is stated only for germs, since that is the context in which it will be used later.

Corollary 6.2.5. *Let η_k be the Lyapunov quantities for system* (6.12a) *with regard to the antisaddle at the origin, let g_{kk} be the focus quantities for the complexification* (6.15) *of* (6.12a)*, and let $g_{kk}^{\mathbb{R}}$ denote the polynomial function defined by* (6.18)*. Let $I=\langle\eta_{2k+1}:k\in\mathbb{N}\rangle=\langle\mathbf{g}_{kk}:k\in\mathbb{N}\rangle\subset\mathscr{G}_{(A^*,B^*)}$. Suppose $\{\eta_{k_1},\ldots,\eta_{k_m}\}$ and $\{\mathbf{g}_{j_1,j_1},\ldots,\mathbf{g}_{j_n,j_n}\}$ are the minimal bases for I with respect to the retention conditon with respect to the ordered sets $\{\eta_3,\eta_5,\eta_7,\ldots\}$ and $\{\mathbf{g}_{11},\mathbf{g}_{22},\ldots\}$, respectively. Then $m=n$ and for $q=1,2,\ldots,m$, $k_q=2j_q+1$.*

Proof. The proof is left to the reader as Exercise 6.11. □

One further consequence of Theorem 6.2.3, which is of independent interest, is that a fine focus of a planar polynomial system is of order k if and only if the first $k-1$ focus quantities vanish. The precise statement is given in the following proposition.

Proposition 6.2.6. *Let η_k be the Lyapunov quantities for system* (6.12a) *with regard to the antisaddle at the origin, let g_{kk} be the focus quantities for the complexification* (6.15) *of* (6.12a)*, and let $g_{kk}^{\mathbb{R}}$ denote the polynomial function defined by* (6.18)*. The system corresponding to a specific choice of the real parameters (A,B) has a kth-order fine focus at the origin if and only if $g_{11}^{\mathbb{R}}(A,B)=\cdots=g_{k-1,k-1}^{\mathbb{R}}(A,B)=0$ but $g_{kk}^{\mathbb{R}}(A,B)\neq 0$.*

Proof. The proof is left to the reader as Exercise 6.12. □

Before continuing with our development of the theory that enables us to estimate the cyclicity of a center in a family (6.12), we pause for a moment to show how Theorem 6.2.3 and Proposition 6.2.6 combine to estimate the cyclicity of a kth-order fine focus in such a family. The bound is universal, in the sense that whatever the nature of the nonlinearities, say perhaps maximal degree two or maximal degree three, a kth-order fine focus has cyclicity at most k.

Theorem 6.2.7. *A fine focus of order k has cyclicity at most $k-1$ for perturbation within family* (6.12) *and at most k for perturbation within family* (6.11).

Proof. By Theorem 6.2.3, for any system of the form (6.12), the difference function $\mathscr{P}$ may be written

$$\begin{aligned}\mathscr{P}(\rho) = {}& \pi g_{11}^{\mathbb{R}}\rho^3 + h_{41}g_{11}\rho^4 \\ & + (h_{51}g_{11}^{\mathbb{R}} + \pi g_{22}^{\mathbb{R}})\rho^5 + (h_{61}g_{11}^{\mathbb{R}} + h_{62}g_{22}^{\mathbb{R}})\rho^6 \\ & + \cdots \\ & + (h_{2k-1,1}g_{11}^{\mathbb{R}} + \cdots + \pi g_{k-1,k-1}^{\mathbb{R}})\rho^{2k-1} + (h_{2k,1}g_{11}^{\mathbb{R}} + \cdots + h_{2k,k}g_{k-1,k-1}^{\mathbb{R}})\rho^{2k} \\ & + (h_{2k+1,1}g_{11}^{\mathbb{R}} + \cdots + \pi g_{kk}^{\mathbb{R}})\rho^{2k+1} + \eta_{2k+2}\rho^{2k+2} + \eta_{2k+3}\rho^{2k+3} + \cdots .\end{aligned}$$

Suppose a system of the form (6.12) corresponding to parameter value (A^*, B^*) has a fine focus of order k at the origin, so that by Proposition 6.2.6, $g_{11}^{\mathbb{R}}$ through $g_{k-1,k-1}^{\mathbb{R}}$ vanish at (A^*, B^*) but $g_{kk}^{\mathbb{R}}(A^*, B^*) \neq 0$. Then because $g_{kk}^{\mathbb{R}}$ is nonzero on a neighborhood of (A^*, B^*), when we factor out $\pi\rho^3$ and collect on the $g_{jj}^{\mathbb{R}}$ (but for simplicity keep the same names for the polynomial weighting functions), we may write

$$\begin{aligned}\mathscr{P}(\rho) = {}& \pi\rho^3 \big[g_{11}^{\mathbb{R}}(1 + h_{41}\rho + \cdots + h_{2k+1,1}\rho^{2k-2}) \\ & \quad + g_{22}^{\mathbb{R}}(1 + h_{62}\rho + \cdots + h_{2k+1,2}\rho^{2k-4})\rho^2 \\ & \quad + g_{33}^{\mathbb{R}}(1 + h_{83}\rho + \cdots + h_{2k+1,3}\rho^{2k-6})\rho^4 \\ & \quad + \cdots \\ & \quad + g_{k-1,k-1}^{\mathbb{R}}(1 + h_{2k,k-1}\rho + h_{2k+1,2k-1}\rho^2)\rho^{2k-4} \\ & \quad + g_{kk}^{\mathbb{R}}\rho^{2k-2} + \eta_{2k+2}\rho^{2k-1} + \eta_{2k+3}\rho^{2k} + \cdots\big] \\ = {}& \pi\rho^3 \big[g_{11}^{\mathbb{R}}(1 + \psi_1(\rho)) + g_{22}^{\mathbb{R}}\rho^2(1 + \psi_2(\rho)) + \cdots + g_{kk}^{\mathbb{R}}\rho^{2k-2}(1 + \psi_k(\rho))\big],\end{aligned}$$

valid on a neighborhood of (A^*, B^*). If a perturbation is made within family (6.12), then by Proposition 6.1.2 $\mathscr{P}$ has at most $k-1$ isolated zeros in a small interval $0 < \rho < \varepsilon$.

For the remainder of the proof, the case that the perturbation is made within family (6.11), see Exercise 6.14. □

We have indicated earlier that we hope to use the focus quantities to treat the cyclicity problem. The focus quantities arise from the complexification of system

(6.12), but bifurcations to produce limit cycles naturally take place in the larger family (6.11). We have connected the focus quantities and their ideals to the Lyapunov quantities in the restricted context of family (6.12). The next result shows how the minimal basis with respect to the retention condition of the ideal generated by the Lyapunov quantities for the restricted family (6.12) is related to the minimal basis of the ideal generated by the Lyapunov quantities of the larger family (6.11) (with the same indexing set S, of course). We will distinguish between the two sets of Lyapunov quantities by using the notation η_k for those depending on just the parameters (A,B) and $\eta_k(\lambda)$ for those depending on the parameters $(\lambda,(A,B))$, although, of course, $\eta_k(0,(A,B)) = \eta_k(A,B)$. Because the functions $\eta_k(\lambda)$ are not polynomials in the parameters $(\lambda,(A,B))$, we must work in the ring of germs in order to handle domains of convergence. Note also that we treat the η_k merely as analytic functions in the first hypothesis of the theorem, although they are actually polynomials in (A,B).

Lemma 6.2.8. *Fix families* (6.11) *and* (6.12) *with the same indexing set S. Let* $\{\eta_k(\lambda)\}_{k=1}^{\infty}$ *be the Lypaunov quantities for family* (6.11) *and let* $\{\eta_k\}_{k=1}^{\infty}$ *be the Lyapunov quantities for family* (6.12). *Fix* (A^*,B^*) *in* $E(A,B)$ *and suppose that the minimal basis with respect to the retention condition of the ideal* $\langle \eta_1, \eta_2, \ldots \rangle$ *in* $\mathscr{G}_{(A^*,B^*)}$, *is* $\{\eta_{k_1},\ldots,\eta_{k_m}\}$, $k_1 < \cdots < k_m$. *Then* $\{\eta_1(\lambda),\eta_{k_1},\ldots,\eta_{k_m}\}$ *is the minimal basis with respect to the retention condition with respect to the ordered set* $\{\eta_1(\lambda),\eta_2(\lambda),\eta_3(\lambda),\ldots\}$ *of the ideal* $\langle \eta_1(\lambda),\eta_2(\lambda),\eta_3(\lambda),\ldots\rangle$ *in* $\mathscr{G}_{(0,(A^*,B^*))}$.

Proof. The functions $\eta_k(\lambda,(A,B))$ are analytic in a neighborhood of $(0,(A^*,B^*))$, hence by Abel's Lemma their power series expansions converge absolutely there, so we may rearrange the terms in these expansions. Thus, for any $k \in \mathbb{N}$ in a neighborhood of $(0,(A^*,B^*))$, $\eta_k(\lambda,(A,B))$ can be written

$$\eta_k(\lambda,(A,B)) = \check{\eta}_k(\lambda,(A,B)) + \check{\eta}_k(A,B), \tag{6.20}$$

where $\check{\eta}_k(0,(A,B)) \equiv 0$. When $\lambda = 0$, $\eta_k(\lambda,(A,B))$ reduces to $\eta_k(A,B)$, so it must be the case that $\eta_k(0,(A,B)) = 0 + \check{\eta}_k(A,B) = \eta_k(A,B)$ and (6.20) becomes

$$\eta_k(\lambda,(A,B)) = \check{\eta}_k(\lambda,(A,B)) + \eta_k(A,B). \tag{6.21}$$

Since

$$\eta_1(\lambda,(A,B)) = e^{2\pi\lambda} - 1 = 2\pi\lambda(1 + \tfrac{1}{2!}(2\pi\lambda) + \cdots),$$

there exists a function $u_k(\lambda,(A,B))$ that is real analytic on a neighborhood of $(0,(A^*,B^*))$ in $E(\lambda,(A,B))$ such that

$$\check{\eta}_k(\lambda,(A,B)) = u_k(\lambda,(A,B))\eta_1(\lambda,(A,B)).$$

Thus (6.21) becomes, suppressing the (A,B) dependence in the notation,

$$\eta_k(\lambda) = u_k(\lambda)\eta_1(\lambda) + \eta_k. \tag{6.22}$$

Let L denote the set $\{\eta_{k_1},\dots,\eta_{k_m}\}$. Because L is the minimal basis with respect to the retention condition of the ideal $\langle \eta_k : k \in \mathbb{N}\rangle$ in $\mathscr{G}_{(A^*,B^*)}$, (6.22) implies that for all $k \in \mathbb{N}$, the identity

$$\begin{aligned}&\eta_k(\lambda,(A,B))\\ &= u_k(\lambda,(A,B))\eta_1(\lambda,(A,B)) + h_{k,1}(A,B)\eta_{k_1}(A,B) + \cdots + h_{k,m}(A,B)\eta_{k_m}(A,B)\end{aligned}$$

holds on a neighborhood of $(0,(A^*,B^*))$ in $E(\lambda,(A^*,B^*))$ for functions $h_{k,q}$ that are defined and real analytic on that neighborhood, albeit without λ dependence. The same equation is therefore true at the level of germs in $\mathscr{G}_{(0,(A^*,B^*))}$. Thus

$$M = \{\eta_1(\lambda),\eta_{k_1},\dots,\eta_{k_m}\}$$

is a basis of the ideal $\langle \eta_1(\lambda),\eta_2(\lambda),\dots\rangle \subset \mathscr{G}_{(0,(A^*,B^*))}$. We must show that it is minimal among all bases that satisfy the retention condition with respect to the set $\{\eta_1(\lambda),\eta_2(\lambda),\dots\}$. Hence let

$$N = \{\eta_1(\lambda),\eta_{j_1}(\lambda),\dots,\eta_{j_n}(\lambda)\}$$

be the unique minimal basis with respect to the retention condition (which must contain $\eta_1(\lambda)$, since $\eta_1(\lambda)$ is first on the list and is not $\mathbf{0}$), with the labelling chosen so that $j_1 < \cdots < j_n$, and suppose, contrary to what we wish to show, that it is not the basis M. There are four ways this can happen, which we treat in turn.

Case 1: There exists $p \in \{1,2,\dots,\min\{m,n\}\}$ such that for $q \in \{1,2,\dots,p-1\}$, $k_q = j_q$ and $\eta_{k_q} = \eta_{j_q}(\lambda)$ but $\eta_{k_p} \neq \eta_{j_p}(\lambda)$ and $j_p < k_p$.
Then $k_{p-1} = j_{p-1} < j_p < k_p$, so because L is minimal $\eta_{j_p} = \mathbf{h}_1\eta_{k_1} + \cdots + \mathbf{h}_{p-1}\eta_{k_{p-1}}$ for $\mathbf{h}_1,\dots,\mathbf{h}_{p-1} \in \mathscr{G}_{(A^*,B^*)}$. Applying the corresponding equality of functions that holds on a neighborhood of (A^*,B^*) to (6.22) implies that

$$\begin{aligned}\eta_{j_p}(\lambda) &= u_{j_p}(\lambda)\eta_1(\lambda) + \eta_{j_p}\\ &= u_{j_p}(\lambda)\eta_1(\lambda) + h_1\eta_{k_1} + \cdots + h_{p-1}\eta_{k_{p-1}}\\ &= u_{j_p}(\lambda)\eta_1(\lambda) + h_1\eta_{j_1}(\lambda) + \cdots + h_{p-1}\eta_{j_{p-1}}(\lambda)\end{aligned}$$

is valid on a neighborhood of $(0,(A,B))$ in $E(\lambda,(A,B))$ (although h_q is independent of λ), so the corresponding equality of germs contradicts the fact that N is minimal.

Case 2: There exists $p \in \{1,2,\dots,\min\{m,n\}\}$ such that for $q \in \{1,2,\dots,p-1\}$, $k_q = j_q$ and $\eta_{k_q} = \eta_{j_q}(\lambda)$ but $\eta_{k_p} \neq \eta_{j_p}(\lambda)$ and $j_p > k_p$.
Then $j_{p-1} = k_{p-1} < k_p < j_p$, so $\eta_{k_p}(\lambda) \notin N$, hence, because N is minimal,

$$\begin{aligned}\eta_{k_p}(\lambda) &= \mathbf{h}_0\eta_1(\lambda) + \mathbf{h}_1\eta_{j_1}(\lambda) + \cdots + \mathbf{h}_{p-1}\eta_{j_{p-1}}(\lambda)\\ &= \mathbf{h}_0\eta_1(\lambda) + \mathbf{h}_1\eta_{k_1} + \cdots + \mathbf{h}_{p-1}\eta_{k_{p-1}}\end{aligned}$$

for $\mathbf{h}_1,\dots,\mathbf{h}_{p-1} \in \mathscr{G}_{(0,(A^*,B^*))}$. The corresponding equality of functions that holds on a neighborhood of $(0,(A^*,B^*))$ in $E(\lambda,(A,B))$, when evaluated on $\lambda = 0$, implies

that $\eta_{k_p} = \tilde{h}_1 \eta_{k_1} + \cdots + \tilde{h}_{p-1} \eta_{k_{p-1}}$ on a neighborhood of (A^*, B^*) in $E(A,B)$, where for $q = 1, \ldots, p-1$, $\tilde{h}(A,B) = h(0,(A,B))$. The corresponding equality of germs in $\mathscr{G}_{(A^*,B^*)}$ contradicts the fact that L is minimal.
Case 3: $n < m$ and for $q \in \{1,\ldots,n\}$: $k_q = j_q$ and $\eta_{k_q} = \eta_{j_q}(\lambda)$.
Then $k_m > j_n$ and $\eta_{k_m}(\lambda) \notin N$, so

$$\eta_{k_m}(\lambda) = \mathbf{h}_0 \eta_1(\lambda) + \mathbf{h}_1 \eta_{j_1}(\lambda) + \cdots + \mathbf{h}_n \eta_{j_n}(\lambda) = \mathbf{h}_0 \eta_1(\lambda) + \mathbf{h}_1 \eta_{k_1} + \cdots + \mathbf{h}_n \eta_{k_n}$$

for $\mathbf{h}_1, \ldots, \mathbf{h}_n \in \mathscr{G}_{(0,(A^*,B^*))}$. Since $k_m > k_n$, the corresponding equality of functions that holds on a neighborhood of $(0,(A^*,B^*))$ in $E(\lambda,(A,B))$, when evaluated at $\lambda = 0$, gives the same contradiction as in the previous case.
Case 4: $n > m$ and for $q \in \{1,\ldots,m\}$: $k_q = j_q$ and $\eta_{k_q} = \eta_{j_q}(\lambda)$.
Then $j_n > k_m$ and $\eta_{j_n} \notin L$, so

$$\eta_{j_n} = \mathbf{h}_1 \eta_{k_1} + \cdots + \mathbf{h}_m \eta_{k_m} = \mathbf{h}_1 \eta_{j_1}(\lambda) + \cdots + \mathbf{h}_m \eta_{j_m}(\lambda)$$

in $\mathscr{G}_{(0,(A^*,B^*)}$ (although h_q has no λ dependence), so an application of the corresponding equality of functions that holds on a neighborhood of $(0,(A^*,B^*))$ to (6.22) implies that

$$\eta_{j_n}(\lambda) = u_{j_n}(\lambda)\eta_1(\lambda) + \eta_{j_n} = u_{j_n}(\lambda)\eta_1(\lambda) + h_1 \eta_{j_1}(\lambda) + \cdots + h_m \eta_{j_m}(\lambda)$$

on a neighborhood of $(0,(A,B))$ in $E(\lambda,(A,B))$. Thus, because $j_n > j_m$, the corresponding equality of germs contradicts the fact that N is minimal. □

Theorem 6.2.9. *Suppose that for* $(A^*,B^*) \in E(A,B)$*, the minimal basis* M *with respect to the retention condition of the ideal* $J = \langle \mathbf{g}_{11}^{\mathbb{R}}, \mathbf{g}_{22}^{\mathbb{R}}, \ldots \rangle$ *in* $\mathscr{G}_{(A^*,B^*)}$ *for the corresponding system of the form* (6.12) *consists of* m *polynomials. Then the cyclicity of the origin of the system of the form* (6.11) *that corresponds to the parameter string* $(0,(A^*,B^*)) \in E(\lambda,(A,B))$ *is at most* m.

Proof. As stated in the discussion surrounding (6.14), the cyclicity of the origin of an element of family (6.11) with respect to the parameter space $E(\lambda,(A,B))$ is equal to the multiplicity of the function

$$\mathscr{P}(\rho) = \eta_1(\lambda)\rho + \eta_2(\lambda)\rho^2 + \eta_3(\lambda)\rho^3 + \cdots.$$

By the hypothesis and Corollary 6.2.5, the minimal basis with respect to the retention condition of the ideal $\langle \eta_3, \eta_5, \eta_5, \ldots \rangle$ in $\mathscr{G}_{(A^*,B^*)}$ has m elements, hence, by Lemma 6.2.8, in $\mathscr{G}_{(0,(A^*,B^*))}$ the minimal basis with respect to the retention condition of the ideal $\langle \eta_1(\lambda), \eta_2(\lambda), \eta_3(\lambda), \ldots \rangle$ has $m+1$ elements. Then by Theorem 6.1.7 the multiplicity of the function $\mathscr{P}(\rho)$ is at most m. □

The following corollary to the theorem is the result that connects the cyclicity of the simple antisaddle at the origin of system (6.11) (perturbation within $E(\lambda,(A,B))$) to the focus quantities of the complexification of the companion system (6.12).

Corollary 6.2.10. *Fix a family of real systems of the form* (6.11), *with parameter set* $E(\lambda,(A,B))$ *or* $E(\lambda,a)$. *For the associated family* (6.12) *and parameter set* $E(A,B)$ *or* $E(a)$ *consider the complexification* (6.15),

$$\dot{x} = i\left(x - \sum_{(p,q)\in S} a_{pq}x^{p+1}y^{q}\right), \qquad \dot{y} = -i\left(y - \sum_{(p,q)\in S} b_{qp}x^{q}y^{p+1}\right), \tag{6.23}$$

and the associated focus quantities $\{g_{kk}\}_{k=1}^{\infty} \subset \mathbb{C}[a,b]$. *Suppose* $\{g_{k_1,k_1},\ldots,g_{k_m,k_m}\}$ *is a collection of focus quantities for* (6.23) *having the following properties, where we set* $K=\{k_1,\ldots,k_m\}$:
(a) $g_{kk}=0$ *for* $1\le k<k_1$;
(b) $g_{k_q,k_q}\neq 0$ *for* $k_q\in K$;
(c) for $k_q\in K$, $q>1$, *and* $k\in\mathbb{N}$ *satisfying* $k_{q-1}<k<k_q$, $g_{kk}\in\mathscr{B}_{k_{q-1}}$, *the ideal in* $\mathbb{C}[a,b]$ *generated by the first* k_{q-1} *focus quantities;*
(d) the ideal $J=\langle g_{k_1,k_1},\ldots,g_{k_m,k_m}\rangle$ *in* $\mathbb{C}[a,b]$ *is radical; and*
(e) $\mathbf{V}(J)=\mathbf{V}(\mathscr{B})$, *where* $\mathscr{B}$ *is the Bautin ideal* $\langle g_{kk}:k\in\mathbb{N}\rangle$ *in* $\mathbb{C}[a,b]$.
Then the cyclicity of the singularity at the origin of any system in family (6.11), *with respect to the parameter space* $E(\lambda,(A,B))$, *is at most m.*

Proof. We first note that since $g_{kk}(a,\bar{a})=g_{kk}^{\mathbb{R}}(A(a,\bar{b}),\bar{B}(a,\bar{b}))\in\mathbb{R}$ for all $k\in\mathbb{N}$, the observation made at the end of the proof of Lemma 6.1.6 implies that for any collection $\{j_1,\ldots,j_n\}\subset\mathbb{N}$, any $k\in\mathbb{N}$, and any $f_1,\ldots,f_n\in\mathbb{C}[a,b]$,

$$\begin{gathered} g_{kk}=f_1g_{j_1,j_1}+\cdots+f_ng_{j_n,j_n} \\ \text{implies} \\ g_{kk}^{\mathbb{R}}=(\operatorname{Re}f_1)g_{j_1,j_1}^{\mathbb{R}}+\cdots+(\operatorname{Re}f_n)g_{j_n,j_n}^{\mathbb{R}}\,. \end{gathered} \tag{6.24}$$

Since $\mathbf{V}(J)=\mathbf{V}(\mathscr{B})$, for any $k\in\mathbb{N}$ the kth focus quantity g_{kk} vanishes on $\mathbf{V}(J)$, so $g_{kk}\in\mathbf{I}(\mathbf{V}(J))$ (Definition 1.1.10). But because J is a radical ideal, by the Strong Hilbert Nullstellensatz (Theorem 1.3.14), $\mathbf{I}(\mathbf{V}(J))=J$, hence $g_{kk}\in J$. Thus $\mathscr{B}\subset J$, so $J=\mathscr{B}$, and by the additional hypotheses on the set $\{g_{k_1,k_1},\ldots,g_{k_m,k_m}\}$ it is the minimal basis with respect to the retention condition of the Bautin ideal $\mathscr{B}$ with respect to the set $\{g_{11},g_{22},\ldots\}$. Thus, for any $k\in\mathbb{N}$, there exist $f_{k,1},\ldots,f_{k,m}\in\mathbb{C}[a,b]$ such that $g_{kk}=f_1g_{k_1,k_1}+\cdots+f_mg_{k_m,k_m}$. By (6.24) for any (A^*,B^*) in $E(A,B)$, $L:=\{\mathbf{g}_{k_1,k_1}^{\mathbb{R}},\ldots,\mathbf{g}_{k_m,k_m}^{\mathbb{R}}\}$ is then a basis of the ideal $I=\langle\mathbf{g}_{kk}^{\mathbb{R}}:k\in\mathbb{N}\rangle$ in $\mathscr{G}_{(A^*,B^*)}$. Clearly hypothesis (a) implies that for $k<k_1$, $\mathbf{g}_{kk}^{\mathbb{R}}=\mathbf{0}$ in $\mathscr{G}_{(A^*,B^*)}$. By hypothesis (b) and (6.24) for any $k_q\in K$, $q>1$, and any $k\in\mathbb{N}$ satisfying $k_1<k<k_q$, $\mathbf{g}_{kk}^{\mathbb{R}}\in\langle\mathbf{g}_{11},\ldots,\mathbf{g}_{k_{q-1},k_{q-1}}\rangle$ in $\mathscr{G}_{(A^*,B^*)}$. Thus it is apparent that even if L is not the minimal basis with respect to the retention condition M of the ideal $I=\langle\mathbf{g}_{11}^{\mathbb{R}},\mathbf{g}_{22}^{\mathbb{R}},\ldots\rangle$ in $\mathscr{G}_{(A^*,B^*)}$ (because of possible collapsing of g_{k_q,k_q} to $\mathbf{g}_{k_q,k_q}^{\mathbb{R}}=\mathbf{0}$), it nevertheless contains M, which therefore can have at most m elements. The conclusion of the corollary thus follows from Theorem 6.2.9. □

6.3 The Cyclicity of Quadratic Systems and a Family of Cubic Systems

The final result in the previous section, Corollary 6.2.10, tells us that if the Bautin ideal $\mathscr{B} = \langle g_{11}, g_{22}, \ldots \rangle$ corresponding to the complexification of system (6.12) is radical, then the cardinality of the minimal basis of $\mathscr{B}$ with respect to the retention condition is an upper bound on the cyclicity of the simple singularity at the origin of system (6.11), with respect to the parameter set $E(\lambda, (A, B))$, the parameter set corresponding to family (6.11). This result enables us to derive a version of Bautin's Theorem on the cyclicity of quadratic foci and centers. A simple but important result needed for the proof is the fact that any focus or center of a quadratic system is simple. Before we state and prove a lemma to that effect, however, we wish to state one more property of foci of real quadratic systems that is a consequence of previous results, but that has not been mentioned so far, and that will nicely complement the facts about quadratic foci and centers given in this section.

Proposition 6.3.1. *A fine focus of a real quadratic system of differential equations is of order at most three.*

Proof. Consider any quadratic system on $\mathbb{R}^2$ with a singularity at which the eigenvalues of the linear part are nonzero pure imaginary numbers. After a translation of coordinates to move the focus to the origin, a nonsingular linear transformation, and a time rescaling, the complexification of the system has the form (3.129) with some specific set (a^*, b^*) of coefficients. If g_{11}, g_{22}, and g_{33} are all zero at (a^*, b^*), which holds if and only if $g^{\mathbb{R}}_{11}$, $g^{\mathbb{R}}_{22}$, and $g^{\mathbb{R}}_{33}$ are all zero at the corresponding original real parameter string $(A(a^*, b^*), B(a^*, b^*))$, then by Theorem 3.7.1, $(a^*, b^*) \in \mathbf{V}(\mathscr{B}_3) = V_{\mathscr{C}}$, so the singularity is a center, not a focus. Thus at most the first two focus quantities can vanish at a fine focus of a quadratic system. But then Proposition 6.2.6 implies that the focus is of order at most three. □

We now continue our work on the cyclicity of quadratic systems, beginning with the following important fact.

Lemma 6.3.2. *Any focus or center of a quadratic system is simple.*

Proof. We prove the contrapositive, hence consider a quadratic system, which we write as $\dot{\mathbf{u}} = \mathbf{f}(\mathbf{u})$, with a nonsimple isolated singularity at $\mathbf{u_0} \in \mathbb{R}^2$. If $\operatorname{Tr} \mathbf{df}(\mathbf{u_0}) \neq 0$, then by Theorem 65 of §21 of [12], the singularity is a node, saddle, or saddle-node. If $\operatorname{Tr} \mathbf{df}(\mathbf{u_0}) = 0$, then by a translation to move the singularity to the origin, followed by an invertible linear transformation and a time rescaling, we may place the system in one of the two forms

$$\dot{u} = P_2(u, v), \quad \dot{v} = Q_2(u, v) \tag{6.25}$$

or

$$\dot{u} = v + P_2(u, v), \quad \dot{v} = Q_2(u, v), \tag{6.26}$$

where P_2 is either zero or a homogeneous polynomial of degree two, Q_2 is either zero or a homogeneous polynomial of degree two, and $\max\{\deg P_2, \deg Q_2\} = 2$. In the case of (6.25), any line that is composed of solutions of the cubic homogeneous equation $uQ_2(u,v) - vP_2(u,v) = 0$, of which there is at least one, is an invariant line through the origin in the phase portrait of (6.1), so $\mathbf{u_0}$ is not a focus or center. (This can be seen by writing the system in polar coordinates.) In the case of (6.26), by the Implicit Function Theorem, the equation $v + P_2(u,v) = 0$ defines an analytic function $v = \varphi(u) = \alpha_2 u^2 + \alpha_3 u^3 + \cdots$. Write $Q_2(u,v) = au^2 + buv + cv^2$ and define $\psi(u) := Q_2(u, \varphi(u)) = au^2 + \cdots$. If $a = 0$, then v factors out of the $\dot{v}$ equation so the line $v = 0$ is an invariant line through the origin in the phase portrait of $\dot{\mathbf{u}} = \mathbf{f}(\mathbf{u})$, so $\mathbf{u_0}$ is not a focus or a center. If $a \neq 0$, then by Theorem 67 in §22 of [12] (or the simplified scheme described later in that section), the singularity either is a saddle-node or has exactly two hyperbolic sectors, hence is not a focus or a center. □

Theorem 6.3.3. *The cyclicity of any center or focus in a quadratic system is at most three. There exist both foci and centers with cyclicity three in quadratic systems.*

Proof. Let a quadratic system with a singularity that is a focus or a center be given. By Lemma 6.3.2, the system can be written in the form (6.11a), whose complex form (6.11b) in this case is

$$\dot{x} = \lambda x + i(x - a_{10}x^2 - a_{01}x\bar{x} - a_{-12}\bar{x}^2). \tag{6.27}$$

The corresponding system with $\lambda = 0$ has complexification (3.129), or equivalently (3.131), but with $b_{10} = \bar{a}_{01}$, $b_{01} = \bar{a}_{10}$, and $b_{2,-1} = \bar{a}_{-12}$, since it arises from a real family. We will work with the complex parameters a_{10}, a_{01}, and a_{-12} rather than the original real parameters, since it is easier to do so. This is permissible, and all the estimates mentioned below continue to hold for the real parameters, since there is a linear isomorphism between the two parameter sets.

The first three focus quantities for a general complex quadratic system (3.131) (not necessarily the complexification of a real system) are listed in (3.133), (3.134), and (3.135). When we examine the five hypotheses of Corollary 6.2.10 in regard to the collection $\{g_{11}, g_{22}, g_{33}\}$, hypothesis (a) holds vacuously, (b) and (c) hold by inspection, and (e) is (3.137), which we established in the proof of Theorem 3.7.1. Thus the upper bound of three on the cyclicity will follow from the corollary if we can show that $\mathscr{B}_3 = \langle g_{11}, g_{22}, g_{33} \rangle$ is a radical ideal in $\mathbb{C}[a,b]$. One way to do so is to appeal to one of the advanced algorithms that are implemented in such special-purpose computer algebra systems as Macaulay or Singular (see Exercise 6.15). We will follow a different approach using the results derived in earlier chapters, and which has the advantage of illustrating techniques for studying polynomial ideals. We consider the four ideals J_1, J_2, J_3, and J_4 listed in Theorem 3.7.1. Since for any two ideals I and J, $\mathbf{V}(I \cap J) = \mathbf{V}(I) \cup \mathbf{V}(J)$ (Proposition 1.3.18), the identity $\mathbf{V}(\mathscr{B}_3) = \cup_{j=1}^4 \mathbf{V}(J_j)$ (Theorem 3.7.1) and (3.137) suggest that perhaps $\mathscr{B}_3 = \cap_{j=1}^4 J_j$. If this is so, and if each ideal J_j is prime, then because an intersection of prime ideals is radical (Proposition 1.4.4), it follows that $\mathscr{B}_3$ is radical. We will verify both of these conjectures.

We know already that the ideal J_4 is prime because it is I_{sym} (third paragraph of the proof of Theorem 3.7.1), which by Theorem 3.5.9 is always prime. For another approach to showing that J_1 and J_2 are prime, see the proof of Proposition 6.3.4. Here we will prove that they are prime using Gröbner Bases and the Multivariable Division Algorithm.

J_1 is prime. Under the ordering $a_{10} > a_{01} > a_{-12} > b_{2,-1} > b_{10} > b_{01}$, the generators listed for J_1 in Theorem 3.7.1 form a Gröbner basis G for J_1 with respect to lex. Let any $f, g \in \mathbb{C}[a,b] = \mathbb{C}[a_{10}, a_{01}, a_{-12}, b_{2,-1}, b_{10}, b_{01}]$ be given. Then, applying the Multivariable Division Algorithm, they can be written

$$\begin{aligned} f &= f_1(2a_{10} - b_{10}) + f_2(2b_{01} - a_{01}) + r_1 \\ g &= g_1(2a_{10} - b_{10}) + g_2(2b_{01} - a_{01}) + r_2, \end{aligned}$$

where f_1, f_2, g_1, g_2, r_1, and r_2 are in $\mathbb{C}[a,b]$ and r_1 and r_2 are reduced with respect to G, so that for each $j \in \{1,2\}$, either $r_j = 0$ or $r_j \neq 0$ but no monomial in r_j is divisible by $\text{LT}(2a_{10} - b_{10}) = 2a_{10}$ or by $\text{LT}(2b_{01} - a_{01}) = -a_{01}$. Suppose neither f nor g is in J_1, so that by the solution of the Ideal Membership Problem given on page 18 of Section 1.2, $r_j \neq 0$, $j = 1,2$. Recall the notation $[\nu] = a_{10}^{\nu_1} a_{01}^{\nu_2} a_{-12}^{\nu_3} b_{2,-1}^{\nu_4} b_{10}^{\nu_5} b_{01}^{\nu_6}$ for $\nu \in \mathbb{N}_0^6$. Setting $S_j = \text{Supp}(r_j)$, r_j can then be written in the form $r_j = \sum_{\nu \in S_j} r_j^{(\nu)}[\nu]$, where because r_1 and r_2 are reduced, $\nu \in S_1 \cup S_2$ implies that $\nu_1 = \nu_2 = 0$. Clearly the product $r_1 r_2$ has the same form, hence, because fg has the form $h + r_1 r_2$ for some $h \in J_1$, $r_1 r_2$ is the remainder of the product fg upon division by G. Since $r_1 r_2 \neq 0$, $fg \notin J_1$, so J_1 is prime.

J_2 is prime. The same argument as for J_1 applies.

J_3 is prime. Ordering the coefficients $a_{10} > a_{01} > a_{-12} > b_{2,-1} > b_{10} > b_{01}$, the set $G = \{2a_{01} + b_{01}, a_{10} + 2b_{10}, 2a_{-12}b_{2,-1} + b_{10}b_{01}\}$ is a Gröbner basis for J_3 with respect to lex. First observe that the ideal $J = \langle 2a_{-12}b_{2,-1} + b_{10}b_{01} \rangle$ is prime. There are two ways to see this, both based on the fact that $2a_{-12}b_{2,-1} + b_{10}b_{01}$ is irreducible. In general terms, $\mathbb{C}[a,b]$ is a unique factorization domain, and any ideal in a unique factorization domain that is generated by a prime (here an irreducible polynomial) is a prime ideal. In simpler terms, if $h_1 h_2 = f \cdot (2a_{-12}b_{2,-1} + b_{10}b_{01})$, then one or the other of h_1 and h_2 must be divisible by $2a_{-12}b_{2,-1} + b_{10}b_{01}$, since it is impossible that each contribute a factor.

Let any $f, g \in \mathbb{C}[a,b]$ be given, and write them as

$$\begin{aligned} f &= f_1(2a_{01} + b_{01}) + f_2(a_{10} + 2b_{10}) + f_3(2a_{-12}b_{2,-1} + b_{10}b_{01}) + r_1 \\ g &= g_1(2a_{01} + b_{01}) + g_2(a_{10} + 2b_{10}) + g_3(2a_{-12}b_{2,-1} + b_{10}b_{01}) + r_2, \end{aligned}$$

where r_1 and r_2 are reduced with respect to G. Suppose that neither f nor g is in J_3. Then for $j = 1,2$, $r_j \neq 0$, $r_j \notin J$, and r_j has the form $r_j = \sum_{\nu \in S_j} r_j^{(\nu)}[\nu]$ in which ν in $S_1 \cup S_2$ implies that $\nu_1 = \nu_2 = \nu_3 \nu_4 = 0$. The product of f and g can be written in the form $fg = h_1(2a_{01} + b_{01}) + h_2(a_{10} + 2b_{10}) + h_3(2a_{-12}b_{2,-1} + b_{10}b_{01}) + r_1 r_2$. From the form of r_1 and r_2 neither $2a_{01} = \text{LT}(2a_{01} + b_{01})$ nor $a_{10} = \text{LT}(a_{10} + 2b_{10})$ divides $r_1 r_2$, so $r_1 r_2$ reduces to zero modulo G only if $2a_{-12}b_{2,-1} + b_{10}b_{01}$ divides it. But this is impossible, since it would imply that $r_1 r_2 \in J$ in the face of our assumption that neither r_1 nor r_2 is in J and the preliminary observation that J is prime. Thus fg is not in J_3, which is therefore prime.

To show that the ideals $\mathscr{B}_3$ and $\cap_{j=1}^4 J_j$ are the same, we order the coefficients $a_{10} > a_{01} > a_{-12} > b_{2,-1} > b_{10} > b_{01}$, compute the unique reduced Gröbner basis of each with respect to lex under this ordering, and verify that they are identical. The reduced Gröbner basis of $\mathscr{B}_3$ can be quickly computed by any general-purpose symbolic manipulator; it is the six-element set shown in Table 6.1, where in order to avoid fractions each (monic) polynomial in the basis except the quadratic one was doubled before being listed. To find the reduced Gröbner basis of $\cap_{j=1}^4 J_j$, we apply the algorithm given in Table 1.5 on page 37 for computing a generating set for $I \cap J$ from generating sets of I and J three times: first for $J_1 \cap J_2$, then for $(J_1 \cap J_2) \cap J_3$, and finally for $(J_1 \cap J_2 \cap J_3) \cap J_4$. This, too, is easily accomplished using a computer algebra system and yields the same collection of polynomials as for $\mathscr{B}_3$.

$$a_{10}a_{01} - b_{01}a_{10}$$

$$2a_{01}^3 b_{2,-1} - 2a_{-12}b_{10}^3 + 3a_{10}a_{-12}b_{10}^2 - 3a_{01}^2 b_{2,-1}b_{01} - 2a_{01}b_{2,-1}b_{01}^2 + 2a_{10}^2 a_{-12}b_{10}$$

$$2a_{10}a_{-12}b_{10}^2 b_{01} + 2a_{01}^4 b_{2,-1} - 2a_{01}a_{-12}b_{10}^3 - 3a_{01}^3 b_{2,-1}b_{01} - 2a_{01}^2 b_{2,-1}b_{01}^2 + 3a_{-12}b_{10}^3 b_{01}$$

$$2a_{-12}b_{10}^3 b_{01}^2 + 2a_{01}^5 b_{2,-1} - 2a_{01}^2 a_{-12}b_{10}^3 - 3a_{01}^4 b_{2,-1}b_{01} - 2a_{01}^3 b_{2,-1}b_{01}^2 + 3a_{01}a_{-12}b_{10}^3 b_{01}$$

$$\begin{aligned}&2a_{10}a_{-12}^2 b_{2,-1}b_{10}^2 - a_{01}^4 b_{2,-1}b_{10} + a_{01}^3 a_{-12}b_{2,-1}^2 + 2a_{01}^3 b_{2,-1}b_{10}b_{01}\\ &\qquad - 2a_{01}^2 a_{-12}b_{2,-1}^2 b_{01} + a_{01}a_{-12}b_{10}^4 - a_{-12}^2 b_{2,-1}b_{10}^3 - 2a_{-12}b_{10}^4 b_{01}\end{aligned}$$

$$\begin{aligned}&4a_{-12}^2 b_{2,-1}b_{10}^3 b_{01} - 4a_{01}^3 a_{-12}b_{2,-1}^2 b_{01} - 2a_{01}^3 b_{2,-1}b_{10}b_{01}^2 + 2a_{01}^4 a_{-12}b_{2,-1}^2\\ &\qquad - 2a_{01}a_{-12}^2 b_{2,-1}b_{10}^3 + a_{01}^4 b_{2,-1}b_{10}b_{01} - a_{01}a_{-12}b_{10}^4 b_{01} + 2a_{-12}b_{10}^4 b_{01}^2\end{aligned}$$

Table 6.1 Reduced Gröbner Basis of $\mathscr{B}_3$ for System (3.129) (All except the quadratic polynomial were doubled before listing in order to eliminate fractional coefficients.)

Finally, we show that the bound on the cyclicity is sharp. We will denote a real system by the single letter X (usually subscripted) and will continue to specify a real system X by a complex triple $(a_{10}, a_{01}, a_{-12})$. We will construct concrete quadratic systems X_0 and X_1, the first with a center of cyclicity three and the second with a third-order fine focus of cyclicity three. To begin, we observe that if in the discussion in the first few paragraphs of Section 3.1, leading up to the definition of the function $\mathscr{P}$ in (3.14), we regard the right-hand sides of system (3.4) (which we have now replaced by the simpler form (6.11a)) as analytic functions in u, v, λ, and the coefficients of P and Q, then $\mathscr{P}$ is an analytic function of r_0 and the coefficients. Thus we will be able to freely rearrange series expressions for $\mathscr{P}$ and to assert convergence of expansions of $\mathscr{P}$ in powers of r_0 on specified intervals under sufficiently small perturbation of λ and the remaining coefficients. Next we observe that the equality $\mathbf{V}(\mathscr{B}) = \mathbf{V}(\mathscr{B}_3)$ of varieties (Theorem 3.7.1) implies the equality $\sqrt{\mathscr{B}} = \sqrt{\mathscr{B}_3}$ of ideals (Proposition 1.3.16).Hence, by the definition of the radical of an ideal and the fact that $\mathscr{B}_3$ is radical, $\mathscr{B}_3 \subset \mathscr{B} \subset \sqrt{\mathscr{B}} = \sqrt{\mathscr{B}_3} = \mathscr{B}_3$. Thus $\langle g_{jj} : j \in \mathbb{N} \rangle = \langle g_{11}, g_{22}, g_{33} \rangle$. Combining this fact with Theorem 6.2.3, when $\lambda = 0$ in (6.27), the difference function $\mathscr{P}$ of (3.14), whose zeros pick out cycles surrounding the antisaddle at the origin, can be written

$$\begin{aligned}\mathscr{P}(\rho) = \pi g_{11}\rho^3 &+ h_{41}g_{11}\rho^4 \\ &+ (h_{51}g_{11} + \pi g_{22})\rho^5 + (h_{61}g_{11} + h_{62}g_{22})\rho^6 \\ &+ (h_{71}g_{11} + h_{72}g_{22} + \pi g_{33})\rho^7 + (h_{81}g_{11} + h_{82}g_{22} + h_{83}g_{33})\rho^8 \\ &+ (h_{91}g_{11} + h_{92}g_{22} + h_{93}g_{33})\rho^9 + (h_{10,1}g_{11} + h_{10,2}g_{22} + h_{10,3}g_{33})\rho^{10} + \cdots\end{aligned}$$

for some $h_{jk} \in \mathbb{C}[a,b]$. Rearranging terms and rewriting, the right-hand side becomes

$$\pi\rho^3\left[g_{11}(1 + \widetilde{h}_{41}\rho + \cdots) + g_{22}(1 + \widetilde{h}_{62}\rho + \cdots)\rho^2 + g_{33}(1 + \widetilde{h}_{83}\rho + \cdots)\rho^4\right]. \quad (6.28)$$

Because the focus quantities are being computed from the complexification of a real system (so that $b_{jk} = \bar{a}_{kj}$), by (3.133)–(3.135) they are

$$\begin{aligned}g_{11} &= -i[a_{10}a_{01} - \bar{a}_{01}\bar{a}_{10}] = 2\,\mathrm{Im}(a_{10}a_{01}) \\ g_{22} &= -i[a_{10}\bar{a}_{01}^2 a_{-12} - \bar{a}_{10}a_{01}^2\bar{a}_{-12} - \tfrac{2}{3}(\bar{a}_{01}^3 a_{-12} - a_{01}^3\bar{a}_{-12}) \\ &\quad - \tfrac{2}{3}(\bar{a}_{10}^2 a_{01}\bar{a}_{-12} - a_{10}^2\bar{a}_{01}a_{-12})] \\ g_{33} &= \quad i\tfrac{5}{8}[-a_{01}\bar{a}_{01}^4 a_{-12} + 2\bar{a}_{10}\bar{a}_{01}^4 a_{-12} + a_{01}^4\bar{a}_{01}\bar{a}_{-12} - 2\bar{a}_{10}a_{01}^3\bar{a}_{01}\bar{a}_{-12} \\ &\quad - 2a_{10}\bar{a}_{01}^2 a_{-12}^2\bar{a}_{-12} + \bar{a}_{01}^3 a_{-12}^2\bar{a}_{-12} - a_{01}^3 a_{-12}\bar{a}_{-12}^2 + 2\bar{a}_{10}a_{01}^2 a_{-12}\bar{a}_{-12}^2].\end{aligned}$$

To simplify the discussion, we introduce a complex parameter c such that $a_{10} = c\bar{a}_{01}$. Then $g_{11} = 2|a_{01}|^2\,\mathrm{Im}(c)$ and for $c \in \mathbb{R}$ we have

$$g_{22} = \tfrac{2}{3}(2c-1)(c+2)\,\mathrm{Im}(\bar{a}_{01}^3 a_{-12}), \; g_{33} = \tfrac{4}{5}(2c-1)(|a_{-12}|^2 - |a_{01}|^2)\,\mathrm{Im}(\bar{a}_{01}^3 a_{-12}).$$

Choose $\lambda = 0$, choose any pair of nonzero complex numbers a_{01} and a_{-12} satisfying the two conditions (a) $|a_{-12}| - |a_{01}| = 0$ and (b) $\mathrm{Im}(\bar{a}_{01}^3 a_{-12}) > 0$, and choose $a_{10} = -2\bar{a}_{01}$, corresponding to $c = -2$. Let X_0 denote the corresponding real quadratic system. Then $g_{11} = g_{22} = g_{33} = 0$ so X_0 has a center at the origin. Let $\rho_0 > 0$ be such that the power series expansion of $\mathscr{P}$ in powers of ρ converges on the interval $(-4\rho_0, 4\rho_0)$ and the disk D_0 of radius $4\rho_0$ about the origin lies wholly within the period annulus of the center of X_0 at the origin.

Let positive real constants r and ε be given.

Make an arbitrarily small perturbation in a_{-12} so that convergence of $\mathscr{P}(\rho)$ holds on the interval $(-3\rho_0, 3\rho_0)$, conditions (b) and (c) continue to hold, but now $|a_{-12}|^2 - |a_{01}|^2 < 0$. Let X_1 denote the corresponding real quadratic system, which may be chosen so that its coefficients are arbitrarily close to those of X_0. Then for X_1 we have that $g_{11} = g_{22} = 0 < g_{33}$ (since $2c - 1 < 0$), so that the origin is a third-order fine focus for X_1 and is unstable (Remark 3.1.6 and Theorem 6.2.3). There are no periodic orbits of X_1 wholly contained in the disk D_1 of radius $3\rho_0$ about the origin.

Now change a_{10} corresponding to the change in c from -2 to $-2+\delta_1$ for $\delta_1 \in \mathbb{R}^+$. For δ_1 sufficiently small, convergence of $\mathscr{P}(\rho)$ holds on the interval $(-2\rho_0, 2\rho_0)$, g_{33} changes by an arbitrarily small amount, g_{11} remains zero, and g_{22} becomes negative but arbitrarily close to zero. Change a_{10} a second time, now corresponding to the change in c from $-2+\delta_1$ to $(-2+\delta_1)+\delta_2 i$ for $\delta_2 \in \mathbb{R}^+$. For δ_2 sufficiently small, convergence of $\mathscr{P}(\rho)$ holds on the interval $(-\rho_0, \rho_0)$, g_{22} and g_{33} change by arbitrarily small amounts, while g_{11} becomes positive but arbitrarily close to zero. Let X_2 denote the resulting real quadratic system. Because for X_2 the first three focus quantities satisfy $0 < g_{11} \ll -g_{22} \ll g_{33}$ by the expression (6.28) for $\mathscr{P}(\rho)$ and the quadratic formula, for suitably chosen g_{11}, g_{22}, and g_{33} there are exactly two zeros of $\mathscr{P}(\rho)$ in the interval $(0, \rho_0]$, the larger one less than $\sqrt{-g_{22}/g_{33}}$, which can be made to be less than ρ_0. Moreover, $\mathscr{P}(\rho)$ is positive on a neighborhood of 0 in $(0, \rho_0]$. But then by (3.14) and the identity $\eta_1 = e^{2\pi\lambda} - 1$ (see the sentence that follows Definition 3.1.3), if λ is made negative but $|\lambda|$ is sufficiently small, a third isolated zero of $\mathscr{P}(\rho)$ appears in $(0, \rho_0)$ for the corresponding real quadratic system X_3. System X_3 can be chosen to be within ε of X_0 and X_1 and have its three small limit cycles wholly within the disk about the origin of radius r. Thus the center of X_0 and the fine focus of X_1 each have cyclicity three. □

We have easily obtained a proof of the celebrated Bautin theorem (which is the solution of the local 16th Hilbert problem for system (6.1) with $n = 2$) with the help of the methods of computational algebra. Simple good fortune entered in, however, since the ideal $\mathscr{B}_3$ defining the center variety was radical, and it is most likely that it will very seldom be true that the ideal of the first few focus quantities defining the center variety will be radical. We devote the remainder of this section to a study of the real system that has complex form

$$\dot{x} = i(x - a_{10}x^2 - a_{01}x\bar{x} - a_{-13}\bar{x}^3), \tag{6.29}$$

which will illustrate the difficulties encountered with systems of higher order and some of the techniques for working with them. As usual we adjoin to equation (6.29) its complex conjugate and consider $\bar{x}$ and $\bar{a}_{jk}$ as new independent complex variables. We thus obtain system (3.130), which we examined in Section 3.7 (Theorem 3.7.2) and which for reference we reproduce here as

$$\begin{aligned} \dot{x} &= \quad i(x - a_{10}x^2 - a_{01}xy - a_{-13}y^3) \\ \dot{y} &= -i(y - b_{10}xy - b_{01}y^2 - b_{3,-1}x^3). \end{aligned} \tag{6.30}$$

The first nine focus quantities for system (6.30) are listed in the first paragraph of the proof of Theorem 3.7.2. The first five focus quantities determine the center variety of system (6.30) and, therefore, the radical of the ideal of focus quantities of this system; that is, $\sqrt{\mathscr{B}_5} = \sqrt{\mathscr{B}}$. However, in contrast with what is the case for quadratic systems, $\mathscr{B}_5$ is not all of the Bautin ideal $\mathscr{B}$ (Exercise 6.18), nor is it radical:

Proposition 6.3.4. *The ideal* $\mathscr{B}_5 = \langle g_{11}, g_{22}, g_{33}, g_{44}, g_{55} \rangle$ *generated by the first five focus quantities of system* (6.30) *is not radical in* $\mathbb{C}[a_{10}, a_{01}, a_{-13}, b_{3,-1}, b_{10}, b_{01}]$.

Proof. We claim that each of the ideals J_j, $j = 1, \ldots, 8$, in the statement of Theorem 3.7.2 is prime. J_8 is prime by Theorem 3.5.9. For a proof that the remaining seven ideals are prime that involves the theory developed in Chapter 1, see Exercise 6.17. Here we give a purely algebraic proof. To begin with, it is clear that in general for $m, n \in \mathbb{N}$ with $m \leq n$, the ideal $\langle x_1, \ldots, x_m \rangle$ in $\mathbb{C}[x_1, \ldots, x_n]$ is prime. Thus for $c_j \in \mathbb{C}$, $j = 1, \ldots, n$, because the rings $\mathbb{C}[x_1, \ldots, x_n]$ and $\mathbb{C}[x_1 - c_1, \ldots, x_n - c_n]$ are isomorphic, the ideal $\langle x_1 - c_1, \ldots, x_m - c_m \rangle$ in $\mathbb{C}[x_1, \ldots, x_n]$ is prime. Thus if, for $r, s \in \mathbb{N}$, we view $\mathbb{C}[x_1, \ldots, x_r, y_1, \ldots, y_s]$ as $\mathbb{C}[x_1, \ldots, x_r][y_1, \ldots, y_s]$, we have that the ideal $\langle x_{j_1} - y_{k_1}, \ldots, x_{j_p} - y_{k_p} \rangle$ is prime. Applying these ideas to J_1 through J_7 shows that they are prime.

Using the algorithm of Table 1.5 on page 37, we can compute a Gröbner basis of the ideal $\mathscr{J} = \cap_{j=1}^{8} J_j$, which by Proposition 1.4.4 is radical, and obtain from it the unique reduced Gröbner basis of $\mathscr{J}$. Using the known finite set of generators of each of $\mathscr{B}_5$ and $\mathscr{J}$, we can apply the Radical Membership Test (Table 1.4 on page 33) and Exercise 1.35 to determine that $\sqrt{\mathscr{B}_5} = \sqrt{\mathscr{J}} = \mathscr{J}$. (This identity also follows directly from the identity $\mathbf{V}(\mathscr{B}_5) = \mathbf{V}(\mathscr{B}) = \cup_{j=1}^{8} \mathbf{V}(J_j) = \mathbf{V}(\cap_{j=1}^{8} J_j)$ (Theorems 3.7.2 and 1.3.18(2)) and Propositions 1.3.16 and 1.4.4.) But when we compute the reduced Gröbner basis of $\mathscr{B}_5$ we do not get the reduced Gröbner basis for $\mathscr{J}$, so $\mathscr{B}_5 \neq \mathscr{J}$ and we conclude that $\mathscr{B}_5$ is not radical. □

As we have shown above, the first five focus quantities define the center variety of system (6.30), but the corresponding ideal $\mathscr{B}_5$ is not radical. Similarly, the ideals $\mathscr{B}_7$ and $\mathscr{B}_9$ generated by the first seven and the first nine focus quantities are not radical either (Exercise 6.19). By Exercise 6.18, neither $\mathscr{B}_5$ nor $\mathscr{B}_7$ is all of the Bautin ideal $\mathscr{B}$. In order to make progress on the cyclicity problem for family (6.29), we are led to attempt to make the problem more tractable by means of a coordinate transformation. Turn back to Theorem 3.7.2 and examine the generators of the eight ideals that determine the irreducible components of the center variety of family (3.130) (our family (6.30)), and also consider the first focus quantity g_{11}, which is given in the first paragraph of the proof of that theorem. If we replace a_{10} by a multiple of b_{10} and replace b_{01} by a multiple of a_{01}, then the first focus quantity and many of the generators factor as products of distinct variables, so that a decoupling of variables occurs. This suggests that we define a mapping $G : \mathbb{C}^6 \to \mathbb{C}^6$ by

$$\begin{aligned}(a_{10}, a_{01}, a_{-13}, b_{3,-1}, b_{10}, b_{01}) \\ = G(s_1, a_{01}, a_{-13}, b_{3,-1}, b_{10}, s_2) = (s_1 b_{10}, a_{01}, a_{-13}, b_{3,-1}, b_{10}, s_2 a_{01}).\end{aligned} \tag{6.31}$$

That is, we introduce new variables s_1 and s_2 by setting

$$a_{10} = s_1 b_{10}, \qquad b_{01} = s_2 a_{01}. \tag{6.32}$$

This process induces a homomorphism of the ring of polynomials over $\mathbb{C}$ with six indeterminates, defined for any $f \in \mathbb{C}[a, b] = \mathbb{C}[a_{10}, a_{01}, a_{-13}, b_{3,-1}, b_{10}, b_{01}]$ by

$$W : \mathbb{C}[a_{10}, a_{01}, a_{-13}, b_{3,-1}, b_{10}, b_{01}] \quad \to \quad \mathbb{C}[s_1, a_{01}, a_{-13}, b_{3,-1}, b_{10}, s_2]$$
$$: \sum_{\nu \in \mathrm{Supp}(f)} f^{(\nu)} a_{10}^{\nu_1} a_{01}^{\nu_2} a_{-13}^{\nu_3} b_{3,-1}^{\nu_4} b_{10}^{\nu_5} b_{01}^{\nu_6} \mapsto$$
$$\sum_{\nu \in \mathrm{Supp}(f)} f^{(\nu)} s_1^{\nu_1} a_{01}^{\nu_2+\nu_6} a_{-13}^{\nu_3} b_{3,-1}^{\nu_4} b_{10}^{\nu_5+\nu_1} s_2^{\nu_6} = \sum_{\sigma \in \Sigma} f^{(\alpha(\sigma))} s_1^{\sigma_1} a_{01}^{\sigma_2} a_{-13}^{\sigma_3} b_{3,-1}^{\sigma_4} b_{10}^{\sigma_5} s_2^{\sigma_6},$$

where Σ is the image in $\mathbb{N}_0^6$ of $\mathrm{Supp}(f) \subset \mathbb{N}_0^6$ under the invertible linear map

$$\begin{aligned} \omega : \mathbb{R}^6 \to \mathbb{R}^6 : (\nu_1, \nu_2, \nu_3, \nu_4, \nu_5, \nu_6) &\mapsto (\sigma_1, \sigma_2, \sigma_3, \sigma_4, \sigma_5, \sigma_6) \\ &= (\nu_1, \nu_2 + \nu_6, \nu_3, \nu_4, \nu_5 + \nu_1, \nu_6) \end{aligned} \tag{6.33}$$

and $\alpha : \mathbb{R}^6 \to \mathbb{R}^6$ is the inverse of ω. Although the mapping G is not a true change of coordinates (it is neither one-to-one nor onto), the ring homomorphism W is one-to-one (Exercise 6.20), hence is an isomorphism of $\mathbb{C}[a,b]$ with its image, call it C. We denote the image of f under W by $\check{f}$. Similarly, we denote the ideal in C that is the image of an ideal I in $\mathbb{C}[a,b]$ by $\check{I}$.

The image C is not all of $\mathbb{C}[s_1, a_{01}, a_{-13}, b_{3,-1}, b_{10}, s_2]$, so in addition to the ideals $\check{\mathscr{B}}_k$ and $\check{\mathscr{B}}$ *within* C, we have the larger ideals $\langle \check{g}_{11}, \ldots \check{g}_{kk} \rangle$ and $\langle \check{g}_{jj} : j \in \mathbb{N} \rangle$ in $\mathbb{C}[s_1, a_{01}, a_{-13}, b_{3,-1}, b_{10}, s_2]$, in which the weights in the finite linear combinations are not restricted to lie in C. We denote these larger ideals by $\check{\mathscr{B}}_k^+$ and $\check{\mathscr{B}}^+$, respectively. In essence, by means of the transformation G, we have placed the Bautin ideal in a larger ring, and it is here that we are able to prove that the first nine focus quantities suffice to generate the whole ideal.

Proposition 6.3.5. *With the notation of the preceding paragraph,* $\check{\mathscr{B}}^+ = \check{\mathscr{B}}_9^+$. *That is, the polynomials* $\check{g}_{11}$, $\check{g}_{33}$, $\check{g}_{44}$, $\check{g}_{55}$, $\check{g}_{77}$, *and* $\check{g}_{99}$ *form a basis of the ideal of focus quantities of system* (6.30) *in the ring* $\mathbb{C}[s_1, a_{01}, a_{-13}, b_{3,-1}, b_{10}, s_2]$.

Proof. As usual, we write just $\mathbb{C}[a,b]$ for $\mathbb{C}[a_{10}, a_{01}, a_{-13}, b_{3,-1}, b_{10}, b_{01}]$ and for $\nu \in \mathbb{N}_0^6$ let $[\nu]$ denote the monomial $a_{10}^{\nu_1} a_{01}^{\nu_2} a_{-13}^{\nu_3} b_{3,-1}^{\nu_4} b_{10}^{\nu_5} b_{01}^{\nu_6} \in \mathbb{C}[a,b]$. The focus quantities $g_{kk} \in \mathbb{C}[a,b]$ have the form (3.82),

$$g_{kk} = \tfrac{1}{2} \sum_{\{\nu : L(\nu) = (k,k)\}} g_{kk}^{(\nu)} ([\nu] - [\hat{\nu}]), \tag{6.34}$$

where $i g_{kk}^{(\nu)} \in \mathbb{Q}$ and L is defined by (3.71), which in the present situation is the map $L : \mathbb{N}_0^6 \to \mathbb{Z}^2$ given by

$$\begin{aligned} L(\nu) &= \nu_1(1,0) + \nu_2(0,1) + \nu_3(-1,3) + \nu_4(3,-1) + \nu_5(1,0) + \nu_6(0,1) \\ &= (-\nu_3 + 3\nu_4 + (\nu_5 + \nu_1), (\nu_2 + \nu_6) + 3\nu_3 - \nu_4). \end{aligned} \tag{6.35}$$

The caret accent denotes the involution on $\mathbb{C}[a,b]$ given by Definition 3.4.3 and the paragraph following it. For the remainder of this proof we modify it by not taking the complex conjugate of the coefficients in forming the conjugate. The new involution agrees with the old one on monomials, but now, for example, in the expression for g_{kk}, $\widehat{g}_{kk}^{(\nu)} = g_{kk}^{(\nu)}$ since it is a constant polynomial.

For any μ and σ in $\mathbb{N}_0^6$,

$$\begin{aligned}[\mu+\sigma] - [\widehat{\mu+\sigma}] &= \tfrac{1}{2}\Big([\sigma]+[\widehat{\sigma}]\Big)\Big([\mu]-[\widehat{\mu}]\Big) + \tfrac{1}{2}\Big([\mu]+[\widehat{\mu}]\Big)\Big([\sigma]-[\widehat{\sigma}]\Big) \\ &= f_1\Big([\mu]-[\widehat{\mu}]\Big) + f_2\Big([\sigma]-[\widehat{\sigma}]\Big),\end{aligned} \tag{6.36}$$

where $f_j \in \mathbb{Q}[a,b]$ and $\widehat{f_j} = f_j$ for $j = 1, 2$. Every string ν appearing in the sum in (6.34) is an element of the monoid $\mathscr{M}$. Using the algorithm of Table 5.1 on page 235, we construct a Hilbert basis $\mathscr{H}$ of $\mathscr{M}$; the result is

$$\begin{aligned}\mathscr{H} = \{&(100001),(010010),(001\,100),(110000),(000011),\\ &(040\,100),(001\,040),(030\,101),(101\,030),(020\,102),(201\,020),\\ &(010\,103),(301\,010),(000\,104),(401\,000)\}.\end{aligned}$$

For $j = 1,\dots,15$, let μ_j denote the jth element of $\mathscr{H}$ as listed here, so that, for example, $\mu_3 = (001\,100)$. For any $k \in \mathbb{N}$, the sum in (6.34) is finite. Expressing each $\nu \in \operatorname{Supp}(g_{kk})$ in terms of elements of $\mathscr{H}$, applying (6.36) repeatedly, and collecting terms yields an expression for g_{kk} of the form

$$g_{kk} = f_1\Big([\mu_1]-[\widehat{\mu}_1]\Big) + \dots + f_{15}\Big([\mu_{15}]-[\widehat{\mu}_{15}]\Big),$$

where $if_j \in \mathbb{Q}[a,b]$ and $\widehat{f_j} = f_j$ for $j = 1,\dots,15$. It is apparent that f_j is a polynomial in the monomials $[\mu_1],\dots,[\mu_{15}]$. Since $\widehat{\mu}_j = \mu_j$ for $j = 1, 2, 3$, the first three terms do not actually appear. Also, $[\mu_4]-[\widehat{\mu}_4] = g_{11}$ and $[\mu_5]-[\widehat{\mu}_5] = -g_{11}$. Since we are attempting to show that $\breve{g}_{kk} \in \mathscr{B}_9$, which contains $\breve{g}_{11}$ as a generator, we may replace g_{kk} with the polynomial that results when these multiples of g_{11} are removed. That is, we reduce g_{kk} modulo g_{11}. So as not to complicate the notation, we retain the same name g_{kk} for the reduced focus quantity. The remaining elements of $\mathscr{H}$ occur in conjugate pairs, $\mu_{j+1} = \widehat{\mu}_j$ for $j = 6$, 8, 10, 12, and 14. Thus we may use the identity $[\mu_{j+1}]-[\widehat{\mu}_{j+1}] = -\Big([\mu_j]-[\widehat{\mu}_j]\Big)$ for these values of j to finally express the reduced g_{kk} as

$$\begin{aligned}g_{kk} &= (f_6-f_7)\Big([\mu_6]-[\widehat{\mu}_6]\Big) + \dots + (f_{14}-f_{15})\Big([\mu_{14}]-[\widehat{\mu}_{14}]\Big)\\ &= \sum_{j=3}^{7} h_{2j}\Big([\mu_{2j}]-[\widehat{\mu}_{2j}]\Big),\end{aligned} \tag{6.37}$$

where $ih_{2j} \in \mathbb{Q}[a,b]$, $\widehat{h}_{2j} = h_{2j}$ for $j = 3,\dots,7$, and again h_{2j} is a polynomial in the monomials $[\mu_1],\dots,[\mu_{15}]$.

Define an involution on $\mathbb{C}[s_1,a_{01},a_{-13},b_{3,-1},b_{10},s_2]$, also denoted by a caret accent, by

$$s_1 \to s_2, \quad s_2 \to s_1, \quad a_{01} \to b_{10}, \quad a_{-13} \to b_{3,-1}, \quad b_{3,-1} \to a_{-13}, \quad b_{10} \to a_{01}\,.$$

Thus, for example, if

$$\begin{aligned} f &= (2+3i)s_1^2 a_{01} a_{-13}^3 b_{10}^3 + (5-7i)s_1^3 a_{01}^4 b_{3,-1} b_{10}^3 s_2^4 \\ &= f_1(s_1,s_2)\, a_{01} a_{-13}^3 b_{10}^3 + f_2(s_1,s_2)\, a_{01}^4 b_{3,-1} b_{10}^3 \end{aligned}$$

then

$$\begin{aligned} \widehat{f} &= (2+3i)s_2^2 b_{10} b_{3.-1}^3 a_{01}^3 + (5-7i)s_2^3 b_{10}^4 a_{-13} a_{01}^3 s_1^4 \\ &= \widehat{f_1}(s_1,s_2)\, b_{10} b_{3.-1}^3 a_{01}^3 + \widehat{f_2}(s_1,s_2)\, b_{10}^4 a_{-13} a_{01}^3\,. \end{aligned}$$

For any $f \in \mathbb{C}[a,b]$, the conjugate of the image $\check{f}$ of f under the homomorphism W is the image of the conjugate. In particular, f is invariant under the involution on $\mathbb{C}[a,b]$ if and only if its image $\check{f}$ is invariant under the involution on the isomorphic copy $C = W(\mathbb{C}[a,b])$ of $\mathbb{C}[a,b]$ lying in $\mathbb{C}[s_1, a_{01}, a_{-13}, b_{3,-1}, b_{10}, s_2]$.

In order to make the next display easier to read, we extend the bracket notation to $\mathbb{C}[s_1, a_{01}, a_{-13}, b_{3,-1}, b_{10}, s_2]$, expressing the monomial $a_{01}^{\nu_1} a_{-13}^{\nu_2} b_{3,-1}^{\nu_3} b_{10}^{\nu_4}$ as $[\nu_1, \nu_2, \nu_3, \nu_4]$. With this notation,

$$W([\nu_1,\nu_2,\nu_3,\nu_4,\nu_5,\nu_6]) = s_1^{\nu_1} s_2^{\nu_6}[\nu_2+\nu_6, \nu_3, \nu_4, \nu_5+\nu_1].$$

Then the images of the monomials corresponding to the 15 elements of the Hilbert basis $\mathscr{H}$ of $\mathscr{M}$ are (note the ordering of the last two columns compared to the ordering in the display for $\mathscr{H}$)

$$\begin{array}{lll} [100001] \to s_1 s_2[1001] & [040100] \to \quad [4010] & [001040] \to \quad [0104] \\ [010010] \to \quad [1001] & [030101] \to s_2[4010] & [101030] \to s_1[0104] \\ [001100] \to \quad [0110] & [020102] \to s_2^2[4010] & [201020] \to s_1^2[0104]. \\ [110000] \to \quad s_1[1001] & [010103] \to s_2^3[4010] & [301010] \to s_1^3[0104] \\ [000011] \to \quad s_2[1001] & [000104] \to s_2^4[4010] & [401000] \to s_1^4[0104] \end{array} \tag{6.38}$$

We name the monomials that appear as images as

$$u = [0110] = a_{-13} b_{3,-1}, \qquad v = [1001] = a_{01} b_{10}, \qquad w = [0104] = a_{-13} b_{10}^4.$$

The first two are self-conjugate and $\widehat{w} = [4010] = a_{01}^4 b_{3,-1}$.

Using (6.37) and the last two columns of (6.38), the image of the reduced g_{kk} is

$$\begin{aligned} \check{g}_{kk} &= \check{h}_6(\widehat{w} - w) + \check{h}_8(s_2\widehat{w} - s_1 w) + \check{h}_{10}(s_2^2\widehat{w} - s_1^2 w) \\ &\qquad + \check{h}_{12}(s_2^3\widehat{w} - s_1^3 w) + \check{h}_{14}(s_2^4\widehat{w} - s_1^4 w) \\ &= hw - \widehat{h}\widehat{w}. \end{aligned} \tag{6.39}$$

Knowing that the functions h_{2j} are polynomials in the monomials $[\mu_1],\ldots,[\mu_{15}]$ and using the first column in (6.38), we see that in fact h can be regarded as a polynomial function of s_1, s_2, u, v, w, and $\widehat{w}$. Note, however, that by regrouping the terms in the monomials that compose h, it could be possible to express h in terms of s_1, s_2, u, v, w, and $\widehat{w}$ in diffferent ways. An important example of this phenomenon is the identity $w\widehat{w} = a_{01}^4 a_{-13} b_{3,-1} b_{10}^4 = uv^4$. We will now show that by means of this identity and a possible reduction modulo $\breve{g}_{11}$, we can eliminate all $\widehat{w}$ dependence from h.

For $m,n \in \mathbb{N}_0$, $n \geq 1$, let $w^m \widehat{w}^n f_{m,n}(s_1,s_2,u,v)$ denote the sum of all terms in h that contain $w^m\widehat{w}^n$. For $n \leq m$, we can factor out a power of w to rewrite this expression as $w^{m-n}(w\widehat{w})^n f_{m,n} = w^{m-n}(uv^4)^n f_{m,n}$. For $n > m$, we must move terms between the expressions hw and $\widehat{hw}$ in (6.39). Beginning as in the previous case, we write

$$\begin{aligned}
&\left[\widehat{w}^{n-m}(uv^4)^m f_{m,n}\right]w - \left[w^{n-m}(uv^4)^m \widehat{f}_{m,n}\right]\widehat{w} \\
&\quad = \widehat{w}^{n-m-1}(uv^4)^{m+1} f_{m,n} - w^{n-m-1}(uv^4)^{m+1}\widehat{f}_{m,n} \\
&\quad = (-\widehat{f}_{m,n})(uv^4)^{m+1} w^{n-m-1} - (-f_{m,n})(uv^4)^{m+1}\widehat{w}^{n-m-1} \\
&\quad = \left[-\widehat{f}_{m,n}(uv^4)^{m+1} w^{n-m-2}\right]w - \left[-f_{m,n}(uv^4)^{m+1}\widehat{w}^{n-m-2}\right]\widehat{w},
\end{aligned}$$

which eliminates $\widehat{w}$ dependence from h provided $n \geq m+2$. For $n = m+1$, the last line in the display is not present, and the line that precedes it has the form

$$\begin{aligned}
&(uv^4)^{m+1} f(s_1,s_2,u,v) - (uv^4)^{m+1}\widehat{f}(s_1,s_2,u,v) \\
&\qquad\qquad = (uv^4)^{m+1}\left[f(s_1,s_2,u,v) - \widehat{f}(s_1,s_2,u,v)\right]. \qquad (6.40)
\end{aligned}$$

Any term $A\, s_1^{n_1} s_2^{n_2} u^{n_3} v^{n_4}$ in f produces in (6.40) the binomial

$$A\, u^{n_3+m+1} v^{n_4+4m+4}(s_1^{n_1} s_2^{n_2} - s_1^{n_2} s_2^{n_1}) = A\, u^p v^q (s_1^s s_2^t - s_1^t s_2^s),$$

where $q > 0$. The term does not appear if $s = t$. If $s \neq t$, say $s > t$ (the case $t > s$ is similar), such a binomial can be written

$$A\, u^p v^q (s_1 s_2)^t (s_1^{s-t} - s_2^{s-t}) = A\, u^p v^{q-1} (s_1 s_2)^t (s_1^{s-t-1} - \cdots - s_2^{s-t-1})(s_1 - s_2)v.$$

But $(s_1 - s_2)v = (s_1 - s_2)a_{01}b_{10} = \breve{g}_{11}$, so any such term can be discarded in a further reduction of $\breve{g}_{kk}$ modulo $\breve{g}_{11}$, although as before we will retain the same notation for the reduced focus quantity. We thus have expression (6.39), where, with respect to the current grouping of terms in the monomials that are present in h, $ih \in \mathbb{Q}[s_1,s_2,u,v,w]$.

In general, for any element f of $\mathbb{C}[a,b]$ and any point $(a^*,b^*) \in \mathbb{C}^6$ such that $f(a^*,b^*) = 0$, if $a_{01}^* b_{10}^* \neq 0$, then we can define two numbers $s_1^* = a_{10}^*/b_{10}^*$ and $s_2^* = b_{01}^*/a_{01}^*$, and know that $\breve{f}(s_1^*, a_{10}^*, a_{-13}^*, b_{3,-1}^*, b_{10}^*, s_2^*) = 0$. By Theorem 3.7.2(7), if (a,b) is such that $b_{10} = 2a_{10}$ and $a_{01} = 2b_{01}$, corresponding to $s_1 = s_2 = 1/2$, then $g_{kk}(a,b) = 0$ for all $k \in \mathbb{N}$. This means that if we write the h of (6.39) as $h = \sum_{k=0}^{T} H_k(s_1,s_2,u,v)w^k$, then

$$\sum_{k=0}^{T} H_k(\tfrac{1}{2},\tfrac{1}{2},u,v)w^{k+1} - \sum_{k=0}^{T} \widehat{H}_k(\tfrac{1}{2},\tfrac{1}{2},u,v)\widehat{w}^{k+1}$$
$$= \sum_{k=0}^{T} H_k(\tfrac{1}{2},\tfrac{1}{2},u,v)(w^{k+1} - \widehat{w}^{k+1}) = 0 \quad (6.41)$$

since u and v are self-conjugate. Although the H_k are polynomials in u and v, the identity (6.41) does not automatically imply that $H_k(1/2,1/2,u,v)$ is the zero polynomial in $\mathbb{C}[u,v]$, for u, v, w, and $\widehat{w}$ are not true indeterminates, but only shorthand expressions for monomials in (a,b). As stated, identity (6.41) holds subject to the condition $uv^4 - w\widehat{w} = 0$. To obtain the conclusion that $H_k(1/2,1/2,u,v)$ is the zero polynomial in $\mathbb{C}[u,v]$, we must show that no term in $H_k w^{k+1}$ can cancel with any term in $H_j w^{j+1}$ or $H_j \widehat{w}^{j+1}$ for $j \neq k$. Any term in $H_k(1/2,1/2,u,v)w^{k+1}$ has the form $A a_{01}^{s} b_{10}^{s+4k+4} a_{-13}^{r+k+1} b_{3,-1}^{r}$; any term in $H_j(1/2,1/2,u,v)w^{j+1}$ has the form $B a_{01}^{n} b_{10}^{n+4j+4} a_{-13}^{m+j+1} b_{3,-1}^{m}$. The powers on a_{01} agree only if $n = s$ and the powers on $b_{3,-1}$ agree only if $m = r$, hence the powers on a_{-13} agree only if $j = k$. The other cases are similar. Thus, for each k, as an element of $\mathbb{C}[a_{01},a_{-13},b_{3,-1},b_{10}]$ $H_k(1/2,1/2,u,v)w^{k+1}$ is zero, hence $H_k(1/2,1/2,u,v)$ is also the zero polynomial in $\mathbb{C}[a_{01},a_{-13},b_{3,-1},b_{10}]$. The map from $\mathbb{C}^4$ to $\mathbb{C}^2$ that carries $(a_{01},a_{-13},b_{3,-1},b_{10})$ to (u,v) is onto, so this means that the polynomial function F on $\mathbb{C}^2$ defined by $F(u,v) = H_k(1/2,1/2,u,v)$ is identically zero. By Proposition 1.1.1, $H_k(1/2,1/2,u,v)$ must be the zero polynomial in $\mathbb{C}[u,v]$. Thus there exist polynomials h_a and h_b in $\mathbb{C}[s_1,a_{10},a_{-13},b_{3,-1},b_{10},s_2]$ such that $h = h_a(2s_1 - 1) + h_b(2s_2 - 1)$. Inserting this expression into (6.39) yields

$$\breve{g}_{kk} = h_a(2s_1 - 1)w + h_b(2s_2 - 1)w - \widehat{h}_a(2s_2 - 1)\widehat{w} - \widehat{h}_b(2s_1 - 1)\widehat{w}, \qquad (6.42)$$

where, with respect to the current grouping of terms in the monomials that are present in h_a and h_b, each of ih_a and ih_b lies in $\mathbb{Q}[s_1,s_2,u,v,w]$, $j = 1, 2$. With respect to that grouping, now reorder the terms in $\breve{g}_{kk}$ so that it is expressed as a sum of three polynomials $\breve{g}_{kk} = \breve{g}_{kk}^{(1)} + \breve{g}_{kk}^{(2)} + \breve{g}_{kk}^{(3)}$, where $\breve{g}_{kk}^{(1)}$ is the sum of all terms that contain v, $\breve{g}_{kk}^{(2)}$ is the sum of all remaining terms that contain u, and $\breve{g}_{kk}^{(3)}$ is the sum of all remaining terms. Thus

(1) $\breve{g}_{kk}^{(1)}$ is a sum of polynomials of the form

$$v^c\left[(2s_1 - 1)fw - (2s_2 - 1)\widehat{f}\widehat{w}\right] \qquad (6.43a)$$

and

$$v^c\left[(2s_2 - 1)fw - (2s_1 - 1)\widehat{f}\widehat{w}\right], \qquad (6.43b)$$

where $c \in \mathbb{N}$ and $if \in \mathbb{Q}[s_1,s_2,u,w]$ (there is no v dependence in f because v is self-conjugate, hence, decomposing any polynomial f that is currently present into a sum of polynomials all of whose terms contain v to the same power, we may assume that v is present to the same power in f and $\widehat{f}$ and can be factored out);

(2) $\breve{g}_{kk}^{(2)}$ is a sum of polynomials of the form

$$u^c\left[(2s_1-1)fw-(2s_2-1)\widehat{f}\widehat{w}\right] \tag{6.44a}$$

and

$$u^c\left[(2s_2-1)fw-(2s_1-1)\widehat{f}\widehat{w}\right], \tag{6.44b}$$

where $c \in \mathbb{N}$ and $if \in \mathbb{Q}[s_1,s_2,w]$ (there is no u dependence in f for the same reason: u is self-conjugate); and

(3) $\breve{g}_{kk}^{(3)}$ is a sum of polynomials of the form

$$(2s_1-1)fw-(2s_2-1)\widehat{f}\widehat{w} \tag{6.45a}$$

and

$$(2s_2-1)fw-(2s_1-1)\widehat{f}\widehat{w}, \tag{6.45b}$$

where $if \in \mathbb{Q}[s_1,s_2,w]$.

The motivation for this decomposition is the following observation, which we will have occasion to use several times in what follows: if a monomial $[\mu]$ either (i) contains s_1 but not b_{10} or (ii) contains s_2 but not a_{01}, or stated in terms of u, v, w, and $\widehat{w}$, if $[\mu]$ either (i) contains s_1 but neither v nor w or (ii) contains s_2 but neither v nor $\widehat{w}$, then it cannot appear in $\breve{g}_{kk}$. For in case (i) $\mu = (n,*,*,*,0,*)$, $n>0$, which has as its image under $\alpha = \omega^{-1}$ the 6-tuple $(*,*,*,*,-n,*)$, so that because of the negative entry μ is not in the image of ω, while in case (ii) $\mu = (*,0,*,*,*,n)$, $n>0$, which has α-image the 6-tuple $(*,-n,*,*,*,*)$, so again μ is not in the image of ω.

We will show that all polynomials of the forms (6.43) and (6.45) that appear in $\breve{g}_{kk}$ lie in $\breve{\mathscr{B}}_5^+$ and that all those of the form (6.44) that appear in $\breve{g}_{kk}$ and are relevant lie in $\breve{\mathscr{B}}_9^+$, where $\breve{\mathscr{B}}_j^+ = \langle \breve{g}_{11},\ldots,\breve{g}_{jj}\rangle$ *as an ideal in the full ring* $\mathbb{C}[s_1,a_{01},a_{-13},b_{3,-1},b_{10},s_2]$, for $j=5,9$. To start we compute the reduced Gröbner basis of $\breve{\mathscr{B}}_5^+$ with respect to lex with $s_1>s_2>a_{01}>b_{10}>a_{-13}>b_{3,-1}$. It contains (among others) the polynomials

$$\begin{aligned}
u_1 &= v(s_1-s_2)\\
u_2 &= v(2s_2-1)(w-\widehat{w})\\
u_3 &= -a_{01}u(2s_2-1)(w-\widehat{w})\\
u_4 &= -b_{10}u[(2s_1-1)w-(2s_2-1))\widehat{w}]\\
u_5 &= a_{01}(2s_2-1)(s_2-3)(s_2+3)(w-\widehat{w})\\
u_6 &= (2s_1-1)(s_1-3)(s_1+3)w-(2s_2-1)(s_2-3)(s_2+3)\widehat{w}
\end{aligned}$$

which therefore also lie in $\breve{\mathscr{B}}_9^+$.

$\breve{g}_{kk}^{(1)} \in \breve{\mathscr{B}}_5^+$. Consider a polynomial of the form specified in (6.43a). A monomial $s_1^{n_1}s_2^{n_2}u^{n_3}v^{n_4}w^s \in f$ gives rise to the binomial

$$v^c\big[(2s_1-1)s_1^{n_1}s_2^{n_2}u^{n_3}v^{n_4}w^{s+1}-(2s_2-1)s_1^{n_2}s_2^{n_1}u^{n_3}v^{n_4}\widehat{w}^{s+1}\big]$$
$$=v\big[(2s_1-1)s_1^{n_1}s_2^{n_2}w^{s+1}-(2s_2-1)s_1^{n_2}s_2^{n_1}\widehat{w}^{s+1}\big]u^{n_3}v^{n_4+c-1}$$

in $\breve{g}_{kk}^{(1)}$. If $n_1 \geq n_2$, then the product $(s_1s_2)^{n_1}$ can be factored out while if $n_1 < n_2$, then the product $(s_1s_2)^{n_2}$ can be factored out, in each case expressing the binomial as a monomial times a binomial of the form

$$v\big[(2s_1-1)s_j^r w^t-(2s_2-1)\widehat{s}_j^r\,\widehat{w}^t\big] \tag{6.46}$$

for $j\in\{1,2\}$ and where $r\in\mathbb{N}_0$ and $t\in\mathbb{N}$. Thus polynomials of the form (6.43a) lie in $\breve{\mathscr{B}}_5^+$ if all monomials of the form (6.46) do. But

$$\begin{aligned}s_j^r(2s_1-1)w^t-\widehat{s}_j^r(2s_2-1)\widehat{w}^t &\\ =(s_j+\widehat{s}_j)\big[s_j^{r-1}(2s_1-1)w^t-\widehat{s}_j^{r-1}(2s_2-1)\widehat{w}^t\big]&\\ -s_j\widehat{s}_j\big[s_j^{r-2}(2s_1-1)w^t-\widehat{s}_j^{r-2}(2s_2-1)\widehat{w}^t\big]&,\end{aligned}$$

so this will follow by mathematical induction on r if it is true for $r=0$ and 1. That it is follows from the fact that for $t=1$, $f_1:=v(2s_2-1)(w-\widehat{w})=u_2$, and for $t>1$,

$$\begin{aligned}f_t:=v(2s_2-1)(w^t-\widehat{w}^t)&=v(2s_2-1)(w-\widehat{w})(w^{t-1}-\cdots-\widehat{w}^{t-1})\\&=u_2(w^{t-1}-\cdots-\widehat{w}^{t-1})\end{aligned}$$

(so they contain u_2 as a factor) and the identities

$$\begin{aligned}v[(2s_1-1)w^t-(2s_2-1)\widehat{w}^t]&=2w^tu_1+f_t\\ v[s_1(2s_1-1)w^t-s_2(2s_2-1)\widehat{w}^t]&=(2s_1+2s_2-1)w^tu_1+s_2f_t\\ v[s_2(2s_1-1)w^t-s_1(2s_2-1)\widehat{w}^k]&=(2s_2(w^t-\widehat{w}^t)+\widehat{w}^t)u_1+s_2f_t,\end{aligned}$$

which are easily verified simply by expanding each side. The demonstration that a polynomial of the form specified in (6.43b) lies in $\breve{\mathscr{B}}_5^+$ is similar.

$\breve{g}_{kk}^{(2)}\in\breve{\mathscr{B}}_9^+$. For this case we will need the identity $\breve{\mathscr{B}}_{10}^+=\breve{\mathscr{B}}_{11}^+=\breve{\mathscr{B}}_{12}^+=\breve{\mathscr{B}}_9^+$. It is established by computing a Gröbner basis G_9 of $\breve{\mathscr{B}}_9^+$, then computing $g_{10,10}$, $g_{11,11}$, and $g_{12,12}$ and reducing each one modulo G_9; the remainder is zero each time. This is sufficient since W is an isomorphism onto its image C.

Our next observation is that no polynomial of the form of (6.44b) actually appears in $\breve{g}_{kk}$. For any monomial $As_1^{n_1}s_2^{n_2}w^{n_3}$ in f yields

$$\begin{aligned}2As_1^{n_1}s_2^{1+n_2}a_{-13}^{1+c+n_3}b_{3,-1}^{c}b_{10}^{4+4n_3}-As_1^{n_1}s_2^{n_2}a_{-13}^{1+c+n_3}b_{3,-1}^{c}b_{10}^{4+4n_3}&\\ -2As_1^{1+n_2}s_2^{n_1}a_{01}^{4+4n_3}a_{-13}^{c}b_{3,-1}^{1+c+n_3}+As_1^{n_2}s_2^{n_1}a_{01}^{4+4n_3}a_{-13}^{c}b_{3,-1}^{1+c+n_3}&.\end{aligned}$$

The first term contains a power of s_1 but not of b_{10}, hence must be cancelled by some other term in $\breve{g}_{kk}^{(2)}$. But no term coming from a polynomial of the form in (6.44b) with a different exponent c' on u can cancel it, for if $c' \neq c$ there will be a mismatch of the exponents on $b_{3,-1}$ in the first and second terms and a mismatch of the exponents on a_{-13} in the third and fourth terms. No term coming from a polynomial of the form in (6.44b) with the same exponent c on u can cancel it either, because of a mismatch of the exponents on a_{-13} in the second term and of a_{01} in the third and fourth terms. By writing out completely a general term coming from a monomial in f in a polynomial of the form in (6.44a) with exponent c' on u, we find similarly that for no choice of c' can we obtain a term to make the cancellation.

The same kind of argument shows that for no polynomial of the form of (6.44a) can the polynomial f contain a monomial $As_1^{n_1} s_2^{n_2} w^{n_3}$ in which $n_2 > 0$. Thus we reduce to consideration of polynomials of the form

$$p_{c,d} = u^c\left[(2s_1 - 1)s_1^r w^d - (2s_2 - 1)s_2^r \widehat{w}^d\right], \tag{6.47}$$

where $c \geq 1$, $d \geq 1$, and $r \geq 0$.

The two situations $d = 1$ and $d \geq 2$ call for different treatments. If $d = 1$, then there is an important shift in the logic of the proof. Heretofore we have implicitly used the fact that if a polynomial f is a sum of polynomials that are in an ideal I, then $f \in I$. But it is by no means necessary that every polynomial in such a sum be in I in order for f to be in I. In particular, it is not necessary that each polynomial $p_{c,1}$ be in $\breve{\mathscr{B}}_9^+$ in order for $\breve{g}_{kk}^{(2)}$ to be there. Indeed, polynomials of the form $p_{1,1}$ and $p_{2,1}$ are not. What is sufficient, and what we will show, is that any polynomial $p_{c,1}$ that is not definitely in $\breve{\mathscr{B}}_{12}^+ = \breve{\mathscr{B}}_9^+$ can be in $\breve{g}_{kk}$ only if k is so low ($k \leq 12$) that $\breve{g}_{kk}^{(2)}$ is a generator of $\breve{\mathscr{B}}_{12}^+$, hence certainly in it.

The argument that was just applied to narrow the scope of possible forms of f in (6.44a) shows that r can be at most 3, where now we use the fact that if s_1 occurs in a term with exponent n, then b_{10} must also occur in that term with exponent at least n (and cancellation of terms that fail this condition is impossible). Since $\deg p_{c,1} = 2c + r + 6$, $\max\deg p_{c,1} = 2c + 9$, hence, for $c \leq 2$, $\deg p_{c,1} \leq 13$. By Exercise 6.21, any monomial present in g_{kk} has degree at least k if k is even and at least $k+1$ if k is odd. Since W preserves degree, the same lower bounds apply to $\breve{g}_{kk}$. Thus, for $c \leq 2$, if $p_{c,1}$ lies in any $\breve{g}_{kk}$, it can do so only for $k \leq 12$, so that $\breve{g}_{kk}$ is in $\breve{\mathscr{B}}_{12}^+ = \breve{\mathscr{B}}_9^+$. For $c = 3$, reduce $p_{3,1}$ modulo a Gröbner basis of $\breve{\mathscr{B}}_9^+$ to obtain remainder zero and conclude that $p_{3,1} \in \breve{\mathscr{B}}_9^+$. For $c \geq 4$, $p_{c,1} = u^{c-3} p_{3,1}$, hence is in $\breve{\mathscr{B}}_9^+$.

For the case $d \geq 2$, we factor out and discard a factor of u^{c-1} from (6.47). Define the polynomial $p_{1,d}^+$ by

$$p_{1,d}^+ = u\left[(2s_1 - 1)s_1^r w^d + (2s_2 - 1)s_2^r \widehat{w}^d\right](w - \widehat{w}).$$

The identity

$$p_{1,d}^+ = -(s_1^r w^{d-1} + s_2^r \widehat{w}^{d-1}) a_{01}^3 b_{3,-1} u_3 + (w - \widehat{w}) s_1^r a_{-13} b_{10}^3 w^{d-2} u_4$$

shows that $p^+_{1,d} \in \breve{\mathscr{B}}^+_5$ for all $d \geq 2$. We will prove that any polynomial of the form $p_{1,d}$ is an element of $\breve{\mathscr{B}}^+_9$ by induction on k.

Basis step. For $d = 2$ we have

$$p_{1,2} = -s_2^r a_{01}^3 b_{3,-1} u_3 - s_1^u a_{-13} b_{10}^3 u_4 + (2s_2 - 1)(uv^4)^2 (s_1^r - s_2^r).$$

Using the binomial theorem if necessary factor $v(s_1 - s_2) = u_1$ out of the last term to see that $p_{1,2}$ is in $\breve{\mathscr{B}}^+_5$.

Inductive step. Suppose $p_{1,d-1} \in \breve{\mathscr{B}}^+_9$. By the identity (see (6.36))

$$\begin{aligned} p_{1,d} &= \tfrac{1}{2} u\big[(2s_1 - 1) s_1^r w^{d-1} - (2s_2 - 1) s_2^r \widehat{w}^{d-1}\big](w + \widehat{w}) \\ &\quad + \tfrac{1}{2} u\big[(2s_1 - 1) s_1^r w^{d-1} - (2s_2 - 1) s_2^r \widehat{w}^{d-1}\big](w - \widehat{w}) \\ &= \tfrac{1}{2} p_{1,d-1}(w + \widehat{w}) + \tfrac{1}{2} p^+_{1,d-1}(w - \widehat{w}) \end{aligned}$$

and the fact that $p^+_{1,j} \in \breve{\mathscr{B}}^+_9$ for all $j \geq 2$, it follows that $p_{1,d} \in \breve{\mathscr{B}}^+_9$.

$\breve{g}^{(3)}_{kk} \in \breve{\mathscr{B}}^+_5$. Just as we applied part (7) of Theorem 3.7.2 to obtain the factorization of the polynomial h in (6.39), we wish to apply parts (2) and (4) now. To do so we extend the fact noted earlier that for any element f of $\mathbb{C}[a,b]$ and any point (a^*, b^*) in $\mathbb{C}^6$ such that $f(a^*, b^*) = 0$, $\breve{f}$ vanishes at the point $(s_1^*, a_{10}^*, a_{-13}^*, b_{3,-1}^*, b_{10}^*, s_2^*)$ for s_1^* and s_2^* defined by $s_1^* = a_{10}^*/b_{10}^*$ and $s_2^* = b_{01}^*/a_{01}^*$, which is possible provided $a_{01}^* b_{10}^* \neq 0$. The extension comes from the observation that if $a_{01}^* = 0$ but $b_{01}^* = 0$ as well, then we may choose *any* value for s_2^* and $\breve{f}$ will still vanish at $(s_1^*, a_{10}^*, a_{-13}^*, b_{3,-1}^*, b_{10}^*, s_2^*)$, provided only that $b_{10} \neq 0$. Thus, by Theorem 3.7.2(2), $\breve{g}_{kk}(3, 0, a_{-13}, 0, b_{10}, s_2) = 0$ holds for all a_{-13}, s_2, and nonzero b_{10}, hence by Exercise 6.22 holds for all a_{-13}, s_2, and b_{10}. Similarly, it follows from part (4) of Theorem 3.7.2 that $\breve{g}_{kk}(-3, 0, a_{-13}, 0, b_{10}, s_2) = 0$ holds for all a_{-13}, s_2, and b_{10}. Since the polynomial f in (6.45) contains only sums of products of s_1, s_2, and $w = a_{-13} b_{10}^4$, and a_{01} and $b_{3,-1}$ do not appear, we conclude that it must contain $s_1 + 3$ and $s_1 - 3$ as factors, so that the polynomials in (6.45) must have the form

$$(2s_1 - 1)(s_1 - 3)(s_1 + 3) f w - (2s_2 - 1)(s_2 - 3)(s_2 + 3) \widehat{f w}$$

and

$$(2s_2 - 1)(s_2 - 3)(s_2 + 3) f w - (2s_1 - 1)(s_1 - 3)(s_1 + 3) \widehat{f w},$$

where $f \in \mathbb{C}[s_1, s_2, w]$. The latter forms do not actually appear since every term that they contain has either s_1 without w or s_2 without $\widehat{w}$, and cancellation of such terms is impossible. Thus we reduce to an examination of polynomials of the form

$$p_k := s_1^r (2s_1 - 1)(s_1 - 3)(s_1 + 3) w^d - s_2^r (2s_2 - 1)(s_2 - 3)(s_2 + 3) \widehat{w}^d,$$

where $r \geq 0$ and $d \geq 1$.

For $d = 1$, an argument as in the case $d = 1$ in the proof that $\breve{g}^{(2)}_{kk}$ is in $\mathscr{B}^+_9$ shows that r is at most 1 and that p_1 can be in g_{kk} only if k is so low that g_{kk} is a generator of $\mathscr{B}^+_9$.

For $d = 2$, we have the identity

$$p_2 = -a_{01}^3 b_{3,-1} s_2^r u_5 + s_1^r w u_6 + (s_2 - 3)(s_2 + 3)(2s_2 - 1)(s_1^r - s_2^r) w \widehat{w}.$$

Using the binomial theorem if necessary factor $v(s_1 - s_2) = u_1$ out of the last term to see that p_2 is in $\breve{\mathscr{B}}_5^+$.

For $d \geq 2$, define a polynomial p_d^+ by

$$p_d^+ = \left[s_1^r(2s_1 - 1)(s_2 - 3)(s_2 + 3)w^d + s_2^r(2s_1 - 1)(s_2 - 3)(s_2 + 3)\widehat{w}^d\right](w - \widehat{w}).$$

Then

$$p_d^+ = -s_1^r (s_2 - 3)(s_2 + 3) a_{01}^3 b_{10}^4 w^{d-2} u_3 + s_2^r a_{01}^3 b_{3,-1} \widehat{w}^{d-1} u_5 - s_1^r w^{k-1} (w - \widehat{w}) u_6,$$

which shows that $p_d^+ \in \breve{\mathscr{B}}_5^+$. Following the pattern from the previous case, the reader is invited to combine these identities with (6.36) to show by mathematical induction that $p_d \in \breve{\mathscr{B}}_5^+$ for all $d \geq 1$. □

Using this result, we obtain the following estimate for the cyclicity of the singularity at the origin for system (6.29).

Proposition 6.3.6. *Fix a particular system in family* (6.29) *corresponding to a parameter value a for which $a_{01} \neq 0$. Then the cyclicity of the focus or center at the origin of this system with respect to perturbation within family* (6.29) *is at most five. If perturbation of the linear part is allowed, then the cyclicity is at most six.*

Proof. For the general family (6.30) of differential equations on $\mathbb{C}^2$, Proposition 6.3.5 implies that for any $k \in \mathbb{N}$, there exist polynomials h_1, h_3, h_4, h_5, h_7, and h_9 in six indeterminates with coefficients in $\mathbb{C}$ such that

$$\breve{g}_{kk} = h_1 \breve{g}_{11} + h_3 \breve{g}_{33} + h_4 \breve{g}_{44} + h_5 \breve{g}_{55} + h_7 \breve{g}_{77} + h_9 \breve{g}_{99}. \tag{6.48}$$

For any point $(a,b) \in \mathbb{C}^6$, this identity can be evaluated at (a,b), but that tells us nothing about $g_{kk}(a,b)$. The transformation G defined by equation (6.31) maps the open set $D = \{(z_1, z_2, z_3, z_4, z_5, z_6) : z_2 z_5 \neq 0\}$ one-to-one onto itself and has the global inverse

$$F(a_{10}, a_{01}, a_{-13}, b_{3,-1}, b_{10}, b_{01}) = \left(\tfrac{a_{10}}{b_{10}}, a_{01}, a_{-13}, b_{3,-1}, b_{10}, \tfrac{b_{01}}{a_{01}}\right)$$

on D. Thus if $(a,b) \in D$, that is, if $a_{01} b_{10} \neq 0$, then F is defined at (a,b) and when (6.48) is evaluated at $F(a,b)$, the left-hand side is $g_{kk}(a,b)$ and the right-hand side is a sum of expressions of the form (letting $T_j = \operatorname{Supp}(h_j)$ and $S_j = \operatorname{Supp}(g_{jj})$)

$$\sum_{\nu\in T_j} h_j^{(\nu)} \left(\tfrac{a_{10}}{b_{10}}\right)^{\nu_1} a_{01}^{\nu_2} a_{-13}^{\nu_3} b_{3,-1}^{\nu_4} b_{10}^{\nu_5} \left(\tfrac{b_{01}}{a_{01}}\right)^{\nu_6}$$
$$\times \sum_{\nu\in S_j} g_{jj}^{(\nu)} \left(\tfrac{a_{10}}{b_{10}}\right)^{\nu_1} a_{01}^{\nu_2+\nu_6} a_{-13}^{\nu_3} b_{3,-1}^{\nu_4} b_{10}^{\nu_5+\nu_1} \left(\tfrac{b_{01}}{a_{01}}\right)^{\nu_6}$$
$$= \frac{f_j(a,b)}{a_{01}^{r_j} b_{10}^{s_j}} g_{jj}(a,b)$$

for some polynomials f_j and constants $r_j,\ s_j \in \mathbb{N}_0$. Clearly the resulting identity

$$g_{kk} = \frac{f_1(a,b)}{a_{01}^{r_1} b_{10}^{s_1}} g_{11} + \frac{f_3(a,b)}{a_{01}^{r_3} b_{10}^{s_3}} g_{33} + \frac{f_4(a,b)}{a_{01}^{r_4} b_{10}^{s_4}} g_{44} + \frac{f_5(a,b)}{a_{01}^{r_5} b_{10}^{s_5}} g_{55} + \frac{f_6(a,b)}{a_{01}^{r_6} b_{10}^{s_7}} g_{77} + \frac{f_9(a,b)}{a_{01}^{r_9} b_{10}^{s_9}} g_{99}$$

of analytic functions holds on D. Hence, for any $(a,b) \in D$, the set of germs $\{\mathbf{g}_{11}, \mathbf{g}_{33}, \mathbf{g}_{44}, \mathbf{g}_{55}, \mathbf{g}_{77}, \mathbf{g}_{99}\}$ in the ring $\mathscr{G}_{(a,b)}$ of germs of complex analytic functions at (a,b) is a minimal basis with respect to the retention condition of the ideal $\langle \mathbf{g}_{kk} : k \in \mathbb{N}\rangle$. Since parameter values of interest are those for which (6.30) is the complexification of the real system (in complex form) (6.29), they satisfy $b_{10} = \bar{a}_{01}$, so the condition $a_{01} \neq 0$ ensures that (a,b) is in D, and the set of germs $\{\mathbf{g}_{11}^{\mathbb{R}}, \mathbf{g}_{33}^{\mathbb{R}}, \mathbf{g}_{44}^{\mathbb{R}}, \mathbf{g}_{55}^{\mathbb{R}}, \mathbf{g}_{77}^{\mathbb{R}}, \mathbf{g}_{99}^{\mathbb{R}}\}$ in the ring $\mathscr{G}_{(A(a,b),B(a,b))}$ of complex analytic functions at $(A(a,b),B(a,b))$ is a minimal basis with respect to the retention condition of the ideal $\langle \mathbf{g}_{kk}^{\mathbb{R}} : k \in \mathbb{N}\rangle$. Consequently, Theorem 6.2.9 implies that the cyclicity of the origin with respect to perturbation within the family of the form (6.11) that corresponds to (6.29) (that is, allowing perturbation of the linear terms) is at most six. If perturbations are constrained to remain within family (6.29), then, by Lemma 6.2.8, the cyclicity is at most five. □

Our treatment of the cyclicity problem for family (6.29) is incomplete in two ways: there remain the question of the sharpness of the bounds given by Proposition 6.3.6 (see Exercise 6.23 for an outline of how to obtain at least three small limit cycles, or four when perturbation of the linear part is permitted) and the question of what happens at all in the situation $a_{01} = 0$. Since family (6.29) was intended only as an example, we will not pursue it any further, except to state and prove one additional result concerning it, a result that we have chosen because we wish to again bring to the reader's attention the difference between bifurcation from a focus and bifurcation from a center.

Proposition 6.3.7. *Fix a particular system in family* (6.29) *corresponding to a parameter value a. If* $a_{01} = 0$ *but* $a_{10}^4 a_{-13} - \bar{a}_{10}^4 \bar{a}_{-13} \neq 0$, *then the cyclicity of the origin with respect to perturbation within family* (6.29) *is at most three. If perturbation of the linear part is allowed, then the cyclicity is at most four.*

Proof. Form the complexification of (6.29), system (6.30) with $b_{qp} = \bar{a}_{pq}$. The focus quantities for family (6.30) are listed at the beginning of the proof of Theorem 3.7.2. When $a_{01} = 0$ (hence $b_{10} = \bar{a}_{01} = 0$) the first four of them are zero and

$$g_{55} = -i\tfrac{7}{60}a_{-13}b_{3,-1}(a_{10}^4 a_{-13} - b_{01}^4 b_{3,-1}) = -i\tfrac{7}{60}|a_{-13}|^2(a_{10}^4 a_{-13} - \bar{a}_{10}^4 \bar{a}_{-13}).$$

Since $a_{10}^4 a_{-13} - \bar{a}_{10}^4 \bar{a}_{-13} \neq 0$ only if $|a_{-13}| \neq 0$, the second condition in the hypothesis ensures that $g_{55} \neq 0$, so that the origin is not a center but a fine focus of order five. An argument similar to the proof of Corollary 6.1.3 gives the result. □

6.4 Bifurcations of Critical Periods

Suppose that a particular element of the family of systems of real differential equations (6.12), corresponding to a specific string a^* of the coefficients of the complex form (6.12b), has a center at the origin. The series on the right-hand side of the expansion (4.7) for the period function $T(r)$ (Definition 4.1.1) for (6.12) defines a real analytic function on a neighborhood of 0 in $\mathbb{R}$, whose value for $r > 0$ is the least period of the trajectory of (4.1) (which is the same as (6.12a)) through the point with coordinates $(r,0)$. We will still call this extended function the *period function* and continue to denote it by $T(r)$. By Proposition 4.2.4, $r = 0$ is a *critical point* of T, a point at which T' vanishes. In this context any value $r > 0$ for which $T'(r) = 0$ is called a *critical period*. The question that we address in this section is the maximum number of zeros of T' that can bifurcate from the zero at $r = 0$, that is, the maximum number of critical periods that can lie in a sufficiently small interval $(0, \varepsilon_0)$, when the coefficient string a is perturbed from a^* but remains within the center variety $V_{\mathscr{C}}^{\mathbb{R}}$ of the original family (6.12). This is called the *problem of bifurcations of critical periods*, which was first considered by C. Chicone and M. Jacobs ([45]), who investigated it for quadratic systems and for some general Hamiltonian systems. In this section we show how this problem can be examined by applying the techniques and results that we have developed in earlier sections to the complexification (6.15) of (6.12b). From the standpoint of the methods used, the problem of bifurcation of critical periods is analogous to the problem of small limit cycle bifurcations when the parameter space is not $\mathbb{R}^{n^2+3n}$ but an affine variety in $\mathbb{R}^{n^2+3n}$. To be specific, the problem of the cyclicity of antisaddles of focus or center type in polynomial systems, studied in the previous two sections, is the problem of the multiplicity of the function $\mathscr{F}(z,\theta)$ of (6.2) when this function is the difference function $\mathscr{P}$ of Definition 3.1.3 and $\mathscr{E} = \mathbb{R}^{n^2+3n}$ is the space of parameters of a two-dimensional system of differential equations whose right-hand sides are polynomials of degree n without constant terms. The problem of bifurcations of critical periods is the problem of the multiplicity of the function $\mathscr{F}(r,(a,\bar{a})) = T'(r)$ (there is now the parameter pair $(a,b) = (a,\bar{a})$, since we have complexified), where $T(r)$ is the period function given by equations (4.7) and (4.26), and $\mathscr{E}$ is not the full space $\mathbb{R}^{n^2+3n}$ but the center variety $V_{\mathscr{C}}^{\mathbb{R}}$ of family (6.12).

At the outset we make the following simplifying observation. As presented so far, the problem is to investigate the multiplicity of the function $T'(r)$, which, by (4.26), is $T'(r,(a,\bar{a})) = \sum_{k=1}^{\infty} 2kp_{2k}(a,\bar{a})r^{2k-1}$. Using Bautin's method of Section 6.1, this amounts to analyzing the ideal $\langle 2kp_{2k} : k \in \mathbb{N}\rangle \subset \mathbb{C}[a,b]$. But this is the same ideal as $\langle p_{2k} : k \in \mathbb{N} \rangle \subset \mathbb{C}[a,b]$, which arises from the study of the multiplicity of the function

$$\mathscr{T}(r,(a,\bar{a})) = T(r) - 2\pi = \sum_{k=1}^{\infty} p_{2k}(a,\bar{a})r^{2k}, \tag{6.49}$$

whose zeros count the number of small cycles that maintain their original period 2π after perturbation. Fix an initial parameter string $(a^*,b^*) = (a^*,\overline{a^*}) \in V_{\mathscr{C}}^{\mathbb{R}}$. For any specific parameter string $(a,\bar{a}) \in V_{\mathscr{C}}^{\mathbb{R}}$ near (a^*,b^*), the number of critical periods and the number of cycles of period 2π, in a neighborhood of $(0,0)$ of radius $\varepsilon_0 > 0$, may differ, but the upper bounds on the number of critical periods and on the number of period-2π cycles that arise from examining the ideal presented with generators $\{2kp_{2k}\}$ on the one hand and examining the same ideal but presented with generators $\{p_{2k}\}$ on the other are the same. Thus we can avoid taking derivatives and work with the function $\mathscr{T}$ of (6.49) when we use Bautin's method to obtain an upper bound on the number of critical periods.

According to Proposition 6.1.2, if we can represent the function $\mathscr{T}(r,(a,\bar{a}))$ in the form (6.4), then the multiplicity of $\mathscr{T}$ at any point is at most $s-1$. We are interested in bifurcations of critical periods for systems from $V_{\mathscr{C}}^{R}$; however, we consider this variety as enclosed in $V_{\mathscr{C}}$ and look for representation (6.4) on components of the center variety $V_{\mathscr{C}}$. Because the p_{2k} are polynomials with real coefficients, if such a representation exists in $\mathbb{C}[a,b]$, then it also exists in $\mathbb{R}[a,b]$.

We will build our discussion here around systems of the form $\dot{u} = -v + U(u,v)$, $\dot{v} = u + V(u,v)$, where U and V are homogeneous cubic polynomials (or one of them may be zero). We will find that a sharp upper bound for the number of critical periods that can bifurcate from any center in this family (while remaining in this family and still possessing a center at the origin) is three (Theorem 6.4.10). We begin with the statement and proof of a pair of lemmas that are useful in general.

Lemma 6.4.1. *Suppose system $(a^*,b^*) = (a^*,\overline{a^*})$ corresponds to a point of the center variety $V_{\mathscr{C}}^{\mathbb{R}}$ of family* (6.15) *that lies in the intersection of two subsets A and B of $V_{\mathscr{C}}^{\mathbb{R}}$ and that there are parametrizations of A and B in a neighborhood of $(a^*,\overline{a^*})$ such that the function $\mathscr{T}(r,(a,\bar{a}))$ can be written in the form* (6.4) *with $s = m_A$ and $s = m_B$, respectively. Then the multiplicity of $\mathscr{T}(r,(a,\bar{a}))$ with respect to $A \cup B$ is equal to* $\max(m_A - 1, m_B - 1)$.

Proof. Apply Proposition 6.1.2 to $\mathscr{T}(r,(a,\bar{a}))$ with $\mathscr{E} = A$ and to $\mathscr{T}(r,(a,\bar{a}))$ with $\mathscr{E} = B$ separately. □

The second lemma, a slight modification of a result of Chicone and Jacobs, provides a means of bounding the number of bifurcating critical periods in some cases without having to find a basis of the ideal of isochronicity quantities at all.

Lemma 6.4.2. *Let $\mathscr{F}(z,\theta)$ be defined by (6.2) and convergent in a neighborhood U of the origin in $\mathbb{R} \times \mathbb{C}^n$ and such that each function $f_j(\theta)$ is a homogeneous polynomial satisying* $\deg(f_j) > \deg(f_0)$ *for $j > 0$. Suppose there exists a closed set B in $\mathbb{C}^n$ that is closed under rescaling by nonnegative real constants and is such that $|f_0(\theta)| > 0$ for all $\theta \in B \setminus \mathbf{0}$. Then there exist $\varepsilon > 0$ and $\delta > 0$ such that for* *each fixed $\theta \in B$ that satisfies $0 < |\theta| < \delta$, the equation $\mathscr{F}(z,\theta) = 0$, regarded as an equation in z alone, has no solutions in the interval $(0,\varepsilon)$.*

Proof. For any θ for which $f_0(\theta) \neq 0$, we can write

$$\mathscr{F}(z,\theta) = f_0(\theta)\left[1 + \left(\frac{f_1(\theta)}{f_0(\theta)}\right)z + \left(\frac{f_2(\theta)}{f_0(\theta)}\right)z^2 + \cdots\right].$$

The hypotheses on the f_j imply that $\left|f_j(\theta)/f_0(\theta)\right| \to 0$ as $|\theta| \to 0$, which suggests the truth of the lemma, at least when $B = \mathbb{C}^n$. Thus let the function $\mathscr{G}$ be defined on $U \cap (\mathbb{R} \times B)$ by

$$\mathscr{G}(z,\theta) = \begin{cases} 1 & \text{if } \theta = \mathbf{0} \\ \dfrac{\mathscr{F}(z,\theta)}{f_0(\theta)} & \text{if } \theta \neq \mathbf{0}. \end{cases}$$

We will show that $\mathscr{G}$ is continuous on a neighborhood of the origin in $\mathbb{R} \times B$. Since for any $(z,\theta) \in U \cap (\mathbb{R} \times B)$, $\mathscr{G}(z,\theta) = 1$ if $z = 0$, this implies that ε and δ as in the conclusion of the lemma exist for $\mathscr{G}$. But also for any $(z,\theta) \in U \cap (\mathbb{R} \times B)$, for $\theta \neq \mathbf{0}$, $\mathscr{F}(z,\theta) = 0$ only if $\mathscr{G}(z,\theta) = 0$, so the conclusion of the lemma then holds for $\mathscr{F}$ with the same ε and δ. Clearly continuity of $\mathscr{G}$ is in question only at points in the set $\{(z,\theta) : (z,\theta) \in U \cap (\mathbb{R} \times B) \text{ and } \theta = \mathbf{0}\}$.

Let $D = \{\theta : |\theta| \leq R\}$, where R is a fixed real constant chosen such that $\mathscr{F}$ converges on the compact set $[-R,R] \times D$, and let M denote the supremum of $|\mathscr{F}|$ on D. The Cauchy Inequalities give $|f_j(\theta)| \leq M/R^j$ for all $\theta \in D$.

By the hypothesis that B is closed under rescaling by nonnegative real constants, any $\theta \in D \cap B$ can be written as $\theta = \rho\theta'$ for a unique $\rho \in [0,1]$ and $\theta' \in B$ satisfying $|\theta'| = R$, which is unique for nonzero θ. Thus if, for $j \in \mathbb{N}_0$, f_j is homogeneous of degree $m_j \in \mathbb{N}_0$, then

$$|f_j(\theta)| = |f_j(\rho\theta')| = \rho^{m_j}|f_j(\theta')| \leq \rho^{m_j}\frac{M}{R^j}.$$

Since $f_0(\theta) \neq 0$ for $\theta \in B \setminus \{\mathbf{0}\}$ and $\{\theta : |\theta| = R\} \cap B$ is compact, we have that $m := \inf\{|f_0(\theta')| : |\theta'| = R \text{ and } \theta' \in B\}$ is positive, and for all $\theta \in (D \cap B) \setminus \{\mathbf{0}\}$,

$$|f_0(\theta)| = |f_0(\rho\theta')| = \rho^{m_0}|f(\theta')| \geq m\rho^{m_0}.$$

For $j \in \mathbb{N}$, set $m'_j = m_j - m_0 > 0$. Combining the displayed estimates gives, for $\theta \in (D \cap B) \setminus \{\mathbf{0}\}$,

$$\left|\frac{f_j(\theta)}{f_0(\theta)}z^j\right| \leq \frac{\rho^{m_j}\frac{M}{R^j}}{m\rho^{m_0}}2^{-j}R^j \leq \frac{M}{m}\rho^{m'_j}2^{-j} \leq \frac{M}{m}2^{-j}\rho.$$

Thus

$$\left|\sum_{j=0}^{\infty}\frac{f_j(\theta)}{f_0(\theta)}z^j - 1\right| \leq \frac{M}{m}\rho\left(\sum_{j=1}^{\infty}2^{-j}\right) = \frac{M}{m}\rho$$

so that $|\mathscr{G}(z,\theta)-1|<\eta$ for $|\theta|=\rho R$, hence ρ, sufficiently small, and $\mathscr{G}$ is continuous on $[-R/2,R/2]\times(D\cap B)$. □

The first step in investigating the bifurcation of critical periods of a center of a real family (6.12) is to write the family in complex form (6.12b) and form the complexification (6.15). The complex form of (6.12a) when the nonlinearities are homogeneous cubics polynomials (perhaps one of them zero) is

$$\dot{x}=i(x-a_{20}x^3-a_{11}x^2\bar{x}-a_{02}x\bar{x}^2-a_{-13}\bar{x}^3) \tag{6.50}$$

for $a_{pq}\in\mathbb{C}$. The complexification of this family is

$$\begin{aligned}\dot{x}&=\quad i(x-a_{20}x^3-a_{11}x^2y-a_{02}xy^2-a_{-13}y^3)\\ \dot{y}&=-i(y-b_{3,-1}x^3-b_{20}x^2y-b_{11}xy^2-b_{02}y^3)\,.\end{aligned} \tag{6.51}$$

In general, the center and isochronicity varieties must be found next. It would be too lengthy a process to derive them for system (6.51) here, so we will simply describe them in the following two theorems (but see Example 1.4.13). Characterization of centers in family (6.15) was first given by Sadovskii in [163], where a derivation can be found, although his conditions differ from those listed here. (The real case was first completely solved by Malkin; see [133].) The conditions in Theorem 6.4.4 are close to those derived in [58]; see also [141]. The conditions in both theorems are for family (6.51) without the restriction $b_{qp}=\bar{a}_{pq}$ that holds when (6.51) arises as the complexification of a real family.

Theorem 6.4.3. *The center variety* $V_{\mathscr{C}}$ *of family* (6.51) *consists of the following three irreducible components:*
1. $\mathbf{V}(C_1)$, *where* $C_1=\langle a_{11}-b_{11},\,3a_{20}-b_{20},\,3b_{02}-a_{02}\rangle$;
2. $\mathbf{V}(C_2)$, *where* $C_2=\langle a_{11},\,b_{11},\,a_{20}+3b_{20},\,b_{02}+3a_{02},\,a_{-13}b_{3,-1}-4a_{02}b_{20}\rangle$;
3. $\mathbf{V}(C_3)$, *where* $C_3=\langle a_{20}^2a_{-13}-b_{3,-1}b_{02}^2,\,a_{20}a_{02}-b_{20}b_{02},\ a_{20}a_{-13}b_{20}-a_{02}b_{3,-1}b_{02},\,a_{11}-b_{11},a_{02}^2b_{3,-1}-a_{-13}b_{20}^2\rangle$.

Systems from C_1 are Hamiltonian, systems from C_2 have a Darboux first integral, and systems from C_3 are reversible.

Theorem 6.4.4. *The isochronicity and linearizability variety* $V_{\mathscr{I}}=V_{\mathscr{L}}$ *of family* (6.51) *consists of the following seven irreducible components:*
1. $\mathbf{V}(a_{11},a_{02},a_{-13},b_{02},b_{11})$;
2. $\mathbf{V}(a_{20},a_{11},b_{11},b_{20},b_{3,-1})$;
3. $\mathbf{V}(a_{11},a_{02},a_{-13},b_{11},b_{20},b_{3,-1})$;
4. $\mathbf{V}(a_{11},a_{02},b_{02},b_{11},b_{3,-1},a_{20}+3b_{20})$;
5. $\mathbf{V}(a_{20},a_{11},a_{-13},b_{11},b_{20},3a_{02}+b_{02})$;
6. $\mathbf{V}(a_{11},a_{-13},b_{11},b_{3,-1},a_{20}+b_{20},a_{02}+b_{02})$;
7. $\mathbf{V}(a_{11},b_{11},3a_{20}+7b_{20},7a_{02}+3b_{02},\,112b_{20}^3+27b_{3,-1}^2b_{02},\ 49a_{-13}b_{20}^2-9b_{3,-1}b_{02}^2,\,21a_{-13}b_{3,-1}+16b_{20}b_{02},\,343a_{-13}^2b_{20}+48b_{02}^3)$.

Combining these theorems, we can identify all isochronous centers in family (6.50) (Exercise 6.29), which will be important at one point in the analysis.

In general, the center variety of a family of systems is presented, as is the case with Theorem 6.4.3 (and Theorems 3.7.1 and 3.7.2), as a union of subvarieties (typically irreducible), $V_{\mathscr{C}} = \cup_{j=1}^{s} V_j$. If $(a^*, b^*) \in V_1 \setminus \cup_{j=2}^{s} V_j$, then because $\cup_{j=2}^{s} V_j$ is a closed set, any sufficiently small perturbation of (a^*, b^*) that remains in $V_{\mathscr{C}}$ must lie in $\mathbf{V}(C_1)$, and similarly if (a^*, b^*) lies in exactly one of the subvarieties V_{j_0} for any other value of j_0. This observation motivates the following definition of a *proper* perturbation and shows why we may restrict our attention to proper perturbations in the investigation of the bifurcation of critical periods.

Definition 6.4.5. Suppose the center variety of a family (6.15) decomposes as a union of subvarieties $V_{\mathscr{C}} = \cup_{j=1}^{s} V_j$ and that $(a^*, b^*) \in V_{j_0}$. A perturbation (a, b) of (a^*, b^*) is a *proper* perturbation with respect to V_{j_0} if the perturbed system (a, b) also lies in V_{j_0} (same index j_0).

The isochronicity quantities of the complexification (6.15) of a general system (6.12) are defined implicitly by equation (4.26) in terms of the quantities $\widetilde{H}_{2k+1}$, which in turn are defined in terms of the coefficients $Y_1^{(k+1,k)}$ and $Y_2^{(k,k+1)}$ of the distinguished normalization of (6.15). The ideal in $\mathbb{C}[a,b]$ that they generate is denoted by $\mathscr{Y}$ (see (4.28)); we let $\mathscr{Y}_k$ denote the ideal generated by just the first k pairs: $\mathscr{Y}_k = \langle Y_1^{(2,1)}, Y_2^{(1,2)}, \ldots, Y_1^{(k+1,1)}, Y_2^{(k,k+1)} \rangle$. Similarly, we let $P = \langle p_{2k} : k \in \mathbb{N} \rangle$ denote the ideal generated by the isochronicity quantities and $P_k = \langle p_2, \ldots, p_{2k} \rangle$ the ideal generated by just the first k of them. Implementing the Normal Form Algorithm (Table 2.1 on page 75) on a computer algebra system it is a straightforward matter to derive the normal form of (6.15) through low order. For our model system (6.50) and (6.51), the first few nonlinear terms for which we will need explicit expressions are listed in Table 6.2 on page 292. Using these expressions, we can obtain by means of (4.42) expressions for p_2, p_4, p_6, and p_8, which are listed in Table 6.3 on page 293. (The reader can compute p_{10}, which will also be needed, but is too long to list.)

In the coordinate ring $\mathbb{C}[V_{\mathscr{C}}]$, $P_4 = P_5 = P_6$, which suggests the following result.

Lemma 6.4.6. *For system* (6.51),

$$V_{\mathscr{L}} = \mathbf{V}(\mathscr{Y}_4) = V_{\mathscr{C}} \cap \mathbf{V}(P_4) = V_{\mathscr{I}}. \tag{6.52}$$

Proof. Using Singular ([89]), we computed the primary decomposition of $\mathscr{Y}_4$ and found that the associated primes are given by the polynomials defining the varieties (1)–(7) of Theorem 6.4.4, which implies that $V_{\mathscr{L}} = \mathbf{V}(\mathscr{Y}_4)$. The next equality follows by Proposition 4.2.14. The last equality is Proposition 4.2.7. □

We now consider bifurcation of critical periods under proper perturbation with respect to each of the three components of the center variety of family (6.51) identified in Theorem 6.4.3. Our treatment of the component $\mathbf{V}(C_1)$ of $V_{\mathscr{C}}$ illustrates techniques for finding a basis of the ideal P in the relevant coordinate ring. See Exercise 6.30 for a different approach that leads to an upper bound of one, which is sharp (Exercise 6.31).

$$Y_1^{(2,1)} = -ia_{11}$$

$$Y_2^{(1,2)} = ib_{11}$$

$$Y_1^{(3,2)} = i(-4a_{02}a_{20} - 4a_{02}b_{20} - 3a_{-13}b_{3,-1})/4$$

$$Y_2^{(2,3)} = i(4a_{02}b_{20} + 4b_{02}b_{20} + 3a_{-13}b_{3,-1})/4$$

$$\begin{aligned} Y_1^{(4,3)} = {} & i(-6a_{-13}a_{20}^2 + 4a_{11}a_{20}b_{02} - 16a_{02}a_{20}b_{11} - 20a_{02}a_{11}b_{20} - 24a_{-13}a_{20}b_{20} \\ & - 8a_{02}b_{11}b_{20} - 18a_{-13}b_{20}^2 - 24a_{02}^2b_{3,-1} - 11a_{11}a_{-13}b_{3,-1} - 8a_{02}b_{02}b_{3,-1} \\ & - 6a_{-13}b_{11}b_{3,-1})/16 \end{aligned}$$

$$\begin{aligned} Y_2^{(3,4)} = {} & i(-4a_{20}b_{02}b_{11} + 8a_{02}a_{11}b_{20} + 8a_{-13}a_{20}b_{20} + 16a_{11}b_{02}b_{20} + 20a_{02}b_{11}b_{20} \\ & + 24a_{-13}b_{20}^2 + 18a_{02}^2b_{3,-1} + 6a_{11}a_{-13}b_{3,-1} + 24a_{02}b_{02}b_{3,-1} + 6b_{02}^2b_{3,-1} \\ & + 11a_{-13}b_{11}b_{3,-1})/16 \end{aligned}$$

$$\begin{aligned} Y_1^{(5,4)} = {} & i(144a_{02}a_{11}^2a_{20} + 144a_{11}a_{-13}a_{20}^2 - 96a_{11}^2a_{20}b_{02} + 96a_{02}a_{11}a_{20}b_{11} - 216a_{-13}a_{20}^2b_{11} \\ & + 240a_{11}a_{20}b_{02}b_{11} - 432a_{02}a_{20}b_{11}^2 - 288a_{02}a_{11}^2b_{20} - 288a_{02}^2a_{20}b_{20} \\ & - 72a_{11}a_{-13}a_{20}b_{20} - 192a_{02}a_{20}b_{02}b_{20} - 240a_{02}a_{11}b_{11}b_{20} - 576a_{-13}a_{20}b_{11}b_{20} \\ & - 192a_{02}^2b_{20}^2 - 516a_{11}a_{-13}b_{20}^2 - 96a_{02}b_{02}b_{20}^2 - 234a_{-13}b_{11}b_{20}^2 - 582a_{02}^2a_{11}b_{3,-1} \\ & - 132a_{11}^2a_{-13}b_{3,-1} - 660a_{02}a_{-13}a_{20}b_{3,-1} - 192a_{02}a_{11}b_{02}b_{3,-1} \\ & - 144a_{-13}a_{20}b_{02}b_{3,-1} - 336a_{02}^2b_{11}b_{3,-1} - 120a_{11}a_{-13}b_{11}b_{3,-1} + 24a_{02}b_{02}b_{11}b_{3,-1} \\ & - 18a_{-13}b_{11}^2b_{3,-1} - 1120a_{02}a_{-13}b_{20}b_{3,-1} - 300a_{-13}b_{02}b_{20}b_{3,-1} \\ & - 81a_{-13}^2b_{3,-1}^2)/192 \end{aligned}$$

$$\begin{aligned} Y_2^{(4,5)} = {} & i(96a_{20}b_{02}b_{11}^2 - 240a_{11}a_{20}b_{02}b_{11} + 96a_{02}^2a_{20}b_{20} - 24a_{11}a_{-13}a_{20}b_{20} + 432a_{11}^2b_{02}b_{20} \\ & + 192a_{02}a_{20}b_{02}b_{20} + 240a_{02}a_{11}b_{11}b_{20} + 192a_{-13}a_{20}b_{11}b_{20} - 96a_{11}b_{02}b_{11}b_{20} \\ & + 288a_{02}b_{11}^2b_{20} - 144b_{02}b_{11}^2b_{20} + 192a_{02}^2b_{20}^2 + 336a_{11}a_{-13}b_{20}^2 + 288a_{02}b_{02}b_{20}^2 \\ & + 582a_{-13}b_{11}b_{20}^2 + 234a_{02}^2a_{11}b_{3,-1} + 18a_{11}^2a_{-13}b_{3,-1} + 300a_{02}a_{-13}a_{20}b_{3,-1} \\ & + 576a_{02}a_{11}b_{02}b_{3,-1} + 144a_{-13}a_{20}b_{02}b_{3,-1} + 216a_{11}b_{02}^2b_{3,-1} + 516a_{02}^2b_{11}b_{3,-1} \\ & + 120a_{11}a_{-13}b_{11}b_{3,-1} + 72a_{02}b_{02}b_{11}b_{3,-1} - 144b_{02}^2b_{11}b_{3,-1} + 132a_{-13}b_{11}^2b_{3,-1} \\ & + 1120a_{02}a_{-13}b_{20}b_{3,-1} + 660a_{-13}b_{02}b_{20}b_{3,-1} + 81a_{-13}^2b_{3,-1}^2)/192. \end{aligned}$$

Table 6.2 Normal Form Coefficients for System (6.51)

Lemma 6.4.7. *For family* (6.51) *and the decomposition of $V_{\mathscr{C}}$ given by Theorem 6.4.3, the first four isochronicity quantities form a basis of the ideal $P = \langle p_{2k} : k \in \mathbb{N} \rangle$ in $\mathbb{C}[\mathbf{V}(C_1)]$. At most three critical periods bifurcate from centers of the component $\mathbf{V}(C_1)$ (the Hamiltonian centers) under proper perturbations.*

Proof. On the component $\mathbf{V}(C_1)$ of Hamiltonian systems we have

$$b_{11} = a_{11}, \quad a_{20} = b_{20}/3, \quad b_{02} = a_{02}/3, \tag{6.53}$$

$$
\begin{aligned}
p_2 &= \tfrac{1}{2}(a_{11}+b_{11})\\
p_4 &= -\tfrac{1}{4}(a_{11}+b_{11})^2+\tfrac{1}{4}(2a_{02}a_{20}+4a_{02}b_{20}+2b_{02}b_{20}+3a_{-13}b_{3,-1})\\
p_6 &= \tfrac{1}{8}(a_{11}+b_{11})^3\\
&\quad+\tfrac{1}{32}(-4a_{11}a_{20}b_{02}+24a_{02}a_{20}b_{11}+44a_{02}a_{11}b_{20}+32a_{-13}a_{20}b_{20}\\
&\quad+44a_{02}b_{11}b_{20}+29a_{11}a_{-13}b_{3,-1}+32a_{02}b_{02}b_{3,-1}+29a_{-13}b_{11}b_{3,-1}\\
&\quad-4a_{20}b_{02}b_{11}+24a_{11}b_{02}b_{20}+8a_{11}a_{02}a_{20}+8b_{11}b_{02}b_{20}\\
&\quad+6a_{-13}a_{20}^2+42a_{-13}b_{20}^2+42a_{02}^2b_{3,-1}+6b_{02}^2b_{3,-1})\\
p_8 &= \tfrac{1}{16}(a_{11}+b_{11})^4\\
&\quad+\tfrac{1}{192}(-36a_{11}a_{-13}a_{20}^2+24a_{11}^2a_{20}b_{02}+192a_{02}a_{11}a_{20}b_{11}\\
&\quad-288a_{11}a_{20}b_{02}b_{11}+216a_{11}a_{-13}a_{20}b_{20}+288a_{02}a_{20}b_{02}b_{20}\\
&\quad+864a_{02}a_{11}b_{11}b_{20}+576a_{-13}a_{20}b_{11}b_{20}+624a_{02}a_{-13}a_{20}b_{3,-1}\\
&\quad+576a_{02}a_{11}b_{02}b_{3,-1}+144a_{-13}a_{20}b_{02}b_{3,-1}+540a_{11}a_{-13}b_{11}b_{3,-1}\\
&\quad+216a_{02}b_{02}b_{11}b_{3,-1}+1408a_{02}a_{-13}b_{20}b_{3,-1}+624a_{-13}b_{02}b_{20}b_{3,-1}\\
&\quad+192a_{11}b_{02}b_{11}b_{20}+189a_{-13}^2b_{3,-1}^2+384a_{02}^2b_{20}^2+144a_{-13}a_{20}^2b_{11}\\
&\quad+384a_{02}a_{20}b_{11}^2+456a_{02}a_{11}^2b_{20}+384a_{02}^2a_{20}b_{20}+678a_{11}a_{-13}b_{20}^2\\
&\quad+384a_{02}b_{02}b_{20}^2+660a_{-13}b_{11}b_{20}^2+660a_{02}^2a_{11}b_{3,-1}+285a_{11}^2a_{-13}b_{3,-1}\\
&\quad+678a_{02}^2b_{11}b_{3,-1}+285a_{-13}b_{11}^2b_{3,-1}+24a_{20}b_{02}b_{11}^2\\
&\quad+384a_{11}^2b_{02}b_{20}+456a_{02}b_{11}^2b_{20}+144a_{11}b_{02}^2b_{3,-1}\\
&\quad-36b_{02}^2b_{11}b_{3,-1}+48a_{02}^2a_{20}^2+48b_{02}^2b_{20}^2)
\end{aligned}
$$

Table 6.3 Isochronicity Quantities for System (6.51)

so that when we make these substitutions in p_{2k}, we obtain a polynomial $\widetilde{p}_{2k}$ that is in the same equivalence class $[[p_{2k}]]$ in $\mathbb{C}[\mathbf{V}(C_1)]$ as p_{2k}. For example, by Table 6.3, $\widetilde{p}_2 = a_{11}$ and $\widetilde{p}_4 = -a_{11}^2 + (4/3)a_{02}b_{20} + (3/4)a_{-13}b_{3,-1}$. The polynomials $\widetilde{p}_{2k}$ so obtained may be viewed as elements of $\mathbb{C}[a_{11}, a_{02}, a_{-13}, b_{3,-1}, b_{20}]$. Working in this ring and recalling Remark 1.2.6, we apply the Multivariable Division Algorithm to obtain the equivalences

$$
\begin{aligned}
\widetilde{p}_4 &\equiv \tfrac{4}{3}a_{02}b_{20}+\tfrac{3}{4}a_{-13}b_{3,-1} \bmod \langle \widetilde{p}_2\rangle\\
\widetilde{p}_6 &\equiv \tfrac{5}{3}a_{-13}b_{20}^2+\tfrac{5}{3}a_{02}^2b_{3,-1} \bmod \langle \widetilde{p}_2,\widetilde{p}_4\rangle\\
\widetilde{p}_8 &\equiv -\tfrac{105}{32}a_{-13}^2b_{3,-1}^2 \bmod \langle \widetilde{p}_2,\widetilde{p}_4,\widetilde{p}_6\rangle\\
\widetilde{p}_{10} &\equiv 0 \bmod \langle \widetilde{p}_2,\widetilde{p}_4,\widetilde{p}_6,\widetilde{p}_8\rangle.
\end{aligned}
\tag{6.54}
$$

(The actual computation each time is to compute a Gröbner basis G of the ideal in question and reduce $\widetilde{p}_{2k}$ modulo G.)

To prove the first statement of the lemma, we must show that $\widetilde{p}_{2k} \in \langle \widetilde{p}_2, \widetilde{p}_4, \widetilde{p}_6, \widetilde{p}_8\rangle$ for all $k \geq 6$. By Proposition 4.2.11,

$$\widetilde{p}_{2k} = \frac{1}{2} \sum_{\{v : L(v)=(k,k)\}} p_{2k}^{(v)}([v] + [\widehat{v}]). \tag{6.55}$$

In each summand the string v that appears thus lies in the monoid $\mathcal{M}$ associated to family (6.51). We select the order $a_{20} > a_{11} > a_{02} > a_{-13} > b_{3,-1} > b_{20} > b_{11} > b_{02}$ on the coefficients and use the algorithm of Table 5.1 on page 235 to construct a Hilbert basis $\mathcal{H}$ of the monoid $\mathcal{M}$ of system (6.51). The result is

$$\begin{aligned}\mathcal{H} = \{&(2001\,0000), (0000\,1002), (1010\,0000), (0000\,0101),\\ &(1001\,0100), (0010\,1001), (0020\,1000), (0001\,0200), (0100\,0000),\\ &(0000\,0010), (1000\,0001), (0010\,0100), (0001\,1000)\}.\end{aligned}$$

Thus for example $[0020\,1000] = a_{02}^2 b_{3-1}$. We name this string α and similarly set $\beta = (0010\,0100)$ and $\gamma = (0001\,1000)$. Under the substitutions (6.53) every element of $\mathcal{H}$ reduces to a rational constant times one of α, $\widehat{\alpha}$, β, or γ. Thus any v appearing in (6.55) with $k \geq 6$ can be written in the form (see Remark 5.1.18 for the notation) $v = r\alpha + s\widehat{\alpha} + m\beta + n\gamma$, so $[v] = [\alpha]^r[\widehat{\alpha}]^s[\beta]^m[\gamma]^n$. In fact, $\widetilde{p}_{2k}$ can be written as a sum of terms $p_{2k}^{(v)}([v] + [\widehat{v}])$ in such a way that $\widehat{\alpha}$ does not appear in v, this time using the identity $[\alpha][\widehat{\alpha}] = [\beta]^2[\gamma]$. For if $r \geq s$, then we can factor out a power of $[\alpha]$ to form $[v] = [\alpha]^{r-s}([\alpha][\widehat{\alpha}])^s[\beta]^m[\gamma]^n = [\alpha]^{r-s}[\beta]^{m+2s}[\gamma]^{n+s}$. If $r < s$, then we simply replace the summand $p_{2k}^{(v)}([v] + [\widehat{v}])$ with $p_{2k}^{(v)}([\mu] + [\widehat{\mu}])$, where $\mu = \widehat{v}$, altered so as to eliminate $\widehat{\alpha}$ (this is really just a matter of changing the description of the indexing set on the sum in (6.55)). But since β and γ are self-conjugate we have then that for $k \geq 6$, $\widetilde{p}_{2k}$ is a sum of terms of the form a rational constant times

$$[v] + [\widehat{v}] = ([\alpha]^r + [\widehat{\alpha}]^r)[\beta]^m[\gamma]^n. \tag{6.56}$$

Moreover, since $L(\alpha) = (3,3)$ and $L(\beta) = L(\gamma) = (2,2)$, $L_1(v) = 3r + 2m + 2n$, from which it follows that either r or $m + n$ is at least two, since $L_1(v) \geq 6$.

Suppose $r \geq 2$. Then using the formula

$$[\mu + \sigma] + [\widehat{\mu + \sigma}] = \tfrac{1}{2}\Big([\sigma] + [\widehat{\sigma}]\Big)\Big([\mu] + [\widehat{\mu}]\Big) + \tfrac{1}{2}\Big([\mu] - [\widehat{\mu}]\Big)\Big([\sigma] - [\widehat{\sigma}]\Big)$$

we obtain

$$\begin{aligned}[v] + [\widehat{v}] &= [(r-1)\alpha + \alpha] + [(r-1)\widehat{\alpha} + \widehat{\alpha}]\\ &= \tfrac{1}{2}([\alpha] + [\widehat{\alpha}])([(r-1)\alpha] + [(r-1)\widehat{\alpha}])\\ &\qquad + \tfrac{1}{2}([(r-1)\alpha] - [(r-1)\widehat{\alpha}])([\alpha] - [\widehat{\alpha}])\\ &= \tfrac{1}{2}([\alpha] + [\widehat{\alpha}])([\alpha]^{r-1} + [\widehat{\alpha}]^{r-1}) + \tfrac{1}{2}([\alpha]^{r-1} - [\widehat{\alpha}]^{r-1})([\alpha] - [\widehat{\alpha}]).\end{aligned}$$

The second congruence in (6.54) implies that the first factor in the first term in this sum is in $\langle \widetilde{p}_2, \widetilde{p}_4, \widetilde{p}_6 \rangle$. If $r = 2$, the second term in the sum is $\frac{1}{2}([\alpha] - [\widehat{\alpha}])^2$, which a computation shows to be in $\langle \widetilde{p}_2, \widetilde{p}_4, \widetilde{p}_6, \widetilde{p}_8 \rangle$. Otherwise, the second term can be written

$$\tfrac{1}{2}([\alpha]-[\widehat{\alpha}])^2([\alpha]^{r-2}+[\alpha]^{r-3}[\widehat{\alpha}]+\cdots+[\alpha][\widehat{\alpha}]^{r-3}+[\widehat{\alpha}]^{r-2}),$$

which is in $\langle \widetilde{p}_2, \widetilde{p}_4, \widetilde{p}_6, \widetilde{p}_8\rangle$ because $([\alpha]-[\widehat{\alpha}])^2$ is.

Suppose now that in (6.56) $m+n \geq 2$. If $m=0$, then

$$[\gamma]^n = (a_{-13}b_{3,-1})^n = (a_{-13}b_{3,-1})^2(a_{-13}b_{3,-1})^{n-2} \equiv -\tfrac{32}{105}\widetilde{p}_8 \bmod \langle \widetilde{p}_2, \widetilde{p}_4, \widetilde{p}_6\rangle$$

by the third congruence in (6.54). If $n=0$, then

$$[\beta]^m = (a_{02}b_{20})^m = (a_{02}b_{20})^2(a_{02}^2b_{20}^2)^{m-2} \equiv 0 \bmod \langle \widetilde{p}_2, \widetilde{p}_4, \widetilde{p}_6, \widetilde{p}_8\rangle$$

since a computation shows that $a_{02}^2b_{20}^2 \equiv 0 \bmod \langle \widetilde{p}_2, \widetilde{p}_4, \widetilde{p}_6, \widetilde{p}_8\rangle$. If $mn \neq 0$, then

$$\begin{aligned}\beta^m\gamma^n &= (a_{02}b_{20})^m(a_{-13}b_{3,-1})^n \\ &= (a_{02}b_{20}a_{-13}b_{3,-1})(a_{02}b_{20})^{m-1}(a_{-13}b_{3,-1})^{n-1} \equiv 0 \bmod \langle \widetilde{p}_2, \widetilde{p}_4, \widetilde{p}_6, \widetilde{p}_8\rangle\end{aligned}$$

since another computation shows that $a_{02}b_{20}a_{-13}b_{3,-1} \equiv 0 \bmod \langle \widetilde{p}_2, \widetilde{p}_4, \widetilde{p}_6, \widetilde{p}_8\rangle$.

This proves the first statement of the lemma. Because we have been working with the $\widetilde{p}_{2k}$ as elements of $\mathbb{C}[a_{11}, a_{02}, a_{-13}, b_{3,-1}, b_{20}]$ without any conditions on a_{11}, a_{02}, a_{-13}, $b_{3,-1}$, or b_{20}, so that they are true indeterminates, the second statement of the lemma follows from what we have just proven and Theorem 6.1.7. □

For the component $\mathbf{V}(C_2)$ of the center variety for family (6.51) we do not use the Bautin method but apply Lemma 6.4.2 when the original system is isochronous and a simple analysis otherwise.

Lemma 6.4.8. *No critical periods of system* (6.50) *bifurcate from centers of the component* $\mathbf{V}(C_2)$ *of the center variety (centers that have a Darboux first integral) under proper perturbation.*

Proof. Since the first two generators of C_2 are just a_{11} and b_{11}, we may restrict attention to the copy of $\mathbb{C}^6$ lying in $\mathbb{C}^8$ that corresponds to $a_{11} = b_{11} = 0$, replace p_{2k} by $\widetilde{p}_{2k}(a_{20}, a_{02}, a_{-13}, b_{3,-1}, b_{20}, b_{02}) = p_{2k}(a_{20}, 0, a_{02}, a_{-13}, b_{3,-1}, b_{20}, 0, b_{02})$ for $k \geq 2$, and dispense with p_2 altogether. In the remainder of the proof of the lemma we will let (a,b) denote $(a_{20}, a_{02}, a_{-13}, b_{3,-1}, b_{20}, b_{02})$. By Proposition 4.2.12, p_{2k} is a homogeneous polynomial of degree k, and clearly the same is true for $\widetilde{p}_{2k}$, regarded as an element of $\mathbb{C}[a,b]$.

Since the generators of C_2 are homogeneous polynomials, $\mathbf{V}(C_2)$, now regarded as lying in $\mathbb{C}^6$, is closed under rescaling by elements of $\mathbb{R}$ and is, of course, topologically closed. The same is clearly true of $\mathbf{V}(C_2)^{\mathbb{R}} = \mathbf{V}(C_2) \cap \{(a,b) : b = \bar{a}\}$.

The value of $\widetilde{p}_4$, viewed as a function, at any point of $\mathbf{V}(C_2)^{\mathbb{R}}$ is $|a_{02}|^2$, which we obtained by replacing a_{20} by $-3b_{20}$, b_{02} by $-3a_{02}$, $a_{-13}b_{3,-1}$ by $4a_{02}b_{20}$, and then replacing b_{20} by $\bar{a}_{02}$. The polynomial function $\widetilde{p}_4$ takes the value zero only on the string $(a,\bar{a}) = (0,0)$, since in $\mathbf{V}(C_2)^{\mathbb{R}}$ $|a_{-13}|^2 = |a_{02}|^2$. If the initial string $(a^*, \overline{a^*})$ of coefficients in $\mathbf{V}(C_2)^{\mathbb{R}}$ is zero, meaning that it corresponds to the linear system in family (6.50) (which, by Exercise 6.29(b), is the only isochronous center in $\mathbf{V}(C_2)$), the conditions of Lemma 6.4.2 are satisfied (with a change of indices) with $B = \mathbf{V}(C_2)^{\mathbb{R}}$ and $\mathscr{F} = T'$, hence for some small ε the equation $\mathscr{T}'(r,(a,\bar{a})) = 0$ has no solutions for $0 < r < \varepsilon$ for $(a,\bar{a}) \in \mathbf{V}(C_2)^{\mathbb{R}}$ with $|(a,\bar{a})| > 0$ sufficiently small. If the initial string $(a^*,\overline{a^*})$ of coefficients in $\mathbf{V}(C_2)^{\mathbb{R}}$ is nonzero, then $\widetilde{p}_4$ is nonzero on a neighborhood U of $(a^*,\overline{a^*})$ in $\mathbf{V}(C_2)^{\mathbb{R}}$, and for any string $(a,\bar{a})$ that lies in U, $T'(r,(a,\bar{a})) = 4r^3(\widetilde{p}_4(a,\bar{a}) + \cdots)$ has no zeros in $0 < r < \varepsilon$. In either case no bifurcation of critical periods is possible on $\mathbf{V}(C_2)$. □

The third component of the center variety for family (6.51) has such a complicated expression that we analyze the ideal P as an ideal in the coordinate ring $\mathbb{C}[\mathbf{V}(C_3)]$ only in conjunction with several parametrizations of $\mathbf{V}(C_3)$.

Lemma 6.4.9. *At most three critical periods of system* (6.50) *bifurcate from centers of the component* $\mathbf{V}(C_3)$ *of the center variety (reversible systems) under proper perturbation, and this bound is sharp.*

Proof. Refer to Example 1.4.19. Proceeding as we did there, we derive a mapping $F_1 : \mathbb{C}^5 \to \mathbb{C}^8$ defined by

$$\begin{aligned} a_{20} &= sw & b_{3,-1} &= tw^2 \\ a_{11} &= u & b_{20} &= w \\ a_{02} &= v & b_{11} &= u \\ a_{-13} &= tv^2 & b_{02} &= sv \end{aligned} \tag{6.57}$$

and readily verify that $F_1(\mathbb{C}^5)$, the image of $\mathbb{C}^5$ under F_1, lies in $\mathbf{V}(C_3)$. To use Theorem 1.4.14 to show that F_1 defines a polynomial parametrization of $\mathbf{V}(C_3)$, we compute a Gröbner basis of the ideal

$$\langle a_{11}-u,\ b_{11}-u,\ a_{20}-sw,\ b_{02}-sv,\ a_{-13}-tv^2,\ b_{3,-1}-tw^2,\ a_{02}-v,\ b_{20}-w\rangle \tag{6.58}$$

with respect to lex with

$$t > s > v > w > u > a_{11} > b_{11} > a_{20} > b_{02} > a_{02} > b_{20} > a_{-13} > b_{3,-1},$$

and obtain

$$\begin{aligned}\{&a_{-13}b_{20}^2 - a_{02}^2b_{3,-1}, -a_{-13}a_{20}b_{20} + a_{02}b_{02}b_{3,-1}, -a_{02}a_{20} + b_{02}b_{20},\\ &-a_{-13}a_{20}^2 + b_{02}^2b_{3,-1}, -a_{11} + b_{11}, b_{11} - u, b_{20} - w, a_{02} - v, -a_{20} + b_{20}s,\\ &-b_{02} + a_{02}s, b_{3,-1} - b_{20}^2t, a_{-13} - a_{02}^2t, -a_{-13}s + a_{02}b_{02}t,\\ &a_{-13}s^2 - b_{02}^2t, -b_{3,-1}s + a_{20}b_{20}t, b_{3,-1}s^2 - a_{20}^2t\}.\end{aligned}$$

The polynomials in the basis that do not depend on t, s, v, w, and u are exactly the generating polynomials of the ideal C_3. By Theorem 1.4.14, this means that (6.57) is a polynomial parametrization of $\mathbf{V}(C_3)$ (and incidentally confirms, by Corollary 1.4.18, that $\mathbf{V}(C_3)$ is irreducible). This proves only that $\mathbf{V}(C_3)$ is the Zariski closure of $F_1(\mathbb{C}^5)$, that is, the smallest variety in $\mathbb{C}^8$ that contains it, not that $F_1(\mathbb{C}^5)$ covers all of $\mathbf{V}(C_3)$. In fact, by Exercise 6.32, $F_1(\mathbb{C}^5)$ is a proper subset of $\mathbf{V}(C_3)$. Also by that exercise, setting $M = \{(0,u,0,0,0,0,u,0) : u \in \mathbb{C}\} \subset \mathbf{V}(C_3)$, F_1 defines an analytic coordinate system on $F_1(\mathbb{C}^5) \setminus M$ as an embedded submanifold of $\mathbb{C}^8$ ([97, §1.3]). Based on this fact, we make the substitution (6.57) in p_{2k} and regard p_{2k} as an element of $\mathbb{R}[u,v,w,s,t]$ (recall that its coefficients are real), but without a change of name. From Table 6.3 we obtain $p_2 = u$, $p_4 = -u + \frac{1}{4}vw(4+4s+3vwt^2)$, and so on. We will show that in $\mathbb{C}[u,v,w,s,t]$, $p_{2k} \in \langle p_2, p_4, p_6, p_8 \rangle$ for $k \geq 5$, which implies the corresponding inclusion in the ring of germs at every point of $F_1(\mathbb{C}^5) \setminus M$. Since in $\mathbb{C}[u,v,w,s,t]$ $p_2 = u$, it is enough to show that if $\widetilde{p}_{2k}$ denotes the polynomial obtained from p_{2k} by replacing u by zero, then $\widetilde{p}_{2k} \in \langle \widetilde{p}_4, \widetilde{p}_6, \widetilde{p}_8 \rangle$ for $k \in \mathbb{N}$, $k \geq 5$. Computations yield

$$\begin{aligned} \widetilde{p}_4 &= \tfrac{1}{4}vw(4+4s+3vwt^2) \\ \widetilde{p}_6 &= \tfrac{1}{8}v^2w^2t(3+s)(7+3s) \\ \widetilde{p}_8 &\equiv \tfrac{5}{36}v^2w^2(1+s)(7+3s) \bmod \langle \widetilde{p}_4, \widetilde{p}_6 \rangle \end{aligned} \tag{6.59}$$

and $\widetilde{p}_{10} \in \langle \widetilde{p}_4, \widetilde{p}_6, \widetilde{p}_8 \rangle$.

Lemma 6.4.6 says that $V_{\mathscr{I}} \cap \mathbf{V}(u) = \mathbf{V}(\widetilde{p}_4, \widetilde{p}_6, \widetilde{p}_8)$ (intersection with $V_{\mathscr{C}}$ drops out because we are now working in analytic coordinates on $\mathbf{V}(C_3) \setminus M$) so that $\widetilde{p}_{2k}$ vanishes everywhere on $\mathbf{V}(\widetilde{p}_4, \widetilde{p}_6, \widetilde{p}_8)$, hence $\widetilde{p}_{2k} \in \mathbf{I}(\mathbf{V}(\widetilde{p}_4, \widetilde{p}_6, \widetilde{p}_8))$. By the Strong Hilbert Nullstellensatz (Theorem 1.3.14), this latter ideal is $\sqrt{\langle \widetilde{p}_4, \widetilde{p}_6, \widetilde{p}_8 \rangle}$. If the ideal $\langle \widetilde{p}_4, \widetilde{p}_6, \widetilde{p}_8 \rangle$ were radical, then we would have the result that we seek. This is not the case however, because of the squares on v and w in $\widetilde{p}_6$ and $\widetilde{p}_8$. If we define $\tilde{\widetilde{p}}_6 = \widetilde{p}_6/(vw)$ and $\tilde{\widetilde{p}}_8 = \widetilde{p}_8/(vw)$, then a computation shows that

$$\sqrt{\langle \widetilde{p}_4, \widetilde{p}_6, \widetilde{p}_8 \rangle} = \langle \widetilde{p}_4, \tilde{\widetilde{p}}_6, \tilde{\widetilde{p}}_8 \rangle. \tag{6.60}$$

The way forward is to recognize that $\widetilde{p}_{2k}$ also contains v^2w^2 as a factor when $k \geq 6$. That is, we claim that for each $k \geq 6$ there exists $f_k \in \mathbb{C}[v,w,s,t]$ such that

$$\widetilde{p}_{2k} = v^2w^2 f_k. \tag{6.61}$$

To see this, observe that for every element ν of the Hilbert Basis $\mathscr{H}$ displayed in the proof of Lemma 6.4.7, by inspection $[\nu]$ contains (vw) as a factor. Thus if $[\theta] + [\widehat{\theta}]$ appears in p_{2k} (recall (6.55)) with $k \geq 6$, then because $L(\nu) = (r,r)$ with $r \leq 3$ for all $\nu \in \mathscr{H}$, $\theta = m_1\nu_1 + \cdots + m_s\nu_s$ with $m_j \in \mathbb{N}$ where $s \geq 2$. The important point is that s is at least two, since it means that $[\theta] = [m_1\nu_1 + \cdots + m_s\nu_s] = [\nu_1]^{m_1} \cdots [\nu_s]^{m_s}$ must contain v^2w^2.

We know that $\widetilde{p}_{2k}$ vanishes at every point of $\mathbf{V}(\widetilde{p}_4, \widetilde{p}_6, \widetilde{p}_8)$. But now because $\widetilde{p}_{2k} = vw(vwf_k)$, this implies that vwf_k vanishes at every point of $\mathbf{V}(\widetilde{p}_4, \widetilde{p}_6, \widetilde{p}_8)$, too. Furthermore, by (6.60) and Proposition 1.3.16, $\mathbf{V}(\widetilde{p}_4, \widetilde{p}_6, \widetilde{p}_8) = \mathbf{V}(\widetilde{p}_4, \tilde{\widetilde{p}}_6, \tilde{\widetilde{p}}_8)$, so that vwf_k vanishes at every point of $\mathbf{V}(\widetilde{p}_4, \tilde{\widetilde{p}}_6, \tilde{\widetilde{p}}_8)$, hence is in $\mathbf{I}(\mathbf{V}(\widetilde{p}_4, \tilde{\widetilde{p}}_6, \tilde{\widetilde{p}}_8))$, which, by the Strong Nullstellensatz (Theorem 1.3.14) and (6.60), is $\langle \widetilde{p}_4, \tilde{\widetilde{p}}_6, \tilde{\widetilde{p}}_8 \rangle$. That is, there exist h_1, h_2, $h_3 \in \mathbb{C}[v,w,s,t]$ such that $vwf_k(v,w,s,t) = h_1\widetilde{p}_4 + h_2\tilde{\widetilde{p}}_6 + h_3\tilde{\widetilde{p}}_8$, hence $\widetilde{p}_{2k} = v^2w^2f_k = (vwh_1)\widetilde{p}_4 + h_2\widetilde{p}_6 + h_3\widetilde{p}_8$, that is, $\widetilde{p}_{2k} \in \langle \widetilde{p}_4, \widetilde{p}_6, \widetilde{p}_8 \rangle$, as required.

In order to handle as many points in $\mathbf{V}(C_3)$ that were not in the image of the mapping F_1 as possible, we define a second, similar mapping $F_2 : \mathbb{C}^5 \to \mathbb{C}^8$ by

$$\begin{aligned} a_{20} &= v & b_{3,-1} &= tv^2 \\ a_{11} &= u & b_{20} &= sv \\ a_{02} &= sw & b_{11} &= u \\ a_{-13} &= tw^2 & b_{02} &= w. \end{aligned} \tag{6.62}$$

Since $F_2 = \eta \circ F_1$, where $\eta : C^8 \to \mathbb{C}^8$ is the involution defined by

$$a_{20} \to a_{02}, \quad a_{02} \to a_{20}, \quad a_{-13} \to b_{3,-1}, \quad b_{3,-1} \to a_{-13} \quad b_{20} \to b_{02}, \quad b_{02} \to b_{20},$$

which preserves $\mathbf{V}(C_3)$ and $\mathbf{V}(C_3)^{\mathbb{R}}$, F_2 is a parametrization of $\mathbf{V}(C_3)$. By Exercise 6.33, F_2 forms an analytic coordinate system on $F_2(\mathbb{C}^5) \setminus M$ as an embedded submanifold of $\mathbb{C}^8$, where $M \subset \mathbf{V}(C_3)$ is as above.

We now proceed just as we did with the parametrization F_1. We define $\widetilde{p}_{2k}$ as before and similarly set $\tilde{\tilde{p}}_6 = \widetilde{p}_6/(vw)$ and $\tilde{\tilde{p}}_8 = \widetilde{p}_8/(vw)$. Computations now give

$$\begin{aligned} \widetilde{p}_4 &= \tfrac{1}{4} vw(4s + 4s^2 + 3vwt^2) \\ \tilde{\tilde{p}}_6 &= \tfrac{1}{8} vwt(1+3s)(3+7s) \\ \tilde{\tilde{p}}_8 &\equiv -\tfrac{5}{108} svw(1+s)(3+7s) \bmod \langle \widetilde{p}_4, \widetilde{p}_6 \rangle, \end{aligned}$$

$\widetilde{p}_{10} \in \langle \widetilde{p}_4, \widetilde{p}_6, \widetilde{p}_8 \rangle$, and establish the identity (6.60) in this setting as well. We thus obtain the inclusion $p_{2k} \in \langle p_2, p_4, p_6, p_8 \rangle$ for $k \geq 5$ in $\mathbb{C}[u,v,w,s,t]$, which implies the corresponding inclusion in the ring of germs at every point of $F_2(C^5) \setminus M$.

The points of $\mathbf{V}(C_3)$ that are not covered by either parametrization are covered by the parametrization F_3 of $\mathbf{V}(C_3) \cap \mathbf{V}(a_{20}, a_{02}, b_{20}, b_{02})$ defined by

$$a_{11} = b_{11} = u, \quad a_{20} = a_{02} = b_{20} = b_{02} = 0, \quad a_{-13} = w, \quad b_{3,-1} = v. \tag{6.63}$$

With this parametrization $p_2 = u$ and $p_4 = -u^2 + \frac{3}{4}vw$. The Hilbert Basis displayed in the proof of Lemma 6.4.7 reduces to $\{u, vw\}$ so that now for every θ that is in $\mathscr{M}$ $[\theta] + [\widehat{\theta}] = 2[u]^r[vw]^s$, hence (6.55) implies that $P = \langle p_2, p_4 \rangle$.

Let (a^*, b^*) be a point of $\mathbf{V}(C_3)^{\mathbb{R}}$ and let $U \subset \mathbf{V}(C_3)$ be a sufficiently small neighborhood of a^*. Obviously U is covered by the union of the images of the parametrizations (6.57), (6.62), and (6.63). Therefore, using Lemma 6.4.1, we conclude that at most three small critical periods can appear under proper perturbation of system (a^*, b^*).

Finally, to show that the bound is sharp we produce a bifurcation of three critical periods from the center of a system in $\mathbf{V}(C_3)$. We will use the parametrization (6.57) with initial values $u = 0$, $v = w = 1$, $s = -3$, and $t = 2\sqrt{2/3}$. We will not change v or w but will successively change s to $-3 + \alpha$, then t to $2\sqrt{(2-\alpha)/3} + \beta$, and finally u to a nonzero value that we still denote by u. The expressions (6.59) for the isochronicity quantities become, replacing $\widetilde{p}_8$ by the polynomial to which it is equivalent, without changing the notation,

$$\begin{aligned}
\widetilde{p}_4 &= u \\
\widetilde{p}_4 &= \tfrac{1}{4}\beta(4\sqrt{3}\sqrt{2-\alpha}+3\beta) \\
\widetilde{p}_6 &= \tfrac{1}{24}\alpha(-2+3\alpha)(2\sqrt{3}\sqrt{2-\alpha}+3\beta) \\
\widetilde{p}_8 &\equiv \tfrac{5}{36}(\alpha-2)(3\alpha-2)\,.
\end{aligned}$$

By the theory that we have developed, it is enough to show that, replacing r^2 by R, the polynomial $T(R) = p_8R^4 + p_6R^3 + p_4R^2 + p_2R$ has three arbitrarily small positive roots for suitably chosen values of α, β, and u that are arbitrarily close to zero.

Keeping β and u both zero, we move α to a positive value that is arbitrarily close to 0. Then $p_8 \approx 5/9$ and $p_6 \approx -\alpha/\sqrt{6} < 0$. Moving β to a positive value that is sufficiently small the quadratic factor in $T(R) = R^2(p_8R^2 + p_6R + p_4)$ has two positive roots, the larger one less than $-p_6/p_8$, which is about $9\alpha/(5\sqrt{6})$. If we now move u to a sufficiently small positive value, the two positive roots move slightly and a new positive root of T bifurcates from 0, giving the third critical period. □

By the discussion preceding Definition 6.4.5, Lemmas 6.4.7, 6.4.8, and 6.4.9 combine to give the following result.

Theorem 6.4.10. *At most three critical periods bifurcate from centers of system* (6.50), *and there are systems of the form* (6.50) *with three small critical periods.*

6.5 Notes and Complements

Bautin's theorem on the cyclicity of antisaddles in quadratic systems is a fundamental result in the theory of polynomial systems of ordinary differential equations, important not only because of the bound that it provides, but also because of the approach it gives to the study of the problem of cyclicity in any polynomial system. Specifically, Bautin showed that the cyclicity problem in the case of a simple focus or center could be reduced to the problem of finding a basis for the ideal of focus quantities. By way of contrast, a full description of the bifurcation of singularities of quadratic systems cannot be expressed in terms of algebraic conditions on the coefficients ([66]).

In his work Bautin considered quadratic systems with antisaddles in the normal form of Kapteyn. Such systems look simpler in that form because it contains only six real parameters. However, the ideal of focus quantities is not radical when the system is written this way, so that Bautin's method of constructing that ideal was rather complicated. A simpler way to construct a basis for such systems was suggested by Yakovenko ([201]). In [206] Żołądek considered the focus quantities of system (6.12b) in the ring of polynomials that are invariant under the action of the rotation group, and observed that the ideal of focus quantities is radical in that ring.

As to a generalization of Bautin's theorem to the case of cubic systems, as of now only a few partial results are known. In particular, Żołądek ([209]) and Christopher ([51]) have shown that there are cubic systems with 11 small limit cycles bifurcating from a simple center or focus, and Sibirsky ([176]) (see also [207]) has shown that the cyclicity of a linear center or focus perturbed by homogeneous polynomials of the third degree is at most five. For studies of bifurcations of small limit cycles of cubic Liénard systems the reader can consult [56] and for studies of simultaneous bifurcations of such cycles in cubic systems [203].

At present we have no reason to believe that it is possible to find a bound on the cyclicity of a simple center or focus of system (6.1) as a function of n. We can look at the problem from another point of view, however, asking whether it is possible to perform the steps outlined above algorithmically. That is, for a given polynomial system with a singular point that is a simple center in the linear approximation we can ask:

- Does there exist an algorithm for finding the center variety in a finite number of steps?
- Given the center variety, is it possible to find a basis of the Bautin ideal in a finite number of steps?

Affirmative answers to these questions would constitute one sort of solution to the local 16th Hilbert problem.

In this chapter we considered the cyclicity problem only for singularities of focus or center type. For an arbitrary singularity of system (6.1), even for quadratic systems this problem is still only partially solved, although much is known.

An important problem from the point of view of the theory of bifurcations and the study of the Hilbert problem is the investigation of the cyclicity of more general limit periodic sets than singularities, of which the next most complicated is the homoclinic loop mentioned in the introduction to this chapter. Since rescaling the system by a nonzero constant does not alter the phase portrait of (6.1), the parameters may be regarded as lying in the compact sphere $S^{(n+1)(n+2)-1}$ rather than in $\mathbb{R}^{(n+1)(n+2)}$. There is also a natural extension of the vector field $\mathscr{X}$ that corresponds to (6.1) to the sphere, which amounts to a compactification of the phase space from $\mathbb{R}^2$ to the "Poincaré sphere" S^2. If every limit periodic set in this extended context has finite cyclicity, then finitude of $H(n)$ for any n follows ([155]). Although it is still unknown whether even $H(2)$ is finite, considerable progress in the study of the cyclicity of limit periodic sets of quadratic systems has been achieved. Quadratic systems are simple enough that it is possible to list all limit periodic sets that can occur among them. There are 121 of them on the Poincaré sphere, not counting those that consist of a single singular point. See [68] and [156] for a full elaboration of these ideas. There is hope that the finitude of $H(2)$ can be proved in this way, but even for cubic systems this approach is unrealistic. To find out more about methods for studying bifurcations of general limit periodic sets the reader is referred to the works of Dumortier, Li, Roussarie, and Rousseau ([67, 68, 112, 156, 157]), Han and his collaborators ([91, 92, 93, 94]), and Żołądek ([206]). For the state of the art concerning the 16th Hilbert problem and references the reader can consult the recent surveys by Chavarriga and Grau ([37]) and Li ([113]), the books by Christopher and Li ([53]), Roussarie ([156]), Ye ([202]), and Zhang ([205]), and the bibliography of Reyn ([147]).

In Section 6.4 we presented one approach to the investigation of bifurcations of critical periods of polynomial systems as we applied it to system (6.50). The proof given there, based as it is on the use of the center and isochronicity varieties in $\mathbb{C}^8$ of system (6.51), is essentially different from the original proof in [158], which, however, also includes sharp bounds on the maximum number of critical periods that can bifurcate in terms of the order of vanishing of the period function at $r = 0$.

Exercises

6.1 For the purpose of this exercise, for a function f that is defined and analytic in a neighborhood of $\theta^* \in k^n$, in addition to $\mathbf{f}$ let $[f]$ also denote the germ determined by f at θ^*. If $\mathbf{f}$ and $\mathbf{g}$ are germs, show that for any $f_1, f_2 \in \mathbf{f}$ and any $g_1, g_2 \in \mathbf{g}$, $f_1 + g_1 = f_2 + g_2$ and $f_1 g_1 = f_2 g_2$ on a neighborhood of θ^*, so that addition and multiplication of germs are well-defined by $[f] + [g] = [f + g]$ and $[f][g] = [fg]$.

6.2 Show that if $I = \langle \mathbf{f}_1, \ldots, \mathbf{f}_s \rangle \subset \mathscr{G}_{\theta^*}$, then $\mathbf{f} \in I$ if and only if for any $f \in \mathbf{f}$ and $f_j \in \mathbf{f}_j$, $j = 1, \ldots, s$, there exist an open neighborhood U of θ^* and functions h_j, $j = 1, \ldots, s$, such that f and all f_j and h_j are defined and analytic on U and $f = h_1 f_1 + \cdots + h_s f_s$ holds on U.

6.3 Show in two different ways that if system (6.1) has a simple singularity at $(0,0)$, then the singularity is isolated.

a. Interpret the condition

$$\det \begin{pmatrix} \widetilde{U}_u & \widetilde{U}_v \\ \widetilde{V}_u & \widetilde{V}_v \end{pmatrix} \neq 0$$

geometrically and give a geometric argument.

b. Give an analytic proof.

Hint. By a suitable affine change of coordinates and the Implicit Function Theorem near $(0,0)$, $\widetilde{V}(u,v) = 0$ is the graph of a function $v = f(u)$. Consider $g(u) = \widetilde{U}(u, f(u))$.

6.4 Show that if system (6.1) has a simple singularity at $(0,0)$, if the coefficient topology is placed on the set of pairs $(\widetilde{U}, \widetilde{V})$ (each pair is identified with a point of $\mathbb{R}^{(n+1)(n+2)}$ under the usual topology by the $(n+1)(n+2)$ coefficients of $\widetilde{U}$ and $\widetilde{V}$), and if N is a neighborhood of $(0,0)$ that contains no singularity of (6.1) besides $(0,0)$, then for $(\widetilde{U}', \widetilde{V}')$ sufficiently close to $(\widetilde{U}, \widetilde{V})$, the system

$$\dot{u} = \widetilde{U}'(u,v), \qquad \dot{v} = \widetilde{V}'(u,v) \tag{6.64}$$

has exactly one singularity in N.

6.5 Give an example of a system (6.1) with an isolated nonsimple singularity at $(0,0)$ with the property that given any neighborhood N of $(0,0)$, there is a pair

$(\widetilde{U}',\widetilde{V}')$ arbitrarily close to $(\widetilde{U},\widetilde{V})$ in the coefficient topology (Exercise 6.4) such that system (6.64) has more than one singularity in N.

6.6 Same as Exercise 6.5 but with the conclusion that system (6.64) has no singularities in N.

6.7 Show the Lyapunov quantities specified by Definition 3.1.3 for family (6.11a) are finite sums of polynomials in (A,B) and rational functions in λ whose denominators have no real zero besides $\lambda = 0$, all with rational coefficients, times exponentials of the form $e^{m\pi\lambda}$, $m \in \mathbb{N}$.
Hint. Two key points are as follows: the initial value problem that determines w_j is of the form $w_j'(\varphi) - \lambda w_j(\varphi) = S_j(\varphi)e^{m\lambda\varphi}$, $w_j(0) = 0$, where the constant m is at least two; and the antiderivative of a function of the form $e^{\lambda u}\sin^r au \cos^s bu$, $r,s \in \mathbb{N}_0$, is a sum of terms of similar form times a rational function of λ whose denominator has either the form λ, which occurs only if $r = s = 0$, or the form $\lambda^2 + c^2$, some $c \in \mathbb{Z}$ depending on a, b, r, and s.

6.8 Use (3.5)–(3.7) and (3.10) and a computer algebra system to compute the first few Lyapunov quantities at an antisaddle of the general quadratic system (2.74) written in the form (6.11a). (Depending on your computing facility, even η_3 could be out of reach. First do the case $\lambda = 0$.)

6.9 The general quadratic system of differential equations with a simple nonhyperbolic antisaddle at the origin is

$$\dot{u} = -v + A_{20}u^2 + A_{11}uv + A_{02}v^2, \quad \dot{v} = u + B_{20}u^2 + B_{11}uv + B_{02}v^2.$$

a. Compute the complexification

$$\begin{aligned}\dot{x} &= \ \ i\left(x - a_{10}x^2 - a_{01}xy - a_{-12}y^2\right)\\ \dot{y} &= -i\left(y - b_{2,-1}x^2 - b_{10}xy - b_{01}y^2\right),\end{aligned}$$

which is just (3.129) in Section 3.7. That is, express each coefficient of the complexification in terms of the coefficients of the original real system.

b. The first focus quantity is given by (3.133) as $g_{11} = -i(a_{10}a_{01} - b_{10}b_{01})$. Compute $g_{11}^{\mathbb{R}}$, the expression for the first focus quantity in terms of the original coefficients.

6.10 Prove Corollary 6.2.4.

6.11 Use Corollary 6.2.4 to prove Corollary 6.2.5.

6.12 Prove Proposition 6.2.6.

6.13 Find an example of two polynomials f and g and a function h that is real analytic on an open interval U in $\mathbb{R}$ but is not a polynomial such that $f = hg$. Can U be all of $\mathbb{R}$? (This shows why it is necessary to work with the germs $\eta_1, \eta_2, \ldots$ to obtain a contradiction in cases 2 and 3 of the proof of Lemma 6.2.8.)

6.14 a. Explain why the following conclusion to the proof of Theorem 6.2.7 is invalid. "If the perturbation is made within family (6.11), then we first perturb without changing λ from zero to obtain up to $k-1$ zeros in $(0,\varepsilon)$ as well as a zero at $\rho = 0$. Since $\mathscr{P}'(0) = \eta_1 = e^{\lambda} - 1$, we may then change λ to an arbitrarily small nonzero value, whose sign is the opposite of that of $\mathscr{P}(\rho)$

for ρ small, to create an additional isolated zero in $(0,\varepsilon)$."
Hint. Study the first line in the proof of the theorem.

b. Validly finish the proof of Theorem 6.2.7 using the theory developed in the remainder of Section 6.2.

6.15 Prove that the ideal $\mathscr{B}_3$ for a general complex quadratic system (3.131) is radical using a computer algebra system that has a routine for computing radicals of polynomial ideals (such as Singular) as follows:
a. compute the radical of $\mathscr{B}_3$;
b. apply the Equality of Ideals Criterion on page 24.

6.16 The proof that the bound in Theorem 6.3.3 is sharp is based on the same logic and procedure as the faulty proof of part of Theorem 6.2.7 in Exercise 6.14. Explain why the reasoning is valid in the situation of Theorem 6.3.3.

6.17 Show that the ideals J_1 through J_7 listed in Theorem 3.7.2 are prime using Theorem 1.4.17 and the Equality of Ideals Criterion on page 24.

6.18 Show that for family (6.30), $g_{77} \notin \mathscr{B}_5$ and $g_{99} \notin \mathscr{B}_7$.

6.19 a. Show that the ideal $\mathscr{B}_9$ for family (6.30) is not radical.
b. How can you conclude automatically that $\mathscr{B}_5$ cannot be radical either?

6.20 Show that the mapping W that appear in the proof of Theorem 6.3.5 is one-to-one, either directly or by showing that $\ker(W) = \{0\}$.

6.21 [Referenced in Proposition 6.3.5, proof.] Show that any monomial in the kth focus quantity g_{kk} of family (6.30) has degree at least k if k is even and at least $k+1$ if k is odd.
Hint. Any monomial $[\nu]$ in g_{kk} is an $\mathbb{N}_0$-linear combination of elements of the Hilbert basis $\mathscr{H}$ of the monoid $\mathscr{M}$ listed on page 277 in the proof of Proposition 6.3.5 and satisfies $L(\nu) = (k,k)$.

6.22 [Referenced in Proposition 6.3.5, proof.] Let k be $\mathbb{R}$ or $\mathbb{C}$.
a. Let f and g be polynomials in $k[x_1,\dots,x_n]$, let $V \subsetneq k^n$ be a variety, and suppose that if $\mathbf{x} \in k^n \setminus V$, then $f(\mathbf{x}) = g(\mathbf{x})$. Show that $f = g$.
Hint. Recall Proposition 1.1.1.
b. Generalize the result to the "smallest" set on which f and g agree that you can, and to other fields k, if possible.

6.23 Consider family (6.29). We wish to generate as many limit cycles as possible by imitating the technique used at the end of the proof of Theorem 6.3.3.
a. Explain why a fine focus in family (6.29) is of order at most five.
b. Begin with a system that has a fifth-order fine focus at the origin, and adjust the coefficients in order to change g_{11}, g_{33}, and g_{44} in such a way as to produce three limit cycles in an arbitrarily small neighborhood of the origin. By an appropriate change in the linear part, obtain a system arbitrarily close to the original system with four arbitrarily small limit cycles about the origin.

6.24 Determine the cyclicities of systems from $\mathbf{V}(J_j)^{\mathbb{R}}$, where J_j are the ideals defined in the statement of Theorem 3.7.1.

6.25 Suppose the coefficients $f_j(\theta)$ of the series (6.2) are polynomials over $\mathbb{C}$ and that the ideal $I_s = \langle f_1,\dots,f_s\rangle$ is radical. Denote by $\widetilde{\mathscr{F}}(z,\theta) = \sum_{j=s+1}^{\infty} \tilde{f}_j(\theta) z^k$ the function $\mathscr{F}(z,\theta)|_{\mathbf{V}(I_s)}$, by $\tilde{I}$ the ideal generated by the coefficients of $\widetilde{\mathscr{F}}(z,\theta)$

in $\mathbb{C}(\mathbf{V}(I_s))$, and assume that the minimal basis $M_{\tilde{I}}$ of $\tilde{I}$ consists of t polynomials. Prove that the multiplicity of $\mathscr{F}$ (over $\mathbb{R}$) is at most $s+t$.

6.26 Investigate cyclicity of the origin for the system

$$\dot{x} = x - a_{20}x^3 - a_{11}x^2\bar{x} - a_{02}x\bar{x}^2 - a_{-13}\bar{x}^3.$$

6.27 Prove Theorem 6.4.3.

6.28 Prove Theorem 6.4.4.

6.29 Use Theorems 6.4.3 and 6.4.4 to show that for the *real* family (6.50),

a. no nonlinear center corresponding to a point lying in the component $\mathbf{V}(C_1)$ of the center variety $V_{\mathscr{C}}^{\mathbb{R}}$ is isochronous;

b. no nonlinear center corresponding to a point lying in the component $\mathbf{V}(C_2)$ of the center variety $V_{\mathscr{C}}^{\mathbb{R}}$ is isochronous; and

c. the centers corresponding to a points lying in the component $\mathbf{V}(C_3)$ of the center variety $V_{\mathscr{C}}^{\mathbb{R}}$ that are isochronous correspond to precisely the parameter strings $(a_{20}, a_{11}, a_{02}, a_{-13}, b_{3,-1}, b_{20}, b_{11}, b_{02}) \in \mathbb{C}^8$ of the form

(i) $(a, 0, -\bar{a}, 0, 0, -a, 0, \bar{a})$, $a \in \mathbb{C}$;

(ii) $(-\frac{28}{9r}e^{is}, 0, \frac{4}{3r}e^{-is}, \frac{16}{9r}e^{-2is}, \frac{16}{9r}e^{2is}, \frac{4}{3r}e^{is}, 0, -\frac{28}{9r}e^{-is})$, $r, s \in \mathbb{R}$;

(iii) $(a, 0, 0, 0, 0, 0, 0, 0, \bar{a})$, $a \in \mathbb{C}$; and

(iv) $(\frac{4}{7}e^{-is}, 0, -\frac{12}{49}e^{is}, \frac{16}{49}e^{2is}, \frac{16}{49}e^{-2is}, -\frac{12}{49}e^{-is}, 0, \frac{4}{7}e^{is})$, $s \in \mathbb{R}$.

Hint. Systems (i) and (ii) arise from the first parametrization of $\mathbf{V}(C_3)$ in the proof of Lemma 6.4.9; systems (iii) and (iv) arise from the second parametrization. No point covered by the third parametrization corresponds to an isochronous center.

6.30 Reduce the upper bound in Lemma 6.4.7 from three to one as follows. Assume that reduction to $\widetilde{p}_{2k} \in \mathbb{C}[a_{11}, a_{02}, a_{-13}, b_{3,-1}, b_{20}]$ as in the proof of the lemma has been made.

a. Each $\widetilde{p}_{2k}$ is a homogeneous polynomial of degree k.

b. The polynomial function $\widetilde{p}_2$ vanishes at points in $\mathbf{V}(C_1)^{\mathbb{R}}$ other than $\mathbf{0}$, but $\widetilde{p}_4$ does not.
Hint. $\mathbf{V}(C_1)^{\mathbb{R}}$ is obtained from $\mathbf{V}(C_1)$ by making the substitutions $b_{kj} = \bar{a}_{jk}$. The coefficient a_{11} is forced to be real (recall (6.53)).

c. Make a change of variables $z = r^2$ and show that the number of critical periods in $(0, \varepsilon)$ is the number of zeros of the function $\mathscr{F}(z) = \sum_{k=1}^{\infty} 2k\widetilde{p}_{2k}z^{k-1}$ in $(0, \varepsilon^2)$.
Hint. $\mathscr{F}$ corresponds to $\mathscr{T}'(r)/r$.

d. $\mathscr{F}(z)$ has at most one more zero in $(0, \varepsilon^2)$ than does its derivative $\mathscr{F}'$.

e. $\mathscr{F}'$ has no zeros in $(0, \varepsilon^2)$. The result thus follows from parts (c) and (d).
Hint. By parts (a) and (b), Lemma 6.4.2 applies to $\mathscr{F}'$ with $B = \mathbf{V}(C_1)^{\mathbb{R}}$. You must use Exercise 6.29(a).

6.31 Show that the bound in the previous exercise is sharp by finding a perturbation from the linear center into $\mathbf{V}(C_1)^{\mathbb{R}}$ that produces a system with an arbitrarily small critical period.

6.32 Consider the mapping $F_1 : \mathbb{C}^5 \to \mathbb{C}^8$ defined by (6.57).

a. Given $(a_{20},\dots,b_{02}) \in \mathbb{C}^8$, show that equations (6.57) can be solved for (u,v,w,s,t) in terms of $(a_{20},\dots,b_{02})$ if $(a_{02},b_{20}) \neq (0,0)$ and that there is a solution if and only if $a_{20} = a_{-13} = b_{3,-1} = b_{02} = 0$ when $(a_{02},b_{20}) = (0,0)$.

b. Use part (a) to show that the image of F_1 is

$$\mathbf{V}(C_3) \setminus \{(a,r,0,b,c,0,r,d) : (a,b,c,d) \neq (0,0,0,0) \text{ and } a^2b - cd^2 = 0\}$$

(but a, b, c, d, and r are otherwise arbitrary elements of $\mathbb{C}$).

c. Let $K = \{(u,0,0,s,t)\} \subset \mathbb{C}^5$ and $M = \{(0,u,0,0,0,0,u,0) : u \in \mathbb{C}\} \subset \mathbf{V}(C_3)$. Show that F_1 maps $\mathbb{C}^5 \setminus K$ one-to-one onto $F_1(\mathbb{C}^5) \setminus M$ and maps K onto M.

d. Show that the derivative DF_1 of F_1 has maximal rank at every point of $\mathbb{C}^5 \setminus K$.

6.33 Consider the mapping $F_2 : \mathbb{C}^5 \to \mathbb{C}^8$ defined by (6.62).

a. Give an alternate proof that F_2 is a parametrization of $\mathbf{V}(C_3)$ from that given in the text by computing a Gröbner basis of a suitably chosen ideal, as was done for F_1.

b. Proceeding along the lines of the previous exercise, show that the image of F_2 is

$$\mathbf{V}(C_3) \setminus \{(0,r,a,b,c,d,r,0) : (a,b,c,d) \neq (0,0,0,0) \text{ and } a^2c - bd^2 = 0\}$$

(but a, b, c, d, and r are otherwise arbitrary elements of $\mathbb{C}$).

c. Show that F_2 maps $\mathbb{C}^5 \setminus K$ one-to-one onto $F_2(\mathbb{C}^5) \setminus M$ and collapses K onto M, where K and M are the sets defined in part (b) of the previous exercise.

d. Show that the derivative DF_2 of F_2 has maximal rank at every point of $\mathbb{C}^5 \setminus K$.

6.34 The example at the end of the proof of Lemma 6.4.9 produced a bifurcation of three critical periods from a nonlinear center. Modify it to obtain a bifurcation of three critical periods from a linear center.

Appendix

The algorithms presented in this book were written in pseudocode so that they would not be tied to any particular software. It is a good exercise to program them in a general-purpose computer algebra system such as Maple or Mathematica. The algorithms of Chapter 1 have been implemented in a number of computer algebra systems, including Macaulay, Singular, some packages of Maple, and REDUCE. The interested reader can find a short overview of available computer algebra systems with routines for dealing with polynomial ideals in [60].

We present here two Mathematica codes that we used to study the general quadratic system and systems with homogeneous cubic nonlinearities. The codes can be easily modified to compute the focus and linearizability quantities and normal forms of other polynomial systems.

In Figures 6.1 and 6.2 we give Mathematica code for computing the first three focus quantities of system (3.131). Setting $\alpha = 1$ and $\beta = 0$, it is possible to use this code to compute the linearizability quantities I_{kk}. Set $\alpha = 0$ and $\beta = 1$ to compute the quantities J_{kk}. In the code `a12` and `b21` stand for a_{-12} and $b_{2,-1}$.

In Figures 6.3 and 6.4 we present the Mathematica code that we used to compute the first four pairs of the resonant coefficients in the normal form of system (6.51). Since the output of the last two pairs in the presented notation is relatively large, we do not present these quantities in the output, but they are given in Table 6.2.

Note also that we hide the output or large expressions using the symbol "`;`" at the end of the input expressions.

■ The operator (3.71) for system (3.131)

```
In[1]:= l1[nu1_,nu2_,nu3_,nu4_,nu5_,nu6_]:=  1 nu1 + 0 nu2  -  nu3 + 2 nu4 +1 nu5 + 0 nu6;
        l2[nu1_,nu2_,nu3_,nu4_,nu5_,nu6_]:=  0 nu1 + 1 nu2 + 2 nu3  - nu4 + 0 nu5 + 1 nu6;
```

■ Set α=β=1 to compute the focus quantities

■ Set α=1, β=0 to compute the linearizability quantities I_{kk}

■ Set α=0, β=1 to compute J_{kk}

```
In[3]:= α=1;  β=1;
```

■ Definition of function (4.52)

```
In[4]:= v[k1_,k2_,k3_,k4_,k5_,k6_]:=v[k1,k2,k3,k4,k5,k6]=
        Module[{us,coef},coef=l1[k1,k2,k3,k4,k5,k6]-l2[k1,k2,k3,k4,k5,k6];   us=0;
        v[0,0,0,0,0,0]=1;
        If[k1>0,us=us+(l1[k1-1,k2,k3,k4,k5,k6]α+)*v[k1-1,k2,k3,k4,k5,k6]];
        If[k2>0,us=us+(l1[k1,k2-1,k3,k4,k5,k6]α+)*v[k1,k2-1,k3,k4,k5,k6]];
        If[k3>0,us=us+(l1[k1,k2,k3-1,k4,k5,k6]α+)*v[k1,k2,k3-1,k4,k5,k6]];
        If[k4>0,us=us-(l2[k1,k2,k3,k4-1,k5,k6]β-)*v[k1,k2,k3,k4-1,k5,k6]];
        If[k5>0,us=us-(l2[k1,k2,k3,k4,k5-1,k6]β-)*v[k1,k2,k3,k4,k5-1,k6]];
        If[k6>0,us=us-(l2[k1,k2,k3,k4,k5,k6-1]β-)*v[k1,k2,k3,k4,k5,k6-1]];
        If [coef!=0,   us=us/coef]; If [coef==0, gg[k1,k2,k3,k4,k5,k6]=us;   us=0];  us]
```

■ gmax is the number of the focus or linearizability quantity to be computed

```
In[5]:= gmax=3;
```

■ Computing the quantities q[1], q[2], ... up to the order "gmax"

```
In[6]:= Do[k= sc; num=k;  q[num]=0;
        For [i1=0,i1<=2 k,i1++ ,
        For[i2=0,i2<=(2 k-i1),i2++  ,
        For[i3=0,i3<=(2 k-i1-i2),i3++  ,
        For[i4=0,i4<=(2 k-i1-i2-i3),i4++,
        For[i5=0,i5<=(2 k-i1-i2-i3-i4),i5++,
        For[i6=0,i6<=(2 k-i1-i2-i3-i4-i5),i6++,
        If[(l1[i1,i2,i3,i4,i5,i6]==k)  &&(l2[i1,i2,i3,i4,i5,i6]==k),  v[i1,i2,i3,i4,i5,i6];
        q[num]=q[num]+gg[i1,i2,i3,i4,i5,i6]TT[i1,i2,i3,i4,i5,i6]]]]]]]],
        {sc,1,gmax}]
```

Fig. 6.1 Mathematica code for computing focus and linearizability quantities of a quadratic system

■ Definition of monomials of system (3.131)

```
In[7]:= TT[l1_,l2_,l3_,l4_,l5_,l6_]:=a10^l1  a01^l2 a12^l3  b21^l4   b10^l5  b01^l6
```

■ Output of focus or linearizability quantities

```
In[8]:= Do[{g[i] = Factor[q[i]], Print[g[i]]}, {i, 1, gmax}]
```

$$a01\, a10 - b01\, b10$$

$$\frac{1}{3}(24\, a01^2\, a10^2 - 18\, a01\, a10^2\, b01 - 27\, a01^2\, a10\, b10 + 2\, a10^2\, a12\, b10 + 18\, a10\, b01^2\, b10 + 3\, a10\, a12\, b10^2 + 27\, a01\, b01\, b10^2 - 24\, b01^2\, b10^2 - 2\, a12\, b10^3 + 2\, a01^3\, b21 - 3\, a01^2\, b01\, b21 - 2\, a01\, b01^2\, b21)$$

$$\frac{1}{72}(7236\, a01^3\, a10^3 + 696\, a01\, a10^4\, a12 - 10476\, a01^2\, a10^3\, b01 + 3888\, a01\, a10^3\, b01^2 - 13824\, a01^3\, a10^2\, b10 + 254\, a01\, a10^3\, a12\, b10 + 16380\, a01^2\, a10^2\, b01\, b10 - 1212\, a10^3\, a12\, b01\, b10 - 3888\, a10^2\, b01^3\, b10 + 6768\, a01^3\, a10\, b10^2 + 399\, a01\, a10^2\, a12\, b10^2 + 238\, a10^2\, a12\, b01\, b10^2 - 16380\, a01\, a10\, b01^2\, b10^2 + 10476\, a10\, b01^3\, b10^2 - 2222\, a01\, a10\, a12\, b10^3 - 6768\, a01^2\, b01\, b10^3 + 1440\, a10\, a12\, b01\, b10^3 + 13824\, a01\, b01^2\, b10^3 - 7236\, b01^3\, b10^3 + 621\, a01\, a12\, b10^4 + 2\, a12\, b01\, b10^4 - 2\, a01^4\, a10\, b21 + 890\, a01^2\, a10^2\, a12\, b21 - 1440\, a01^3\, a10\, b01\, b21 - 864\, a01\, a10^2\, a12\, b01\, b21 - 238\, a01^2\, a10\, b01^2\, b21 + 1212\, a01\, a10\, b01^3\, b21 - 621\, a01^4\, b10\, b21 - 1754\, a01^2\, a10\, a12\, b10\, b21 + 28\, a10^2\, a12^2\, b10\, b21 + 2222\, a01^3\, b01\, b10\, b21 - 399\, a01^2\, b01^2\, b10\, b21 + 864\, a10\, a12\, b01^2\, b10\, b21 - 254\, a01\, b01^3\, b10\, b21 - 696\, b01^4\, b10\, b21 + 132\, a10\, a12^2\, b10^2\, b21 + 1754\, a01\, a12\, b01\, b10^2\, b21 - 890\, a12\, b01^2\, b10^2\, b21 - 73\, a12^2\, b10^3\, b21 + 73\, a01^3\, a12\, b21^2 - 132\, a01^2\, a12\, b01\, b21^2 - 28\, a01\, a12\, b01^2\, b21^2)$$

Fig. 6.2 Mathematica code for computing focus and linearizability quantities of a quadratic system (continued)

■ Input of system (6.51)

```
In[1]:= xdot = i (x - Sum[a[2 - k, k] x^(3 - k) y^k, {k, 0, 3}])

Out[1]= i (x - y^3 a[-1, 3] - x y^2 a[0, 2] - x^2 y a[1, 1] - x^3 a[2, 0])

In[2]:= ydot = -i (y - Sum[b[k, 2 - k] x^k y^(3 - k), {k, 0, 3}])

Out[2]= -i (y - y^3 b[0, 2] - x y^2 b[1, 1] - x^2 y b[2, 0] - x^3 b[3, -1])
```

■ Normal form (4.22) of (6.51) up to the 11th order

```
In[3]:= x1dot = i x1 + x1 Sum[Y1[2 k] x1^k y1^k, {k, 1, 5}]

Out[3]= i x1 + x1 (x1 y1 Y1[2] + x1^2 y1^2 Y1[4] + x1^3 y1^3 Y1[6] + x1^4 y1^4 Y1[8] + x1^5 y1^5 Y1[10])

In[4]:= y1dot = -i y1 + y1 Sum[Y2[2 k] x1^k y1^k, {k, 1, 5}]

Out[4]= -i y1 + y1 (x1 y1 Y2[2] + x1^2 y1^2 Y2[4] + x1^3 y1^3 Y2[6] + x1^4 y1^4 Y2[8] + x1^5 y1^5 Y2[10])
```

■ Transformation (4.21)

```
In[5]:= xsub = x1 + Sum[Sum[(h1[k - 1, s - k]  x1^(k) y1^(s - k) ), {k, 0, s}], {s, 3, 11}] // Expand;

In[6]:= ysub = y1 + Sum[Sum[(h2[k, s - k - 1] ( x1^(k) y1^(s - k) )), {k, 0, s}], {s, 3, 11}] // Expand;
```

■ Choose the distinguished transformation

```
In[7]:= Do[{h1[k, k] = 0; h2[k, k] = 0}, {k, 1, 6}]
```

■ Create equation (2.35)

```
In[8]:= v1 = (D[xsub, x1]  x1dot + D[xsub, y1]  y1dot - (xdot /. {x → xsub, y → ysub})) // Expand;

In[9]:= v2 = (D[ysub, x1]  x1dot + D[ysub, y1]  y1dot - (ydot /. {x → xsub, y → ysub})) // Expand;
```

Fig. 6.3 Mathematica code for computing the distinguished normal form of system (6.51)

■ sh[s] is a code to solve (2.35) in the space H_s of vector homogeneous functions

```
In[10]:= sh[s_] :=
   Module[{g}, t1[s - 1, 0] = First[Solve[Coefficient[v1 /. y1 → 0, x1^s] == 0, h1[s - 1, 0]]];
    h1[s - 1, 0] = h1[s - 1, 0] /. t1[s - 1, 0];
    t1[-1, s] = First[Solve[Coefficient[v1 /. x1 → 0, y1^s] == 0, h1[-1, s]]];
    h1[-1, s] = h1[-1, s] /. t1[-1, s]; Do[If[s - 1 - i ≠ i,
      {t1[s - 1 - i, i] = First[Solve[Coefficient[v1, x1^(s - i) y1^i] == 0, h1[s - 1 - i, i]]];
       h1[s - 1 - i, i] = h1[s - 1 - i, i] /. t1[s - 1 - i, i]},
      {t1[s - 1 - i, i] = First[Solve[Coefficient[v1, x1^(s - i) y1^i] == 0, Y1[2 i]]];
       Y1[2 i] = Y1[2 i] /. t1[s - 1 - i, i]}], {i, 1, s - 1}];
    t2[s, -1] = First[Solve[Coefficient[v2 /. y1 → 0, x1^s] == 0, h2[s, -1]]];
    h2[s, -1] = h2[s, -1] /. t2[s, -1];
    t2[0, s - 1] = First[Solve[Coefficient[v2 /. x1 → 0, y1^s] == 0, h2[0, s - 1]]];
    h2[0, s - 1] = h2[0, s - 1] /. t2[0, s - 1];
    Do[If[s - 1 - i ≠ i, {t2[i, s - 1 - i] = First[Solve[Coefficient[v2, x1^(i) y1^(s - i) ] == 0,
          h2[i, s - 1 - i]]]; h2[i, s - 1 - i] = h2[i, s - 1 - i] /. t2[i, s - 1 - i]},
      {t2[s - 1 - i, i] = First[Solve[Coefficient[v2, x1^(i) y1^(s - i) ] == 0, Y2[2 i]]];
       Y2[2 i] = Y2[2 i] /. t2[s - 1 - i, i]}], {i, 1, s - 1}] ]
```

■ Calculations using sh

```
In[11]:= Do[sh[k], {k, 3, 9}]
```

■ Output of the coefficients of the normal form

```
In[12]:= Do[Print[{Y1[2 k], Y2[2 k]}], {k, 1, 4}]
```

$\{-i\,a[1,1],\ i\,b[1,1]\}$

$\{\frac{1}{4}\,i\,(-4\,a[0,2]\,a[2,0] - 4\,a[0,2]\,b[2,0] - 3\,a[-1,3]\,b[3,-1]),$

$\frac{1}{4}\,i\,(4\,a[0,2]\,b[2,0] + 4\,b[0,2]\,b[2,0] + 3\,a[-1,3]\,b[3,-1])\}$

Fig. 6.4 Mathematica code for computing the distinguished normal form of system (6.51) (continued)

References

1. W. W. Adams and P. Loustaunau. *An Introduction to Gröbner Bases.* Graduate Studies in Mathematics, Vol. 3. Providence, RI: American Mathematical Society, 1994.
2. A. Algaba, E. Freire, and E. Gamero. Isochronicity via normal form. *Qual. Theory Dyn. Sys.* *1* (2000), no. 2, 133–156.
3. M. I. Al'muhamedov. On conditions for the existence of singular point of the center type. (Russian) *Izv. Fiz.-Mat. Obs.* (Kazan) **9** (1937) 105–121.
4. V. V. Amel'kin. Strong isochronicity of the Liénard system. *Differ. Uravn.* **42** (2006) 579–582; *Differ. Equ.* **42** (2006) 615–618.
5. V. V. Amel'kin and K. S. al'Khaïder. Strong isochronism of polynomial differential systems with a center. *Differ. Uravn.* **35** (1999) 867–873; *Differ. Equ.* **35** (1999) 873–879.
6. V. V. Amel'kin and C. Dang. Isochronicity of the Cauchy-Riemann system in the focus case. (Russian) *Vestsi Nats. Akad. Navuk Belarusi Ser. Fiz. Mat.-Navuk* (1993) 28–31.
7. V. V. Amel'kin and O. B. Korsantiya. Isochronous and strongly isochronous oscillations of two-dimensional monodromic holomorphic dynamical systems. *Differ. Uravn.* **42** (2006) 147–152; *Differ. Equ.* **42** (2006) 159–164.
8. V. V. Amel'kin, N. A. Lukashevich, and A. P. Sadovskii. *Nonlinear Oscillations in Second Order Systems.* (Russian) Minsk: Belarusian State University, 1982.
9. A. F. Andreev. Solution of the problem of the center and the focus in one case. (Russian) *Akad. Nauk SSSR. Prikl. Mat. Meh.* **17** (1953) 333–338.
10. A. F. Andreev. *Singular Points of Differential Equations.* (Russian) Minsk: Vysh. Shkola, 1979.
11. A. F. Andreev, A. P. Sadovskii, and V. A. Tsikalyuk. The center-focus problem for a system with homogeneous nonlinearities in the case of zero eigenvalues of the linear part. *Differ. Uravn.* **39** (2003) 147–153; *Differ. Equ.* **39** (2003) 155–164.
12. A. A. Andronov, E. A. Leontovitch, I. I. Gordon, and A. G. Maier. *Qualitative Theory of Second-Order Dynamic Systems.* Israel Program for Scientific Translations. New York: John Wiley and Sons, 1973.
13. A. A. Andronov, E. A. Leontovitch, I. I. Gordon, and A. G. Maier. *Theory of Bifurcations of Dynamic Systems on a Plane.* Israel Program for Scientific Translations. New York: John Wiley and Sons, 1973.
14. R. Bamón. Quadratic vector fields in the plane have a finite number of limit cycles. *Publ. Math. Inst. Hautes Etudes Sci.* **64** (1986) 111–142.
15. V. V. Basov. *The Normal Forms Method in the Local Qualitative Theory of Differential Equations: Formal Theory of Normal Forms.* (Russian) Saint Petersburg: Izd. Saint Petersburg University, 2001.
16. V. V. Basov. *The Normal Forms Method in the Local Qualitative Theory of Differential Equations: Analytical Theory of Normal Forms.* (Russian) Saint Petersburg: Izd. Saint Petersburg University, 2002.

17. N. N. Bautin. On the number of limit cycles which appear with the variation of coefficients from an equilibrium position of focus or center type. *Mat. Sb.* **30** (1952) 181–196; *Amer. Math. Soc. Transl.* **100** (1954) 181–196.
18. T. Becker and V. Weispfenning. *Gröbner Bases: A Computational Approach to Commutative Algebra.* Graduate Texts in Mathematics, Vol. 141. New York: Springer-Verlag, 1993.
19. Y. N. Bibikov. *Local Theory of Nonlinear Analytic Ordinary Differential Equations.* Lecture Notes in Mathematics, Vol. 702. New York: Springer-Verlag, 1979.
20. R. I. Bogdanov. Versal deformations of a singular point on the plane in the case of zero eigenvalues. *Funct. Anal. Appl.* **9** (1975) 144–145.
21. Yu. L. Bondar. Solution of the problem of the isochronicity of the center for a cubic system. (Russian) *Vestn. Beloruss. Gos. Univ. Ser. 1 Fiz. Mat. Inform.* (2005) 73–76.
22. Yu. L. Bondar and A. P. Sadovskii. Solution of the center and focus problem for a cubic system that reduces to the Liénard system. *Differ. Uravn.* **42** (2006) 11–22, 141; *Differ. Equ.* **42** (2006) 10–25.
23. J. Brainen and R. Laubenbacher. *Lectures Notes of the Summer School, Laramie, Wyoming, 2000.*
24. A. D. Brjuno. Analytic form of differential equations. I, II. (Russian) *Tr. Mosk. Mat. Obs.* **25** (1971) 119–262; **26** (1972) 199–239.
25. A. D. Brjuno. *A Local Method of Nonlinear Analysis for Differential Equations.* Moscow: Nauka, 1979; *Local Methods in Nonlinear Differential Equations.* Translated from the Russian by William Hovingh and Courtney S. Coleman. Berlin: Springer-Verlag, 1989.
26. B. Buchberger. Ein Algorithmus zum Auffinden der Basiselemente des Restklassenringes nach einem nulldimensionalen Polynomideal. PhD Thesis, Mathematical Institute, University of Innsbruck, Austria, 1965; An algorithm for finding the basis elements of the residue class ring of a zero dimensional polynomial ideal. *J. Symbolic Comput.* **41** (2006) 475–511.
27. B. Buchberger. Gröbner bases: an algorithmic method in polynomial ideals theory. In: *Multidimensional Systems: Theory and Applications* (N. K. Bose, Ed.), 184–232. Dordrecht, The Netherlands: D. Reidel, 1985.
28. B. Buchberger. Introduction to Gröbner bases. In: *Gröbner Bases and Applications* (Linz, 1998), London Math. Soc. Lecture Note Ser. **251**, 3–31. Cambridge: Cambridge University Press, 1998.
29. A. Campillo and M. M. Carnicer. Proximity inequalities and bounds for the degree of invariant curves by foliations of $\mathbf{P}^2_{\mathbf{C}}$. *Trans. Amer. Math. Soc.* **349** (1997) 2211–2228.
30. D. Cerveau and A. Lins Neto. Holomorphic foliations in CP(2) having an invariant algebraic curve. *Ann. Inst. Fourier (Grenoble)* **41** (1991) 883–903.
31. J. Chavarriga, I. A. García, and J. Giné. On Lie's symmetries for planar polynomial differential systems. *Nonlinearity* **14** (2001) 863–880.
32. J. Chavarriga, I. A. García, and J. Giné. Isochronicity into a family of time-reversible cubic vector fields. *Appl. Math. Comput.* **121** (2001) 129–145.
33. J. Chavarriga, H. Giacomini, J. Giné, and J. Llibre. Local analytic integrability for nilpotent centers. *Ergodic Theory Dynam. Systems* **23** (2003) 417–428.
34. J. Chavarriga and J. Giné. Integrability of a linear center perturbed by fourth degree homogeneous polynomial. *Publ. Mat.* **40** (1996) 21–39.
35. J. Chavarriga and J. Giné. Integrability of a linear center perturbed by fifth degree homogeneous polynomial. *Publ. Mat.* **40** (1996) 335–356.
36. J. Chavarriga, J. Giné, and I. A. García. Isochronous centers of a linear center perturbed by fourth degree homogeneous polynomial. *Bull. Sci. Math.* **123** (1999) 77–96.
37. J. Chavarriga and M. Grau. Some open problems related to 16th Hilbert problem. *Sci. Ser. A Math. Sci. (N. S.)* **9** (2003) 1–26.
38. J. Chavarriga and M. Sabatini. A survey of isochronous centers. *Qual. Theory Dyn. Syst.* **1** (1999) 1–70.
39. L. Chen and M. Wang. The relative position and number of limit cycles of a quadratic differential system. *Acta Math. Sinica* **22** (1979) 751–758.
40. X. Chen, V. G. Romanovski, and W. Zhang. Linearizability conditions of time-reversible quartic systems having homogeneous nonlinearities. *Nonlinear Anal.* **69** (2008) 1525–1539.

41. L. A. Cherkas. On the conditions for a center for certain equations of the form $yy' = P(x) + Q(x)y + R(x)y^2$. *Differ. Uravn.* **8** (1972) 1435–1439; *Differ. Equ.* **8** (1972) 1104–1107.
42. L. A. Cherkas. Conditions for a Liénard equation to have a center. *Differ. Uravn.* **12** (1976) 292–298; *Differ. Equ.* **12** (1976) 201–206.
43. L. A. Cherkas. Conditions for a center for the equation $yy' = \sum_{i=0}^{3} p_i(x)y^i$. *Differ. Uravn.* **14** (1978) 1594–1600, 1722; *Differ. Equ.* **14** (1978) 1133–1137.
44. C. Chicone. *Ordinary Differential Equations with Applications.* New York: Springer-Verlag, 1999.
45. C. Chicone and M. Jacobs. Bifurcation of critical periods for plane vector fields. *Trans. Amer. Math. Soc.* **312** (1989) 433–486.
46. C. Chicone and D. S. Shafer. Separatrix and limit cycles of quadratic systems and Dulac's theorem. *Trans. Amer. Math. Soc.* **278** (1983) 585–612.
47. A. R. Chouikha. Isochronous centers of Liénard type equations and applications. *J. Math. Anal. Appl.* **331** (2007) 358–376.
48. A. R. Chouikha, V. G. Romanovski, and X. Chen. Isochronicity of analytic systems via Urabe's criterion. *J. Phys. A* **40** (2007) 2313–2327.
49. C. Christopher. Invariant algebraic curves and conditions for a centre. *Proc. Roy. Soc. Edinburgh Sect. A* **124** (1994) 1209–1229.
50. C. Christopher. An algebraic approach to the classification of centers in polynomial Liénard systems. *J. Math. Anal. Appl.* **229** (1999) 319–329.
51. C. Christopher. Estimating limit cycles bifurcations. In: *Trends in Mathematics, Differential Equations with Symbolic Computations* (D. Wang and Z. Zheng, Eds.), 23–36. Basel: Birkhäuser-Verlag, 2005.
52. C. Christopher and J. Devlin. On the classification of Liénard systems with amplitude-independent periods. *J. Differential Equations* **200** (2004) 1–17.
53. C. Christopher and C. Li. *Limit Cycles of Differential Equations.* Basel: Birkhäuser-Verlag, 2007.
54. C. Christopher and J. Llibre. Algebraic aspects of integrability for polynomial systems. *Qual. Theory Dyn. Syst.* **1** (1999) 71–95.
55. C. Christopher, J. Llibre, C. Pantazi, and X. Zhang. Darboux integrability and invariant algebraic curves for planar polynomial systems. *J. Phys. A* **35** (2002) 2457–2476.
56. C. Christopher and S. Lynch. Small-amplitude limit cycle bifurcations for Liénard systems with quadratic or cubic damping or restoring forces. *Nonlinearity* **12** (1999) 1099–1112.
57. C. Christopher, P. Mardešić, and C. Rousseau. Normalizable, integrable, and linearizable saddle points for complex quadratic systems in $\mathbb{C}^2$. *J. Dyn. Control Sys.* **9** (2003) 311–363.
58. C. Christopher and C. Rousseau. Nondegenerate linearizable centres of complex planar quadratic and symmetric cubic systems in $\mathbb{C}^2$. *Publ. Mat.* **45** (2001) 95–123.
59. A. Cima, A. Gasull, V. Mañosa, and F. Mañosas. Algebraic properties of the Liapunov and periodic constants. *Rocky Mountain J. Math.* **27** (1997) 471–501.
60. D. Cox, J. Little, and D. O'Shea. *Ideals, Varieties, and Algorithms*, 2nd edition. New York: Springer-Verlag, 1997.
61. G. Darboux. Mémoire sur les équations différentielles algébriques du premier ordre et du premier degré. *Bull. Sci. Math. Sér.* 2 **2** (1878) 60–96, 123–144, 151–200.
62. W. Decker, G. Pfister, H. Schönemann, and S. Laplagne. A SINGULAR 3.0 library for computing the primary decomposition and radical of ideals, `primdec.lib`, 2005.
63. J. Devlin, N. G. Lloyd, and J. M. Pearson. Cubic systems and Abel equations. *J. Differential Equations.* **147** (1998) 435–454.
64. H. Dulac. Détermination et intégration d'une certaine classe d'équations différentielles ayant pour point singulier un centre. *Bull. Sci. Math.* (2) **32** (1908) 230–252.
65. F. Dumortier. Singularities of vector fields on the plane. *J. Differential Equations* **23** (1977) 53–106.
66. F. Dumortier and P. Fiddelaers. Quadratic models for generic local 3-parameter bifurcations on the plane. *Trans. Amer. Math. Soc.* **326** (1991) 101–126.

67. F. Dumortier and C. Li. Perturbation from an elliptic Hamiltonian of degree four I, II, III, IV. *J. Differential Equations* **175** (2001) 209–243; **176** (2001) 114–175; **188** (2003) 473–511, 512–554.
68. F. Dumortier, R. Roussarie, and C. Rousseau. Hilbert's 16th problem for quadratic vector fields *J. Differential Equations* **110** (1994) 95–123.
69. J. Ecalle. *Introduction aux fonctions analysables et preuve constructive de la conjecture de Dulac*. Paris: Hermann, 1992.
70. V. F. Edneral. Computer evaluation of cyclicity in planar cubic system. In: *Proceedings of the 1997 International Symposium on Symbolic and Algebraic Computation*, 305–309. New York: ACM Press, 1997.
71. V. F. Edneral. Looking for periodic solutions of ODE systems by the normal form method. In: *Trends in Mathematics, Differential Equations with Symbolic Computations* (D. Wang and Z. Zheng, Eds.), 173–200. Basel: Birkhäuser-Verlag, 2005.
72. W. W. Farr, C. Li, I. S. Laboriau, and W. F. Langford. Degenerate Hopf bifurcation formulas and Hilbert's 16th problem. *SIAM J. Math. Anal.* **20** (1989) 13–30.
73. J.-P. Françoise and A. Fronville. Computer algebra and bifurcation theory of vector fields of the plane. *Proceedings of the 3rd Catalan Days on Applied Mathematics (Lleida, 1996)*, 57–63. Lleida: Univ. Lleida, 1997.
74. M. Frommer. Über das Auftreten von Wirbeln und Strudeln (geschlossener und spiraliger Integralkurven) in der Umgebung rationaler Unbestimmtheitsstellen. *Math. Ann.* **109** (1934) 395–424.
75. A. Fronville, A. P. Sadovski, and H. Żołądek. The solution of the $1 : -2$ resonant center problem in the quadratic case. *Fund. Math.* **157** (1998) 191–207.
76. A. Gasull, A. Guillamon, and V. Mañosa. Centre and isochronicity conditions for systems with homogeneous nonlinearities. *Proceedings of the 2nd Catalan Days on Applied Mathematics* (Odeillo, 1995), 105–116. Perpiniá: Presses Universitaires de Perpignan, 1995.
77. A. Gasull, A. Guillamon, and V. Mañosa. An analytic-numerical method for computation of the Liapunov and period constants derived from their algebraic structure. *SIAM J. Numer. Anal.* **36** (1999) 1030–1043.
78. A. Gasull, J. Llibre, V. Mañosa, and F. Mañosas. The focus-centre problem for a type of degenerate system. *Nonlinearity* **13** (2000) 699–729.
79. A. Gasull and J. Torregrosa. Center problem for several differential equations via Cherkas' method. *J. Math. Anal. Appl.* **228** (1998) 322–343.
80. A. Gasull and J. Torregrosa. A new approach to the computation of the Lyapunov constants. The geometry of differential equations and dynamical systems. *Comput. Appl. Math.* **20** (2001) 149–177.
81. P. Gianni, B. Trager, and G. Zacharias. Gröbner bases and primary decomposition of polynomials. *J. Symbolic Comput.* **6** (1988) 146–167.
82. J. Giné and M. Grau. Linearizability and integrability of vector fields via commutation. *J. Math. Anal. Appl.* **319** (2006) 326–332.
83. J. Giné and M. Grau. Characterization of isochronous foci for planar analytic differential systems. *Proc. Roy. Soc. Edinburgh Sect. A* **135** (2005) 985–998.
84. J. Giné and J. Llibre. A family of isochronous foci with Darboux first integral. *Pacific J. Math.* **218** (2005) 343–355.
85. H.-G. Gräbe. CALI–a REDUCE Package for Commutative Algebra, Version 2.2, 1995. `http://www.informatik.uni-leipzig.de/~graebe/ComputerAlgebra/Software/Cali.`
86. R. L. Graham, D. E. Knuth, and O. Patashnik. *Concrete Mathematics*. New York: Addison-Wesley, 1994.
87. H. Grauert and K. Fritzsche. *Several Complex Variables*. New York: Springer-Verlag, 1976.
88. D. Grayson and M. Stillman. Macaulay2: a software system for algebraic geometry and commutative algebra. `http://www.math.uiuc.edu/Macaulay2.`
89. G.-M. Greuel, G. Pfister, and H. Schönemann. SINGULAR 3.0. A Computer Algebra System for Polynomial Computations. Centre for Computer Algebra, University of Kaiserslautern, 2005. `http://www.singular.uni-kl.de/`

90. R. C. Gunning and H. Rossi. *Analytic Functions of Several Complex Variables.* Englewood Cliffs, NJ: Prentice Hall, 1965.
91. M. Han, Y. Lin, and P. Yu. A study on the existence of limit cycles of a planar system with third-degree polynomials. *Internat. J. Bifur. Chaos Appl. Sci. Engrg.* **14** (2004) 41–60.
92. M. Han, Y. Wu, and P. Bi. Bifurcation of limit cycles near polycycles with n vertices. *Chaos Solitons Fractals* **22** (2004) 383–394.
93. M. Han and C. Yang. On the cyclicity of a 2-polycycle for quadratic systems. *Chaos Solitons Fractals* **23** (2005) 1787–1794.
94. M. Han and T. Zhang. Some bifurcation methods of finding limit cycles. *Math. Biosci. Eng.* **3** (2006) 67–77.
95. P. Hartman. *Ordinary Differential Equations.* New York: John Wiley & Sons, 1964; Boston: Birkhäuser, 1982 (reprint of the second edition).
96. M. Hervé. *Several Complex Variables: Local Theory.* London: Oxford University Press, 1963.
97. M. Hirsch. *Differential Topology.* New York: Springer-Verlag, 1976.
98. Yu. Il'yashenko. Algebraic nonsolvability and almost algebraic solvability of the center-focus problem. *Funktsion. Anal. i Prilozhen.* **6** (1972) 197–202; *Funct. Anal. Appl.* **6** (1972) 30–37.
99. Yu. Il'yashenko. *Finiteness Theorems for Limit Cycles.* Providence, RI: American Mathematical Society, 1991.
100. Yu. Ilyashenko. Centennial history of Hilbert's 16th problem. *Bull. Amer. Math. Soc. (N. S.)* **39** (2002) 301–354.
101. J.-P. Jouanolou. *Equations de Pfaff algébriques.* Lecture Notes in Mathematics, Vol. 708. New York: Springer-Verlag, 1979.
102. A. S. Jarrah, R. Laubenbacher, and V. Romanovski. The Sibirsky component of the center variety of polynomial differential systems. Computer algebra and computer analysis (Berlin, 2001). *J. Symbolic Comput.* **35** (2003) 577–589.
103. W. Kapteyn. On the centra of the integral curves which satisfy differential equations of the first order and the first degree. *Proc. Kon. Akad. Wet., Amsterdam* **13** (1911) 1241–1252.
104. W. Kapteyn. New researches upon the centra of the integrals which satisfy differential equations of the first order and the first degree. *Proc. Kon. Acad. Wet., Amsterdam* **14** (1912) 1185–1185; **15** (1912) 46–52.
105. I. S. Kukles. Sur les conditions nécessaires et suffisantes pour l'existence d'un centre. *Dokl. Akad. Nauk SSSR* **42** (1944) 160–163.
106. I. S. Kukles. Sur quelques cas de distinction entre un foyer et un centre. *Dokl. Akad. Nauk SSSR* **42** (1944) 208–211.
107. N. N. Ladis. Commuting vector fields and isochrony. (Russian) *Vestn. Beloruss. Gos. Univ. Ser. I Fiz. Mat. Inform.* (1976) 21–24, 93.
108. J. S. W. Lamb and J. A. G. Roberts. Time-reversal symmetry in dynamical systems: a survey. Time-reversal symmetry in dynamical systems (Coventry, 1996). *Phys. D* **112** (1998) 1–39.
109. J. P. La Salle. *The Stability of Dynamical Systems.* Philadelphia: Society for Industrial and Applied Mathematics, 1976.
110. S. Lefschetz. *Differential Equations: Geometric Theory.* New York: Wiley Interscience, 1963; New York: Dover Publications, 1977 (reprint).
111. V. L. Le and A. P. Sadovskii. The center and focus problem for a cubic system in the case of zero eigenvalues of the linear part. (Russian) *Vestn. Beloruss. Gos. Univ. Ser. 1 Fiz. Mat. Inform.* (2002) 75–80.
112. C. Li and R. Roussarie. The cyclicity of the elliptic segment loops of the reversible quadratic Hamiltonian systems under quadratic perturbations. *J. Differential Equations* **205** (2004) 488–520.
113. J. Li. Hilbert's 16th problem and bifurcations of planar polynomial vector fields. *Internat. J. Bifur. Chaos Appl. Sci. Engrg.* **13** (2003) 47–106.
114. A. Liapounoff. Problème général de la stabilité du mouvement. *Annales de la Faculté des Sciences de Toulouse* Sér. 2 **9** (1907) 204–474. Reproduction in *Annals of Mathematics Studies* **17**, Princeton: Princeton University Press, 1947, reprinted 1965, Kraus Reprint Corporation, New York.

115. A. M. Liapunov. *Stability of Motion*. With a contribution by V. Pliss. Translated by F. Abramovici and M. Shimshoni. New York: Academic Press, 1966.
116. Y. Liu. Formulas of focal values, center conditions and center integrals for the system $(E_3^{(3)})$. *Kexue Tongbao (English Ed.)* **33** (1988) 357–359.
117. Y. Liu. Formulas of values of singular point and the integrability conditions for a class of cubic system, $M(3) \geq 7$. *Chinese Sci. Bull.* **35** (1990) 1241–1245.
118. Y. Liu and J. Li. Theory of values of singular point in complex autonomous differential systems. *Sci. China Ser. A* **33** (1990) 10–23.
119. J. Llibre. Integrability of polynomial differential systems. *Handbook of Differential Equations, Volume 1: Ordinary Differential Equations* (A. Cañada, P. Drábek, and A. Fonda, Eds.). Amsterdam: Elsevier, 2004.
120. J. Llibre and G. Rodriguez. Invariant hyperplanes and Darboux integrability for d-dimensional polynomial differential systems. *Bull. Sci. Math.* **124** (2000) 599–619.
121. N. G. Lloyd and J. M. Pearson. REDUCE and the bifurcation of limit cycles. *J. Symbolic Comput.* **9** (1990) 215–224.
122. N. G. Lloyd and J. M. Pearson. Computing centre conditions for certain cubic systems. *J. Comput. Appl. Math.* **40** (1992) 323–336.
123. N. G. Lloyd and J. M. Pearson. Bifurcation of limit cycles and integrability of planar dynamical systems in complex form. *J. Phys. A* **32** (1999) 1973–1984.
124. N. G. Lloyd and J. M. Pearson. Symmetry in planar dynamical systems. *J. Symbolic Comput.* **33** (2002) 357–366.
125. N. G. Lloyd, J. M. Pearson, and C. Christopher. Algorithmic derivation of centre conditions. *SIAM Rev.* **38** (1996) 619–636.
126. N. G. Lloyd, J. M. Pearson, and V. G. Romanovsky. Centre conditions for cubic systems in complex form. Preprint, 1998.
127. W. S. Loud. Behavior of the period of solutions of certain plane autonomous systems near centers. *Contributions to Differential Equations* **3** (1964) 21–36.
128. V. A. Lunkevič and K. S. Sibirskii. Conditions for a center in the case of homogeneous nonlinearities of third degree. *Differ. Uravn.* **1** (1965) 1482–1487; *Differ. Equ.* **1** (1965) 1164–1168.
129. S. Lynch. *Dynamical Systems with Applications Using Maple*, 2nd edition. Boston: Birkhäuser, 2008.
130. S. Lynch. *Dynamical Systems with Applications Using Mathematica.* Boston: Birkhäuser, 2007.
131. S. Lynch. Symbolic computation of Lyapunov quantities and the second part of Hilbert's sixteenth problem. In: *Trends in Mathematics, Differential Equations with Symbolic Computations* (D. Wang and Z. Zheng, Eds.), 1–22. Basel: Birkhäuser-Verlag, 2005.
132. S. MacLane and G. Birkhoff. *Algebra.* New York: Macmillan, 1967.
133. K. E. Malkin. Criteria for the center for a certain differential equation. (Russian) *Volž. Mat. Sb. Vyp.* **2** (1964) 87–91.
134. K. E. Malkin. Conditions for the center for a class of differential equations. (Russian) *Izv. Vysš. Učebn. Zaved. Matematika* **50** (1966) 104–114.
135. P. Mardešić, L. Moser-Jauslin, and C. Rousseau. Darboux linearization and isochronous centers with a rational first integral. *J. Differential Equations* **134** (1997) 216–268.
136. P. Mardešić, C. Rousseau, and B. Toni. Linearization of isochronous centers. *J. Differential Equations* **121** (1995) 67–108.
137. L. Mazzi and M. Sabatini. A characterization of centres via first integrals. *J. Differential Equations* **76** (1988) 222–237.
138. N. B. Medvedeva. Analytic solvability of the center–focus problem in some classes of vector fields with a complex monodromic singular point. *Proc. Steklov Inst. Math.* **2002** Algebra, Topology, Mathematical Analysis, Suppl. 2, S120–S141.
139. J. Murdock. *Normal Forms and Unfoldings for Local Dynamical Systems.* New York: Springer-Verlag, 2003.
140. L. Perko. *Differential Equations and Dynamical Systems*, 3rd edition. New York: Springer-Verlag, 2001.

141. I. I. Pleshkan. A new method of investigation on the isochronism of a system of differential equations. *Dokl. Akad. Nauk SSSR* **182** (1968) 768–771; *Soviet Math. Dokl.* **9** (1968) 1205–1209.
142. V. A. Pliss. On the reduction of an analytic system of differential equations to linear form. *Differ. Uravn.* **1** (1965) 153–161; *Differ. Equ.* **1** (1965) 111–118.
143. H. Poincaré. Mémoire sur les courbes définies par une équation différentielle. *J. Math. Pures et Appl.* (Sér. 3) **7** (1881) 375–422; (Sér. 3) **8** (1882) 251–296; (Sér. 4) **1** (1885) 167–244; (Sér. 4) **2** (1886) 151–217.
144. M. J. Prelle and M. F. Singer. Elementary first integrals of differential equations. *Trans. Amer. Math. Soc.* **279** (1983) 215–229.
145. N. B. Pyzhova, A. .P. Sadovskii, and M. L. Hombak. Isochronous centers of a reversible cubic system. (Russian) *Tr. Inst. Mat. Natl. Akad. Nauk Belarusi, Inst. Mat., Minsk* **4** (2000) 120–127.
146. R. H. Rand and D. Armbruster. *Perturbation Methods, Bifurcation Theory and Computer Algebra.* New York: Springer-Verlag, 1987.
147. J. W. Reyn. *A Bibliography of the Qualitative Theory of Quadratic Systems of Differential Equations in the Plane.* 3rd edition. Report 94-02, Delft University of Technology, 1994.
148. V. G. Romanovski, X. Chen, and Z. Hu. Linearizability of linear systems perturbed by fifth degree homogeneous polynomials. *J. Phys. A* **40** (2007) 5905–5919.
149. V. Romanovski and M. Robnik. The center and isochronicity problems for some cubic systems. *J. Phys. A* **34** (2001) 10267–10292.
150. V. G. Romanovski and D. S. Shafer. On the center problem for $p:-q$ resonant polynomial vector fields. *Bull. Belg. Math. Soc. Simon Stevin* **15** (2008) (in press).
151. V. G. Romanovski and D. S. Shafer. Time-reversibility in two-dimensional polynomial systems. In: *Trends in Mathematics, Differential Equations with Symbolic Computations* (D. Wang and Z. Zheng, Eds.), 67–84. Basel: Birkhauser Verlag, 2005.
152. V. G. Romanovskii. *On the Number of Limit Cycles of a Second Order System of Differential Equations.* (Russian) PhD thesis, Leningrad State University, 1986.
153. V. G. Romanovskii. Calculation of Lyapunov numbers in the case of two pure imaginary roots. *Differ. Uravn.* **29** (1993) 910–912; *Differ. Equ.* **29** (1993) 782–784.
154. V. Romanovsky and A. Şubă. On Bautin ideal of the quadratic system. In: *Proceedings of the Third International Conference "Differential Equations and Applications," Saint Petersburg, June 12–17, 2000. Mathematical Research* **7** 21–25. St. Petersburg: St. Petersburg State Technical University Press, 2000.
155. R. Roussarie. A note on finite cyclicity property and Hilbert's 16th problem. *Lecture Notes in Mathematics, Vol. 1331.* New York: Springer-Verlag, 1988.
156. R. Roussarie. Bifurcations of planar vector fields and Hilbert's sixteenth problem. *Progress in Mathematics* **164**. Basel: Birkhäuser, 1998.
157. C. Rousseau. Bifurcation methods in polynomial systems. In: *Bifurcations and Periodic Orbits of Vector Fields* (Montreal). *NATO Adv. Sci. Inst. Ser. C* **405** 383–428. Dordrecht: Kluwer Acad. Publ., 1993.
158. C. Rousseau and B. Toni. Local bifurcation of critical periods in vector fields with homogeneous nonlinearities of the third degree. *Canad. Math. Bull.* **36** (1993) 473–484.
159. C. Rousseau and B. Toni. Local bifurcations of critical periods in the reduced Kukles system. *Canad. J. Math.* **49** (1997) 338–358.
160. M. Sabatini. On the period function of Liénard systems. *J. Differential Equations* **152** (1999) 467–487.
161. M. Sabatini. Characterizing isochronous centers by Lie brackets. *Differential Equations Dynam. Systems* **5** (1997) 91–99.
162. M. Sabatini. On the period function of $x'' + f(x)x'^2 + g(x) = 0$. *J. Differential Equations* **196** (2004) 151–168.
163. A. P. Sadovskii. Holomorphic integrals of a certain system of differential equations. *Differ. Uravn.* **10** (1974) 558–560; *Differ. Equ.* **10** (1974) 425–427.
164. A. P. Sadovskii. The problem of the center and focus for analytic systems with a zero linear part. I. *Differ. Uravn.* **25** (1989) 790–799; *Differ. Equ.* **25** (1989) 552–560.

165. A. P. Sadovskii. The problem of the center and focus for analytic systems with a zero linear part. II. *Differ. Uravn.* **25** (1989) 950–956; *Differ. Equ.* **25** (1989) 682–687.
166. A. P. Sadovskii. Solution of the center and focus problem for a cubic system of nonlinear oscillations. *Differ. Uravn.* **33** (1997) 236–244, 286; *Differ. Equ.* **33** (1997) 236–244.
167. A. P. Sadovskii. Centers and foci of a class of cubic systems. *Differ. Uravn.* **36** (2000) 1652–1657; *Differ. Equ.* **36** (2000) 1812–1818.
168. N. A. Saharnikov. On Frommer's center conditions. (Russian) *Akad. Nauk SSSR. Prikl. Mat. Meh.* **12** (1948) 669–670.
169. N. A. Saharnikov. On conditions for the existence of a center and a focus. (Russian) *Akad. Nauk SSSR. Prikl. Mat. Meh.* **14** (1950) 513–526.
170. D. Schlomiuk. Elementary first integrals and algebraic invariant curves of differential equations. *Exposition. Math.* **11** (1993) 433–454.
171. D. S. Shafer. Weak singularities under weak perturbation. *J. Dynam. Differential Equations* **16** (2004) 65–90.
172. S. L. Shi. A concrete example of the existence of four limit cycles for plane quadratic system. *Sci. Sinica Ser. A* **23** (1980) 153–158.
173. T. Shimoyama and K. Yokoyama. Localization and primary decomposition of polynomial ideals. *J. Symbolic Comput.* **22** (1996) 247–277.
174. K. S. Sibirskii. On the conditions for existence of a center and a focus. (Russian) *Uč. Zap. Kišinevsk. Univ.* **11** (1954) 115–117.
175. K. S. Sibirskii. The principle of symmetry and the problem of the center. (Russian) *Kišinev. Gos. Univ. Uč. Zap.* **17** (1955) 27–34.
176. K. S. Sibirskii. On the number of limit cycles in the neighborhood of a singular point. *Differ. Uravn.* **1** (1965) 53–66; *Differ. Equ.* **1** (1965) 36–47.
177. K. S. Sibirskii. *Algebraic Invariants of Differential Equations and Matrices.* (Russian) Kishinev: Shtiintsa, 1976.
178. K. S. Sibirsky. *Introduction to the Algebraic Theory of Invariants of Differential Equations.* Kishinev: Shtiintsa, 1982; Nonlinear Science: Theory and Applications. Manchester: Manchester University Press, 1988.
179. K. S. Sibirskii and A. S. Shubè. Coefficient conditions for a center in the sense of Dulac for a differential system with one zero characteristic root and cubic right-hand sides. *Dokl. Akad. Nauk SSSR* **303** (1988) 799–803; *Soviet Math. Dokl.* **38** (1989) 609–613.
180. C. L. Siegel. Über die normalform analytischer Differential–Gleichungen in der Nähe einer Gleichgewichtslösung. *Nachr. der Akad. Wiss. Göttingen Math.–Phys. Kl. IIa* (1952) 21–30.
181. C. L. Siegel and J. K. Moser. *Lectures on Celestial Mechanics.* Berlin: Springer-Verlag, 1971, 1995.
182. M. F. Singer. Liouvillian first integrals of differential equations. *Trans. Amer. Math. Soc.* **333** 673–688.
183. S. Sternberg. On the structure of local homeomorphisms of Euclidean n-space, II. *Amer. J. Math.* **80** (1958) 623–631.
184. B. Sturmfels. Gröbner bases and convex polytopes. University Lecture Series, Vol. 8. Providence, RI: American Mathematical Society, 1996.
185. B. Sturmfels. *Algorithms in Invariant Theory*. New York: Springer-Verlag, 1993.
186. A. Şubă and D. Cozma. Solution of the problem of the centre for cubic differential system with three invariant straight lines in generic position. *Qual. Theory Dyn. Syst.* **6** (2005) 45–58.
187. F. Takens. Singularities of vector fields. *Publ. Math. Inst. Hautes Etudes Sci.* **43** (1974) 47–100.
188. A. Tsygvintsev. Algebraic invariant curves of plane polynomial differential systems. *J. Phys. A* **34** (2001) 663–672.
189. M. Urabe. Potential forces which yield periodic motions of a fixed period. *J. Math. Mech.* **10** (1961) 569–578.
190. M. Urabe. The potential force yielding a periodic motion whose period is an arbitrary continuous function of the amplitude of the velocity. *Arch. Ration. Mech. Anal.* **11** (1962) 27–33.

191. J. V. Uspensky. *Theory of Equations*. New York: McGraw-Hill, 1948.
192. W. V. Vasconcelos. *Computational Methods in Commutative Algebra and Algebraic Geometry*. Algorithms and Computation in Mathematics, Vol. 2. Berlin: Springer-Verlag, 1998.
193. E. P. Volokitin and V. V. Ivanov. Isochronicity and commutability of polynomial vector fields. *Sibirsk. Mat. Zh.* **40** (1999) 30–48; *Siberian Math. J.* **40** (1999) 23–38.
194. A. P. Vorob'ev. On periodic solutions in the case of a center. *Dokl. Akad. Nauk BSSR* **6** (1962) 281–284.
195. B. L. van der Waerden. *Modern Algebra*. New York: Frederick Ungar Publishing, 1969.
196. D. Wang. Mechanical manipulation for a class of differential systems. *J. Symbolic Comput.* **12** (1991) 233–254.
197. D. Wang. Irreducible decomposition of algebraic varieties via characteristic sets and Gröbner bases. *Comput. Aided Geom. Design* **9** (1992) 471–484.
198. D. Wang. *Elimination Methods*. New York: Springer-Verlag, 2001.
199. D. Wang. *Elimination Practice: Software Tools and Applications*. London: Imperial College Press, 2004.
200. S. Willard. *General Topology*. Reading, MA: Addison-Wesley, 1970.
201. S. Yakovenko. A geometric proof of the Bautin theorem. Concerning the Hilbert Sixteenth Problem. Advances in Mathematical Sciences, Vol. 23; *Amer. Math. Soc. Transl.* **165** (1995) 203–219.
202. Y.-Q. Ye. *Theory of Limit Cycles*. Transl. Math. Monographs, Vol. 66. Providence, RI: American Mathematical Society, 1986.
203. P. Yu and M. Han. Twelve limit cycles in a cubic case of the 16th Hilbert problem. *Internat. J. Bifur. Chaos Appl. Sci. Engrg.* **15** (2005) 2191–2205.
204. A. Zegeling. Separatrix cycles and multiple limit cycles in a class of quadratic systems. *J. Differential Equations* **113** (1994) 355–380.
205. Z. Zhang, T. Ding, W. Huang, and Z. Dong. *Qualitative Theory of Differential Equations*. Transl. Math. Monographs, Vol. 101. Providence, RI: American Mathematical Society, 1992.
206. H. Żołądek. Quadratic systems with center and their perturbations. *J. Differential Equations* **109** (1994) 223–273.
207. H. Żołądek. On a certain generalization of Bautin's theorem. *Nonlinearity* **7** (1994) 273–279.
208. H. Żołądek. The classification of reversible cubic systems with center. *Topol. Methods Nonlinear Anal.* **4** (1994) 79–136.
209. H. Żołądek. Eleven small limit cycles in a cubic vector field. *Nonlinearity* **8** (1995) 843–860.
210. H. Żołądek. The problem of center for resonant singular points of polynomial vector fields. *J. Differential Equations* **135** (1997) 94–118.
211. H. Żołądek. Algebraic invariant curves for the Liénard equation. *Trans. Amer. Math. Soc.* **350** (1998) 1681–1701.
212. H. Żołądek. New examples of holomorphic foliations without algebraic leaves. *Studia Math.* **131** (1998) 137–142.

Index of Notation

$\overset{(s+1)}{\sim}$	[binary relation] agreement of series through order $s+1$
$\prec$	[binary relation] $u(\mathbf{z}) \prec v(\mathbf{z})$: series $u(\mathbf{z})$ majorizes series $v(\mathbf{z})$
$\setminus$	set difference $A \setminus B$: elements of A that are not in B
$\lvert\alpha\rvert$	for $\alpha = (\alpha_1, \dots, \alpha_n) \in \mathbb{N}_0^n$, $\lvert\alpha\rvert = \alpha_1 + \cdots + \alpha_n$
(α, κ)	for $\alpha \in \mathbb{N}_0^N$ and $\kappa \in \mathbb{C}^n$, the scalar $\sum_{j=1}^n \alpha_j \kappa_j$
η_j	jth Lyapunov number
$[\nu]$	for $\nu \in \mathbb{N}_0^{2\ell}$, $[\nu] = a_{p_1,q_1}^{\nu_1} \cdots a_{p_\ell,q_\ell}^{\nu_\ell} b_{q_\ell,p_\ell}^{\nu_{\ell+1}} \cdots b_{q_1,p_1}^{\nu_{2\ell}} \in \mathbb{C}[a,b]$
σ^ζ	for $\sigma \in \mathbb{R} \setminus \{0\}$, $\zeta \in \mathbb{Z}^{2\ell}$, $\sigma^\zeta = (\sigma^{\zeta_1}, \dots, \sigma^{\zeta_{2\ell}})$
$\sigma^{-\zeta}\mathbf{c}$	for $\sigma \in \mathbb{R} \setminus \{0\}$, $\zeta \in \mathbb{Z}^{2\ell}$, $\mathbf{c} \in E(a,b)$, $\sigma^{-\zeta}\mathbf{c} = (\sigma^{-\zeta_1} c_1, \dots, \sigma^{-\zeta_{2\ell}} c_{2\ell})$
$\mathbf{a}(t)$	for $\mathbf{a} \in k^n$, the unique solution of (2.1) with initial condition $\mathbf{a}(0) = \mathbf{a}$, on its maximal interval of existence
(a,b)	The vector of coefficients of system (3.3), ordered $(a_{p_1,q_1}, a_{p_2,q_2}, \dots, a_{p_\ell,q_\ell}, b_{q_\ell,p_\ell}, \dots, b_{q_2,p_2}, b_{q_1,p_1})$
$\mathscr{B}$	the Bautin ideal, generated by the focus quantities
$\mathscr{B}_k$	the ideal $\langle g_{11}, g_{22}, \dots, g_{kk} \rangle$ generated by the first k focus quantities
$\breve{\mathscr{B}}$, $\breve{\mathscr{B}}_k$	the images of $\mathscr{B}$ and $\mathscr{B}_k$ under the homomorphism W of Proposition 6.3.5
$\breve{\mathscr{B}}^+$	the ideal $\langle \breve{g}_{kk} : k \in \mathbb{N} \rangle$ described in the paragraph preceding Proposition 6.3.5
$\breve{\mathscr{B}}_k^+$	the ideal $\langle \breve{g}_{11}, \dots, \breve{g}_{kk} \rangle$ described in the paragraph preceding Proposition 6.3.5
$(\hat{b}, \hat{a})$	the involution of (a,b) defined by (5.29)
$\mathbb{C}$	the field of complex numbers
$\mathbb{C}[a,b]$	the polynomial ring in the variables a_{pq}, b_{qp}, $(p,q) \in S$
$\mathbb{C}[V]$	the coordinate ring of a variety V in $\mathbb{C}^n$
$\mathbf{c}^\nu$	for $\mathbf{c} \in E(a,b)$, $\nu \in \mathbb{N}_0^{2\ell}$, $\mathbf{c}^\nu = [\nu]\rvert_{\mathbf{c}} = c_1^{\nu_1} \cdots c_{2\ell}^{\nu_{2\ell}}$
$\operatorname{div} \mathscr{X}$	divergence of the vector field $\mathscr{X}$
$E(a)$	the parameter space $\mathbb{C}^\ell$ of the complex form (3.51) of the real polynomial system (3.4)
$E_P(a)$	subset of $E(a)$ for which $\widetilde{P}$ and $\widetilde{Q}$ are relatively prime polynomials
$E(a,b)$	the parameter space $\mathbb{C}^{2\ell}$ of system (3.3)

F	the function defined by (3.151) (Section 3.8)
$\mathbf{f}$	a mapping into $\mathbb{R}^n$ or $\mathbb{C}^n$ for $n \geq 2$
$\mathbf{f}$	for an analytic function f, the germ at θ^* induced by f (Chapter 6)
$\widehat{f}$	the conjugate of the polynomial f defined in Definition 3.4.3 (except: omitting complex conjugation of the coefficients in the proof of Theorem 6.3.5)
$f^\natural(\mathbf{z})$	the trivial majorant of $f(\mathbf{z}) = \sum_{\alpha \in \mathbb{N}^n} f^{(\alpha)} \mathbf{z}^\alpha$: $f^\natural(\mathbf{z}) = \sum_{\alpha \in \mathbb{N}^n} \lvert f^{(\alpha)} \rvert \mathbf{z}^\alpha$
$f \mid g$	polynomial f divides polynomial g
$f \xrightarrow{F} h$	f reduces to h modulo $F = \{f_1, \dots, f_s\}$ (Definition 1.2.5)
$f \equiv g \bmod I$	$f - g \in I$ (f, g polynomials and I an ideal)
$\langle f_1, \dots, f_s \rangle$	ideal generated by the polynomials $f_1, \dots, f_s$
G	the function defined by (3.28b)
G	the function defined by (3.151) (Section 3.8)
$\mathscr{G}_{\theta^*}$	the ring of germs of analytic functions of θ at the point $\theta^* \in k^n$
g_{kk}	the kth focus quantity of (3.3)
g_{k_1,k_2}	the coefficient of $x^{k_1+1} y^{k_2+1}$ in (3.53)
$\breve{g}_{kk}$	the image of g_{kk} under the homomorphism W of Proposition 6.3.5
$\mathbb{H}$	the set of radical ideals in the ring $k[x_1, \dots, x_n]$
$\mathscr{H}$	the ideal $\langle H_{2j+1} : j \in \mathbb{N} \rangle = \langle \widetilde{H}_{2j+1} : j \in \mathbb{N} \rangle$ for $H(w) = \sum_{k=1}^{\infty} H_{2k+1} w^k$ defined by (3.28b)
$\mathscr{H}_k$	the ideal $\langle H_{2j+1} : j = 1, \dots, k \rangle = \langle \widetilde{H}_{2j+1} : j = 1, \dots, k \rangle$
$\mathscr{H}_s$	the set of functions from $\mathbb{R}^n$ to $\mathbb{R}^n$ (or $\mathbb{C}^n$ to $\mathbb{C}^n$) all of whose components are homogeneous polynomial functions of degree s (Chapter 2)
H	the function defined by (3.28b)
$\widetilde{H}$	$(i/2)H$
$\mathbb{I}$	the set of all ideals in the ring $k[x_1, \dots, x_n]$
$\mathbf{I}(S)$	the ideal of a set or variety S
$\sqrt{I}$	the radical of an ideal I
$I + J$	the sum of two ideals I and J
$I : J$	the quotient ideal of two ideals I and J
IJ	the product ideal of two ideals I and J
I_{kk}, J_{kk}	the kth linearizability quantities of (4.27)
I_{Ham}	the ideal of Hamiltonian systems of (3.3) defined by (3.85)
I_{sym}	the symmetry or Sibirsky ideal
$i\mathbb{Q}$	the set of pure imaginary elements of $\mathbb{Q}(i)$, $i\mathbb{Q} = \{iq : q \in \mathbb{Q}\}$
k	a field
$k[V]$	the coordinate ring of a variety V in k^n
$k[x_1, \dots, x_n]$	the ring of polynomials in n indeterminates with coefficients in k
$k(x_1, \dots, x_n)$	the ring of rational functions in n indeterminates with coefficients in k
$k[[x_1, \dots, x_n]]$	the ring of formal power series in n indeterminates with coefficients in k

$\mathscr{L}$	the homological operator on $\mathscr{H}_s$ (Chapter 2)
$\mathscr{L}$	the linearizability ideal $\langle I_{kk}, J_{kk} : k \in \mathbb{N} \rangle$ (Chapter 4)
$L(\nu)$	the linear map $L : \mathbb{N}_0^{2\ell} \to \mathbb{Z}^2$ of (3.71)
$LCM(x^\alpha, x^\beta)$	least common multiple of monomials x^α and x^β
$LC(f)$	leading coefficient of a polynomial f
$LM(f)$	leading monomial of a polynomial f
$LT(f)$	leading term of a polynomial f
$LT(S)$	set of leading terms of a set S of polynomials
$\mathscr{M}$	$\mathscr{M} = \{\nu \in \mathbb{N}_0^{2\ell} : L(\nu) = (j, j) \text{ for some } j \in \mathbb{N}_0\}$, L defined by (3.71)
$\mathbb{N}$	the set of natural numbers $\{1, 2, 3, \ldots\}$
$\mathbb{N}_0$	$\{0\} \cup \mathbb{N}$
$\mathbb{N}_{-n}$	$\{-n, \ldots, -1, 0\} \cup \mathbb{N}$
$\mathscr{P}$	difference function of (3.14)
P	the ideal $\langle p_{2k} : k \in \mathbb{N} \rangle$ (Chapters 4 and 6)
P_k	the ideal $\langle p_2, p_4, \ldots, p_{2k} \rangle$ (Chapters 4 and 6)
p_{2k}	the kth isochronicity quantity
$\mathbb{Q}$	the field of rational numbers
$\mathbb{Q}(i)$	the field of Gaussian rationals, $\mathbb{Q}(i) = \{a + ib : a, b \in \mathbb{Q}\}$
$\mathbb{R}$	the field of real numbers
$R(\alpha)$	for $\alpha \in k^n$, the set of indices of nonzero entries
$\lfloor r \rfloor$	the greatest integer less than or equal to r (floor function)
$\bar{S}$	the Zariski closure of a subset S of k^n
$\mathrm{Supp}(f)$	the set of ν such that the coefficient of $\mathbf{x}^\nu$ in the polynomial $f = \sum f^{(\nu)} \mathbf{x}^\nu$ is nonzero
U	the rotation group of Definition 5.1.4
U_φ	element of the rotation group U
$U_\varphi^{(a)}, U_\varphi^{(b)}$	blocks in the matrix representation of U_φ
$\mathbb{V}$	the set of varieties in k^n, for k a field
$\mathbf{V}(I)$	the variety of an ideal I
$V_{\mathscr{C}}$	the center variety for family (3.3)
$V_{\mathscr{I}}$	the isochronicity variety for family (3.3)
$V_{\mathscr{L}}$	the linearizability variety for family (3.3)
$\mathscr{X}$	the vector field $\mathscr{X} = \sum f_1 \frac{\partial}{\partial x_1} + \cdots + f_n \frac{\partial}{\partial x_n}$ corresponding to the system of differential equations $\dot{x}_1 = f_1(\mathbf{x}), \ldots, \dot{x}_n = f_n(\mathbf{x})$
$X_m^{(\alpha)}$	the coefficient of the monomial $\mathbf{x}^\alpha$ in the mth component X_m of the vector function $\mathbf{X}$
$\{X_m(\mathbf{y} + \mathbf{h}(\mathbf{y}))\}^{(\alpha)}$	the coefficient of $\mathbf{y}^{(\alpha)}$ in the expansion of $X_m(\mathbf{y} + \mathbf{h}(\mathbf{y}))$ in powers of $\mathbf{y}$
x^α	the monomial $x_1^{\alpha_1} \cdots x_n^{\alpha_n}$, $\alpha = (\alpha_1, \ldots, \alpha_n) \in \mathbb{N}_0^n$
$\mathscr{Y}$	the ideal $\left\langle Y_1^{(j+1,j)},\ Y_2^{(j,j+1)} : j \in \mathbb{N} \right\rangle$ defined in (4.28)
$\mathbb{Z}$	the ring of integers

Index